Numerical Methods in Porous Media

This book serves as a practical guide to using numerical methods with MATLAB®/Python implementation for simulating fluid flow and transport in porous media, covering the fundamentals and their application in a way that is easy to understand for those without a technical background. In addition, it covers the governing equations of fluid mechanics and heat transfer, boundary conditions and initial conditions, dimensional analysis and scaling laws, and analytical and numerical methods for solving fluid flow and heat transfer problems.

- Provides a comprehensive introduction to numerical methods for simulating fluid flow and transport in porous media.
- Examines numerical methods and their implementation in porous media simulations using MATLAB® and Python.
- Explains the applications for numerical methods in porous media, including transport, heat transfer, and fluid dynamics.

Numerical Methods in Porous Media

MATLAB® and Python Approaches

Mohammed Fathy El-Amin

CRC Press
Taylor & Francis Group
Boca Raton London New York

CRC Press is an imprint of the
Taylor & Francis Group, an **informa** business

Designed cover image: Shutterstock

First edition published 2026
by CRC Press
2385 NW Executive Center Drive, Suite 320, Boca Raton FL 33431

and by CRC Press
4 Park Square, Milton Park, Abingdon, Oxon, OX14 4RN

CRC Press is an imprint of Taylor & Francis Group, LLC

ISBN: 978-1-032-72279-5 (hbk)
ISBN: 978-1-032-72496-6 (pbk)
ISBN: 978-1-032-72497-3 (ebk)

DOI: 10.1201/9781032724973

Typeset in CMR10 font
by KnowledgeWorks Global Ltd.

Dedicated to the soul of my beloved father,
to my dear mothers, my wife, my children,
and to my brothers and sisters,
whose love, support, and prayers have always been my source of strength.

Contents

List of Figures

List of Tables

Preface

The study of transport phenomena in porous media lies at the intersection of physics, mathematics, and computational engineering. From the extraction of hydrocarbons and the management of groundwater to the design of energy-storage systems and environmental remediation, porous-media flow models underpin many of the critical challenges and opportunities in modern science and technology. Over the past few decades, advances in numerical analysis have transformed our ability to simulate and understand these complex processes, yet practitioners often face steep learning curves when it comes to implementing robust, efficient, and extensible code.

This book was born out of the need for a single, coherent resource that couples the theory of numerical methods with hands-on implementation in two of the most widely used programming environments: MATLAB® and Python. While MATLAB® has long been the workhorse of engineers for rapid prototyping and matrix-based computations, Python, with its rich ecosystem (NumPy, SciPy, Pandas, and beyond), has emerged as a powerful, open-source alternative for production-scale modeling and data science. By presenting parallel MATLAB® and Python code side by side, this book aims to equip readers with institutional resources.

Each chapter begins with a concise review of the governing equations and discretization techniques, followed by step-by-step guidance on translating those methods into working code. For every major algorithm—whether finite difference, finite element, mixed methods, or fractional-order schemes, you will find mainly in MATLAB® scripts and Python modules (mainly for ML computation and a few for numerical computations). We start with fundamentals—single-phase Darcy flow and basic ODE/PDE solvers—and advance through two-phase flow, fractured reservoirs, fractional models, non-isothermal coupling, nanoparticle transport, underground hydrogen storage, and AI-driven surrogate modeling. Each chapter concludes with practical exercises, real-world case studies, and review questions designed to reinforce learning and encourage further exploration. This book is intended for graduate students, researchers, and practicing engineers in petroleum engineering, hydrogeology, environmental engineering, and applied mathematics. A working knowledge of calculus, basic PDE theory, and introductory programming (in any language) will help you dive in immediately.

Chapter 1 lays the theoretical foundations—porosity, permeability, Darcy's law, representative elementary volumes, and scale transitions. **Chapter 2** introduces the programming environments, essential libraries, and basic numerical tools (differentiation, integration, ODE/PDE solvers). **Chapter 3** surveys spatial and temporal discretization techniques: finite difference, finite element, finite volume, spectral, and meshless methods. **Chapter 4** develops analytical solutions and validates numerical implementations for steady and unsteady Darcy flow in one and two dimensions. **Chapter 5** covers multiphase conservation laws, capillary pressure, relative permeability models, and IMPES/IMEX schemes. **Chapter 6** explores single-continuum, dual-continuum, discrete fracture, and hybrid models, including uncertainty quantification. **Chapter 7** introduces fractional calculus for anomalous diffusion, memory effects, and nonlocal interactions, with numerical solvers. **Chapter 8** develops coupled flow–heat transfer models, energy equations, and iterative solvers for stiff, nonlinear systems. **Chapter 9** examines colloid filtration theory, adsorption/desorption kinetics, and numerical validation against laboratory data. **Chapter 10** reviews UHS principles, compositional multiphase flow equations, geochemical interactions, and pilot project case studies. **Chapter 11** surveys machine learning and physics-informed neural networks, data-driven surrogate modeling, and hybrid physics–ML approaches.

Early chapters build the core numerical framework; later chapters can be consulted independently for specialized topics. All MATLAB® scripts and Python files are provided. Experiment by changing physical parameters, grid resolutions, and boundary conditions to deepen your understanding. Each chapter contains review questions to enrich their knowledge and problems to examine their understanding.

Mohammed Fathy El-Amin
Jeddah, Saudi Arabia
April 2025

Biography

Professor Mohammed Fathy El-Amin is a professor of applied mathematics and computational sciences at Effat University, Saudi Arabia. With a career spanning more than 25 years, Dr. El-Amin has established a global reputation for pioneering contributions in computational modeling, fluid dynamics, reservoir simulation, porous media transport, heat and mass transfer, hydrogen energy, and renewable energy technologies. He earned his Ph.D. in applied mathematics. His research journey included prestigious fellowships from the Alexander von Humboldt (AvH) Foundation in Germany and the Japan Society for the Promotion of Science (JSPS) in Japan, as well as research appointments at renowned institutions such as Stuttgart University, Kyushu University, and King Abdullah University of Science and Technology (KAUST). Dr. El-Amin has published over 200 peer-reviewed articles, book chapters, and conference papers, alongside several edited volumes and special journal issues. His recent authored books, including Numerical Modeling of Nanoparticle Transport in Porous Media (Elsevier, 2023) and Fractional Modeling of Fluid Flow and Transport Phenomena (Elsevier, 2025), reflect his leadership in bridging mathematical theory with practical energy and environmental challenges. Currently, Dr. El-Amin leads research projects on atmospheric water generation using desiccant materials and underground hydrogen storage, aiming to support sustainable energy and water security initiatives. An active member of several international scientific societies, including INTERPORE and the Society of Petroleum Engineers (SPE), Dr. El-Amin has been consistently recognized among the World's Top 2% Scientists by Stanford University rankings. His contributions have earned him multiple awards for excellence in research, teaching, and civic engagement. Beyond research, Dr. El-Amin is deeply committed to mentoring graduate students, supervising numerous MSc and Ph.D. theses, and actively participating in university leadership roles, including chairing promotion and research committees.

Acknowledgments

The successful completion of this book would not have been possible without the support, collaboration, and dedication of numerous individuals and organizations.

I am deeply grateful to the publishing team for their expert guidance and unwavering professional support throughout the development and production of this work. In particular, I wish to thank Mr. Joseph Clements, Editorial Manager, for his exceptional coordination, consistent encouragement, and meticulous attention to detail throughout the editorial process. I also thank Ms. Maggie Apostolis for her remarkable assistance and steadfast commitment to ensuring a smooth and timely publication journey.

To the readers, I sincerely hope that the book ahead offers meaningful insights, practical approaches, and inspiration for further exploration of fluid transport phenomena in porous media.

Mohammed Fathy El-Amin
Effat University, Jeddah, Saudi Arabia

1 Introduction to Transport Phenomena in Porous Media

1.1 INTRODUCTION

Transport phenomena in porous media form the foundation of many natural and engineered processes, from groundwater flow and oil recovery to biological systems and materials science. This chapter provides the essential theoretical framework needed to understand and model these complex systems. This chapter aims to enable readers to define and characterize porous media through fundamental properties such as porosity and permeability, understand the physical mechanisms governing fluid transport in porous materials, and apply Darcy's law and its extensions to describe fluid flow in porous media. Readers will learn to distinguish between different flow regimes and their governing equations, recognize the challenges in modeling multiphase flow and non-Newtonian fluids, and appreciate the role of scale in porous media analysis, from pore to continuum levels. The chapter establishes the fundamental concepts and theoretical foundations of transport phenomena in porous media. We begin with basic definitions and properties of porous materials, examining how their structure affects fluid flow and transport processes. The discussion progresses through single-phase flow concepts, building up to more complex phenomena such as multiphase flow and non-Newtonian behavior. Special attention is given to the physical mechanisms governing transport at different scales, from microscopic pore-level interactions to macroscopic continuum behavior. The theoretical principles introduced here are essential for understanding fluid behavior in applications ranging from petroleum engineering and groundwater hydrology to chemical processing and environmental remediation.

Porous media, defined by their interconnected networks of void spaces, exhibit unique characteristics that enable the storage and transmission of fluids. Understanding these phenomena requires an integration of physics, chemistry, and material science, with applications spanning hydrogeology, petroleum engineering, environmental science, and chemical engineering.

1.2 POROUS MATERIALS

Porous materials and media are essential for understanding a range of geological and engineering processes, integrating principles from physics, chemistry,

DOI: 10.1201/9781032724973-1

and material science. These materials are defined by their porosity; the diverse size, shape, and connectivity of these voids contribute to the material's permeability, which is crucial for various fields, including hydrogeology, petroleum geology, and environmental engineering.

Porous media can be categorized into three primary types: artificial, biological, and geological. Artificial media, such as orderly arrays of spheres, fibers, or cylinders, are synthetic structures that are not found naturally. Despite their simplicity, these artificial structures have garnered extensive attention in research, with numerous experimental, theoretical, and numerical studies aiding significantly in pore-scale modeling.

Biological media include structures like bones, tissues, and membranes, all of which are filled with diverse fluids. In certain instances, the configuration of biological media supports pore-scale modeling frameworks, like capillary networks. An example of this is the network of pores and tubes that are integral to plant tissues.

Geological media garner considerable attention due to their practical significance in areas such as water resource management and contamination cleanup [1]. This particular study focuses solely on geological porous media. In environmental engineering, a distinction is made between unconsolidated media, found in the pores of loose sediments, and consolidated media, present in the cracks and gaps within solid rock. For example, a layer of sand particles serves as an excellent representation of unconsolidated media. Sand consists of coarse, nearly spherical silica particles and can be idealized as a cluster of spheres of varying sizes, with typical porosity ranging between 30% and 45%.

Buried sand undergoes a transformation when chemicals in groundwater crystallize within the pore spaces, cementing the sand grains together to form sandstone, a type of consolidated sedimentary rock that constitutes about 25% of the world's sedimentary rocks. Other forms of consolidated porous materials include shales, with porosities between 10% and 25%, and carbonate rocks, primarily composed of calcite, with porosity ranging from 20% to over 50%. Naturally, these rocks display erratic solid/pore interfaces, with a pore distribution that is far more erratic than that found in unconsolidated sands and sandstones.

While the geometry of artificial and biological media is relatively straightforward to determine, the complexity of geological porous media poses significant challenges. Describing the three-dimensional pore morphology of geological substances is a profound and complex subject in the field of modern pore-scale modeling.

1.2.1 ROCK PROPERTIES

In geosciences and reservoir engineering, the physical properties of rocks play a pivotal role in determining their capability to store and transmit fluids. Two of the most fundamental properties are porosity and permeability, which

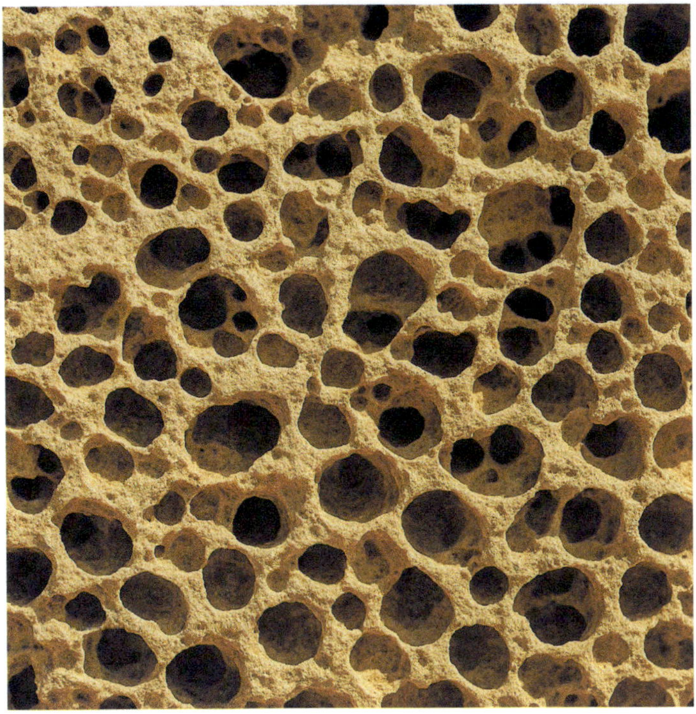

Figure 1.1 A close-up view of porosity of a porous surface.

together dictate the productivity of hydrocarbon reservoirs, aquifers, and the efficiency of geotechnical engineering projects.

POROSITY

Porosity, denoted by the symbol ϕ, is a measure of the void spaces within a geological formation. It is a ratio, typically expressed as a percentage, representing the total volume occupied by the voids (pores) in a given volume of rock (Figure 1.1). Mathematically, porosity is given by:

$$\phi = \frac{V_\mathrm{p}}{V_\mathrm{t}} \times 100\% \tag{1.1}$$

where V_p is the volume of pore spaces and V_t is the total volume of the rock sample.

Geologists use petrographic microscopes, CT scanners, and field porosity maps to visualize and quantify porosity. Figure 1.1 illustrates a typical field porosity map, which shows porosity distribution across a geological formation.

Porosity, the measure of void spaces in a material, is categorized into several types based on characteristics and geological processes. Total porosity encompasses all pore spaces in a rock, regardless of their contribution to fluid transport. Effective porosity, in contrast, accounts only for the interconnected pores that facilitate fluid movement. Secondary porosity arises from geological alterations like fracturing, dissolution, and dolomitization, which modify the initial pore network. Each type of porosity plays a significant role in determining the storage and movement capacity of fluids within the rock structure.

Pore volume compressibility, denoted as c_f, describes how the volume of the pores within a rock changes in response to variations in pressure. It is a critical parameter in reservoir engineering for understanding rock deformation and fluid flow behavior under different pressure conditions. The mathematical expression for pore volume compressibility is given by:

$$c_f = -\frac{1}{V_p}\frac{dV_p}{dp} = -\frac{1}{\phi}\frac{d\phi}{dp} \tag{1.2}$$

where V_p is the pore volume and $\frac{dV_p}{dp}$ represents the rate of change of pore volume with respect to pressure, while $\frac{d\phi}{dp}$ represents the rate of change of porosity with respect to pressure. This equation highlights the inverse relationship between pore volume and pressure changes, making it a fundamental concept for understanding reservoir behavior.

PERMEABILITY

Permeability refers to the ability of a rock to transmit fluids through its pore network. Absolute or intrinsic permeability, denoted as K, is a property of the rock itself, independent of the type of fluid passing through it. It is crucial in the assessment of a reservoir's production potential. Larger and more uniform grain sizes generally increase permeability, as does the size of pore throats—the smaller connections between pores that facilitate fluid movement. The geometry and connectivity of these pores also play a crucial role; complex, tortuous paths tend to reduce permeability [2]. Additionally, the overall structure and fabric of the rock, including aspects such as stratification, bedding, and the presence or absence of fractures, can either improve or impede the flow of fluids. These elements combined dictate the ease with which fluids can traverse the geological material. Permeability is measured in m^2 (or darcy (D)), where one darcy equals $1 \times 10^{-12}\ \mathrm{m}^2$.

In general, higher porosity often indicates the potential for higher permeability because more void spaces are available for fluid storage and flow. However, porosity alone does not guarantee high permeability. The connectivity and alignment of the pore spaces play a critical role. Rocks with well-connected pores tend to exhibit higher permeability than those with isolated or poorly connected pores.

PERMEABILITY AND POROSITY CORRELATIONS

Empirical correlations have been developed to estimate permeability (k) from porosity (ϕ). These correlations are often specific to rock types and require

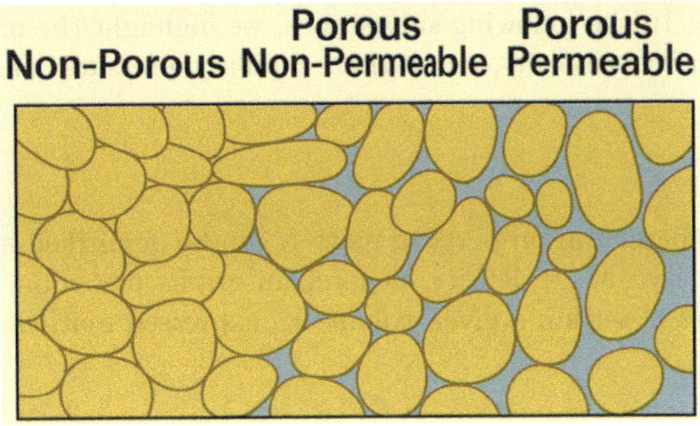

Figure 1.2 Illustration of pore connectivity and its impact on permeability (adapted and redrawn from publicly available internet sources; original creator unknown).

calibration with experimental data. Two commonly used forms, including the power-law relationship:

$$k = a\phi^b, \tag{1.3}$$

where a and b are empirical constants derived from experimental data. And Kozeny-Carman equation is given by:

$$k = \frac{\phi^3}{(1-\phi)^2 S^2} \tag{1.4}$$

where S is the specific surface area of the pores. This equation highlights that permeability is dependent not only on porosity but also on the geometry and surface characteristics of the pores.

Figure 1.2 provides a conceptual representation of how pore connectivity influences permeability. The diagram categorizes rock structures into three distinct zones: (1) non-porous and non-permeable, which have no pore spaces and represent completely solid rocks incapable of storing or transmitting fluids; (2) porous but non-permeable, where the rock has pore spaces but they are unconnected, allowing fluid storage but preventing flow; and (3) porous and permeable, which feature interconnected pore spaces, enabling both fluid storage and flow. The transition from no pore spaces to connected pores highlights the critical role of pore connectivity in determining a rock's permeability. This understanding is essential for evaluating reservoir quality and designing efficient hydrocarbon recovery strategies, emphasizing the importance of detailed rock characterization in reservoir engineering.

1.2.2 FLUID PROPERTIES

Fluids within porous media exhibit distinct properties that significantly influence their behavior and movement. These properties are crucial for understanding and modeling the flow and distribution of fluids in subsurface environments. In the following subsections, we highlight the most basic ones, including density, viscosity, saturation, capillary pressure, and relative permeabilities.

DENSITY

Density, a fundamental property of matter, can be quantified as mass density or molar density. Mass density, denoted in grams per cubic meter $[g/m^3]$, measures mass m within a given volume V, expressed mathematically as:

$$\rho = \frac{m}{V} \tag{1.5}$$

The density of a substance varies with pressure and temperature, typically increasing with pressure and decreasing with temperature. For mixtures, composition further influences density. Density directly affects buoyancy, influencing whether materials sink or float, with interplay involving temperature and species concentration.

VISCOSITY

Viscosity, a measure of a fluid's resistance to deformation, manifests as dynamic viscosity. It correlates the fluid's shear stress τ with the rate of velocity change across a gradient $\left(\frac{\partial u}{\partial y}\right)$, formalized as:

$$\mu = \frac{\tau}{\left(\frac{\partial u}{\partial y}\right)} \tag{1.6}$$

where u is the velocity.

For liquids, viscosity increases with pressure and decreases with temperature. In gases, viscosity increases with pressure but decreases with a drop in temperature. Kinematic viscosity v, used in certain scenarios, is dynamic viscosity divided by density:

$$v = \frac{\mu}{\rho} \tag{1.7}$$

1.3 FLUID FLOW IN POROUS MEDIA

Continuum mechanics provides a framework for analyzing material behavior without focusing on the atomic or molecular origins of the material. In this approach, a material is treated as a continuum, meaning it can be divided and infinitely subdivided into smaller elements. This allows scientists and engineers to study the behavior of matter at scales larger than the distances between individual particles. Instead of tracking the motion of each molecule or atom, continuum mechanics focuses on the average behavior of large numbers of particles, enabling a macroscopic perspective.

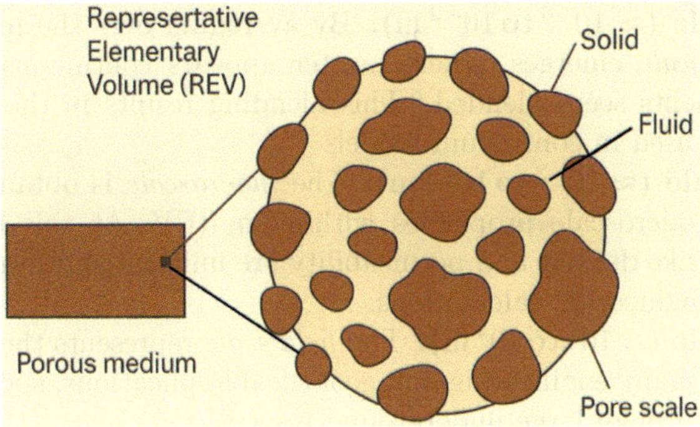

Figure 1.3 Schematic representation of a representative elementary volume (REV) in a porous medium (Adapted from [4]).

This methodology is particularly effective in studying phenomena like *fluid flow*, which encompasses the movement of liquids and gases, as well as flow through porous media such as hydrocarbons moving through underground reservoirs. *Fluid mechanics*, a branch of continuum mechanics, examines the behavior of fluids and applies the continuum assumption to simplify the analysis of complex systems.

The continuum assumption eliminates molecular-level discontinuities by averaging properties such as density, velocity, pressure, and temperature across a small but finite volume. As a result, these properties are treated as continuous functions that vary smoothly from one point to another within the flow. This approach is critical in fluid mechanics because it allows the study of macroscopic parameters without delving into molecular-scale interactions.

Within fluid mechanics, *fluid dynamics* specifically addresses the effects of forces on fluid motion. This field is a cornerstone of modern science and engineering and presents significant mathematical challenges. Many fluid dynamics problems remain unsolved or partially solved, requiring advanced numerical methods and computational simulations for meaningful solutions.

When analyzing flow in porous media, such as underground reservoirs, the continuum approach becomes more complex. The challenge lies in defining a suitable *representative elementary volume* (REV) for averaging. This volume must be large enough to ensure that the averaged properties (e.g., porosity and permeability) are independent of the REV itself, yet small enough to capture localized variations within the system. The scale at which this averaging occurs bridges the gap between microscopic and macroscopic perspectives (Figure 1.3).

Several scales are defined in continuum mechanics, each providing a unique lens for understanding material behavior:

Molecular Scale ($\approx 10^{-10}\,\mathrm{m}$): At this scale, the movement and interactions of individual molecules are described. Molecular behavior is inherently discrete and not treated as continuous.

Microscale ($\approx 10^{-6}$ to 10^{-2} m): By averaging over the molecular scale, the *microscale* emerges, where matter appears continuous, and molecular movements seem blended. This blending results in the smoothed-out properties used in continuum models.

Macroscale ($\approx 10^{-3}$ to 10^{-1} m): The *macroscale* is obtained by further averaging microscale properties within an REV. At this level, material properties like density and permeability are uniform and form the basis for practical engineering calculations.

Field Scale ($\approx 10^{1}$ to 10^{4} m): The *field scale* represents the dimensions of interest in many engineering and geological applications, such as analyzing fluid flow through large underground reservoirs.

A critical aspect of simulating flow through porous media lies in accounting for the varying scales of length.

The transport of fluids through porous media is governed by several physical processes that can affect the movement of fluids such as water, oil, or gas. These processes include advection, diffusion, and dispersion. Advection refers to the bulk movement of fluid driven by pressure gradients, while diffusion is the movement of particles from high to low concentration due to molecular motion. Dispersion combines both advection and diffusion, accounting for the spreading of fluid as it moves through the heterogeneous structure of the porous medium. These transport mechanisms are fundamental to understanding how fluids behave in natural and engineered subsurface environments, influencing key applications such as groundwater flow, oil recovery, and pollutant transport.

In general, the field of fluid mechanics, and particularly, flow in porous media, heavily relies on differential equations to describe properties such as velocity, stress, strain, and temperature fluctuation within materials. Conservation laws, principles like the conservation of mass, momentum, and energy, lay the groundwork for the equations that steer various physical systems. Moving from conventional fluid dynamics to fluid dynamics in porous media, the context changes, but many of the foundational principles remain consistent. Porous media refers to materials that have pores or voids through which fluids can flow, such as rocks, soils, and biological tissues.

The conservation equations are mathematical expressions representing the principles of mass, momentum, and energy conservation. The continuity equation, asserting that mass remains constant, typically connects fluid density changes to velocity variations. The energy conservation principle, within the fluid movement, links internal energy alterations to force-induced work and heat exchange. The Navier-Stokes equations, evolving from Newton's second law, outline the progression of a fluid's velocity field, factoring in viscous and external forces. However, in the case of porous media, these fundamentals require adaptations to consider the effect of the solid matrix's characteristics. Therefore, supplementary equations like Darcy's law and the Brinkman

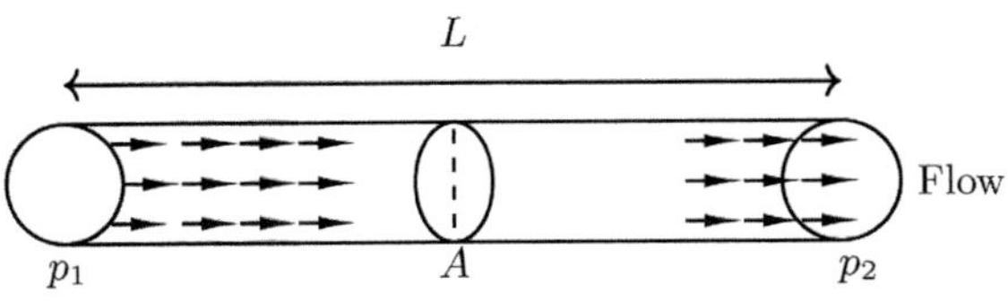

Figure 1.4 A schematic diagram for Darcy's law.

equation depict fluid movement intricacies in such environments. Darcy's law is a foundational principle in this area, which relates the flow rate of a fluid through a porous medium to the pressure gradient and the intrinsic permeability of the medium.

1.3.1 DARCY'S LAW

Darcy's law is a fundamental principle used to describe the flow of a fluid through a porous medium [3]. Named after the French engineer Henry Darcy, who formulated it in the 19th century, based on experiments he conducted to investigate water flow in porous media. Darcy's law is foundational in many subsurface fluid flow modeling, particularly in the fields of groundwater and oil reservoir simulation.

Darcy's law plays the role of the momentum conservation law, which replaces the Navier-Stokes equations in the case of conventional fluid flows. It relates the flow of fluid through a porous medium to the pressure gradient and the permeability of the medium. It looks similar to Fourier's law in the field of heat conduction and Fick's law in diffusion theory. The equation can be written in its simple, single-phase incompressible fluid flow through a porous medium as:

$$Q = -K\frac{A}{\mu}\nabla P \tag{1.8}$$

where Q is the volumetric flow rate per unit area (m^3/s), K is the permeability of the medium (m^2 or darcies/millidarcies), A is the cross-sectional area perpendicular to the flow direction, μ is the fluid viscosity (in Pa·s or cP), ∇P is the pressure gradient (Pa/m). See Figure 1.4. The negative sign denotes that flow occurs from regions of higher pressure to regions of lower pressure. Presenting the Darcy velocity (flux), $\mathbf{u} = \frac{Q}{A}$, the Darcy equation can be written as follows:

$$\mathbf{u} = -\frac{K}{\mu}\nabla P \tag{1.9}$$

The influence of gravity is accounted for within the pressure forces in the following equation:

$$\mathbf{u} = -\frac{K}{\mu}(\nabla P - \rho g \nabla z)$$

In this equation, g denotes the gravitational constant, and z refers to the vertical spatial coordinate.

The continuity equation for incompressible flow in porous media is:

$$\nabla \cdot \mathbf{u} = q \tag{1.10}$$

where q is a source or sink term.

The single-phase pressure equation in porous media combines Darcy's law with the continuity equation. Substituting Darcy's law into the continuity equation, we get:

$$\nabla \cdot \left(-\frac{K}{\mu} (\nabla P - \rho g \nabla z) \right) = q \tag{1.11}$$

This is the single-phase pressure equation in porous media. It describes the balance between the flow of fluid due to pressure gradients and any sources or sinks present in the domain. The equation is a second-order elliptic partial differential equation (PDE) and can be solved using various numerical methods, depending on the specifics of the problem and the boundary conditions.

1.3.2 PEACEMAN MODEL

The Peaceman model [5] is widely used in reservoir simulation to calculate the pressure at a well (p_w) based on the pressure in the grid cell containing the well (p_{grid}). It corrects for the discrepancy caused by representing a well within a discretized grid system. The pressure at the wellbore is given by:

$$p_w = p_{\text{grid}} - \frac{q \cdot B \cdot \mu}{2\pi h k} \ln \left(\frac{r_{\text{e}}}{r_w} \right) \tag{1.12}$$

where p_w is the pressure at the wellbore, p_{grid} is the pressure at the center of the grid cell containing the well, q is the flow rate (positive for production, negative for injection), B is the formation volume factor, which accounts for fluid compressibility and reservoir conditions, μ is the fluid viscosity, h is the reservoir thickness, k is the reservoir permeability, r_w is the wellbore radius (physical radius of the well), and r_{e} is the Effective well radius, defined for a rectangular grid as:

$$r_{\text{e}} = 0.14\sqrt{\Delta x \Delta y} \tag{1.13}$$

For square grid blocks ($\Delta x = \Delta y = \Delta$) is the

$$r_{\text{e}} = 0.14\Delta \tag{1.14}$$

The Peaceman model connects the grid-based simulation pressure (p_{grid}) with the actual well pressure (p_w). In **production wells**, $q > 0$, and the pressure decreases relative to p_{grid}. In **injection wells**, $q < 0$, and the pressure increases relative to p_{grid}.

1.3.3 DISPERSION MODEL

Dispersion in porous media is a complex phenomenon arising from both molecular diffusion and mechanical dispersion. The total dispersion coefficient, D, which encapsulates the cumulative effects of these processes, is expressed as:

$$D = D_{\text{diff}} + D_{\text{disp}}, \tag{1.15}$$

where the flow vector j_x is influenced by these dispersion processes along with the advective transport:

$$j_x = -D\frac{\partial c}{\partial x} + \mathbf{u}c, \tag{1.16}$$

Here, j_x represents the flux of the solute in the x-direction, c is the concentration, and $\mathbf{u}$ is the velocity field.

The molecular diffusion coefficient, D_{diff} $[\text{m}^2.\text{s}^{-1}]$, is a fundamental parameter characterizing the spread of molecules due to random thermal motion. It can be calculated using Stokes' equation:

$$D_{\text{diff}} = \tau\frac{k_B T}{3\pi\mu d}, \tag{1.17}$$

where τ denotes the tortuosity of the porous medium, a factor that accounts for the increased path length due to the medium's structure. k_B is the Boltzmann constant, a fundamental constant relating the average kinetic energy of particles in a gas with the temperature of the gas. T is the absolute temperature, μ is the dynamic viscosity of the fluid, and d represents the particle diameter.

In contrast, the mechanical dispersion coefficient, D_{disp} $[\text{m}^2.\text{s}^{-1}]$, accounts for the spreading of solutes due to variations in velocity at different points within the flow field, particularly in porous media. This coefficient is influenced by the flow conditions and is dependent on Darcy's velocity $|\mathbf{u}|$, as described by:

$$\phi D_{\text{disp}} = d_{l,w}|\mathbf{u}|, \tag{1.18}$$

where $d_{l,w}$ is the longitudinal dispersion coefficient, which quantifies the degree of spreading along the flow direction, and ϕ represents the porosity of the medium. This term reflects the impact of heterogeneities within the medium that cause differential flow velocities and contribute to the overall dispersion process.

Together, these expressions provide a comprehensive understanding of how solutes are transported through porous media, integrating the effects of both microscopic diffusion and macroscopic flow-driven dispersion.

1.3.4 NON-NEWTONIAN FLUIDS FLOW

Non-Newtonian fluid flow in porous media is challenging due to the fluids' variable viscosity, which differs significantly from Newtonian behavior. Unlike Newtonian fluids, non-Newtonian fluids have a viscosity that changes with shear rate, introducing nonlinearity into the flow equations. Rheological models like the power-law, Bingham plastic, and Herschel-Bulkley are used to extend Darcy's law to account for shear-rate-dependent viscosity:

$$\mathbf{u} = -\frac{K}{\mu_{\text{eff}}(\dot{\gamma})}(\nabla P - \rho g \nabla z), \tag{1.19}$$

where $\mu_{\text{eff}}(\dot{\gamma})$ is the effective viscosity dependent on shear rate, and K is the specific permeability.

The Ostwald-de Waele power-law model describes the relationship between shear stress and shear rate:

$$\tau = K\dot{\gamma}^n, \tag{1.20}$$

with τ as shear stress, K as the consistency coefficient, $\dot{\gamma}$ as shear rate, and n as the flow behavior index. Although specific models by Christopher and Middleman, and Dharmadhikari and Kale are not available, permeability relations can be conceptualized as:

$$K = f(n) = \alpha \cdot n^{\beta}, \tag{1.21}$$

$$K = g(n, \phi) = \frac{\phi^n}{1 + \gamma \cdot n}, \tag{1.22}$$

where K is permeability, n is the power-law index, and α, β, and γ are empirical or theoretical coefficients.

1.3.5 BEYOND DARCY'S LAW

It is worth noting that Darcy's law is strictly valid only for laminar, low-velocity flow regimes within porous media. When conditions deviate from the assumptions that underlie Darcy's law, such as higher flow velocities, the onset of inertial effects, or even turbulence, more advanced models are required. These models introduce additional terms or couple different governing equations to capture the complexities of fluid flow beyond the Darcy regime.

Figure 1.5 illustrates the flow regimes of a fluid in a porous medium as characterized by the Reynolds number, a dimensionless quantity used to predict flow patterns. For very low Reynolds numbers ($Re \ll 1$), the flow is generally considered to be within the strict Darcy regime: it is fully laminar, and Darcy's law provides a linear relationship between flux and pressure gradient. As the Reynolds number increases (on the order of $1 < Re < 10$, depending on permeability and other factors), inertial effects begin to emerge,

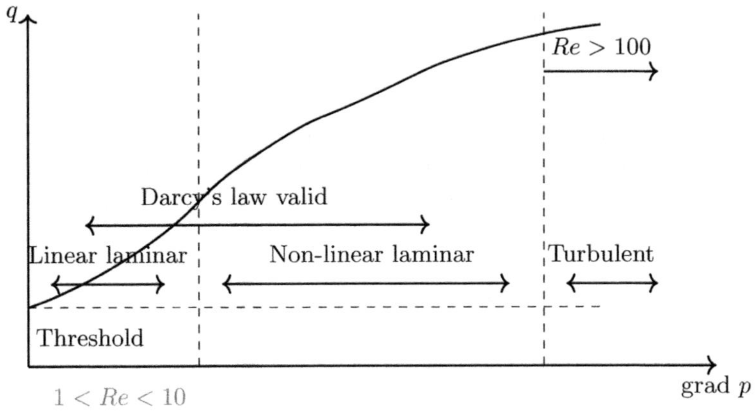

Figure 1.5 A schematic diagram for flows beyond Darcy's law. (Adapted from [6]).

causing deviations from strict linearity. In such a transition regime, Darcy's law alone may not sufficiently capture the flow behavior. At even higher Reynolds numbers (approaching or exceeding $Re \approx 100$), the flow becomes increasingly influenced by inertia and may eventually become turbulent. Under these conditions, Darcy's law is no longer adequate, and non-linear or more comprehensive formulations must be employed.

Forchheimer Model:

The Forchheimer model introduces an additional inertial term to Darcy's law. While Darcy's law can be viewed as a linear relationship between velocity and pressure gradient, the Forchheimer equation adds a velocity-squared term, capturing the non-linear dependence that arises when inertial forces become significant but before fully developed turbulence sets in. In its simplest form, the Forchheimer equation can be written as:

$$\nabla p - \mu \frac{\mathbf{u}}{K} + \beta \rho |\mathbf{u}| \mathbf{u}$$

where μ is the fluid viscosity, K is the permeability, ρ is the fluid density, $\mathbf{u}$ is the velocity field, and β is the Forchheimer coefficient. The second term on the right-hand side introduces the non-linear dependence on velocity, making this model suitable for intermediate Reynolds number flows.

Brinkman Model:

The Brinkman equation serves as a bridging model between Darcy and Navier–Stokes equations. It modifies Darcy's law to include a Laplacian (viscous diffusion) term for the velocity field. This approach captures effects such as boundary layers near solid walls and partial penetration into porous media, extending Darcy's applicability to cases where simple linear flow assumptions break down:

$$-\nabla p + \mu_e \nabla^2 \mathbf{u} = \mu \frac{\mathbf{u}}{K}$$

Here, μ_e is an effective viscosity that may differ from the fluid's bulk viscosity. The Brinkman model is often employed in domains featuring both free fluid flow and flow through porous regions, ensuring a smoother transition between these regimes.

Navier–Stokes–Darcy Coupling:

For configurations involving free flow adjacent to porous media, such as groundwater movement interacting with open channels or rivers, a coupled Navier–Stokes–Darcy approach may be used. In this hybrid formulation, the Navier–Stokes equations govern flow in the open-fluid region, while Darcy's law (or one of its aforementioned modifications) governs flow in the porous medium. Appropriate interface conditions ensure continuous momentum and mass transfer across the boundary. Such coupled models are crucial in environmental and industrial applications, where free and porous flows coexist and influence each other.

Non-Darcy Turbulent Models:

When the flow becomes fully turbulent within the porous structure (at sufficiently high Reynolds numbers), even Forchheimer-type corrections are insufficient. Models that incorporate turbulence closures, such as Reynolds–Averaged Navier–Stokes (RANS) equations adapted for porous media, are required to accurately predict velocity distributions, pressure losses, and mixing processes. These methods are more complex and rely on additional parameters derived from experiments or direct numerical simulations.

1.3.6 COMPRESSIBLE FLUID FLOW IN POROUS MEDIA

Unlike incompressible fluids, studied in the previous chapters, compressible fluids are those whose density changes significantly in response to pressure changes. Gases are the most common example of compressible fluids. As they move through porous media, the pressure, temperature, and density of the gas can vary significantly. Understanding compressible fluid flow in porous media is crucial in several fields. In petroleum engineering, it helps predict and enhance the recovery of oil and gas from reservoirs. In environmental engineering, it's important to assess the migration of gases in soils, such as in landfills or contaminated sites. In chemical engineering, this knowledge is used in designing and operating filters and reactors. Modeling compressible fluid flow in porous media is challenging due to the need to account for variations in fluid properties and complex interactions with the porous matrix.

For an illustration of compressible fluid flow in porous media, let us examine the scenario of a single-phase gas flow within an underground reservoir. The gas is assumed to be ideal, and the flow is considered to be uniform in all directions (isotropic), without any absorption or release of gas. Being the reservoir flow is isotropic means that the permeability $\mathbf{K} = k\mathbf{I}$, where k is a scalar and $\mathbf{I}$ is the identity tensor. The equation governing mass balance in this context is:

$$\frac{\partial(\phi\rho)}{\partial t} + \nabla \cdot (\rho \mathbf{u}) = q \tag{1.23}$$

Using the ideal gas law, the mass density can be written as $\rho = \gamma p$. The velocity may be expressed using Darcy's law; therefore, Eq. (8.33) becomes,

$$\gamma\phi\frac{\partial p}{\partial t} - \nabla \cdot \left(\frac{\gamma k}{\mu} p \nabla p\right) = q \tag{1.24}$$

where $\gamma = \frac{M}{RT}$, R is the gas constant, T is the temperature, and M is the molecular weight.

IDEAL GAS LAW

The simplest equation of state describing pressure-volume-temperature (PVT) behavior for gases in porous media is the ideal gas law:

$$PV = nRT \tag{1.25}$$

where P is the absolute pressure, V is the volume occupied by the gas, n is the moles of gas, R is the gas constant, and T is the absolute temperature

The density ρ_g of gas in porous media can be derived using the ideal gas law:

$$\rho_g = \frac{PM_w}{RT} \tag{1.26}$$

where M_w is the molecular weight of the gas. Specific gravity γ_g is the ratio of the gas density to the density of air under the same conditions:

$$\gamma_g = \frac{\rho_g}{\rho_{\text{air}}} \tag{1.27}$$

REAL GAS LAW

The ideal gas law assumes perfect behavior, but real gases deviate due to intermolecular forces and kinetic motion. At high temperatures and low pressures, gases approximate ideal behavior.

At low pressures, real gases tend to behave similarly to an ideal gas. As pressure increases leading to a decrease in pV, as observed in Figure 1.6, C_o and CH_2H_2. At very high pressures, causing pV to rise sharply. This behavior is evident in gases like helium (He), which have weak intermolecular forces.

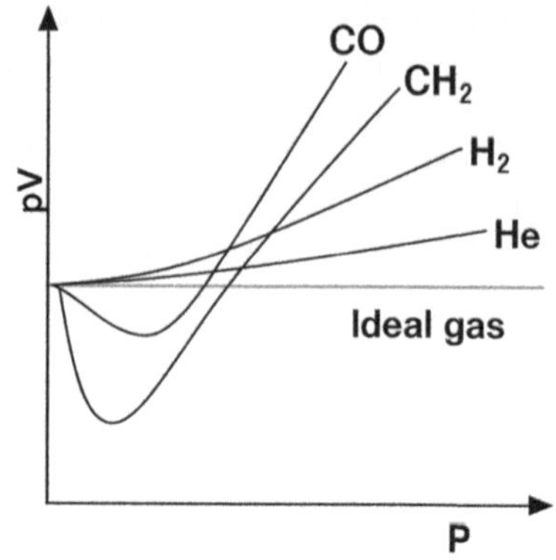

Figure 1.6 Deviation of real gases from ideal gas behavior. The graph compares the pV versus P behavior for different real gases (CO, CH_2H_2, H_2, He) against the ideal gas law (Adapted from [67]).

The gas deviation factor (z) quantifies deviations:

$$z = \frac{V_a}{V_i} \tag{1.28}$$

where V_a is the actual volume and V_i is the ideal volume. The real gas law incorporates z:

$$pV_a = znRT \tag{1.29}$$

Deviations from ideal behavior can lead to errors in calculations if z is omitted. Compressibility quantifies the fractional change in volume with pressure:

$$c_g = -\frac{1}{V}\frac{dV}{dp} \tag{1.30}$$

Gas compressibility is significantly higher than that of oil or water, impacting fluid flow in porous media.

EQUATIONS OF STATE

The Van der Waals equation represents a significant advancement in understanding the behavior of real gases by incorporating molecular size and intermolecular forces into the ideal gas law. It is expressed as:

$$\left[p + a\left(\frac{n}{V}\right)^2\right][V - nb] = nRT \tag{1.31}$$

where p is the pressure of the gas, V is the volume of the gas, n is the number of moles, R is the universal gas constant, T is the absolute temperature, a

is the parameter accounting for attractive forces between molecules, and b is the parameter accounting for the finite size of gas molecules.

The Van der Waals equation serves as the foundation for modern cubic equations of state, which aim to refine predictions of gas behavior across a broader range of thermodynamic conditions.

Several modifications to the Van der Waals equation have been proposed to enhance its predictive accuracy for real gases. These improved cubic equations of state introduce additional temperature- and pressure-dependent parameters:

Redlich-Kwong Equation [8]

$$p = \frac{RT}{V-b} - \frac{a}{V(V+b)\sqrt{T}} \tag{1.32}$$

Introduced to better handle high-temperature and low-pressure conditions, this equation adds a temperature-dependent attractive term.

Soave-Redlich-Kwong Equation [9]

$$p = \frac{RT}{V-b} - \frac{a(T)}{V(V+b)} \tag{1.33}$$

This equation replaces the square root of temperature with a more generalized function $a(T)$, improving accuracy for hydrocarbons and other non-polar fluids.

Peng-Robinson Equation [10]

$$p = \frac{RT}{V-b} - \frac{a(T)}{V(V+b)+b(V-b)} \tag{1.34}$$

The Peng-Robinson equation enhances phase equilibrium calculations by improving predictions for both liquid and vapor phases.

Stryjek-Vera-Peng-Robinson Equation [11]

$$p = \frac{RT}{V-b} - \frac{a(\theta,T)}{V(V+b)+b(V-b)} \tag{1.35}$$

By introducing a temperature correction term θ, this equation accounts for polar and associating fluids, such as alcohols and water.

Patel-Teja Equation [12]

$$p = \frac{RT}{V-b} - \frac{a(T)}{V(V+b)+c(V-b)} \tag{1.36}$$

Designed to accommodate a wider range of fluids, the Patel-Teja equation adds an additional parameter c to improve flexibility in thermodynamic predictions.

Figure 1.7 Anisotropic rock in subsurface formations.

1.4 ANISOTROPIC POROUS MEDIA

Anisotropic porous media are geological formations where key properties, especially permeability, vary with direction. Unlike isotropic media, where permeability is the same in all directions, anisotropic media exhibit directional dependence, meaning fluids flow more easily in certain directions than others. This anisotropy is especially important in subsurface environments such as sedimentary basins, aquifers, and hydrocarbon reservoirs, where stratification, layering, and mineral alignment lead to preferential flow paths. These directional variations directly influence the movement of water, oil, gas, and contaminants, making anisotropy a critical factor in groundwater modeling, reservoir simulation, and environmental remediation (Figure 1.7).

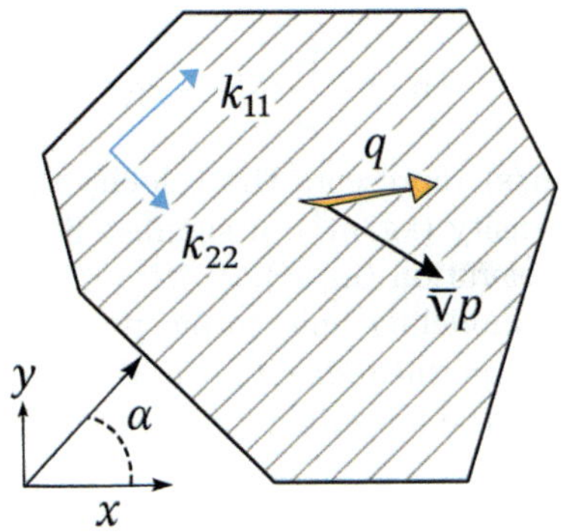

Figure 1.8 Pressure gradient vector against flux vector in anisotropic porous media. Unlike isotropic media, where these vectors are parallel, anisotropy causes them to diverge due to directional variations in permeability, necessitating a tensorial representation of flow (Adapted from [16]).

The development of anisotropy results from long-term physicochemical and mechanical processes such as deposition, compaction, and tectonic stress that shape the structure of porous media over geological timescales. These processes produce formations with permeability that is typically higher along bedding planes and significantly lower in the perpendicular direction.

Permeability in porous media is broadly classified as either isotropic or anisotropic. In isotropic media, permeability is uniform in all directions. In contrast, anisotropic media require a more complex mathematical treatment, where permeability is represented by a second-order tensor that varies with direction. The principal directions of this tensor often align with the geological layering, capturing the inherent directional bias in fluid flow.

The concept of anisotropic permeability dates back to Darcy's 1856 experiments. While his formulation assumed isotropy, later research showed that in anisotropic media, the pressure gradient and flow vectors are not aligned, as illustrated in Figure 1.8. This non-parallel behavior demands a tensor-based approach to accurately describe flow.

In anisotropic media, permeability along the x, y, and z directions typically differs. These components are described by the principal values of the permeability tensor. While heterogeneity can be handled using a spatially varying scalar function $k(x, y, z)$, anisotropy requires a tensor framework to capture its directional nature.

Sedimentary deposition remains the most common origin of anisotropy. When coordinate axes are aligned with the principal directions, the permeability tensor simplifies to a diagonal form. If misaligned, off-diagonal terms appear, indicating cross-directional flow behavior [1, 13].

The momentum conservation equation governing single-phase flow in porous media is expressed through Darcy's law:

$$\mathbf{u} = -\frac{\mathbf{K}}{\mu}(\nabla p - \rho \mathbf{g}) \tag{1.37}$$

In this equation, $\mathbf{u}$ (m/s) represents the velocity vector of the single-phase flow, ρ denotes fluid density (kg/m^3), μ signifies fluid viscosity (Pa.s), and g is the gravitational acceleration (m/s^2).

A critical aspect of modeling anisotropic porous media is the appropriate representation of the permeability tensor $\mathbf{K}$. We treat permeability as a second-order tensor with the following structure:

$$\mathbf{K} = \begin{bmatrix} K_{xx} & K_{xy} & K_{xz} \\ K_{yx} & K_{yy} & K_{yz} \\ K_{zx} & K_{zy} & K_{zz} \end{bmatrix} \tag{1.38}$$

Consequently, neglecting the gravity effect, Eq. (1.37) can be expanded to explicitly show the tensor-vector multiplication:

$$\begin{bmatrix} u_x \\ u_y \\ u_z \end{bmatrix} = -\frac{1}{\mu} \begin{bmatrix} K_{xx} & K_{xy} & K_{xz} \\ K_{yx} & K_{yy} & K_{yz} \\ K_{zx} & K_{zy} & K_{zz} \end{bmatrix} \begin{bmatrix} \frac{\partial p}{\partial x} \\ \frac{\partial p}{\partial y} \\ \frac{\partial p}{\partial z} \end{bmatrix} \tag{1.39}$$

The permeability tensor in Eq. (1.38) possesses several important mathematical properties. It is assumed to be symmetric and positive definite, with potentially non-zero off-diagonal entries. The elements along the principal diagonal (K_{xx}, K_{yy}, and K_{zz}) represent the magnitude of permeability along the principal coordinate axes. The off-diagonal elements play a crucial role in characterizing cross-flow phenomena, where pressure gradients in one direction induce flow components in perpendicular directions. For example, the K_{xz} element contributes to flow in the x-direction resulting from pressure differentials in the z-direction.

For three-dimensional systems, the complete permeability tensor contains nine elements as shown in Eq. (1.38), while two-dimensional systems require four elements. Due to the symmetry assumption ($K_{xy} = K_{yx}$, $K_{xz} = K_{zx}$, $K_{yz} = K_{zy}$), only six independent elements need to be determined in the three-dimensional case. This symmetry property implies that permeability values in opposite directions are equivalent, ensuring that the symmetric tensor transforms principal coordinate axes without causing system deformation [14, 15, 17, 18].

In many subsurface scenarios, porous media exhibit lateral isotropy combined with vertical anisotropy. If we designate the coordinate axes along the principal permeability directions as (x_0, y_0, z_0), the permeability tensor assumes the following simplified form:

$$\mathbf{K_0} = \begin{bmatrix} K_H & 0 & 0 \\ 0 & K_H & 0 \\ 0 & 0 & K_V \end{bmatrix} \tag{1.40}$$

In practical geological settings, however, formation bedding rarely aligns perfectly with the horizontal plane. Instead, these formations typically form an angle θ with the horizontal. When constructing reservoir models using horizontal and vertical orientations, the original permeability tensor $\mathbf{K}$ must undergo rotation through a transformation matrix $R_1(\theta)$:

$$R_1 = \begin{bmatrix} \cos\theta & 0 & \sin\theta \\ 0 & 1 & 0 \\ -\sin\theta & 0 & \cos\theta \end{bmatrix} \tag{1.41}$$

Following this coordinate transformation, the permeability tensor in the new coordinate system (x_1, y_1, z_1) takes the form:

$$K_1 = R_1^T K_0 R_1 = \begin{bmatrix} K_{xx} & 0 & K_{xz} \\ 0 & K_H & 0 \\ K_{xz} & 0 & K_{zz} \end{bmatrix} \tag{1.42}$$

where the components are related to the principal permeabilities and rotation angle by:

$$K_{xx} = K_H \cos^2\theta + K_V \sin^2\theta \tag{1.43}$$

$$K_{xz} = (K_H - K_V) \sin\theta \cos\theta \tag{1.44}$$

$$K_{zz} = K_H \sin^2\theta + K_V \cos^2\theta \tag{1.45}$$

This represents the comprehensive anisotropic permeability model that accounts for both the intrinsic directional properties of the medium and their orientation relative to the chosen coordinate system.

A critical parameter in anisotropy studies is the anisotropy ratio, defined as the ratio between permeability values in the principal directions, as depicted in Figure 1.9. A higher anisotropy ratio indicates a lower k_{22} value for a given k_{11} permeability component, resulting in more pronounced directional flow behavior.

Anisotropic layer flow patterns in porous media of various anisotropy angles: 30°, 45°, and 60°, the permeability tensor is

$$k = \begin{bmatrix} 9.869 & 0 \\ 0 & 9.869 \end{bmatrix} \times 10^{-14}\,\text{m}^2.$$

In the anisotropic layer, the permeability varies with orientation angle θ. At $\theta = 30°$, the permeability tensor is

$$k = \begin{bmatrix} 7.649 & 3.846 \\ 3.846 & 3.208 \end{bmatrix} \times 10^{-14}\,\text{m}^2;$$

at $\theta = 45°$, it is

$$k = \begin{bmatrix} 5.428 & 4.441 \\ 4.441 & 5.428 \end{bmatrix} \times 10^{-14}\,\text{m}^2;$$

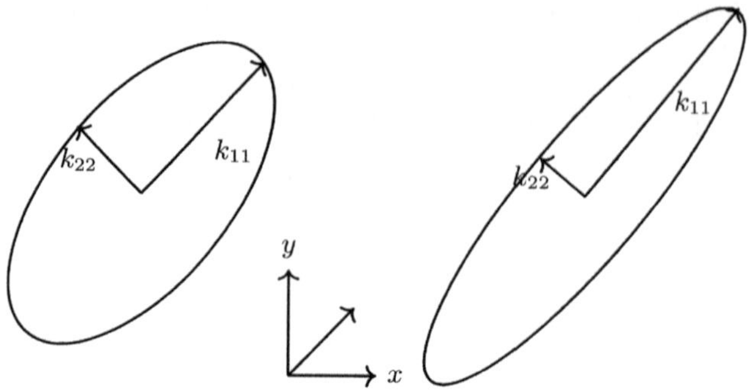

Figure 1.9 Illustration of the anisotropy ratio, defined as the ratio of permeability components in the principal directions (k_{11}/k_{22}). This ratio quantifies the degree of anisotropy, where a larger value reflects a greater disparity between horizontal and vertical permeability, influencing the directional behavior of fluid flow in porous media (Adapted from [16]).

and at $\theta = 60°$, it is

$$k = \begin{bmatrix} 3.208 & 3.846 \\ 3.846 & 7.649 \end{bmatrix} \times 10^{-14}\,\text{m}^2.$$

1.5 CONCLUSION

This chapter lays the groundwork for understanding transport in porous media by establishing principles and models essential to both theoretical and applied studies. It begins by defining porous materials and their structural properties, then progresses through fluid behavior, flow regimes, and the mathematical tools required to model them. The chapter emphasizes the importance of porosity, permeability, fluid properties, and scale in shaping flow behavior. It also introduces key models for dealing with non-Newtonian fluids and compressible gas transport. By combining physical insight with mathematical rigor, this chapter equips readers with the necessary foundation to tackle more advanced topics in subsequent chapters.

REVIEW QUESTIONS

1. Define the term "porous media" and describe the main types of porous materials. Provide examples of each.
2. What are the key properties of porous materials, and how do porosity and permeability influence fluid flow?
3. Define porosity and differentiate between total porosity and effective porosity.
4. Explain Darcy's law and its significance in modeling fluid flow through porous media. What are its limitations?

5. Describe the transport mechanisms in porous media, including advection, diffusion, and dispersion. How do they interact?
6. State the assumptions of the ideal gas law. Why does the ideal gas law fail to describe real gases at low temperatures or high pressures?
7. Define the compressibility factor Z. What does it indicate when Z deviates from 1, and why is this important for real gases?
8. What are the main differences between the Van der Waals equation and improved cubic equations of state, such as the Peng-Robinson equation?

DECLARATION

During the preparation of this work, the author used ChatGPT in order to improve the language and readability. After using this tool/service, the author reviewed and edited the content as needed and takes full responsibility for the content of the publication.

2 MATLAB® and Python Implementation of Numerical Methods

2.1 INTRODUCTION TO MATLAB®

MATLAB®, which stands for MATrix LABoratory, is a high-level programming language and interactive environment designed for numerical computing, data analysis, and algorithm development. It is widely used in engineering, scientific research, and applied mathematics due to its powerful built-in functions and toolboxes that simplify complex mathematical calculations. MATLAB® is particularly strong in matrix manipulations, numerical linear algebra, and data visualization, making it an ideal tool for implementing numerical methods.

MATLAB® has long been a preferred environment for numerical simulations due to its specialized toolboxes and intuitive syntax. Higham and Higham [19] offer a rigorous foundation in numerical analysis with MATLAB®, while Hahn et al. [20] provide a practical guide geared toward engineers and applied scientists.

MATLAB® syntax is straightforward and similar to traditional mathematical notation, making it accessible for both beginners and advanced users. Additionally, MATLAB® provides extensive visualization capabilities, enabling users to create high-quality plots and graphical representations of data with minimal effort. MATLAB® also supports integration with other programming languages and software.

This section covers the basic MATLAB® syntax, control flow structures, functions, and file operations, along with examples of how these features can be applied to numerical methods relevant to porous media simulations, which is the theme of this book.

BASIC MATLAB® SYNTAX AND CONCEPTS

MATLAB® is designed to work primarily with matrices and arrays, which are fundamental to many numerical methods. This section introduces some essential MATLAB® operations and concepts.

Display Text: To display the string of the title of this book "Numerical Methods in Porous Media: MATLAB® and Python Approaches" in the command window:

```
disp('Numerical Methods in Porous Media: MATLAB® and Python
    Approaches')
```

DOI: 10.1201/9781032724973-2

The output is:

```
Numerical Methods in Porous Media: MATLAB® and Python Approaches
```

Strings: Text in MATLAB® is handled as character arrays or strings. Here is how to concatenate and display strings:

```
head = 'The book title is:';
title = 'Numerical Methods in Porous Media';
disp([head ' ' title])
```

In this example, `head` and `title` are concatenated using square brackets `[]` and displayed using `disp`. The output is:

```
The book title is: Numerical Methods in Porous Media
```

Arrays: MATLAB® excels at array and matrix operations, which are central to many numerical methods.

```
Twophase = {'water', 'oil', 'rock'};
disp(Twophase)
```

`Twophase` is a cell array containing three string elements. The output is:

```
    'water'    'oil'    'rock'
```

Structures: MATLAB® structures allow data to be organized in key-value pairs, similar to dictionaries in Python (as we will see later in this chapter).

```
Rock.Property = 'Porosity';
Rock.Value = 0.35;
disp(Rock)
```

The structure `Rock` contains fields `Property` and `Value`. The output is:

```
  struct with fields:
    Property: 'Porosity'
    Value: 0.3500
```

Basic Data Types: MATLAB® supports a variety of data types, including integers, floating-point numbers, strings, and logical values.

```
x = 5;
y = 3.14;
disp([x, y])
```

`x` is an integer, and `y` is a floating-point number. The `disp` function displays:

```
    5    3.1400
```

CONTROL FLOW

Control flow in MATLAB® allows for decision-making, looping, and more complex programming constructs.

Conditional Statements: MATLAB® uses `if`, `elseif`, and `else` statements to execute code based on conditions.

```
x = 5;
if x > 0
    disp('x is positive')
elseif x == 0
    disp('x is zero')
else
    disp('x is negative')
end
```

The output will be:

```
x is positive
```

For Loops: For loops are used to iterate over arrays and other data structures.

```
fluids = {'water', 'oil', 'gas'};
for i = 1:length(fluids)
    disp(['The fluid is ' fluids{i}])
end
```

This loop iterates over the elements in the `fluids` array and displays each one. The output is:

```
The fluid is water
The fluid is oil
The fluid is gas
```

While Loops: While loops repeat code as long as a condition is true.

```
saturation = 0.2;
while saturation <= 1.0
    disp(['Current saturation: ' num2str(saturation)])
    saturation = saturation + 0.1;
end
```

The output will be:

```
Current saturation: 0.2
Current saturation: 0.3
Current saturation: 0.4
Current saturation: 0.5
Current saturation: 0.6
Current saturation: 0.7
Current saturation: 0.8
Current saturation: 0.9
Current saturation: 1.0
```

FUNCTIONS AND SCRIPTS

Functions in MATLAB® allow for code modularity and reuse. Scripts are sequences of MATLAB® commands saved in a file and are useful for automating repetitive tasks.

Defining a Function: Functions are defined using the `function` keyword. Here is an example function to calculate porosity:

```
function porosity = calculate_porosity(pore_volume, bulk_volume)
    porosity = pore_volume / bulk_volume;
end
```

This function can be saved as `calculate_porosity.m` and called as follows:

```
porosity = calculate_porosity(0.25, 1.0);
disp(['The␣porosity␣is␣' num2str(porosity)])
```

The output is:

```
The porosity is 0.25
```

Scripts: MATLAB® scripts are files containing a sequence of commands. They are useful for automating tasks and running multiple commands sequentially.

File Input and Output: Reading from and writing to files is straightforward in MATLAB®, allowing for data to be stored and retrieved easily.

Writing to a File:

```
fileID = fopen('reservoir_data.txt','w');
fprintf(fileID, 'Porosity:␣\%0.2f\n', 0.25);
fprintf(fileID, 'Permeability:␣\%d␣mD\n', 150);
fclose(fileID);
```

Reading from a File:

```
fileID = fopen('reservoir_data.txt','r');
content = fscanf(fileID, '\%s');
fclose(fileID);
disp(content)
```

The output will be the content of the file:

```
Porosity: 0.25Permeability: 150 mD
```

MATLAB® LIBRARIES AND TOOLBOXES

MATLAB® has a rich ecosystem of toolboxes that extend its capabilities in various domains. For numerical modeling and simulation of fluid flow in porous media, the following toolboxes are particularly relevant:

Partial Differential Equation (PDE) Toolbox: Provides functions for solving partial differential equations, commonly used in modeling flow and transport in porous media.

FEATool MATLAB® CFD Toolbox: The FEATool CFD Toolbox is a MATLAB®-based environment designed to simplify the setup and simulation of complex fluid dynamics problems.

Optimization Toolbox: Useful for parameter estimation, inverse problems, and optimizing numerical models in porous media simulations.

Symbolic Math Toolbox: Aids in deriving analytical solutions, symbolic differentiation, and simplifying governing equations for porous flow.

Curve Fitting Toolbox: Helps in analyzing experimental data related to permeability, porosity, and fluid transport properties.
Statistics and Machine Learning Toolbox: Useful for data-driven modeling and analysis of porous media behavior, including regression models and clustering techniques.
Parallel Computing Toolbox: Enhances computational efficiency by enabling parallel processing of large-scale fluid flow simulations.

PROJECT: ANALYZING FLUID FLOW IN A POROUS MEDIUM

We will use MATLAB® to analyze fluid flow in a porous medium. The project involves loading a dataset, performing calculations, and visualizing the results.
Loading Data:

```
data = load('fluid_flow_data.mat');
disp(data)
```

Visualizing Data:

```
depth = data.Depth;
pressure = data.Pressure;
plot(depth, pressure, '-o')
xlabel('Depth (m)')
ylabel('Pressure (kPa)')
title('Pressure vs Depth in a Porous Medium')
grid on
```

The output will be a plot similar to the one generated in Python, showing the relationship between pressure and depth.

MATLAB® Functions for Numerical Methods

MATLAB's extensive library of built-in functions allows for the implementation of various numerical methods, such as:
Numerical Integration:

```
f = @(x) x.^2 + 3*x + 2;
integral_value = integral(f, 0, 1);
disp(['The integral value is ' num2str(integral_value)])
```

This code calculates the definite integral of the function $f(x) = x^2 + 3x + 2$ from 0 to 1. The output is:

```
The integral value is 3.8333
```

Solving Linear Systems:

```
A = [3, 2; 1, 4];
b = [5; 6];
x = A\b;
disp(['Solution: x1 = ' num2str(x(1)) ', x2 = ' num2str(x(2))])
```

This example solves the linear system $Ax = b$ using MATLAB's backslash operator. The output is:

```
Solution: x1 = 0.8462, x2 = 1.3846
```

OBJECT-ORIENTED PROGRAMMING (OOP) IN MATLAB®

MATLAB® supports object-oriented programming (OOP) and has the concept of a "class." MATLAB® classes allow you to define objects with properties and methods, similar to classes in other object-oriented languages like Python, C++, or Java.

Defining a Class:

```
% Define a class named Fluid
classdef Fluid
    properties
        Name          % Property to store the name of the fluid
        Viscosity     % Property to store the viscosity of the
            fluid
    end
    methods
        % Constructor method
        function obj = Fluid(name, viscosity)
            obj.Name = name;
            obj.Viscosity = viscosity;
        end

        % Method to describe the fluid
        function description = describe(obj)
            description = ['The fluid ', obj.Name, ...
                           ' has a viscosity of ', ...
                           num2str(obj.Viscosity), ' cp.'];
        end
    end
end
```

Using the Class:

```
fluid1 = Fluid('Water', 1.0);
disp(fluid1.describe())
```

The output is:

```
The fluid Water has a viscosity of 1.0 cp.
```

Inheritance:

```
classdef ReservoirFluid < Fluid
    properties
        Density
    end
    methods
        function obj = ReservoirFluid(name, viscosity, density)
            obj@Fluid(name, viscosity);
            obj.Density = density;
        end
        function description = describe(obj)
            description = ['The fluid ', obj.Name, ...
                           ' has a viscosity of ', ...
                           num2str(obj.Viscosity), ...
                           ' cp and a density of ', ...
                           num2str(obj.Density), ' g/cm^3.'];
```

```
        end
    end
end
```

Using Inheritance:

```
brine = ReservoirFluid('Brine', 0.8, 1.2);
disp(brine.describe())
```

The output is:

```
The fluid Brine has a viscosity of 0.8 cp and a density of 1.2 g/
    cm^3.
```

2.2 INTRODUCTION TO PYTHON

Guido van Rossum created Python, a popular high-level programming language that was initially made available in 1991. Python's ease of use and versatility have made it extremely popular over the years. It is applied in many different fields, such as mathematical computation (our focus), system scripting, software engineering, and web development. Because of its versatility, Python can be used for a wide range of tasks. For example, it can be used to develop dynamic web applications on servers, integrate with other applications to automate processes, and connect to database systems for data retrieval and manipulation. Additionally, Python is a powerful tool for handling large amounts of data and carrying out intricate mathematical operations, making it a great choice for tasks ranging from rapid prototyping to the development of software that is ready for production. Cross-platform interoperability, which allows Python to function flawlessly on a variety of operating systems, including Windows, Mac, and Linux, is one of its strongest points. Because of the language's simple syntax, which is similar to that of English, developers can write code more quickly and efficiently—often in fewer lines than they can with other programming languages. Because Python uses an interpreter system, code may be run immediately after it is created, allowing for rapid prototyping. Furthermore, Python allows for the support of several programming paradigms, such as procedural, object-oriented, and functional programming, giving programmers freedom in how they tackle addressing problems.

This book will use Python 3, the most recent major version of the language. While Python 2 is still in use, it is no longer being developed and only receives security patches. The Jupyter Notebook platform is used for code implementation and execution, which is accessed via Anaconda. Download and install the Anaconda distribution, which comes with Jupyter Notebook and many other useful packages pre-installed. (`https://www.anaconda.com/products/distribution`). Open Anaconda Navigator and launch Jupyter Notebook from there. Alternatively, we can open a terminal and type `jupyter notebook`. For further reading, key references include McKinney [22] and Oliphant [23].

BASIC PYTHON SYNTAX AND CONCEPTS

Learning Python allows for interactive coding and immediate feedback. Python ends statements with new lines instead of semicolons or parentheses, like many other computer languages do. Furthermore, Python defines the scope of loops, functions, and classes using indentation and whitespace, whereas other programming languages frequently use curly brackets to indicate scope limits. Python's code structure is understandable and clear thanks to this design decision, which also makes it easy to use for both novice and seasoned developers.

Print: In order to print the string "Numerical Methods in Porous Media: MATLAB® and Python Approaches" to the console.

```
print("Numerical␣Methods␣in␣Porous␣Media:␣MATLAB® and Python
    Approaches")
```

The output becomes:

```
Numerical Methods in Porous Media: MATLAB® and Python Approaches
```

Strings: Textual data in Python is represented by strings. A string is a collection of characters surrounded by double quotes (" ") or single quotes (' '). For example:

```
head = "The␣book␣title␣is:"
title = "Numerical␣Methods␣in␣Porous␣Media"
print(head + "␣" + title)
```

`head` and `title` are strings (sequences of characters). The + operator concatenates the strings, resulting in:

```
The book title is: Numerical Methods in Porous Media
```

Lists: One of Python's four built-in data structures for storing collections of data is a list; the other three have different properties and are called tuples, sets, and dictionaries. The collection of data is stored in a list, which is a sequence data type. Other sequence data types include strings and tuples.

```
Twophase = ["water", "oil", "rock"]
print(Twophase)
```

`Twophase` is a list containing three string elements. The `print` function outputs:

```
['water', 'oil', 'rock']
```

Dictionaries: Data values are stored in key:value pairs using dictionaries. A dictionary is an arranged*, modifiable collection that forbids duplication. For example:

```
Rock = {"Property": "Porosity", "Value": 0.35}
print(Rock)
```

`Rock` is a dictionary with keys `Property` and `Value`. The `print` function outputs:

```
{'Property': 'Porosity', 'Value': 0.35}
```

Tuples: Another built-in data structure in Python for storing collections of data is a tuple. Tuples are immutable, which means that once a tuple is generated, its elements cannot be altered, added, or removed, in contrast to lists. When working with non-modifiable data, tuples are frequently utilized.

```
ReservoirProperties = ("permeability", "porosity", "saturation")
print(ReservoirProperties)
```

`ReservoirProperties` is a tuple containing three string elements. The `print` function outputs:

```
('permeability', 'porosity', 'saturation')
```

Sets: In Python, an unordered collection of distinct elements is stored in a set. Sets are used mostly for actions involving membership testing, deduplication, and mathematical operations like unions and intersections. Unlike lists and tuples, sets do not permit duplicate elements.

```
minerals = {"quartz", "feldspar", "mica", "quartz"}
print(minerals)
```

`minerals` is a set containing three unique string elements. Even though "quartz" is listed twice, sets automatically remove duplicates. The `print` function outputs:

```
{'quartz', 'feldspar', 'mica'}
```

Basic Data Types: If `x` is an integer (whole number), and `y` is a float (a number with a decimal point), the `print` function outputs `5` and `3.14`.

```
x = 5
y = 3.14
print(x, y)
```

The output takes the form:

```
5 3.14
```

CONTROL FLOW

The sequence in which certain words, commands, or function calls are carried out or assessed is referred to as the control flow in programming. Python allows developers to create logic that reacts dynamically to various inputs or conditions by using control flow components like loops, conditionals, and function calls to determine the program's route throughout execution.

Conditional Statements are a fundamental control flow mechanism that allow a program to execute certain blocks of code based on whether specific conditions are met. In Python, conditional statements include `if`, `elif`, and `else` clauses, which evaluate expressions and run different pieces of code depending on whether the conditions are true or false. This enables the creation of programs that can make decisions and perform different actions depending on varying circumstances.

```
x = 5
if x > 0:
    print("x␣is␣positive")
elif x == 0:
    print("x␣is␣zero")
else:
    print("x␣is␣negative")
```

In this example, **if**, **elif**, and **else** are used to execute code based on conditions. The code prints a message depending on whether x is positive, zero, or negative.

```
x is positive
```

LOOPS

Loops are a fundamental control flow structure in programming that allow code to be executed repeatedly based on a condition or over a sequence of elements. They are essential for performing repetitive tasks, iterating over collections of data, or automating processes that require repeated actions.

For Loop: A **for** loop in Python is used to iterate over a sequence (such as a list, tuple, or string), executing a block of code for each element in the sequence. It is particularly useful when we need to perform the same operation on each item in a collection.

```
fluids = ["water", "oil", "gas"]
for fluid in fluids:
    print(f"The␣fluid␣is␣{fluid}")
```

This loop iterates over each item in the **fluids** list and prints a statement identifying each fluid. The output becomes:

```
The fluid is water
The fluid is oil
The fluid is gas
```

While Loop: A **while** loop in Python repeatedly executes a block of code as long as a specified condition is true. It is typically used when the number of iterations is not known beforehand and depends on a dynamic condition. In the following example, the **while** loop simulates the change in fluid saturation in a porous medium. The variable **saturation** starts at 0.2 and increases by 0.1 in each iteration until it reaches a maximum value of 1.0.

```
saturation = 0.2
while saturation <= 1.0:
    print(f"Current␣saturation:␣{saturation}")
    saturation += 0.1
```

The output of this code is:

```
Current saturation: 0.2
Current saturation: 0.3
Current saturation: 0.4
```

```
Current saturation: 0.5
Current saturation: 0.6
Current saturation: 0.7
Current saturation: 0.8
Current saturation: 0.9
Current saturation: 1.0
```

FUNCTIONS AND MODULES

Functions in Python are reusable blocks of code that perform a specific task. They allow for code modularity and reuse, making programs more organized and easier to maintain. Functions are defined using the `def` keyword, followed by the function name and any parameters it requires.

```
def calculate_porosity(pore_volume, bulk_volume):
    return pore_volume / bulk_volume

porosity = calculate_porosity(0.25, 1.0)
print(f"The porosity is {porosity}")
```

`def` defines a new function called calculate˙porosity that takes two parameters: `pore_volume` and `bulk_volume`. The function calculates and returns the porosity, which is the ratio of pore volume to bulk volume. `print(f"The porosity is {porosity}")` calls the function with specific values and prints the calculated porosity. The output of this code is:

```
The porosity is 0.25
```

Modules in Python are collections of functions, classes, and variables that are grouped together in a file and can be imported into other programs. Modules provide additional functionality that can be reused across different parts of a program or even across different projects.

```
import math
def calculate_flow_rate(permeability, pressure_gradient,
    viscosity):
    return (permeability * pressure_gradient) / viscosity

flow_rate = calculate_flow_rate(100, 50, 10)
print(f"The flow rate is {flow_rate} units")
```

`import` allows we to use external libraries (modules), such as `math`, to access advanced mathematical functions and constants. The `calculate_flow_rate` function computes the flow rate based on permeability, pressure gradient, and viscosity. `print(f"The flow rate is {flow_rate} units")` calls the function with specific values and prints the calculated flow rate. Therefore, the output of the above module is:

```
The flow rate is 500.0 units
```

FUNCTIONS: DEFAULT ARGUMENTS

We can define functions with optional parameters by using default arguments. Python will use the default value indicated if a default parameter value is not supplied at the time the function is called.

```
def calculate_flow_rate(permeability, pressure_gradient,
    viscosity=1.0):
    return (permeability * pressure_gradient) / viscosity

# With default viscosity
flow_rate_default = calculate_flow_rate(100, 50)
print(flow_rate_default)
# Output: 5000.0

# With specified viscosity
flow_rate_specified = calculate_flow_rate(100, 50, viscosity=2.0)
print(flow_rate_specified)
# Output: 2500.0
```

FUNCTIONS: KEYWORD ARGUMENTS

By giving the names of the parameters when calling a function with keyword arguments, we can make the function call easier to comprehend and modify the argument's order.

```
def calculate_flow_rate(permeability, pressure_gradient,
    viscosity):
    return (permeability * pressure_gradient) / viscosity

flow_rate = calculate_flow_rate(viscosity=1.0, permeability=100,
    pressure_gradient=50)
print(flow_rate)
# Output: 5000.0
```

Variable-Length Arguments (*args and **kwargs)

Variable-length arguments allow a function to accept an arbitrary number of arguments. *args is used for non-keyword arguments, and **kwargs is used for keyword arguments.

```
def calculate_total_flow_rate(*flow_rates):
    return sum(flow_rates)

total_flow = calculate_total_flow_rate(100, 150, 200)
print(total_flow)
# Output: 450

def print_fluid_properties(**properties):
    for key, value in properties.items():
        print(f"{key}: {value}")

print_fluid_properties(Fluid="Oil", Viscosity=0.8, Density=900)
```

```
# Output:
# Fluid: Oil
# Viscosity: 0.8
# Density: 900
```

LAMBDA FUNCTIONS

Lambda functions are defined with the `lambda` keyword and are anonymous short functions. They are frequently employed for quick tasks, especially when filtering or sorting are involved.

```
flow_rate = lambda permeability, pressure_gradient, viscosity: (
    permeability * pressure_gradient) / viscosity

result = flow_rate(100, 50, 1.0)
print(result)
# Output: 5000.0
```

FILE INPUT, OUTPUT, OPENING, WRITING, AND READING OPERATIONS

File input/output (I/O) operations are essential for reading from and writing to files, allowing for permanent data storage and retrieval in Python. These operations are particularly beneficial when working with data from external sources or when saving the results of computations.

Python provides built-in functions to handle file operations efficiently. We can open a file in various modes depending on what we want to do:

Read mode (`'r'`): Opens a file for reading. The file must already exist; otherwise, an error will occur.
Write mode (`'w'`): Opens a file for writing. If the file doesn't exist, it will be created. If the file exists, its contents will be overwritten.
Append mode (`'a'`): Opens a file for writing, but instead of overwriting the file, it appends new content to the end. If the file doesn't exist, it will be created.
Read/Write mode (`'r+'`): Opens a file for both reading and writing. The file must exist; otherwise, an error will occur. This mode allows us to read from and write to the file without overwriting its existing content.

```
# Writing to a file
with open("reservoir_data.txt", "w") as file:
    file.write("Porosity: 0.25\n")
    file.write("Permeability: 150 mD\n")

# Appending to a file
with open("reservoir_data.txt", "a") as file:
    file.write("Saturation: 0.8\n")

# Reading from a file
with open("reservoir_data.txt", "r") as file:
```

```
    content = file.read()
    print(content)
```

open("reservoir_data.txt", "w") opens the file reservoir_data.txt in write mode. If the file doesn't exist, it will be created. Existing content is overwritten. file.write() writes data to the file. In this example, porosity and permeability data are written to the file. open("reservoir_data.txt", "a") opens the file in append mode, allowing additional data to be written at the end of the file without overwriting the existing content. Here, saturation data is appended. open("reservoir_data.txt", "r") opens the file in read mode. file.read() reads the entire content of the file and stores it in the content variable. print(content) prints the content of the file to the console. The output takes the form:

```
Porosity: 0.25
Permeability: 150 mD
Saturation: 0.8
```

Additionally, a reservoir_data.txt file is created with the content shown above. Note that the with statement automatically closes the file after the block of code is executed, even if an error occurs, which is a best practice for file handling.

CSV FILES

CSV (Comma-Separated Values) files are widely used for storing tabular data in a simple text format. Python's built-in csv module and the Pandas library are powerful tools for handling CSV files. For example, for reading CSV files with CSV module, we may have:

```
import csv

with open("fluid_flow_data.csv", "r") as file:
    reader = csv.reader(file)
    for row in reader:
        print(row)
```

import csv imports the csv module. csv.reader(file) reads the CSV file, returning each row as a list of strings. The for row in reader loop iterates over each row in the CSV file and prints it. For writing to CSV files with CSV module, we present this example:

```
import csv

with open("output_flow_data.csv", "w", newline='') as file:
    writer = csv.writer(file)
    writer.writerow(["Depth (m)", "Pressure (kPa)"])
    writer.writerow([100, 200])
    writer.writerow([200, 400])
```

csv.writer(file) creates a writer object for the CSV file. writer.writerow() writes a row to the CSV file. In this case, it writes the header and data rows related to fluid flow.

USING PANDAS FOR CSV OPERATIONS

The Pandas library is particularly suited for working with structured data like CSV files. While the built-in csv module is useful for simple operations, Pandas offers a more comprehensive and efficient approach to handling CSV data, allowing for easy data loading, manipulation, and export. With Pandas, we can perform complex data analysis tasks with just a few lines of code, making it an essential tool for data scientists and engineers. For example:

```
import pandas as pd

df = pd.read_csv("fluid_flow_data.csv")
print(df)

df.to_csv("output_flow_data.csv", index=False)
```

`import pandas as pd` imports the Pandas library. `pd.read_csv("fluid_flow_data.csv")` reads the CSV file into a Pandas DataFrame. `df.to_csv("output_flow_data.csv", index=False)` writes the DataFrame to a new CSV file, without including row indices.

JSON FILES

JSON (JavaScript Object Notation) is a lightweight data interchange format that is easy to read and write for humans and easy to parse and generate for machines. This is an example of reading JSON files:

```
import json

with open("reservoir_properties.json", "r") as file:
    data = json.load(file)
    print(data)
```

`import json` imports the JSON module. `json.load(file)` reads the JSON file and parses it into a Python dictionary. `print(data)` outputs the contents of the dictionary.

While for writing to JSON files, we introduce this example:

```
import json

data = {
    "Porosity": 0.25,
    "Permeability": 150,
    "Fluid": "Water"
}

with open("output_reservoir_properties.json", "w") as file:
    json.dump(data, file, indent=4)
```

`data` is a dictionary containing reservoir properties. `json.dump(data, file, indent=4)` writes the dictionary to a JSON file with an indentation of 4 spaces for readability.

HANDLING FILE PATHS

Python has modules called `os` and `pathlib` that help with file path management, which is important when working with files from multiple operating systems.

`os` Module

Python's `os` module has functions for interacting with directories, file paths, and environment variables, allowing for interaction with the operating system. It is especially helpful for tasks like creating file paths, managing files and directories, and running system commands that need to work with many operating systems. We may make sure that file-related actions are reliable and cross-platform by utilizing the `os` module.

```
import os

file_path = os.path.join("data", "reservoir_data.txt")
print(file_path)
# Output   data/reservoir_data.txt
```

`import os` imports the `os` module. `os.path.join()` joins directory names and filenames into a complete path, ensuring compatibility across different operating systems. `print(file_path)` outputs the complete file path.

`pathlib` Module

The `pathlib` module's object-oriented file path handling technique makes working with file system paths easier to understand. We may quickly and simply conduct file operations, verify that a file exists, and manipulate paths with `pathlib`. This more contemporary method of handling paths is especially helpful for developers who want to create code that is clear, reusable, and compatible with several platforms.

```
from pathlib import Path

file_path = Path("data") / "reservoir_data.txt"
print(file_path)
# Output   data/reservoir data.txt
```

`from pathlib import Path` imports the `Path` class from the `pathlib` module. `Path("data") / "reservoir_data.txt"` constructs a file path using the / operator for easier path handling. `print(file_path)` outputs.

Error Handling

A fault that happens while the program is being executed is called an exception. Except in cases where the code is syntactically accurate but something goes wrong during execution, exceptions arise instead of syntax errors, which occur when the Python interpreter is unable to grasp the code. For example, dividing by zero, trying to access an index that does not exist, or working with a file that cannot be found.

Try, Except, Else, and Finally Blocks

Python provides a structured way to handle exceptions using the `try`, `except`, `else`, and `finally` blocks. This structure allows us to catch and handle errors, execute specific code if no exceptions occur, and clean up resources no matter what.

```
try:
    result = 10 / 0
except ZeroDivisionError:
    print("Error: You can't divide by zero!")
else:
    print("Division successful!")
finally:
    print("This will run no matter what.")

# Ouput: Error: You can't divide by zero!
#        This will run no matter what.
```

The `try` block contains the code that might throw an exception. The `except` block catches the specific `ZeroDivisionError` and handles it by printing an error message. The `else` block contains code that runs only if no exceptions occur in the `try` block. The `finally` block contains code that runs no matter what, often used for cleanup actions like closing files or releasing resources. The output of this program is:

```
Error: You can't divide by zero!
This will run no matter what.
```

Raising and Customizing Exceptions

Sometimes, we may want to raise an exception to signal that something has gone wrong. Python allows us to do this with the `raise` statement. We can also define custom exceptions by creating new classes that inherit from Python's built-in `Exception` class.

```
def calculate_pressure(depth):
    if depth < 0:
        raise ValueError("Depth cannot be negative.")
    return 10 * depth

try:
    pressure = calculate_pressure(-100)
except ValueError as e:
    print(f"Error: {e}")

# Output  Error: Depth cannot be negative.
```

`raise ValueError("Depth cannot be negative.")` raises a `ValueError` exception if an invalid depth is provided. The `try` block attempts to calculate the pressure with an invalid depth. The `except` block catches the `ValueError` and prints a custom error message. The output of this code is:

```
Error: Depth cannot be negative.
```

PYTHON LIBRARIES

This section introduces key Python libraries that are essential for data manipulation, analysis, and visualization. We will cover NumPy, Pandas, and Matplotlib, which are widely used in data science, engineering, and scientific computing.

NumPy: Operations on Arrays

NumPy (Numerical Python) is a powerful library for numerical computing in Python. It provides support for large, multi-dimensional arrays and matrices, along with a collection of mathematical functions to operate on these arrays. NumPy is crucial for efficient data manipulation and numerical computations and serves as the foundation for many other data science libraries. In the context of fluid flow in porous media, NumPy can be used to handle and manipulate large datasets related to properties such as permeability, porosity, and fluid saturation.

```
import numpy as np

# Creating an array of porosity values
porosity = np.array([0.2, 0.25, 0.3, 0.35])
print("Porosity␣array:", porosity)

# Performing mathematical operations
adjusted_porosity = porosity * 1.1
print("Adjusted␣Porosity␣array:", adjusted_porosity)
```

`import numpy as np` imports the NumPy library and aliases it as `np`. `np.array([0.2, 0.25, 0.3, 0.35])` creates a NumPy array of porosity values. The array allows for element-wise operations, such as multiplying each value by 1.1 to adjust the porosity. The output takes the form:

```
Porosity array: [0.2  0.25 0.3  0.35]
Adjusted Porosity array: [0.22 0.275 0.33  0.385]
```

Example:

```
import numpy as np
permeabilities = np.array([100, 150, 200, 250])
print(permeabilities)
```

`np.array([100, 150, 200, 250])` creates a NumPy array representing permeability values in millidarcies. The output is:

```
[100 150 200 250]
```

Pandas: Data Manipulation and Analysis

Pandas is a powerful Python library for data manipulation and analysis, particularly suited for structured data. It provides data structures like data frames, which are ideal for handling and analyzing data from experiments or simulations in fluid flow through porous media. Pandas simplifies the process of loading, manipulating, and analyzing data, making it an essential tool in the data science toolkit.

```
import pandas as pd

# Creating a DataFrame with reservoir properties
data = {
    "Porosity": [0.2, 0.25, 0.3],
    "Permeability": [100, 150, 200],
    "Depth": [1000, 1500, 2000]
}

df = pd.DataFrame(data)
print(df)

# Calculating new column
df["Hydraulic␣Conductivity"] = df["Permeability"] * df["Porosity"
    ]
print(df)
```

`import pandas as pd` imports the Pandas library and aliases it as `pd`. `pd.DataFrame(data)` creates a DataFrame from a dictionary containing reservoir properties like porosity, permeability, and depth. A new column `"Hydraulic Conductivity"` is calculated by multiplying permeability and porosity. The output is:

```
   Porosity  Permeability  Depth
0      0.20           100   1000
1      0.25           150   1500
2      0.30           200   2000

   Porosity  Permeability  Depth  Hydraulic Conductivity
0      0.20           100   1000                    20.0
1      0.25           150   1500                    37.5
2      0.30           200   2000                    60.0
```

Example:

```
import pandas as pd
data = {"Porosity": [0.2, 0.25, 0.3], "Permeability": [100, 150,
    200]}
df = pd.DataFrame(data)
print(df)
```

`data` is a dictionary where keys are property names like "Porosity" and "Permeability", and values are lists of corresponding measurements. `pd.DataFrame(data)` creates a DataFrame from the dictionary, organizing the data in a tabular format. The output is:

```
   Porosity  Permeability
0      0.20           100
1      0.25           150
2      0.30           200
```

Matplotlib: Customizing and Saving Plots

Matplotlib is a comprehensive library for creating static, animated, and interactive visualizations in Python. It is widely used in scientific computing for plotting data, making it an essential tool for visualizing results in fluid flow studies within porous media. With Matplotlib, we can create a wide range of plots, including line graphs, scatter plots, bar charts, and histograms, to better understand the behavior of fluids in porous materials. Let us present this example:

```
import matplotlib.pyplot as plt
pressures = [200, 400, 600, 800, 1000]
depths = [100, 200, 300, 400, 500]
plt.plot(depths, pressures)
plt.xlabel('Depth (m)')
plt.ylabel('Pressure (kPa)')
plt.title('Pressure vs Depth in a Porous Medium')
plt.show()
```

`plt.plot(depths, pressures)` creates a line graph plotting pressure as a function of depth in a porous medium. `plt.xlabel('Depth (m)')` and `plt.ylabel('Pressure (kPa)')` label the x-axis and y-axis, respectively. `plt.title('Pressure vs Depth in a Porous Medium')` adds a title to the plot. `plt.show()` displays the plot.
The output appears as a plot window showing a line graph where pressure increases with increasing depth, demonstrating the relationship between pressure and depth in a porous medium (see Figure 2.1).

SciPy: Working with Mathematical and Statistical Functions

Based on NumPy, SciPy is a library that offers more scientific computing features. SciPy is essential in numerical computations as it contains modules for statistics, eigenvalue problems, integration, optimization, and interpolation. We can carry out a wide range of statistical tests and procedures with SciPy's extensive collection of statistical analysis routines. For example:

```
from scipy import stats

# Generating a dataset of porosity values
porosity_values = [0.2, 0.25, 0.22, 0.24, 0.26, 0.23]

# Calculating mean and standard deviation
mean = stats.tmean(porosity_values)
std_dev = stats.tstd(porosity_values)

print(f"Mean Porosity: {mean}")
print(f"Standard Deviation: {std_dev}")
```

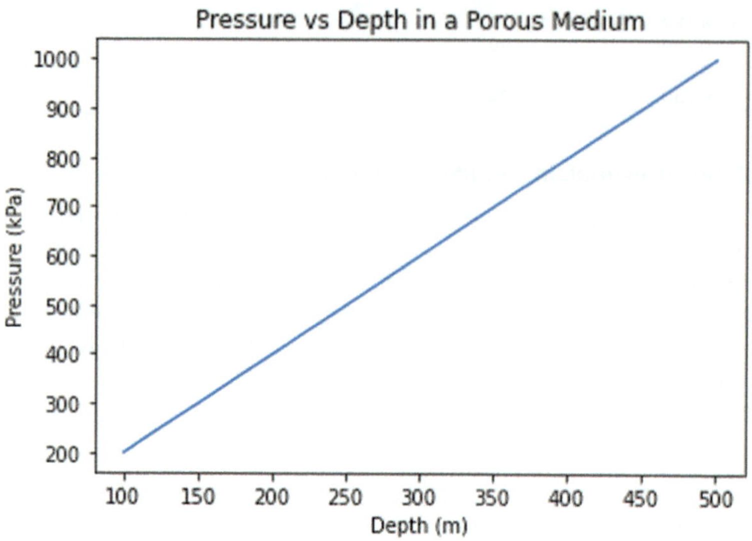

Figure 2.1 Plot output.

`from scipy import stats` imports the statistics module from SciPy. `stats.tmean(porosity_values)` calculates the mean of the porosity values. `stats.tstd(porosity_values)` calculates the standard deviation of the porosity values. The output becomes:

```
Mean Porosity: 0.23333333333333334
Standard Deviation: 0.019531249999999997
```

PROJECT

In this project, we analyze fluid flow data related to a porous medium. The dataset includes properties like porosity, permeability, depth, and pressure. To begin our analysis, we will load the fluid flow dataset using Pandas. The dataset is typically stored in a CSV file and contains columns such as porosity, permeability, depth, and pressure.

Note that this is a hypothetical example, and numbers may not be physically correct.

```
import pandas as pd

# Load the dataset
df = pd.read_csv("fluid_flow_data.csv")

# Display the first few rows of the dataset
print(df.head())

# Display summary statistics
print(df.describe())
```

pd.read_csv("fluid_flow_data.csv") loads the dataset into a Pandas DataFrame. df.head() displays the first few rows of the dataset to give an overview of its structure. df.describe() provides summary statistics, such as mean, standard deviation, and quartiles, for each numerical column in the dataset. The first few rows of the dataset and the summary statistics will be printed to give an initial understanding of the data. The output of this code is:

```
 Porosity  Permeability  Depth  Pressure
0       0.20            100   1000       100
1       0.25            150   1500       150
2       0.30            200   2000       200
3       0.35            250   2500       250
4       0.22            120   1200       120
        Porosity  Permeability       Depth     Pressure
count  10.000000     10.000000    10.00000   10.000000
mean    0.271000    171.000000  1710.00000  171.000000
std     0.046056     46.055522   460.55522   46.055522
min     0.200000    100.000000  1000.00000  100.000000
25\%     0.242500    142.500000  1425.00000  142.500000
50\%     0.270000    170.000000  1700.00000  170.000000
75\%     0.297500    197.500000  1975.00000  197.500000
max     0.350000    250.000000  2500.00000  250.000000
```

After loading and exploring the data, the next step is to perform data manipulation. We will calculate the hydraulic conductivity for each entry in the dataset using the formula: Hydraulic Conductivity = Permeability * Porosity.

```
# Calculate Hydraulic Conductivity
df["Hydraulic␣Conductivity"] = df["Permeability"] * df["Porosity"
    ]

# Add a new column for pressure gradient
df["Pressure␣Gradient"] = df["Pressure"] / df["Depth"]

# Display the updated DataFrame
print(df.head())
```

df["Hydraulic Conductivity"] adds a new column to the DataFrame by calculating the product of permeability and porosity. df["Pressure Gradient"] adds another column that calculates the pressure gradient as the ratio of pressure to depth. print(df.head()) displays the first few rows of the updated DataFrame with the new columns. The output will show the first few rows of the dataset, now including columns for Hydraulic Conductivity and Pressure Gradient. Here is the output:

```
   Porosity  Permeability  Depth  Pressure  Hydraulic
        Conductivity  \
0       0.20            100   1000       100
    20.0
1       0.25            150   1500       150
    37.5
```

```
2       0.30                200    2000        200
     60.0
3       0.35                250    2500        250
     87.5
4       0.22                120    1200        120
     26.4

   Pressure Gradient
0                 0.1
1                 0.1
2                 0.1
3                 0.1
4                 0.1
```

Visualization is key to understanding the relationships between different variables in the dataset. We will use Matplotlib to create plots that highlight these relationships.

```
import matplotlib.pyplot as plt

# Plot Pressure vs Depth
plt.figure(figsize=(8, 6))
plt.plot(df["Depth"], df["Pressure"], marker='o')
plt.xlabel("Depth (m)")
plt.ylabel("Pressure (kPa)")
plt.title("Pressure vs Depth")
plt.grid(True)
plt.show()

# Plot Permeability vs Hydraulic Conductivity
plt.figure(figsize=(8, 6))
plt.scatter(df["Permeability"], df["Hydraulic Conductivity"],
    color='r')
plt.xlabel("Permeability (mD)")
plt.ylabel("Hydraulic Conductivity")
plt.title("Permeability vs Hydraulic Conductivity")
plt.grid(True)
plt.show()
```

The first plot shows the relationship between depth and pressure, indicating how pressure varies with depth in the porous medium. The second plot is a scatter plot that visualizes the relationship between permeability and hydraulic conductivity. `plt.grid(True)` adds a grid to the plots for better readability. Two plots will be generated: one showing the relationship between depth and pressure, and another showing the relationship between permeability and hydraulic conductivity.

Using SciPy, we will perform basic statistical analysis on the dataset to derive insights from the data. This could include calculating means, standard deviations, or conducting hypothesis tests.

```
from scipy import stats

# Calculate mean and standard deviation for porosity
mean_porosity = stats.tmean(df["Porosity"])
```

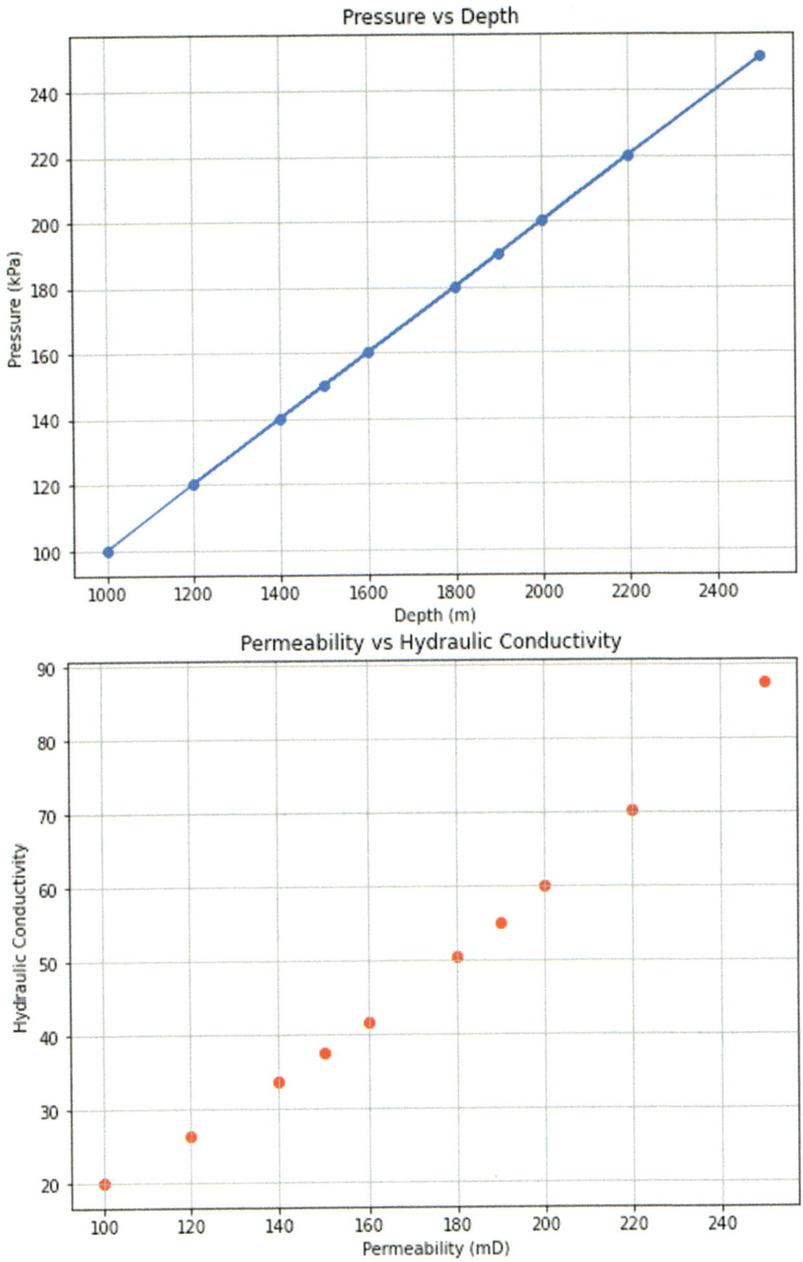

Figure 2.2 Visualization of fluid flow properties: Relationship between pressure and depth, relationship between permeability and hydraulic conductivity.

```
std_porosity = stats.tstd(df["Porosity"])

print(f"Mean␣Porosity:␣{mean_porosity}")
print(f"Standard␣Deviation␣of␣Porosity:␣{std_porosity}")

# Perform a hypothesis test to check if the mean porosity is
    significantly different from a given value
t_stat, p_value = stats.ttest_1samp(df["Porosity"], 0.25)
print(f"T-statistic:␣{t_stat},␣P-value:␣{p_value}")
```

`stats.tmean(df["Porosity"])` calculates the mean porosity of the dataset. `stats.tstd(df["Porosity"])` calculates the standard deviation of porosity. `stats.ttest_1samp(df["Porosity"], 0.25)` performs a onesample ttest to determine if the mean porosity is significantly different from 0.25. The output takes the form:

```
Mean Porosity: 0.271
Standard Deviation of Porosity: 0.04605552204797064
T-statistic: 1.4419081124379998, P-value: 0.18320599515667765
```

The output will include the mean and standard deviation of porosity, as well as the results of the hypothesis test.

OOP CONCEPTS

Object-oriented programming (OOP) is a programming paradigm that organizes software design around data or objects rather than functions and logic. In OOP, objects are instances of classes, which are blueprints for creating objects. Each object can contain both data, known as attributes, and functions, known as methods, that operate on the data. OOP allows developers to model real-world entities and relationships, making the code more modular, reusable, and easier to maintain. By using concepts such as inheritance, polymorphism, and encapsulation, OOP promotes code reusability and the creation of complex systems with greater flexibility and clarity. This paradigm is widely used in software development to create scalable and efficient applications.

Classes and Objects

The key concepts in OOP are classes and objects. An object's behavior (methods) and data (attributes) are contained in an object's blueprint, which is defined by a class. The `class` keyword in Python allows classes to be defined, and objects are made by instantiating these classes.

```
class Fluid:
    def __init__(self, name, viscosity):
        self.name = name
        self.viscosity = viscosity

    def describe(self):
        return f"The␣fluid␣{self.name}␣has␣a␣viscosity␣of␣{self.
            viscosity}␣cp."
```

```
fluid1 = Fluid("Water", 1.0)
print(fluid1.describe())
```

`class` defines a new class named `Fluid`. `__init__` is the constructor method that initializes the object with `name` and `viscosity` attributes. `describe` is a method that returns a description of the fluid. `fluid1 = Fluid("Water", 1.0)` creates an instance of the `Fluid` class. `print(fluid1.describe())` calls the `describe` method on `fluid1` and prints `"The fluid Water has a viscosity of 1.0 cp"`. The output of this code is:

```
The fluid Water has a viscosity of 1.0 cp.
```

Inheritance

Inheritance allows one class (child class) to inherit the attributes and methods of another class (parent class). This helps in reusing code and establishing relationships between classes.

```
class ReservoirFluid(Fluid):
    def __init__(self, name, viscosity, density):
        super().__init__(name, viscosity)
        self.density = density

    def describe(self):
        return f"The fluid {self.name} has a viscosity of {self.
            viscosity} cp and a density of {self.density} g/cm^3.
            "

# Creating an object of the ReservoirFluid class
brine = ReservoirFluid("Brine", 0.8, 1.2)
print(brine.describe())
```

`class ReservoirFluid(Fluid):` defines a new class named `ReservoirFluid` that inherits from `Fluid`. `super().__init__(name, viscosity)` calls the constructor of the parent class `Fluid`. `brine = ReservoirFluid("Brine", 0.8, 1.2)` creates an instance of the `ReservoirFluid` class. The output takes the form:

```
The fluid Brine has a viscosity of 0.8 cp and a density of 1.2 g/
    cm^3.
```

Polymorphism

Polymorphism allows objects of different classes to be treated as objects of a common superclass. It enables the same method to behave differently based on the object that calls it.

```
# Polymorphism Example
for fluid in [water, oil, brine]:
    print(fluid.describe())
```

The for loop iterates through a list of different fluid objects (water, oil, and brine). The describe() method is called on each object, and each object responds differently based on its class definition. The output is:

```
The fluid Water has a viscosity of 1.0 cp.
The fluid Oil has a viscosity of 10.0 cp.
The fluid Brine has a viscosity of 0.8 cp and a density of 1.2 g/
    cm^3.
```

SPECIAL METHODS (MAGIC METHODS)

Magic methods allow us to define how objects of the class behave with built-in operations, such as string representation, comparison, and arithmetic operations.

Implementing __str__, __repr__, and __eq__

```
class Fluid:
    def __init__(self, name, viscosity):
        self.name = name
        self.viscosity = viscosity

    def __str__(self):
        return f"{self.name} with viscosity {self.viscosity} cp"

    def __repr__(self):
        return f"Fluid(name={self.name}, viscosity={self.
            viscosity})"

    def __eq__(self, other):
        return self.viscosity == other.viscosity

# Example Usage
fluid1 = Fluid("Water", 1.0)
fluid2 = Fluid("Oil", 10.0)
fluid3 = Fluid("Brine", 1.0)

print(str(fluid1))  # Output: Water with viscosity 1.0 cp
print(repr(fluid2)) # Output: Fluid(name=Oil, viscosity=10.0)
print(fluid1 == fluid3) # Output: True
```

__str__ defines how an object is printed when passed to print(). __repr__ defines a more detailed string representation, often used in debugging. __eq__ allows comparison between two objects using the == operator. The output is:

```
Water with viscosity 1.0 cp
Fluid(name=Oil, viscosity=10.0)
True
```

PROJECT: BUILDING A FLUID SIMULATION SYSTEM

In this project, we'll build a simple fluid simulation system using the concepts of OOP. The system will include classes for different fluids and reservoirs, and methods to calculate properties like flow rates and pressure drops.

DEFINING THE FLUID AND RESERVOIR CLASSES

```
class Fluid:
    def __init__(self, name, viscosity):
        self.name = name
        self.viscosity = viscosity

    def describe(self):
        return f"{self.name} with viscosity {self.viscosity} cp"

class Reservoir:
    def __init__(self, name, depth, porosity):
        self.name = name
        self.depth = depth
        self.porosity = porosity

    def describe(self):
        return f"{self.name} reservoir with depth {self.depth} m 
            and porosity {self.porosity}"

# Example Usage
fluid = Fluid("Water", 1.0)
reservoir = Reservoir("Reservoir A", 2000, 0.25)

print(fluid.describe())
print(reservoir.describe())
```

`Fluid` class represents a fluid with a name and viscosity. `Reservoir` class represents a reservoir with a name, depth, and porosity. `describe()` methods provide descriptions of the fluid and reservoir. The output is:

```
Water with viscosity 1.0 cp
Reservoir A reservoir with depth 2000 m and porosity 0.25
```

CALCULATING FLOW RATES

Next, we will add a method to the `Fluid` class to calculate the flow rate given a pressure gradient and permeability.

```
class Fluid:
    def __init__(self, name, viscosity):
        self.name = name
        self.viscosity = viscosity

    def calculate_flow_rate(self, permeability, pressure_gradient
        ):
        return (permeability * pressure_gradient) / self.
            viscosity
```

```
# Example Usage
flow_rate = fluid.calculate_flow_rate(permeability=200,
    pressure_gradient=50)
print(f"Flow rate: {flow_rate} units")
```

`calculate_flow_rate()` method calculates the flow rate using the formula: Flow Rate $= \frac{\text{Permeability} \times \text{Pressure Gradient}}{\text{Viscosity}}$. The flow rate is calculated based on the properties of the fluid and the given permeability and pressure gradient. The output is:

```
Flow rate: 10000.0 units
```

2.2.1 EFFECTIVE RESERVOIR ROCKS AND FLUIDS

A wide range of permeabilities for various rock types encountered such as granite of $10^{-5} - 10^{-3}$ [mD], limestone is of $10^{-3} - 10^{-1}$ [mD], sandstone is of $10^{-1} - 10^{1}$ [mD], oil rocks is of $10^{4} - 10^{1}$ [mD], clean sand is of $10^{6} - 10^{4}$ [mD], and clean gravel is of $10^{8} - 10^{6}$.

Listing 2.1: Python code for permeability generation

```
import numpy as np
import matplotlib.pyplot as plt

# Parameters
Lx, Ly = 40, 16          # Domain dimensions
nx, ny = 120, 50         # Discretization
xs = Lx * np.linspace(0, 1, nx+1)  # x-axis
ys = Ly * np.linspace(0, 1, ny+1)  # y-axis

md = 9.86923e-16  # md to m^2

# Initial Permeability
permXX = 1 * md * np.ones((nx, ny))

cxs = 0.5 * (xs[1:] + xs[:-1])
cys = 0.5 * (ys[1:] + ys[:-1])

# Defining low permeability regions
low_perm_regions = [
    ((cxs > 1) & (cxs < 1.5), (cys > 0) & (cys < 4)),
    ((cxs > 10) & (cxs < 10.5), (cys > 4) & (cys < 8)),
    ((cxs > 20) & (cxs < 20.5), (cys > 0) & (cys < 4)),
    ((cxs > 30) & (cxs < 30.5), (cys > 4) & (cys < 8)),
    ((cxs > 38) & (cxs < 38.5), (cys > 0) & (cys < 4)),
    ((cxs > 1) & (cxs < 1.5), (cys > 8) & (cys < 12)),
    ((cxs > 10) & (cxs < 10.5), (cys > 12) & (cys < 16)),
    ((cxs > 20) & (cxs < 20.5), (cys > 8) & (cys < 12)),
    ((cxs > 30) & (cxs < 30.5), (cys > 12) & (cys < 16)),
    ((cxs > 38) & (cxs < 38.5), (cys > 0) & (cys < 4))
]

for x_cond, y_cond in low_perm_regions:
```

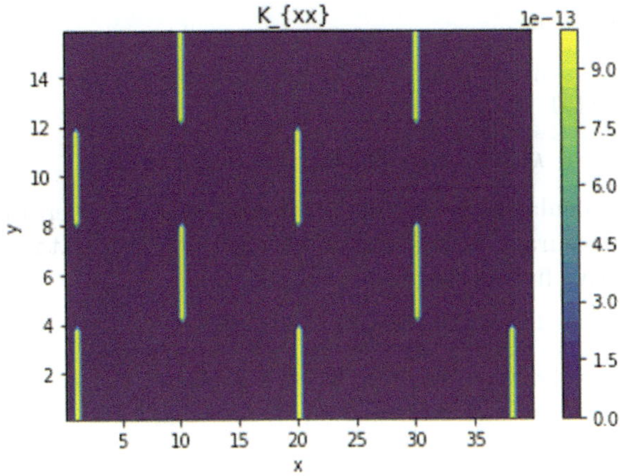

Figure 2.3 Permeability map K_{xx} illustrating the distribution of permeability in the domain.

```
    permXX[np.ix_(x_cond, y_cond)] = 1000 * md

permYY = np.copy(permXX)

# Plotting
plt.figure(1)
plt.contourf(cxs, cys, permXX.T, 20)
plt.colorbar()
plt.title('K_{xx}')
plt.xlabel('x')
plt.ylabel('y')
plt.show()
```

Implementing the above Python code will result in Fig 2.3. The map highlights regions of low permeability, which are intentionally defined in the model to simulate barriers or low-conductivity zones within the system. These regions significantly affect fluid flow and heat transfer and are critical in understanding the overall behavior of the system being modeled.

2.3 NUMERICAL OPERATIONS IN PYTHON

Numerical computation is essential for solving real-world mathematical problems, and Python provides robust libraries like `NumPy` , `SciPy` , and `FiPy` to handle differentiation, integration, and differential equations efficiently. This section covers these numerical operations in detail.

2.3.1 NUMERICAL DIFFERENTIATION

Numerical differentiation approximates the derivative of a function by discretizing it over small intervals. A common approach is the finite difference method, which uses central differences for improved accuracy:

```
import numpy as np
def f(x): return np.sin(x)
x = np.linspace(0, 2 * np.pi, 100)
dx = x[1] - x[0]
df_dx = np.gradient(f(x), dx)
print(df_dx)  # Approximation of the derivative
```

This method calculates the derivative of $f(x) = \sin(x)$ over a uniform grid.

For higher accuracy at specific points, SciPy's `derivative` function provides an adaptive finite difference method:

```
from scipy.misc import derivative
def f(x): return np.sin(x)
df_dx = derivative(f, np.pi / 4, dx=1e-6)
print(df_dx)  # Approximate value of cos(pi/4)
```

NUMERICAL INTEGRATION

Numerical integration calculates the definite integral of a function over a specified interval. Adaptive quadrature is a widely used method, implemented in SciPy's `quad` function:

```
from scipy.integrate import quad
def f(x): return np.sin(x)
result, error = quad(f, 0, np.pi)
print(result)  # Approximation of the integral
```

This example computes the integral of $\sin(x)$ over $[0, \pi]$, resulting in an approximation close to 2.

For higher-dimensional integrals, `dblquad` and `tplquad` can be used:

```
from scipy.integrate import dblquad
def f(x, y): return x * y
result, error = dblquad(f, 0, 1, lambda x: 0, lambda x: 1)
print(result)  # Double integral over [0, 1] x [0, 1]
```

ORDINARY DIFFERENTIAL EQUATIONS (ODEs)

Python's `solve_ivp` function provides an efficient way to solve initial value problems for ODEs. It uses adaptive methods like Runge-Kutta to ensure accuracy:

```
from scipy.integrate import solve_ivp
import numpy as np
def dydx(t, y): return -2 * y + np.sin(t)
t_span = (0, 10)  # Time range
y0 = [1]  # Initial condition
solution = solve_ivp(dydx, t_span, y0, t_eval=np.linspace(0, 10,
    100))
print(solution.t)  # Time points
print(solution.y[0])  # Solution values
```

This example solves the ODE $\frac{dy}{dt} = -2y + \sin(t)$ with an initial condition $y(0) = 1$.

PARTIAL DIFFERENTIAL EQUATIONS (PDEs)

PDEs often require discretization techniques like finite differences, finite elements, or finite volumes. Libraries such as `FiPy` offer tools for solving PDEs numerically:

```
from fipy import CellVariable, Grid1D, DiffusionTerm
nx = 50
dx = 1.0
mesh = Grid1D(nx=nx, dx=dx)
phi = CellVariable(mesh=mesh, value=0.0)
phi.setValue(1.0, where=mesh.x < nx / 4 * dx)
D = 1.0
eq = DiffusionTerm(coeff=D)
for step in range(10):
    eq.solve(var=phi, dt=0.1)
print(phi.value)  # Final solution values
```

This example demonstrates solving a diffusion equation over a 1D domain using finite volume methods.

NUMERICAL EIGENVALUE PROBLEMS

Eigenvalues and eigenvectors can be computed numerically using linear algebra techniques. Python's `NumPy` provides the `eig` function:

```
import numpy as np
A = np.array([[6, 2], [2, 3]])
eigenvalues, eigenvectors = np.linalg.eig(A)
print("Eigenvalues:", eigenvalues)
print("Eigenvectors:\n", eigenvectors)
```

This computes the eigenvalues and eigenvectors of a symmetric matrix A.

SPARSE LINEAR SYSTEMS

Many PDE problems lead to large, sparse linear systems, which can be efficiently solved using `SciPy`'s sparse matrix solvers:

```
from scipy.sparse import diags
from scipy.sparse.linalg import spsolve
n = 100
A = diags([-1, 2, -1], [-1, 0, 1], shape=(n, n)).tocsc()
b = np.ones(n)
x = spsolve(A, b)
print(x)  # Numerical solution of Ax = b
```

EXAMPLE

The pressure equation, derived from Darcy's law and the continuity equation, is given by:

$$\nabla \cdot (k\nabla p) = q$$

where p is the pressure (Pa), k is the permeability (m^2), and q is the source or sink term (e.g., injection or extraction rate, m^3/s).

In this example, the equation is solved numerically on a 2D rectangular domain using finite differences. The domain has pressure $p = 1 \times 10^6$ Pa on the left boundary, pressure $p = 1 \times 10^5$ Pa on the right boundary, and zero flux (Neumann condition) on the top and bottom boundaries. The fluid has a viscosity $\mu = 1 \times 10^{-3}$ Pa.s, and the permeability is constant $k = 1 \times 10^{-12}\,m^2$. A source term is included at the center of the domain to simulate fluid injection.

The governing equation is discretized using finite differences, and the resulting sparse linear system is solved using Python's `scipy.sparse.linalg.spsolve`.

2.3.2 PYTHON IMPLEMENTATION

The Python code used to solve this equation is as follows:

```
import numpy as np
import matplotlib.pyplot as plt
from scipy.sparse import diags
from scipy.sparse.linalg import spsolve

# Define the domain and grid
Lx, Ly = 1.0, 1.0   # Domain size (meters)
Nx, Ny = 50, 50   # Number of grid points
dx = Lx / (Nx - 1)
dy = Ly / (Ny - 1)

# Create grid points
x = np.linspace(0, Lx, Nx)
y = np.linspace(0, Ly, Ny)
X, Y = np.meshgrid(x, y)

# Permeability field (constant for simplicity)
k = 1e-12  # Permeability in m^2 (sandstone-like)

# Viscosity of fluid (water)
mu = 1e-3  # Pa.s

# Source term (e.g., injection at the center)
q = np.zeros((Ny, Nx))
q[Ny // 2, Nx // 2] = 1e-4  # Source term in m^3/s

# Initialize pressure field
p = np.zeros((Ny, Nx))

# Construct the finite difference Laplacian operator (2D)
coeff = k / (mu * dx**2)  # Factor based on permeability and
    viscosity
main_diag = -4 * np.ones(Nx * Ny) * coeff
off_diag = np.ones(Nx * Ny - 1) * coeff
far_diag = np.ones(Nx * Ny - Nx) * coeff
```

```
# Handle grid wrap issues in off-diagonal
for i in range(1, Ny):
    off_diag[i * Nx - 1] = 0

# Sparse matrix construction
diagonals = [main_diag, off_diag, off_diag, far_diag, far_diag]
offsets = [0, 1, -1, Nx, -Nx]
A = diags(diagonals, offsets, shape=(Nx * Ny, Nx * Ny), format="
    csr")

# Flatten the source term
q_flat = q.flatten()

# Apply Dirichlet boundary conditions
for j in range(Ny):
    # Left boundary (p = 100 Pa)
    idx = j * Nx
    A[idx, :] = 0
    A[idx, idx] = 1
    q_flat[idx] = 1e6  # Left boundary pressure

    # Right boundary (p = 0 Pa)
    idx = j * Nx + (Nx - 1)
    A[idx, :] = 0
    A[idx, idx] = 1
    q_flat[idx] = 1e5  # Right boundary pressure

# Apply Neumann conditions (zero flux) on top and bottom
for i in range(Nx):
    # Top boundary (zero flux)
    idx = i
    q_flat[idx] -= coeff * p[0, i] if i > 0 and i < Nx - 1 else 0

    # Bottom boundary (zero flux)
    idx = (Ny - 1) * Nx + i
    q_flat[idx] -= coeff * p[Ny - 1, i] if i > 0 and i < Nx - 1
        else 0

# Solve the linear system
p_flat = spsolve(A, q_flat)

# Reshape the pressure solution back to 2D
p = p_flat.reshape((Ny, Nx))

# Plot the solution
plt.figure(figsize=(8, 6))
plt.contourf(X, Y, p, levels=50, cmap="viridis")
plt.colorbar(label="Pressure (Pa)")
plt.xlabel("x (m)")
plt.ylabel("y (m)")
plt.title("Pressure Distribution in the Domain")
plt.show()
```

The resulting pressure distribution across the domain is shown in the figure below. The pressure is highest on the left boundary (100 Pa) and decreases

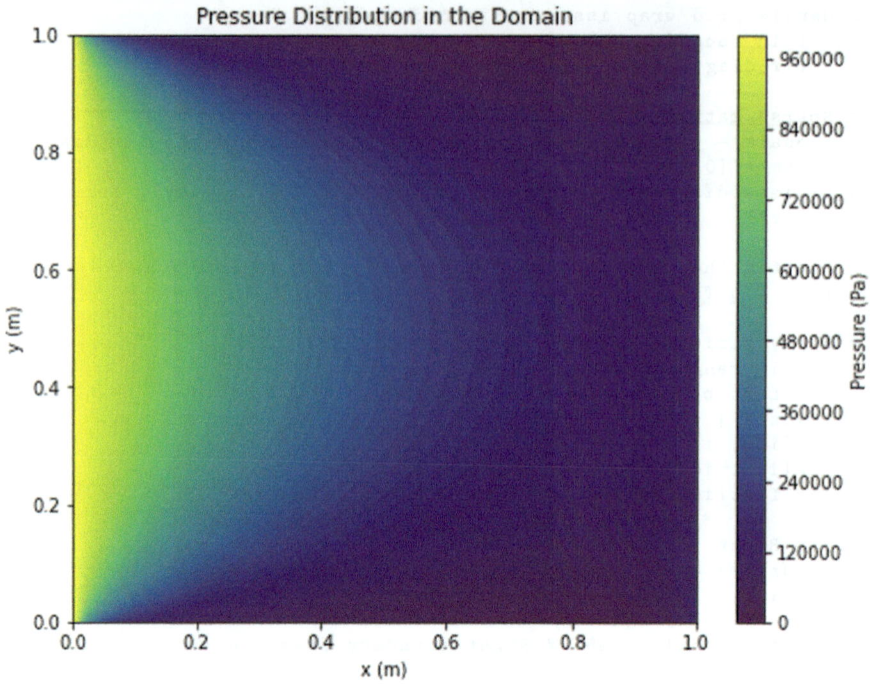

Figure 2.4 Pressure distribution in the domain.

linearly toward the right boundary (0 Pa), with a local perturbation due to the source term at the center.

2.3.3 MATLAB® PDEPE FUNCTION

This section focuses on the employment of MATLAB®'s `pdepe` function to solve one-dimensional PDEs that describe transport phenomena in porous media. While higher-dimensional simulations are often used,1D computations remain highly effective. In many applications, such as aquifers and different types of reservoirs, 1D models capture the depth-dependent behavior accurately, providing valuable insights into system dynamics without the computational complexity of multi-dimensional models. The `pdepe` solver is particularly useful for parabolic and elliptic PDEs, which are commonly used to model diffusion, fluid flow, and heat transfer in porous systems.

The `pdepe` function solves the PDEs by combining spatial discretization with time integration. The Galerkin method was used for spatial discretization. This method projects the solution of the PDE onto a set of basis functions, converting the continuous PDE into a discrete set of algebraic equations. The weak form of the PDE is solved, and the Galerkin method ensures that the error in the solution is orthogonal to the basis functions used for the approximation [24].

Mathematically, the Galerkin method seeks to find an approximate solution $u(x,t)$ to a PDE of the form:

$$\frac{\partial u}{\partial t} = \mathcal{L}(u, x, t), \tag{2.1}$$

where $\mathcal{L}$ is a differential operator, by expanding $u(x,t)$ in terms of a set of basis functions $\phi_i(x)$:

$$u(x,t) \approx \sum_{i=1}^{N} u_i(t)\phi_i(x). \tag{2.2}$$

An adaptive time-stepping algorithm was employed for time integration. This approach dynamically adjusts the time step Δt during the solution process to ensure stability and accuracy. When the solution changes slowly, larger time steps are used, while smaller time steps are chosen when rapid changes occur.

The adaptive time-stepping algorithm ensures that the local truncation error remains within a specified tolerance.

The pdepe function in MATLAB® is designed to solve PDEs that are either parabolic or elliptic in nature. Specifically, it handles PDEs of the following form:

$$c(x,t,u,\frac{\partial u}{\partial x})\frac{\partial u}{\partial t} = \frac{\partial}{\partial x}\left(f(x,t,u,\frac{\partial u}{\partial x})\right) + s(x,t,u,\frac{\partial u}{\partial x}) \tag{2.3}$$

This is a general form that can describe a variety of physical phenomena, such as diffusion, which involves processes like heat flow or changes in chemical concentration; advection, referring to the transport of a substance or quantity through a medium; and reaction-diffusion, where substances spread with the addition of a reaction term, modeling processes such as biological or chemical reactions. This code can also be applied to solve PDEs related to single and two-phase flow in porous media. For instance, in single-phase flow, the PDE can model fluid flow through a porous medium governed by Darcy's law, incorporating factors like pressure and permeability. In two-phase flow, the code can be used to simulate more complex interactions, such as the simultaneous movement of oil and water in a reservoir, accounting for relative permeability and capillary pressure effects [25, 26, 27]. Breaking down the terms in this equation:

$c(x,t,u,\frac{\partial u}{\partial x})$: Coefficient of the time derivative $\frac{\partial u}{\partial t}$, representing how the quantity u changes with respect to time.

$\frac{\partial}{\partial x}\left(f(x,t,u,\frac{\partial u}{\partial x})\right)$: Flux term representing the flow of the quantity u through space. In heat conduction, this would correspond to the flow of heat through the medium.

$s(x,t,u,\frac{\partial u}{\partial x})$: Source term representing any external additions or removals of u, such as internal heat sources in the case of heat conduction.

In general, pdepe solves PDEs that describe processes like diffusion, advection, reaction-diffusion, and fluid flow in a one-dimensional spatial domain with time evolution.

MATLAB® Code Structure

The pdepe function in MATLAB® is used to solve one-dimensional PDEs with time-dependent boundary conditions and initial conditions. The function solves systems of parabolic and elliptic PDEs (2.3). The MATLAB® code for solving the PDEs consists of the main function pdepe and three key sub-functions: pdex4pde, pdex4ic, and pdex4bc. The main function pdepe orchestrates the overall solution process by calling the sub-functions responsible for defining the PDEs, initial conditions, and boundary conditions. It solves the system of equations over the spatial and temporal domains. The sub-function pdex4pde contains the main code that defines the system of PDEs to be solved. It specifies the differential equations in their spatially discretized form. The sub-function pdex4ic defines the initial conditions for the system. These are necessary for solving time-dependent PDEs and provide the starting point for time integration. The sub-function pdex4bc defines the boundary conditions, which are essential for ensuring that the solution remains well-posed and physically realistic. The general syntax of the pdepe function is as follows:

```
sol = pdepe(m, pdefun, icfun, bcfun, xmesh, tspan)
```

This command takes in six parameters:

m: Symmetry parameter (specifies the geometry of the problem).
pdefun: A function handle defining the PDE.
icfun: A function handle defining the initial conditions.
bcfun: A function handle defining the boundary conditions.
xmesh: A vector specifying the spatial points.
tspan: A vector specifying the time points.

Let us now break down each of these parameters in detail:

The symmetry parameter m defines the geometry of the PDE problem. It determines the type of symmetry and, consequently, the dimensionality of the problem. When $m = 0$, it corresponds to the slab or Cartesian symmetry, which describes one-dimensional problems on a flat plane or along a line. This is typically applied when the physical system has no variation in two of the spatial dimensions. For $m = 1$, the symmetry is cylindrical, often used in cases such as radial diffusion in a cylinder, where the problem exhibits rotational symmetry around an axis. Finally, when $m = 2$, the symmetry is spherical, which is commonly used for problems like radial diffusion in a sphere, where the physical system has symmetry in all directions from a central point. For example, if you are solving a PDE in Cartesian coordinates, you will use $m = 0$.

The pdefun function describes the PDE to be solved. The function pdefun must return three values $[c, f, s]$, where:

Example: Heat Equation

For example, consider the heat equation in one dimension:

$$\frac{\partial u}{\partial t} = \alpha \frac{\partial^2 u}{\partial x^2}$$

Here, α is the thermal diffusivity. In this case:

$$c(x, t, u, \frac{\partial u}{\partial x}) = 1$$

$$f(x, t, u, \frac{\partial u}{\partial x}) = \alpha \frac{\partial u}{\partial x}$$

$$s(x, t, u, \frac{\partial u}{\partial x}) = 0$$

The pdefun would be written as:

```
function [c, f, s] = heatPDE(x, t, u, dudx)
    alpha = 1;  % Thermal diffusivity
    c = 1;
    f = alpha * dudx;
    s = 0;
end
```

Initial Condition Function (icfun):

The initial condition function icfun defines the initial state of the solution at $t = 0$. It specifies how the solution $u(x, 0)$ is distributed over the spatial domain at the initial time.

$$u0 = \texttt{icfun}(x)$$

where $u0$ is the initial value of the solution u at position x.

For example, if the initial temperature distribution along a rod is $u(x, 0) = 100 \sin\left(\frac{\pi x}{10}\right)$, the initial condition function can be written as:

```
function u0 = icfun(x)
    u0 = 100 * sin(pi * x / 10);
end
```

Boundary Condition Function (bcfun):

The boundary condition function bcfun specifies the boundary conditions at the edges of the spatial domain. It defines how the solution behaves at the boundaries $x = x_l$ (left boundary) and $x = x_r$ (right boundary).

The boundary conditions are typically expressed in the form:

$$p(x,t) + q(x,t) \cdot u = 0$$

where p and q describe the type of boundary condition.

A Dirichlet boundary condition fixes the value of the solution u at the boundary. For example, if the temperature at the boundaries is fixed at zero, this is expressed as:

$$u(0,t) = 0, \quad u(L,t) = 0$$

For a Dirichlet condition, we set:

$$p = u - u_{\text{fixed}}, \quad q = 0$$

In `bcfun`, it would look like:

```
function [pl, ql, pr, qr] = bcfun(xl, ul, xr, ur, t)
    pl = ul;  % Left boundary: u(0,t) = 0
    ql = 0;
    pr = ur;  % Right boundary: u(L,t) = 0
    qr = 0;
end
```

A Neumann boundary condition specifies the derivative (flux) of the solution at the boundary. For example, if the heat flux at the boundaries is zero, this is expressed as:

$$\frac{\partial u}{\partial x}(0,t) = 0, \quad \frac{\partial u}{\partial x}(L,t) = 0$$

For a Neumann condition, we set:

$$p = 0, \quad q = 1$$

In `bcfun`, it would look like:

```
function [pl, ql, pr, qr] = bcfun(xl, ul, xr, ur, t)
    pl = 0;  % Neumann condition: du/dx(0,t) = 0
    ql = 1;
    pr = 0;  % Neumann condition: du/dx(L,t) = 0
    qr = 1;
end
```

Spatial Mesh Vector (`xmesh`):

The spatial mesh vector **`xmesh`** defines the discrete points in the spatial domain where the solution is computed. For example, if we are solving the PDE over the interval $[0, 10]$, we can define a spatial mesh of 100 points as follows:

```
xmesh = linspace(0, 10, 100);
```

Time Span Vector (`tspan`):

The time span vector **tspan** defines the points in time where the solution is computed. For example, if the solution is required between time $t = 0$ and $t = 5$, and we want 50 time points, we can define:

```
tspan = linspace(0, 5, 50);
```

Output Solution (`sol`):

The output `sol` is a 3D array that contains the computed solution for all spatial and temporal points. The dimensions of `sol` correspond to the number of time points, the number of spatial points, and the number of solution components (in the case of a system of equations). For example, if there are 50 time points and 100 spatial points, the solution `sol` will have a size of 50×100. We can extract and plot the solution as follows:

```
surf(xmesh, tspan, sol)
xlabel('Position (x)')
ylabel('Time (t)')
zlabel('Solution (u)')
title('PDE Solution using pdepe')
```

This will generate a 3D surface plot showing how the solution $u(x, t)$ evolves over space and time.

In upcoming chapters, we present some real applications for flow in porous media using MATLAB® built-in pdepe function, including nanoparticle transport and water harvesting using desiccant materials. Using PDEs, these problems are framed for numerical solutions.

2.4 MRST MATLAB® TOOLBOX

The MATLAB® Reservoir Simulation Toolbox (MRST) [28, 29] is an open-source software platform developed by SINTEF, designed to support research, teaching, and practical applications in reservoir simulation. MRST is a powerful tool for modeling and simulating subsurface flow, focusing on petroleum reservoirs, groundwater management, and carbon storage. MRST provides a wide array of prebuilt models, solvers, and visualization tools that can be easily customized and extended. It is especially beneficial for researchers and practitioners who require rapid prototyping and development of numerical methods or who wish to simulate complex physical phenomena in porous media. MRST includes standard reservoir simulation models like black-oil, compositional models, and advanced thermal and chemical processes. MRST users can modify or extend existing functionalities to suit specific research or industrial needs. It provides 2D and 3D visualization for grids, flow fields, and reservoir properties. It is open source and free. Easily accessible to anyone for research or teaching purposes.

MRST employs a wide range of numerical techniques, algorithms, and solvers tailored for simulating complex subsurface flow phenomena. The toolbox integrates finite volume methods as its primary discretization technique, ensuring mass conservation and robustness across diverse geological configurations. For solving large-scale linear and nonlinear systems, MRST utilizes iterative solvers such as GMRES (Generalized Minimal Residual Method) and preconditioning techniques like algebraic multigrid (AMG) and incomplete LU (ILU) decomposition to enhance computational efficiency. Nonlinear solvers, including Newton-Raphson and its variants, are implemented to handle coupled multiphase and compositional flow equations with high accuracy. MRST's modular framework allows for the integration of advanced numerical methods, such as higher-order discretization schemes and multiscale solvers, enabling users to balance computational cost and precision. These methods are rigorously tested against industry benchmarks and real-world reservoir problems, demonstrating their robustness in handling highly heterogeneous reservoirs, fractured systems, and challenging physical conditions such as capillary pressure and thermal effects. This flexibility and reliability make MRST an indispensable tool for both academic research and industrial applications.

MRST supports a variety of simulation models catering to different reservoir dynamics, including several models such as single-phase flow, black-oil model, compositional model, thermal model, reactive transport models, fractured reservoirs, and carbon Storage. MRST can be applied to simulate and predict subsurface flow and reservoir performance. It can be used to test new numerical methods or physical models and for educational purposes to teach students about reservoir simulation principles and methods. MRST can support CO_2 storage, geothermal energy, and hydrogen storage studies.

MRST DOWNLOAD

Ensure you have a valid MATLAB® installation, as MRST depends on MATLAB® for execution. MRST is freely available and can be downloaded from the official website:

1. Visit *https://www.sintef.no/projectweb/mrst/*.
2. Click on the **Download** section and select the desired version (latest is recommended).
3. Extract the downloaded zip file to a directory on your computer.
4. Add the MRST directory to MATLAB®'s search path using the command:

```
addpath('path_to_mrst');
mrstModule add all;
```

In the upcoming chapters, we present some examples that can be situated using MRST, including Single-phase flow, two-phase oil-water flow, waterflooding in an oil reservoir, fractured reservoir simulation, CO_2 injection and storage, and well placement optimization.

CONCLUSION

This chapter provided a practical foundation for using MATLAB® and Python to implement numerical methods in porous media applications. By walking through programming basics and advanced features in both languages, it demonstrated how numerical tools can be applied to simulate fluid flow, analyze reservoir properties, and solve PDEs. Projects and case studies illustrated the power of combining code with scientific models, offering a reproducible path from theoretical concepts to computational results. The skills and tools introduced here are essential for the simulations and numerical strategies presented in the following chapters.

REVIEW QUESTIONS

1. What are the key physical processes modeled using MATLAB®'s `pdepe` function in porous media simulations, as discussed in this chapter?
2. Why is one-dimensional (1D) modeling effective for porous media applications, despite the availability of higher-dimensional models?
3. Explain the role of the Galerkin method in solving PDEs with MATLAB®'s `pdepe` function.

PROBLEMS

1. Solve the diffusion equation for a chemical in a cylindrical tube of length 5 units. The initial concentration of the chemical is:

$$u(r,0) = \exp(-r^2)$$

The boundary condition is that the concentration at the boundary is held at zero, $u(5,t) = 0$, and there is no source term. The PDE is:

$$\frac{\partial u}{\partial t} = D\left(\frac{1}{r}\frac{\partial}{\partial r}\left(r\frac{\partial u}{\partial r}\right)\right)$$

where D is the diffusion coefficient.

2. Solve the reaction-diffusion equation for a chemical inside a spherical do main of radius 1 unit. The chemical undergoes a reaction of the form $R(u) = u(1-u)$. The initial concentration is $u(r,0) = 0.1$, and the boundary condition at the sphere's surface is $u(1,t) = 1$. The PDE is:

$$\frac{\partial u}{\partial t} = D\left(\frac{1}{r^2}\frac{\partial}{\partial r}\left(r^2\frac{\partial u}{\partial r}\right)\right) + R(u)$$

where D is the diffusion coefficient, and $R(u) = u(1-u)$ is the reaction term.

3. A chemical is diffusing through a porous medium of length $L = 10$ units. The initial concentration is given by:

$$c(x, 0) = \exp\left(-\frac{x^2}{L^2}\right)$$

The boundary condition is that the chemical concentration is zero at both boundaries:

$$c(0, t) = 0, \quad c(L, t) = 0$$

The governing diffusion equation is:

$$\phi \frac{\partial c}{\partial t} = D \frac{\partial^2 c}{\partial x^2}$$

where ϕ is the porosity of the medium and D is the diffusion coefficient of the chemical.

4. Consider heat transfer in a porous rock formation of length $L = 20$ units. The initial temperature distribution is:

$$T(x, 0) = T_0 \sin\left(\frac{\pi x}{L}\right)$$

At both ends, the temperature is held at a constant value T_b:

$$T(0, t) = T_b, \quad T(L, t) = T_b$$

The heat conduction equation in the porous medium is:

$$\phi \rho C_p \frac{\partial T}{\partial t} = k \frac{\partial^2 T}{\partial x^2}$$

where ρ is the density, C_p is the specific heat capacity, k is the thermal conductivity, and ϕ is the porosity.

DECLARATION

During the preparation of this work, the author used ChatGPT in order to improve the language and readability. After using this tool/service, the author reviewed and edited the content as needed and takes full responsibility for the content of the publication.

3 Computational Methods for Transport Phenomena in Porous Media

3.1 INTRODUCTION

The simulation of transport phenomena in porous media, such as the flow of fluids in petroleum reservoirs or the movement of contaminants in groundwater, requires solving complex partial differential equations (PDEs). These equations describe the interplay between spatial variations and temporal dynamics, often coupling variables like pressure, velocity, and concentration. Accurately resolving these interactions is critical for predicting system behavior and making informed engineering decisions. Numerical discretization methods are crucial in translating these continuous equations into solvable algebraic systems. Spatial discretization divides the domain into finite elements or cells, transforming spatial derivatives into algebraic expressions. This step is essential for capturing local variations in properties like permeability, porosity, or saturation. Temporal discretization, on the other hand, involves approximating time derivatives to evolve the solution through discrete time steps. Together, these techniques allow for the numerical solution of PDEs that govern transport in porous media.

The choice of the discretization method significantly impacts the accuracy and stability of the solution. Spatial methods such as finite difference, finite element, and mixed finite element techniques ensure local conservation of mass and momentum. Temporal schemes, ranging from explicit to implicit methods, manage the stability and convergence of time-dependent problems.

Numerical models for transport in porous media often involve strongly coupled and nonlinear equations. For example, the governing equations for two-phase flow or reactive transport incorporate interactions between variables like pressure, saturation, and chemical concentrations. These couplings introduce complexities that can lead to numerical instability or slow convergence during simulations. Addressing these challenges requires advanced discretization techniques and iterative solvers capable of maintaining stability and ensuring convergence, even under highly nonlinear conditions.

Real-world porous media, such as geological formations, are rarely uniform. They often exhibit heterogeneity, with spatial variations in properties like permeability, porosity, and anisotropy, where properties differ depending on the direction. Capturing these local phenomena accurately necessitates

DOI: 10.1201/9781032724973-3

high-resolution spatial discretization methods that can represent fine-scale variations. Failure to account for these complexities can result in oversimplified models and inaccurate predictions, particularly in applications like groundwater remediation or enhanced oil recovery.

Stability and convergence are fundamental concerns in the numerical solution of time-dependent partial differential equations. Explicit time–stepping methods are conditionally stable, requiring the time step to satisfy the Courant–Friedrichs–Lewy (CFL) criterion, which can severely limit the time step size for simulations involving small spatial scales. On the other hand, implicit methods offer unconditional stability but come at a higher computational cost due to the need for iterative solvers. Balancing stability and computational efficiency is critical for reliable and practical simulations.

Simulations of transport in porous media often involve large-scale domains with millions of computational cells, particularly for applications in reservoir engineering or regional groundwater flow modeling. These high-dimensional problems demand computational efficiency to reduce runtime and resource usage. Techniques such as matrix-oriented implementations, parallel computing, and adaptive time-stepping are essential to handle the computational load while maintaining accuracy and resolution.

The accurate implementation of boundary conditions is a crucial aspect of numerical modeling. Depending on the physical scenario, boundary conditions may include Dirichlet conditions (fixed values), Neumann conditions (specified fluxes), or mixed types that combine both. Improper handling of boundary conditions can introduce errors and reduce the reliability of the simulation results. Advanced techniques are often required to adapt boundary conditions to irregular geometries and complex physical processes, ensuring that the numerical model reflects the real-world system accurately.

In petroleum engineering, numerical discretization methods are crucial in reservoir simulations. These simulations are used to model oil and gas recovery, predict reservoir performance, and optimize production strategies. Techniques like the Cell-Centered Finite Difference (CCFD) method are popular for conserving mass locally and handling complex reservoir geometries. The implicit pressure explicit saturation (IMPES) scheme is another commonly used approach, allowing efficient coupling of pressure and saturation equations for two-phase flow simulations. These methods provide critical insights into reservoir dynamics, enabling engineers to make informed decisions to maximize recovery.

Numerical discretization is equally important in groundwater hydrology, where it is used to study groundwater contamination, remediation strategies, and aquifer recharge processes. Mixed Finite Element Methods (MFEM) are frequently employed in these applications, offering accurate flux and pressure calculations essential for understanding subsurface flow behavior. These techniques effectively address challenges posed by heterogeneous aquifers and complex boundary conditions, ensuring that numerical models align closely with real-world observations.

Enhanced Oil Recovery (EOR) methods, such as CO2 injection or thermal recovery, depend heavily on robust numerical discretization techniques. These simulations require accurate spatial and temporal discretization, coupling to model the intricate interactions between injected agents and the reservoir rock and fluids. Advanced methods, including multiscale and adaptive schemes, are often used to capture the fine-scale details of EOR processes while maintaining computational efficiency. The insights gained from these simulations are vital for designing effective EOR strategies and improving recovery rates.

In environmental engineering, numerical discretization methods are employed to model the transport of pollutants and reactive species in porous media. Applications include the simulation of contaminant plume migration, chemical reactions in groundwater, and remediation efforts. Multiscale and adaptive time-stepping methods are particularly useful in these scenarios, as they allow simulations to focus computational resources on regions and timescales where dynamic changes are most significant. These methods provide valuable tools for assessing environmental risks and designing effective mitigation strategies.

By addressing these challenges and leveraging the strengths of spatial and temporal discretization methods, researchers and engineers can better understand and manage the complex dynamics of transport in porous media. These techniques provide the foundation for various applications, from petroleum and groundwater engineering to enhanced recovery processes and environmental protection. This chapter aims to equip readers with a comprehensive understanding of these methods, paving the way for their application in diverse and impactful fields.

3.1.1 NUMERICAL METHODS FOR RESERVOIR SIMULATION

In the context of reservoir simulation, various numerical methods are employed to approximate the solutions to the governing equations that describe fluid flow within the reservoir. One of the primary techniques used is the Finite Difference Method (FDM). This method involves discretizing the reservoir into a grid of blocks or cells and then applying finite difference approximations to estimate the derivatives required to solve the equations. Within FDM, there are different approaches based on where the derivatives are approximated. In the block-centered FDM, the domain is divided into blocks, and derivatives are approximated at the center of each block. Alternatively, the cell-centered FDM also discretizes the domain but focuses on approximating the derivatives at the center of each individual cell, providing a slightly different perspective on the flow behavior.

Another widely used approach is the Finite Element Method (FEM), which is known for its flexibility and accuracy in handling complex geometries and boundary conditions. The standard FEM utilizes continuous functions to approximate the solution across the domain, making it a conventional and reliable method. A more advanced version is the Mixed FEM, which incorporates both displacement and stress fields as primary variables, allowing for a more comprehensive analysis of the system. Additionally, the Discontinuous Galerkin (DG) FEM uses discontinuous basis functions to achieve high accuracy, especially in cases where the solution might have sharp gradients or discontinuities. To further enhance the accuracy, mesh adaptivity is employed, which involves adjusting the mesh during the simulation to better capture the important features of the solution.

The Finite Volume Method (FVM) is another powerful numerical approach used in reservoir simulation. This method divides the domain into finite volumes or control volumes and solves the governing equations by integrating over these volumes. FVM is particularly effective in ensuring the conservation of quantities like mass or energy across the boundaries of the control volumes. Within FVM, various flux formulations, such as the Multi-Point Flux Approximation (MPFA), are utilized to calculate the fluxes at the boundaries of the control volumes, ensuring accurate and stable solutions. Moreover, special techniques like Mixed Finite Element Methods and Orthogonal Collocation on Finite Differences (OCCFD) are incorporated to handle complex boundary conditions and to accommodate variable grid structures, further enhancing the robustness and versatility of the simulation process.

3.2 NUMERICAL ACCURACY AND STABILITY

Numerical accuracy and stability form the foundation of reliable computational modeling, influencing every aspect of numerical methods. In the subsections that follow, we examine precision levels related to round-off and truncation errors, analyze techniques for evaluating these discrepancies, and discuss strategies to ensure that small input variations do not destabilize the overall computation.

3.2.1 NUMBERS PRECISION

Using a more accurate representation of numbers, such as double precision instead of single precision, can help to reduce errors in numerical computations. Single precision uses less memory but has less precision, while double precision uses more memory but has higher precision. The choice between the two depends on the specific requirements of the application. Single precision, also known as "float" in some programming languages, represents a floating-point number using 32 bits of memory. This means that there are 23 bits for the mantissa (or the significant digits) and 8 bits for the exponent. A floating-point number is a way of representing real numbers in computers; it is called "floating" because the decimal point can "float" around the digits of the number. The mantissa represents the digits of the number, while the exponent represents the power of 10 by which the number is multiplied.

The range of a single precision number is from approximately 3.4×10^{-38} to 3.4×10^{38}. This means that single-precision numbers can represent very small and very large numbers, but with less precision than double-precision numbers. Single precision is generally used when memory is a constraint or when the precision of the computation is not critical. On the other hand, double precision, also known as "double" in some programming languages, represents a floating-point number using 64 bits of memory. This means that there are 52 bits for the mantissa and 11 bits for the exponent. The range of a double precision number is from approximately 1.7×10^{-308} to 1.7×10^{308}. This means that double-precision numbers can represent even smaller and even larger numbers than single-precision numbers and with higher precision. Double precision is generally used when high precision is required, such as in scientific and engineering applications. Some languages and libraries offer even more precision than double precision, such as quadruple precision, which uses 128 bits to represent a number. These higher precision formats are often used in scientific and engineering computations where extremely high levels of accuracy are required. It is important to choose the appropriate precision level for a specific application, considering the trade-off between accuracy and performance. Higher precisions, like quadruple precision and so on, are available, but those are not widely supported and are less common.

Modern hardware still adheres to the IEEE 754 standard for floating-point arithmetic, meaning that the distinctions between single precision (32-bit) and double precision (64-bit) continue to be relevant. Even though contemporary systems offer more memory and faster processors, the trade-offs between memory usage, computational speed, and numerical accuracy are still important considerations. Many applications, especially in scientific computing and engineering, rely on double precision for higher accuracy, while single precision is often favored in scenarios where memory is limited or performance is prioritized. Additionally, while higher precision formats like quadruple precision exist, they are less common due to limited hardware support and increased computational costs.

3.2.2 ERROR ANALYSIS

Error analysis and numerical stability are important concepts in numerical analysis because they determine the reliability and accuracy of numerical algorithms. Error analysis is the study of the difference between the true solution of a problem and the approximate solution obtained by a numerical algorithm. Two types of errors can occur in numerical computation: truncation errors and round-off errors. Truncation errors are caused by approximating a continuous problem with a discrete one, while round-off errors are caused by the finite precision of computer arithmetic. Numerical stability refers to the ability of a numerical algorithm to produce accurate results even when small errors are present in the input data. An algorithm is considered stable if small changes in the input data result in small changes in the output. An algorithm is considered unstable if small changes in the input data result in large changes in the output. There are several methods that can be used to improve the stability of a numerical algorithm, such as using a more accurate representation of numbers, such as using double precision instead of single precision. Round-off error is the difference between a computed result and the true result due to the limitations of the computer's arithmetic. It occurs when numbers are rounded to a certain number of decimal places or significant figures. The difference between the precise (exact) solution of an infinite series and the approximation achieved by utilizing a finite number of terms is known as the truncation error. It also gives an illustration of how to figure out the truncation error for a particular series. One of the most common methods for estimating errors is by using Taylor series expansions. The idea is to approximate the function by its Taylor series and then use the remainder term as an estimate of the error. Truncation error is the difference between the exact solution of a problem and an approximate solution obtained by using a finite number of terms or steps. This type of error occurs when a mathematical series is truncated or when an algorithm is approximated. Discretization is the process of approximating a continuous problem by a discrete one usually contains a kind of truncation error. For example, when solving a differential equation using finite difference or finite element methods, the continuous problem is approximated by a discrete one with some error.

3.2.3 ERROR ESTIMATION

Error estimation is the process of determining the difference between the true solution of a problem and the approximate solution obtained by a numerical algorithm. There are different ways to quantify the error, but one of the most common methods is to use the difference between the true solution and the approximate solution at a specific point or over a certain domain.

For example, let u be the true solution of a problem and $\hat{u}$ be the approximate solution obtained by a numerical algorithm. The absolute error, E, at a specific point x_0 is given by:

$$E = |u(x_0) - \hat{u}(x_0)| \tag{3.1}$$

The relative error, e, at a specific point x_0 is given by:

$$e = \left| \frac{u(x_0) - \hat{u}(x_0)}{u(x_0)} \right| \tag{3.2}$$

The global error, E_{global}, over a certain domain D is given by:

$$E_{\text{global}} = \max\{|u(x) - \hat{u}(x)| : x \in D\} \tag{3.3}$$

The global relative error, e_{global}, over a certain domain D is given by:

$$e_{\text{global}} = \max\left\{\left|\frac{u(x) - \hat{u}(x)}{u(x)}\right| : x \in D\right\} \tag{3.4}$$

Global relative error is a measure of the accuracy of an approximation or a numerical method. It is a ratio of the difference between the approximate solution and the true solution to the true solution. It is used to compare the accuracy of different methods or to compare the results obtained with different inputs. The global relative error gives an indication of how well the method approximates the true solution. A small global relative error indicates that the method is accurate, while a large global relative error indicates that the method is not very accurate.

Keep in mind that the global relative error can be affected by the choice of the true solution. For example, if the true solution is not known, a common practice is to compare the results obtained by different methods with each other. This can be useful to compare the relative performance of different algorithms, but it is not always meaningful to compare the global relative error with a fixed threshold.

3.2.4 NUMERICAL STABILITY

Numerical stability refers to the ability of a numerical algorithm to produce accurate results even when small errors are present in the input data. An algorithm is considered stable if small changes in the input data result in small changes in the output. An algorithm is considered unstable if small changes in the input data result in large changes in the output. There are several methods that can be used to improve the stability of a numerical algorithm. For example, using a more accurate representation of numbers, such as using double precision instead of single precision, using a smaller step size or tolerance in iterative methods.

3.3 NUMERICAL METHODS

In general, numerical analysis deals with the design, analysis, and implementation of algorithms for solving mathematical problems. It is the study of algorithms for solving mathematical problems using numerical methods. It is an interdisciplinary field that combines mathematical theory, computer science, and engineering to develop efficient and accurate methods for solving a wide range of problems. The field of numerical analysis is constantly evolving as new techniques and technologies are developed to tackle increasingly complex problems.

Numerical methods are often implemented and executed using computer algorithms, which is necessary for the efficient implementation and execution of numerical methods. The advancement of computer hardware and software has greatly expanded the range of problems that can be solved using numerical methods and has also led to the development of new and more efficient methods. Additionally, the theoretical foundations for analyzing the performance and complexity of numerical algorithms are essential for understanding their limitations and for developing new methods.

3.4 FINITE DIFFERENCE METHODS

Finite difference methods are a cornerstone of numerical analysis, providing a means to approximate the derivatives of functions that are otherwise continuous. By replacing continuous derivatives with discrete approximations, these methods enable the numerical solution of differential equations that arise in various fields of science and engineering. The first difference quotients are the simplest and most fundamental finite difference approximations, offering an initial step toward understanding more complex numerical methods.

Consider a function $F = F(x_1, x_2, t)$, where x_1, x_2, and t are independent variables. The derivative of F with respect to any of these variables can be approximated using finite differences. For instance, the first derivative of F with respect to x_1 can be approximated by a forward finite difference quotient as follows:

$$\frac{\partial F}{\partial x_1} \approx \frac{F(x_1 + h_1, x_2, t) - F(x_1, x_2, t)}{h_1}$$

Here, h_1 represents a small step size along the x_1 direction. This discrete approximation serves as the foundation for many numerical algorithms designed to solve differential equations. This approach is useful when information about the function is more readily available at points ahead of the current position.

The backward difference quotient, in contrast, approximates the derivative by using the function's value at a point behind the current point:

$$\frac{\partial F}{\partial x_1} \approx \frac{F(x_1, x_2, t) - F(x_1 - h_1, x_2, t)}{h_1}$$

This method is often used when past information is available or when the forward difference is not feasible.

The central difference quotient provides a more accurate approximation by considering points on both sides of the current position:

$$\frac{\partial F}{\partial x_1} \approx \frac{F(x_1 + h_1, x_2, t) - F(x_1 - h_1, x_2, t)}{2h_1}$$

This symmetric approach typically results in a higher accuracy compared to forward or backward differences, especially when the function is smooth.

The accuracy of finite difference approximations can be assessed using Taylor series expansions, which express the function F as a sum of its derivatives at a particular point.

For instance, expanding $F(x_1 + h_1, x_2, t)$ using a Taylor series yields:

$$F(x_1 + h_1, x_2, t) = F(x_1, x_2, t) + h_1 \frac{\partial F}{\partial x_1} + \frac{h_1^2}{2} \frac{\partial^2 F}{\partial x_1^2} + \text{higher-order terms}$$

By substituting this expansion into the difference quotient formulas, one can evaluate the truncation error, which is the error introduced by approximating the derivative.

The forward and backward difference methods introduce an error that is proportional to h_1. Specifically:

$$\frac{\partial F}{\partial x_1} \approx \frac{F(x_1 + h_1, x_2, t) - F(x_1, x_2, t)}{h_1} - \frac{h_1}{2} \frac{\partial^2 F}{\partial x_1^2} + \text{higher-order terms}$$

The central difference method, by averaging values on both sides of the point of interest, reduces the error to a second-order term in h_1:

$$\frac{\partial F}{\partial x_1} \approx \frac{F(x_1 + h_1, x_2, t) - F(x_1 - h_1, x_2, t)}{2h_1} + \frac{h_1^2}{6} \frac{\partial^3 F}{\partial x_1^3} + \text{higher-order terms}$$

This higher accuracy is particularly beneficial in simulations requiring precise derivative approximations.

Second difference quotients are employed to approximate the second derivatives of functions, which are essential in solving higher-order differential equations.

$$\frac{\partial^2 F}{\partial x^2} \approx \frac{F(x + h, y, z, t) - 2F(x, y, z, t) + F(x - h, y, z, t)}{h^2} + \mathcal{O}(h^2)$$

This formula, known as the centered second difference quotient, provides a second-order approximation for the second derivative, meaning the error decreases quadratically as the step size h becomes smaller.

The error associated with the second difference quotient can also be derived using Taylor series expansions. For example, expanding $F(x + h, y, z, t)$ and $F(x - h, y, z, t)$ gives, respectively:

$$F(x + h, y, z, t) = F(x, y, z, t) + h \frac{\partial F}{\partial x} + \frac{h^2}{2} \frac{\partial^2 F}{\partial x^2} + \frac{h^3}{6} \frac{\partial^3 F}{\partial x^3} + \mathcal{O}(h^4)$$

and

$$F(x - h, y, z, t) = F(x, y, z, t) - h \frac{\partial F}{\partial x} + \frac{h^2}{2} \frac{\partial^2 F}{\partial x^2} - \frac{h^3}{6} \frac{\partial^3 F}{\partial x^3} + \mathcal{O}(h^4)$$

Subtracting these expansions and dividing by h^2 provides the second derivative approximation:

$$\frac{\partial^2 F}{\partial x^2} \approx \frac{F(x+h,y,z,t) - 2F(x,y,z,t) + F(x-h,y,z,t)}{h^2} + \mathcal{O}(h^2)$$

This confirms that the centered second difference quotient has second-order accuracy, making it an effective tool for numerical differentiation.

3.5 TEMPORAL DISCRETIZATION METHODS

Transport in porous media is governed by partial differential equations (PDEs) that describe the movement of fluids, heat, or chemical species through a porous matrix. These PDEs often take the form of advection–diffusion–reaction equations (or some subset thereof). Solving these equations numerically requires the discretization of both space and time. While spatial discretization methods (e.g., finite differences, finite elements, finite volumes) address how to approximate derivatives in space, temporal discretization methods focus on how the solution evolves in time. This section highlights various numerical schemes for discretizing time and discusses their relative advantages and limitations in the context of transport in porous media.

Transport in porous media is governed by partial differential equations (PDEs) that describe the movement of fluids, heat, or chemical species through a porous matrix. These PDEs often take the form of advection–diffusion–reaction equations. Solving these equations numerically requires the discretization of both space and time.

Temporal discretization methods transform the continuous time variable t into discrete time levels t^n, facilitating the numerical solution of time-dependent PDEs. Between these discrete instances, transport variables, such as pressure, saturation, concentration, or temperature, are advanced using various algorithms that differ in their stability, accuracy, and computational efficiency.

Consider a general partial differential equation for transport phenomena:

$$\frac{\partial c}{\partial t} + \mathcal{L}(c) = 0, \tag{3.5}$$

where $\mathcal{L}$ represents a spatial differential operator that encapsulates phenomena such as advection and diffusion. The manner in which this equation is discretized in time determines the numerical method employed.

Temporal discretization methods can be broadly categorized into three types: explicit, implicit, and mixed methods. Each offers distinct advantages and limitations for modeling transport in porous media.

Explicit Methods

Explicit methods evaluate spatial operators using known values from the previous time step. The Forward Euler method is the classic example:

$$\frac{c^{n+1} - c^n}{\Delta t} + \mathcal{L}(c^n) = 0, \tag{3.6}$$

which can be rearranged to compute the next time step explicitly:

$$c^{n+1} = c^n - \Delta t \cdot \mathcal{L}(c^n). \tag{3.7}$$

Explicit methods are computationally inexpensive per time step, as no system of equations needs to be solved. However, their stability is conditional, typically limited by the CFL condition. They may require prohibitively small time steps for advection-dominated problems or systems with large diffusion coefficients, which can make them inefficient for certain applications in porous media.

Implicit Methods

Implicit methods evaluate spatial operators using unknown values from the new time step. The Backward Euler method exemplifies this approach:

$$\frac{c^{n+1} - c^n}{\Delta t} + \mathcal{L}(c^{n+1}) = 0, \tag{3.8}$$

which can be rearranged as:

$$c^{n+1} + \Delta t \cdot \mathcal{L}(c^{n+1}) = c^n. \tag{3.9}$$

Implicit methods are unconditionally stable for many problems, allowing much larger time steps than explicit methods. This comes at the cost of requiring a system of equations to be solved at each time step, making them computationally expensive per step, especially for nonlinear problems. They are particularly beneficial for problems with stiff terms or large diffusion coefficients, which are common in heterogeneous porous media simulations.

Mixed Methods

Mixed methods combine aspects of both explicit and implicit approaches. The Crank-Nicolson method, for example, uses a time-centered discretization:

$$\frac{c^{n+1} - c^n}{\Delta t} + \frac{1}{2}\mathcal{L}(c^n) + \frac{1}{2}\mathcal{L}(c^{n+1}) = 0, \tag{3.10}$$

Another approach is the IMEX (Implicit-Explicit) family of methods, which treat different terms with different temporal discretizations:

$$\frac{c^{n+1} - c^n}{\Delta t} + \mathcal{L}_{\text{explicit}}(c^n) + \mathcal{L}_{\text{implicit}}(c^{n+1}) = 0, \tag{3.11}$$

where stiff terms are handled implicitly and non-stiff terms explicitly.

Mixed methods provide a balance between stability and computational cost. They are often second-order accurate in time (particularly in the case of Crank-Nicolson) and can selectively apply implicit treatment to terms causing stiffness. This selective approach makes them particularly useful for multi-physics problems with varying time scales, such as coupled flow and transport processes in porous media.

Two-phase flow in porous media (e.g., water and gas) is governed by coupled nonlinear equations for pressure and saturation. The choice of temporal discretization significantly impacts simulation stability and efficiency. Common discretization approaches include IMPES (IMplicit Pressure, Explicit Saturation). This hybrid approach solves the pressure equation implicitly:

$$\nabla \cdot \left[\left(\frac{kk_{rw}}{\mu_w} + \frac{kk_{rg}}{\mu_g} \right)^n \nabla p^{n+1} \right] = q_w + q_g \tag{3.12}$$

Then updates saturations explicitly:

$$\phi \frac{S_w^{n+1} - S_w^n}{\Delta t} + \nabla \cdot \left(\frac{kk_{rw}^n}{\mu_w} \nabla p_w^{n+1} \right) = q_w \tag{3.13}$$

The IMPES method is computationally efficient but can suffer from stability issues, especially in problems with high heterogeneity or large permeability contrasts.

On the other hand, the fully implicit method (FIM) solves pressure and saturation equations simultaneously:

$$\phi \frac{S_w^{n+1} - S_w^n}{\Delta t} + \nabla \cdot \left(\frac{kk_{rw}^{n+1}}{\mu_w} \nabla p_w^{n+1} \right) = q_w \tag{3.14}$$

$$\phi \frac{S_g^{n+1} - S_g^n}{\Delta t} + \nabla \cdot \left(\frac{kk_{rg}^{n+1}}{\mu_g} \nabla p_g^{n+1} \right) = q_g \tag{3.15}$$

The FIM approach offers superior stability but requires solving a larger, coupled nonlinear system, making it computationally intensive.

Another alternative approach, namely, the sequential implicit method (SIM), can be used. This approach first solves pressure implicitly, then solves saturation implicitly:

$$\nabla \cdot \left[\left(\frac{kk_{rw}}{\mu_w} + \frac{kk_{rg}}{\mu_g} \right)^n \nabla p^{n+1} \right] = q_w + q_g \tag{3.16}$$

$$\phi \frac{S_w^{n+1} - S_w^n}{\Delta t} + \nabla \cdot \left(\frac{kk_{rw}^{n+1}}{\mu_w} \nabla p_w^{n+1} \right) = q_w \tag{3.17}$$

SIM offers better stability than IMPES while being more computationally efficient than FIM, providing a middle ground for many applications.

The choice of temporal discretization method for transport in porous media depends on the specific problem characteristics, stability requirements, and computational resources. Explicit methods offer simplicity but face stability constraints, while implicit methods provide stability at increased computational cost. Mixed methods and specialized approaches like IMPES, FIM, and SIM offer various compromises between stability and efficiency.

For highly heterogeneous porous media with significant permeability contrasts, fully implicit or sequential implicit methods often provide the best balance between stability and computational efficiency. The optimal choice ultimately depends on the specific application requirements, including the desired accuracy, available computational resources, and the physical characteristics of the porous medium.

3.5.1 UPWIND SCHEME

The upwind scheme is primarily a spatial discretization technique used for approximating advection terms. However, it interacts strongly with temporal discretization in determining overall stability and accuracy. In advection-dominated transport problems, standard central difference schemes can introduce non-physical oscillations. The upwind scheme combats these by biasing the spatial derivative in the direction of flow:

$$\left.\frac{\partial c}{\partial x}\right|_i \approx \begin{cases} \dfrac{c_i - c_{i-1}}{\Delta x}, & \text{if } u > 0, \\ \dfrac{c_{i+1} - c_i}{\Delta x}, & \text{if } u < 0, \end{cases}$$

where u is the local velocity. The upwind scheme ensures that information is propagated in the correct direction of flow, helping to maintain numerical stability and reduce spurious oscillations.

In two-phase flow simulations, the upwind scheme is particularly important for discretizing the flux terms containing relative permeability functions. The nonlinear relative permeability term $k_{rw}(S_w)$ requires careful treatment. Using a one-dimensional finite volume discretization, the flux between cells i and $i+1$ becomes:

$$F_{i+1/2} = -\frac{k_{i+1/2}}{\mu_w} k_{rw}(S_w^*) \frac{p_{i+1} - p_i}{\Delta x}$$

For upwinding, the saturation-dependent relative permeability $k_{rw}(S_w^*)$ is evaluated at:

$$S_w^* = \begin{cases} S_{w,i}, & \text{if } (p_{i+1} - p_i) < 0 \text{ (flow from } i \text{ to } i+1), \\ S_{w,i+1}, & \text{if } (p_{i+1} - p_i) > 0 \text{ (flow from } i+1 \text{ to } i). \end{cases}$$

This upwind treatment is crucial for capturing shock fronts and maintaining physical saturation values between 0 and 1. The upwind scheme is usually implemented within an IMPES scheme.

3.5.2 STABILITY CONDITIONS

The Courant–Friedrichs–Lewy (CFL) condition provides a fundamental relationship between the time step size Δt, the spatial grid size Δx, and the wave speed (or flow velocity) u. For explicit schemes, the CFL condition essentially states that the information (or fluid parcel) must not travel more than one grid cell within a single time step:

$$\text{CFL} = \frac{u\,\Delta t}{\Delta x} \leq 1,$$

in one dimension (higher dimensions have analogous forms). When this condition is violated, explicit schemes can become unstable. Implicit methods can circumvent the strictness of this constraint, but large CFL numbers can still degrade accuracy or cause slow convergence in nonlinear solvers.

In porous media simulations, the CFL constraint requires careful consideration due to several complexities. The velocity u represents the phase velocity (Darcy flux divided by porosity), which can vary significantly across the domain. This is because heterogeneous permeability fields (common in geological formations), saturation-dependent relative permeability effects, pressure gradients that vary in space and time, and gravitational effects in vertical or inclined systems.

For two-phase flow, the CFL condition must account for the maximum wave speed of the system. In the fractional flow formulation, this is related to the maximum derivative of the fractional flow function:

$$\text{CFL} = \frac{\Delta t}{\Delta x \phi} \max_{S_w} \left| \frac{df_w(S_w)}{dS_w} \frac{q_t}{\phi} \right| \leq 1,$$

where f_w is the fractional flow function, q_t is the total flux, and ϕ is porosity.

To efficiently handle the CFL constraint in highly heterogeneous media, adaptive time-stepping strategies are often employed:

$$\Delta t^{n+1} = \min\left(\Delta t_{\text{max}},\, \alpha \cdot \frac{\Delta x}{u^n_{\text{max}}}\right),$$

where α is a safety factor (typically 0.5–0.9), u^n_{max} is the maximum velocity at the current time step, and Δt_{max} is an upper bound on the time step.

Beyond advection, the complete stability criteria in porous media must also consider diffusion processes, which can be rigorously analyzed using von Neumann stability analysis. This technique examines how numerical errors propagate in Fourier space by decomposing the solution into a sum of Fourier modes. For a diffusion problem:

$$\frac{\partial c}{\partial t} = D \frac{\partial^2 c}{\partial x^2},$$

discretized with a forward-time, central-space scheme:

$$\frac{c_j^{n+1} - c_j^n}{\Delta t} = D\frac{c_{j+1}^n - 2c_j^n + c_{j-1}^n}{(\Delta x)^2},$$

we substitute a Fourier mode $c_j^n = \xi^n e^{ikj\Delta x}$, where $i = \sqrt{-1}$, k is the wavenumber, and ξ is the amplification factor. This yields:

$$\xi = 1 + 2\gamma(\cos(k\Delta x) - 1),$$

where $\gamma = D\Delta t/(\Delta x)^2$ is the diffusion number. For stability, we require $|\xi| \leq 1$ for all wavenumbers, which leads to:

$$\gamma = \frac{D\Delta t}{(\Delta x)^2} \leq \frac{1}{2},$$

for an explicit scheme. This is the von Neumann stability condition for diffusion problems, which translates to:

$$\Delta t \leq \frac{(\Delta x)^2}{2D},$$

In heterogeneous porous media, local diffusion coefficients vary substantially. The effective diffusion coefficient depends on molecular diffusion, mechanical dispersion, and medium properties:

$$D_{\text{eff}} = D_m\tau + \alpha_L|v|,$$

where D_m is the molecular diffusion coefficient, τ is the tortuosity factor, α_L is the longitudinal dispersivity, and $|v|$ is the magnitude of the Darcy velocity. This spatially varying diffusion coefficient means that the stability constraint can be significantly more restrictive in regions of high dispersivity or velocity, potentially overriding the advective CFL condition.

For reactive transport, the characteristic reaction time scale imposes another constraint:

$$\Delta t \leq \frac{1}{k_r},$$

where k_r is the reaction rate constant.

The Society of Petroleum Engineers (SPE) Comparative Solution Project 10 provides a benchmark case with extreme permeability contrasts spanning several orders of magnitude. In explicit simulations of this model, permeability varies from approximately 10^{-5} to 10^4 millidarcies, resulting in velocity variations that can span 6-9 orders of magnitude. The CFL condition is typically governed by a small subset of cells with the highest velocity, and without adaptive time stepping, simulation efficiency is severely compromised. In practice, this benchmark requires either implicit methods or operator splitting techniques that selectively apply implicit treatment to regions with severe

CFL constraints, while using explicit methods elsewhere to balance accuracy and efficiency.

To manage CFL constraints efficiently in complex porous media simulations, several strategies have been developed. Local time-stepping allows different regions of the domain to advance with different time steps based on local CFL constraints. Flux limiting can be employed in addition to upwinding to maintain stability while achieving higher-order accuracy in regions with smooth solutions. Modified IMPES methods with stability enhancement use adaptive implicit regions to dynamically apply implicit treatment only to cells violating the CFL constraint. Multi-rate methods allow different physical processes (e.g., fast reactions vs. slow advection) to be integrated using different time steps within the same simulation. These strategies allow for efficient simulation of complex multiphase flow processes in highly heterogeneous porous media while respecting the fundamental CFL stability constraints.

3.5.3 RELAXATION FACTORS

In iterative methods for solving the discrete equations arising from implicit time discretization, a **relaxation factor** (also called an **under-relaxation factor**) is introduced to stabilize the iteration process. This technique is essential for both linear and nonlinear systems encountered in porous media simulations. The basic relaxation scheme can be expressed as:

$$\phi^{k+1} = \phi^k + \omega(\tilde{\phi} - \phi^k),$$

where ϕ is a generic variable (e.g., pressure, velocity, or saturation), k is the iteration index, $\tilde{\phi}$ is the "raw" solution update from the iterative method, and ω is the relaxation factor (with $0 < \omega \leq 1$). When $\omega < 1$, changes in the solution are damped, making the algorithm more stable but potentially slower to converge.

From a mathematical perspective, relaxation can be understood in terms of spectral radius analysis. For a fixed-point iteration $\phi^{k+1} = G(\phi^k)$, convergence is guaranteed when the spectral radius $\rho(G')$ of the Jacobian matrix G' is less than unity. Relaxation effectively modifies the iteration to:

$$\phi^{k+1} = (1 - \omega)\phi^k + \omega G(\phi^k),$$

which has a spectral radius of $\rho((1 - \omega)I + \omega G')$. For many problems, there exists an optimal ω that minimizes this spectral radius and accelerates convergence.

For strongly coupled, nonlinear systems typical in multiphase flow, the relaxation factor often needs to be determined adaptively:

$$\omega^{k+1} = \min\left(\omega_{\max}, \max\left(\omega_{\min}, \alpha\frac{\|\mathbf{r}^{k-1}\|}{\|\mathbf{r}^k\|}\omega^k\right)\right),$$

where $\mathbf{r}^k$ is the residual vector at iteration k, and α is a scaling parameter typically between 0.5 and 1.0.

For two-phase flow simulations in heterogeneous porous media, relaxation techniques are particularly crucial in several contexts:

Fully Implicit Methods: When solving the coupled pressure-saturation system, nonlinear iterations can diverge due to the rapid changes in relative permeability near residual saturations. A sequential relaxation strategy may be employed:

$$p^{k+1} = p^k + \omega_p(\tilde{p} - p^k),$$
$$S^{k+1} = S^k + \omega_S(\tilde{S} - S^k),$$

with typically $\omega_p > \omega_S$ to prioritize pressure convergence before adjusting saturations. Values of ω_S as low as 0.2-0.3 are common for challenging cases with sharp saturation fronts or in highly heterogeneous domains.

Newton-Raphson Method: For the nonlinear solver commonly employed in implicit schemes, a damped Newton approach implements relaxation as:

$$\mathbf{x}^{k+1} = \mathbf{x}^k + \omega^k \Delta\mathbf{x}^k,$$

where $\Delta\mathbf{x}^k = -(\mathbf{J}^k)^{-1}\mathbf{F}(\mathbf{x}^k)$ is the Newton step, $\mathbf{J}^k$ is the Jacobian matrix, and $\mathbf{F}(\mathbf{x}^k)$ is the residual function. The relaxation factor ω^k is often determined by a line search algorithm to ensure residual reduction.

Variable Switching: In challenging regions of the solution space (e.g., near phase transitions), primary variables may need to be switched (e.g., from pressure-saturation to pressure-capillary pressure). During such transitions, aggressive relaxation ($\omega \approx 0.1$–0.3) helps maintain stability.

Modern reservoir simulators employ sophisticated relaxation strategies:

Component-Wise Relaxation: Different ω values are applied to different equation components based on their sensitivity:

$$\omega_i = \min\left(1.0, \frac{C}{\max(|\Delta\phi_i|)}\right),$$

where C is a user-defined threshold (often 0.2–0.5 for saturations).

Chopping: A non-smooth variant of relaxation, "chopping" limits the maximum change in sensitive variables:

$$\Delta S^{k+1} = \text{sign}(\Delta\tilde{S}) \cdot \min(|\Delta\tilde{S}|, \Delta S_{\max}),$$

where $\Delta S_{\max}$ might be set to 0.2 per iteration.

Trust-Region Methods: Advanced approaches combine relaxation with trust-region techniques, particularly for reservoirs with strong capillary effects or complex wettability behavior:

$$\min_{\mathbf{p}} \|\mathbf{F}(\mathbf{x}^k + \mathbf{p})\|_2^2 \quad \text{subject to} \quad \|\mathbf{p}\| \le \Delta^k,$$

where Δ^k is the trust region radius, adaptively adjusted based on the model's predictive quality.

In the challenging SPE10 model with its extreme permeability contrasts, relaxation factors play a vital role in achieving convergence. Numerical experiments have shown that dynamic relaxation schemes outperform static ones, with optimal values evolving throughout the simulation:

Early time steps (dominated by pressure transients): $\omega_p \approx 0.7$–1.0, $\omega_S \approx 0.5$–0.7
Middle regime (saturation front propagation): $\omega_p \approx 0.8$–1.0, $\omega_S \approx 0.3$–0.5
Late time (approaching steady state): $\omega_p \approx 0.9$–1.0, $\omega_S \approx 0.6$–0.8

For porous media problems that involve strong nonlinearities, sharp fronts, or coupled transport–reaction processes, judicious use of relaxation is not merely beneficial but often critical for convergence. The computational overhead of implementing sophisticated relaxation strategies is typically negligible compared to the benefits gained in robustness and overall simulation efficiency.

3.6 CELL-CENTERED FINITE DIFFERENCE METHOD

The Cell-Centered Finite Difference (CCFD) method is a numerical technique for solving partial differential equations governing transport phenomena in porous media. It is widely used in groundwater flow, contaminant transport, and reservoir modeling in the petroleum industry. It discretizes the computational domain into control volumes with variables placed at cell centers, approximating fluxes at interfaces to enforce the conservation of mass, momentum, and energy. Although fundamentally a finite difference method, CCFD can be interpreted as a finite volume method on structured grids, simplifying implementation. In reservoir simulations, Darcy's law and mass conservation are discretized on a rectangular mesh, and their performance is often comparable to that of the mixed finite element method [30].

We discretize the rectangular domain into $n_x \times n_y$ cells, defining the mesh as:

$$\{(x_i, y_j) \mid i = 0, 1, \ldots, n_x;\ j = 0, 1, \ldots, n_y\}, \tag{3.18}$$

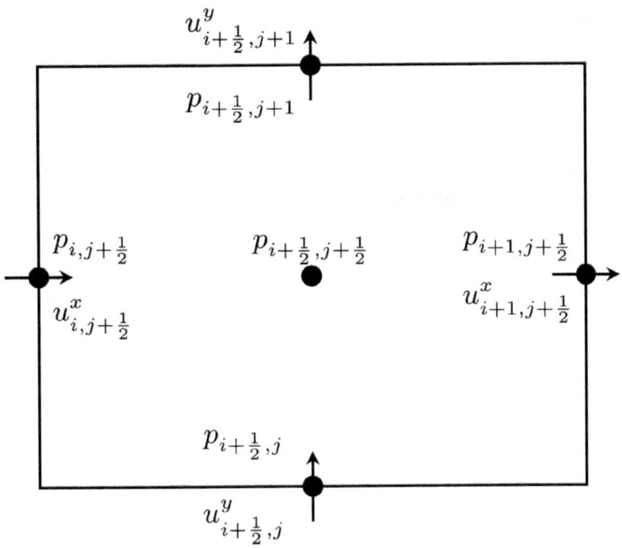

Figure 3.1 Generic CCFD rectangular cell.

where $x_0 = 0 < x_1 < \cdots < x_{n_x} = L_x$ and $y_0 = 0 < y_1 < \cdots < y_{n_y} = L_y$. Each cell is centered at $(x_{i+\frac{1}{2}}, y_{j+\frac{1}{2}})$, where:

$$\begin{aligned} x_{i+\frac{1}{2}} &= \frac{x_i + x_{i+1}}{2}, \quad \Delta x_i = x_{i+1} - x_i, \quad i = 0, 1, \ldots, n_x - 1, \\ y_{j+\frac{1}{2}} &= \frac{y_j + y_{j+1}}{2}, \quad \Delta y_j = y_{j+1} - y_j, \quad j = 0, 1, \ldots, n_y - 1. \end{aligned} \tag{3.19}$$

The pressure is approximated at the center of each cell:

$$p_{i+\frac{1}{2},j+\frac{1}{2}} = p(x_{i+\frac{1}{2}}, y_{j+\frac{1}{2}}). \tag{3.20}$$

The velocities are approximated at the cell faces using Darcy's law. The discrete velocities are defined as:

$$\begin{aligned} u^x_{i,j+\frac{1}{2}} &= u_x(x_i, y_{j+\frac{1}{2}}), \\ u^y_{i+\frac{1}{2},j} &= u_y(x_{i+\frac{1}{2}}, y_j). \end{aligned} \tag{3.21}$$

Detailed numerical discretization techniques will be presented in the upcoming chapters, especially Chapters 4 and 5.

3.7 VECTORIZATION TECHNIQUE

Efficiently solving partial differential equations (PDEs) numerically requires the design of algorithms that are both robust and computationally efficient. By employing such algorithms, solutions can be obtained in significantly less time. MATLAB®, a powerful tool for matrix operations, exhibits inefficiencies when loops are heavily utilized due to the overhead of interpreting each iteration. Consequently, optimizing computations by replacing loops with vectorized operations is essential to achieve better performance.

Vectorization involves restructuring algorithms to operate on entire arrays or matrices simultaneously, thereby eliminating the need for explicit loops. Sun et al. [31] introduced a method that uses vectorization to shift cell-centered data to face-centered positions, and vice versa. This eliminates looping constructs in numerical codes, significantly reducing CPU time. This approach enables the computation of difference equations across all grid cells simultaneously, leveraging MATLAB's optimized matrix operations. Additionally, for post-processing, the face-center information can be efficiently mapped to the physical grid nodes, facilitating contour plotting and streamline visualization. While this implementation uses finite difference methods, the vectorization technique is equally applicable to finite-volume and finite-element frameworks.

To efficiently perform operations on grid data, vectorization can be used to shift cell-centered data vertically (upward and downward) and pad the resulting vectors with zeros. This operation extends the dimensions of the original data while maintaining consistency with boundary conditions.

By shifting the reshaped column vector upward and padding zeros at the new entries, the resulting vector extends to a dimension of $(m+1)n \times 1$. For example:

Original vector of dimension $(mn \times 1)$:

$$p = \begin{bmatrix} p_{1,1} \\ \vdots \\ p_{m,1} \\ p_{1,2} \\ \vdots \\ p_{m,2} \\ \vdots \\ p_{1,n} \\ \vdots \\ p_{m,n} \end{bmatrix} \tag{3.22}$$

After upward shift and zero-padding:

$$p_{\text{shifted-up}} = \begin{bmatrix} p_{1,1} \\ \vdots \\ p_{m,1} \\ 0 \\ p_{1,2} \\ \vdots \\ p_{m,2} \\ 0 \\ \vdots \\ p_{1,n} \\ \vdots \\ p_{m,n} \\ 0 \end{bmatrix} \tag{3.23}$$

The new vector now has the dimension $(m+1)n \times 1$.
Similarly, shifting downward inserts zeros at the top, creating another vector of dimension $(m+1)n \times 1$:

$$p_{\text{shifted-down}} = \begin{bmatrix} 0 \\ p_{1,1} \\ \vdots \\ p_{m,1} \\ 0 \\ p_{1,2} \\ \vdots \\ p_{m,2} \\ \vdots \\ 0 \\ p_{1,n} \\ \vdots \\ p_{m,n} \end{bmatrix} \tag{3.24}$$

These operations are efficiently implemented in MATLAB® using `circshift` or equivalent array-manipulation functions.
Subtracting the upward-shifted vector from the downward-shifted vector produces finite difference terms needed for numerical algorithms:

$$\text{Difference} = p_{\text{shifted-down}} - p_{\text{shifted-up}} \tag{3.25}$$

Adding the shifted vectors yields the average values between adjacent cells:

$$\text{Average} = \frac{p_{\text{shifted-down}} + p_{\text{shifted-up}}}{2} \tag{3.26}$$

Shifting matrices, such as A_E, A_W, A_N, A_S, are used to map cell-centered information to face-centered positions, thereby extending vectors of dimension $mn \times 1$ to a higher dimension of $(m+1)n \times 1$. These matrices enable seamless vectorized operations for the finite difference scheme.

Properties of Shifting Matrices:

The following key properties ensure stability, consistency, and efficiency when these matrices are used in numerical computations:

Orthogonality:

$$A_E^T A_E = A_W^T A_W = A_S^T A_S = A_N^T A_N = I_{mn \times mn}. \tag{3.27}$$

This orthogonality ensures that each shifting matrix A_α (where $\alpha \in \{E, W, S, N\}$) acts as a linear isometry on the discrete vector space of grid values.

Zero Product with Boundary Matrices:

$$A_E^T B_E = A_W^T B_W = 0_{mn \times n}, \quad A_N^T B_N = A_S^T B_S = 0_{mn \times m}. \tag{3.28}$$

The boundary matrices B_E, B_W, B_N, and B_S represent operators that extract or impose conditions at the domain boundaries.

Non-Identity Mappings:

$$A_E A_E^T, A_W A_W^T, A_S A_S^T, A_N A_N^T \neq I, \quad 0. \tag{3.29}$$

Although the shifting matrices are orthogonal, their products with their own transposes do not revert to the identity matrix or vanish entirely. This is a result of how shifting matrices rearrange indexing within a flattened $m \times n$ grid vector.

The transmissibility matrices are defined to account for the relationship between the permeability at face centers and cell centers. For example, the D_{xx} matrix, which represents transmissibility in the x-direction, is defined as:

$$D_{xx}(i,j) = \begin{cases} \frac{2}{\frac{\Delta x(i)}{K_{xx}(i,j)} + \frac{\Delta x(i-1)}{K_{xx}(i-1,j)}} & i \neq 1, m+1 \text{ and } j = 1, n \\ \frac{2(i)}{K_{xx}(i,j)} & i = 1, m+1 \text{ and } j = 1, n \end{cases} \tag{3.30}$$

Similarly, the transmissibility matrix in the y-direction, D_{yy}, is defined as:

$$D_{yy}(i,j) = \begin{cases} \frac{2}{\frac{\Delta y(j)}{K_{yy}(i,j)} + \frac{\Delta y(j-1)}{K_{yy}(i,j-1)}} & j \neq 1, n+1 \text{ and } i = 1, m \\ \frac{2(j)}{K_{yy}(i,j)} & j = 1, n+1 \text{ and } i = 1, m \end{cases} \tag{3.31}$$

Here $D_{xx} \in \mathbb{R}^{(m+1)n\times(m+1)n}$ is a diagonal and square matrix for the x-direction. $D_{yy} \in \mathbb{R}^{m(n+1)\times m(n+1)}$ is a diagonal and square matrix for the y-direction.

To compute the transmissibility matrices, the segment lengths in the x- and y-directions are first calculated as:

$$dx^*(i) = x(i+1) - x(i), \quad dy^*(j) = y(j+1) - y(j) \tag{3.32}$$

These are then extended to cover the entire grid using the MATLAB® `repmat` or equivalent `repeat` function:

$$dx^{**} = \text{repeat}\{dx^*, n\}, \quad dy^{**} = \text{repeat}\{dy^{*T}, m\} \tag{3.33}$$

where $dx^{**}, dy^{**} \in \mathbb{R}^{mn}$, dx^{**} repeats dx^* n times, and dy^{**} repeats dy^{*T} m times.

Element-Wise Operations and Reshaping:

Using dx^{**} and dy^{**}, element-wise division with the permeability matrices ($k_{xx}(i,j)$ and $k_{yy}(i,j)$) is performed. The resulting matrices are reshaped into column vectors:

$$k^*_{xx} = \text{reshape}\left(\frac{dx^{**}}{k_{xx}(i,j)}\right), \quad k^*_{yy} = \text{reshape}\left(\frac{dy^{**}}{k_{yy}(i,j)}\right) \tag{3.34}$$

These vectors are then used to construct the transmissibility matrices D_{xx} and D_{yy} as:

$$(A_W + A_E)k^*_{xx}, \quad (A_S + A_N)k^*_{yy} \tag{3.35}$$

where A_W, A_E, A_S, A_N are the vectorized shifting matrices that map cell-centered information to face-centered information.

The difference equations for the unknown pressure p in the x- and y-directions, $\Delta_x p$ and $\Delta_y p$, respectively, incorporating boundary conditions, are given as:

$$\begin{aligned}\Delta_x p &= [(A_W - A_E)p - B_W p_{BW} + B_E p_{BE}]_{[(m+1)n\times 1]}, \\ \Delta_y p &= [(A_S - A_N)p + B_N p_{BN} - B_S p_{BS}]_{[(m+1)n\times 1]}.\end{aligned} \tag{3.36}$$

Here $p_{BW}, p_{BE}, p_{BS}, p_{BN}$ are the given pressures at the west, east, south, and north boundaries, respectively. p_{BW}, p_{BE} are of dimension $(n+1)\times 1$. p_{BS}, p_{BN} are of dimension $(m+1)\times 1$.

Similarly, the difference equations for the unknown velocities $\Delta_x u_x$ and $\Delta_y u_y$ are expressed as:

$$\begin{aligned}\Delta_x u_x &= \left[(A_E - A_W)^T u_x\right]_{[mn\times 1]}, \\ \Delta_y u_y &= \left[(A_N - A_S)^T u_y\right]_{[mn\times 1]}.\end{aligned} \tag{3.37}$$

Using the adjacency matrices approach, the cell-centered finite difference (CCFD) approximation of Darcy's law is given by:

$$
\begin{aligned}
&[(A_W - A_E)^T D_{xx}(A_W - A_E)]p + [(A_S - A_N)^T D_{yy}(A_S - A_N)]p = q \\
&\quad - [(A_W - A_E)^T D_{xx}(B_W p_{BW} - B_E p_{BE})]p \\
&\quad - [(A_S - A_N)^T D_{yy}(B_N p_{BN} - B_S p_{BS})]p.
\end{aligned} \tag{3.38}
$$

Here $D_{xx} = \text{diag}\,(D_{xx}/dx)$, $D_{yy} = \text{diag}\,(D_{yy}/dy)$. A_W, A_E, A_S, A_N are the adjacency matrices for shifting operations.
The right-hand side of the equation contains two symmetric and positive definite matrices, A_x and A_y, defined as:

$$
\begin{aligned}
A_x &= [(A_W - A_E)^T D_{xx}(A_W - A_E)]_{[mn\times mn]}, \\
A_y &= [(A_S - A_N)^T D_{yy}(A_S - A_N)]_{[mn\times mn]}.
\end{aligned} \tag{3.39}
$$

The system of equations defined above can be efficiently solved in MATLAB® using built-in solvers, such as `mldivide` or iterative solvers for large systems. In the presence of both Neumann and Dirichlet boundary conditions, the system must be appropriately modified to incorporate the known velocities at the boundaries.

The structured procedure for implementing this method is outlined in Algorithm 1, ensuring improved performance and scalability in large-scale numerical simulations.

3.7.1 MATLAB® IMPLEMENTATION

In the CCFD approach, the primary unknowns (pressure and saturation) are defined at cell centers, while fluxes are computed at cell interfaces. This staggered grid arrangement provides natural conservation properties and facilitates handling of heterogeneous porous media.

For a 2D domain discretized into $n_x \times n_y$ cells, with cell widths Δx_i and heights Δy_j, the discretized variables are:

Pressure and saturation defined at cell centers $(i + 1/2, j + 1/2)$
Velocity components defined at cell faces: U_x at $(i, j + 1/2)$ and U_y at $(i + 1/2, j)$
Diffusion coefficients harmonically averaged at interfaces

The cell-centered approach requires special treatment of boundary conditions, as the boundary faces lie on the domain boundary while the primary variables are defined at cell centers. This is handled through appropriate flux calculations at the boundary faces.

3.7.1.1 System Assembly

The following MATLAB® code shows how the adjacency matrices are constructed for the finite difference discretization:

Algorithm 1 Vectorization Technique for Grid-Based Numerical Methods

---

Input: Grid size (m, n), cell-centered data p, permeability field K, grid spacing (dx, dy), boundary conditions (p_B, u_B) **Output:** Vectorized operations for finite differences and transmissibility calculations
Step 1: Reshape and Vectorize Grid Data Reshape the $m \times n$ matrix p into a column vector of size $(mn \times 1)$ Compute grid spacing extensions dx^* and dy^* using repetition functions
Step 2: Shift Cell-Centered Data to Face-Centered Positions Apply upward and downward shifts with zero-padding to extend dimensions Compute horizontal shifts (left and right) similarly
Step 3: Compute Finite Difference Operators Compute vertical differences: $p_{\text{shifted-down}} - p_{\text{shifted-up}}$ Compute horizontal differences: $p_{\text{shifted-right}} - p_{\text{shifted-left}}$ Compute averages for adjacent cells
Step 4: Construct Shifting Matrices Define adjacency matrices A_E, A_W, A_N, A_S for shifting operations Ensure orthogonality: $A_E^T A_E = A_W^T A_W = I$ Enforce boundary constraints: $A_E^T B_E = A_W^T B_W = 0$
Step 5: Compute Transmissibility Matrices Compute diagonal transmissibility matrices D_{xx} and D_{yy} Apply harmonic averaging to permeability values at face centers
Step 6: Formulate and Solve the Discrete System Compute discrete pressure equations using shifting matrices Solve the system using MATLAB® solvers (e.g., `mldivide`)
Return Vectorized pressure and velocity fields

```
% West face to cell adjacency
W_unit = {[sparse(1, nx); speye(nx, nx)]};
W = blkdiag(W_unit{ones(ny, 1)});

% East face to cell adjacency
E_unit = {[speye(nx, nx); sparse(1, nx)]};
E = blkdiag(E_unit{ones(ny, 1)});

% South and North face to cell adjacency
S = [sparse(nx, nx*ny); speye(nx*ny, nx*ny)];
N = [speye(nx*ny, nx*ny); sparse(nx, nx*ny)];

% Boundary adjacency matrices
BW_unit = {[1; sparse(nx, 1)]};
BW = blkdiag(BW_unit{ones(ny, 1)});

BE_unit = {[sparse(nx, 1); 1]};
BE = blkdiag(BE_unit{ones(ny, 1)});

BS = [speye(nx, nx); sparse(nx*ny, nx)];
BN = [sparse(nx*ny, nx); speye(nx, nx)];
```

This sparse matrix representation efficiently encodes the connectivity between cells and faces, enabling compact assembly of the system matrices.

The system matrices are assembled using sparse matrix techniques for computational efficiency. The adjacency matrices representing the connectivity of cells and faces are constructed as follows:

1. West face to cell adjacency (W): Connects west faces to their adjacent cells
2. East face to cell adjacency (E): Connects east faces to their adjacent cells
3. South face to cell adjacency (S): Connects south faces to their adjacent cells
4. North face to cell adjacency (N): Connects north faces to their adjacent cells

Additional adjacency matrices are constructed to handle boundary conditions (BW, BE, BS, BN).

3.7.1.2 Advection Operator

For the advection problem described by:

$$\frac{\partial(\phi c)}{\partial t} + \nabla \cdot (\vec{U} c) = q \tag{3.40}$$

The implementation of the upwind scheme for the advection terms is shown in the following code snippet:

```
% Split velocities into positive and negative parts
UxR = max(0, sparse(Uxbar(:)));
UxM = min(0, sparse(Uxbar(:)));
UyR = max(0, sparse(Uybar(:)));
UyL = min(0, sparse(Uybar(:)));

% System matrix with upwind discretization
A = (W - E)' * (diag(UxR)*W + diag(UxL)*E) ...
    + (S - N)' * (diag(UyR)*S + diag(UyL)*N);

% Right-hand side incorporating boundary conditions
rhs = -(W - E)' * (UxR.*(BW*cBW) + UxL.*(BE*cBE)) ...
    - (S - N)' * (UyR.*(BS*cBS) + UyL.*(BN*cBN));
```

This upwind scheme ensures numerical stability by accounting for the direction of flow when discretizing the advection terms.

The discretized matrix form using upwind scheme is:

$$\mathbf{M}\frac{d\mathbf{c}}{dt} + \mathbf{A}\mathbf{c} = \mathbf{b} \tag{3.41}$$

where:

M is the mass matrix representing cell volumes weighted by porosity
A is the advection (stiffness) matrix incorporating upwind discretization
b is the right-hand side vector accounting for boundary conditions
c is the vector of concentration (or saturation) values

The upwind scheme handles the directional nature of advection by splitting velocities into positive and negative components:

$$U_x^+ = \max(0, U_x) \tag{3.42}$$

$$U_x^- = \min(0, U_x) \tag{3.43}$$

The advection matrix is then constructed as:

$$\begin{aligned}\mathbf{A} &= (\mathbf{W}-\mathbf{E})^T \cdot \big(\mathrm{diag}(U_x^+)\mathbf{W} + \mathrm{diag}(U_x^-)\mathbf{E}\big) + (\mathbf{S}-\mathbf{N})^T \\ &\quad \cdot \big(\mathrm{diag}(U_y^+)\mathbf{S} + \mathrm{diag}(U_y^-)\mathbf{N}\big)\end{aligned} \tag{3.44}$$

3.7.1.3 Diffusion Operator

For the diffusion problem described by:

$$\frac{\partial(\phi c)}{\partial t} - \nabla \cdot (D\nabla c) = q \tag{3.45}$$

The harmonic averaging of diffusion coefficients and assembly of the diffusion matrix are implemented as follows:

```
% Compute harmonic averages for diffusion coefficients at
    interfaces
% X-direction diffusion coefficients at interfaces
Dx_bar = (ones(nx + 1, 1) * hys') .* 2 ./ [
    0 + (hxs(1) * ones(1, ny)) ./ difXX(1, :);
    (hxs(1:end-1) * ones(1, ny)) ./ difXX(1:end-1, :) +
    (hxs(2:end) * ones(1, ny)) ./ difXX(2:end, :);
    (hxs(end) * ones(1, ny)) ./ difXX(end, :) + 0];

% Y-direction diffusion coefficients at interfaces
Dy_bar = (hxs * ones(1, ny + 1)) .* 2 ./ [
    0 + (ones(nx, 1) * hys(1)) ./ difYY(:, 1), ...
    (ones(nx, 1) * hys(1:end-1)') ./ difYY(:, 1:end-1) +
    (ones(nx, 1) * hys(2:end)') ./ difYY(:, 2:end), ...
    (ones(nx, 1) * hys(end)') ./ difYY(:, end) + 0];

% System matrix combining x and y diffusion terms
A = (W - E)' * diag(sparse(Dx_bar(:))) * (W - E) ...
    + (S - N)' * diag(sparse(Dy_bar(:))) * (S - N);
```

The harmonic averaging provides a physically consistent way to handle discontinuous coefficients at cell interfaces, particularly important for heterogeneous porous media.

The discretized matrix form is:

$$\mathbf{M}\frac{d\mathbf{c}}{dt} + \mathbf{A}\mathbf{c} = \mathbf{b} \tag{3.46}$$

where the diffusion stiffness matrix is assembled as:

$$\mathbf{A} = (\mathbf{W}-\mathbf{E})^T \cdot \mathrm{diag}(D_x) \cdot (\mathbf{W}-\mathbf{E}) + (\mathbf{S}-\mathbf{N})^T \cdot \mathrm{diag}(D_y) \cdot (\mathbf{S}-\mathbf{N}) \tag{3.47}$$

The diffusion coefficients D_x and D_y at interfaces are computed using harmonic averages to account for heterogeneity:

$$D_{i-1/2,j} = \frac{2\Delta y_j}{\frac{\Delta x_{i-1}}{D_{i-1,j}} + \frac{\Delta x_i}{D_{i,j}}} \tag{3.48}$$

3.7.1.4 Treatment of Boundary Conditions

The implementation supports both Dirichlet and Neumann boundary conditions:

Dirichlet: Fixed value of the solution variable at the boundary
Neumann: Fixed flux at the boundary

The code below shows how boundary conditions are handled:

```
% Set zero flux for Dirichlet boundaries
UBW(DirW) = 0;
UBE(DirE) = 0;
UBdryS(DirS) = 0;
UBN(DirN) = 0;

% Set zero diffusion coefficient for Neumann boundaries
Dx_bar(1, ~DirW) = 0;
Dx_bar(end, ~DirE) = 0;
Dy_bar(~DirS, 1) = 0;
Dy_bar(~DirN, end) = 0;
```

The boundary conditions are then incorporated into the right-hand side vector:

```
% Right-hand side incorporating boundary conditions
rhs = -(W - E)' * (Dx_bar(:) .* (BW * cBW - BE * cBE) ...
    + BW * (UBW .* hys) + BE * (UBE .* hys)) ...
    - (S - N)' * (Dy_bar(:) .* (BS * cBS - BN * cBN) ...
    + BS * (UBdryS .* hxs) + BN * (UBN .* hxs));
```

For Dirichlet conditions, the corresponding flux terms are set to zero, and the boundary values contribute to the right-hand side. For Neumann conditions, the diffusion coefficients at the corresponding boundaries are set to zero, and the prescribed fluxes are incorporated into the right-hand side.

3.8 FINITE VOLUME METHODS

Finite volume methods are a cornerstone of numerical modeling in subsurface flow, particularly in oil and gas reservoir simulations. These methods discretize the conservation laws over a mesh of control volumes, ensuring that fluxes and pressures are approximated in a manner that respects local conservation principles. Two widely adopted approaches in this context are the Two-Point Flux Approximation (TPFA) and the Multipoint Flux Approximation (MPFA). In the following discussion, we outline the mathematical formulation of finite volume methods through the discretization of a general transport equation and describe the techniques used to approximate its various terms.

Consider the general transport equation:

$$\frac{\partial F}{\partial t} + \nabla \cdot (\mathbf{u}F) - \nabla \cdot (\Gamma \nabla F) = S, \tag{3.49}$$

where F is a scalar variable (e.g., concentration or temperature), $\mathbf{u}$ is the velocity vector, Γ denotes the diffusion coefficient, and S represents the source term. Integrating this equation over a control volume $V_{i,j,k}$ centered at (i, j, k) yields:

$$\int_{V_{i,j,k}} \frac{\partial F}{\partial t}\, dV + \int_{\partial V_{i,j,k}} (\mathbf{u}F) \cdot \mathbf{n}\, dA - \int_{\partial V_{i,j,k}} \Gamma \nabla F \cdot \mathbf{n}\, dA = \int_{V_{i,j,k}} S\, dV. \tag{3.50}$$

Each term in the above equation is discretized as follows:

Time Derivative:

$$\int_{V_{i,j,k}} \frac{\partial F}{\partial t}\, dV \approx \frac{F_{i,j,k}^{n+1} - F_{i,j,k}^{n}}{\Delta t} V_{i,j,k}, \tag{3.51}$$

where $F_{i,j,k}^{n+1}$ and $F_{i,j,k}^{n}$ denote the scalar values at the current and previous time levels, respectively.

Advection Term:

$$\int_{\partial V_{i,j,k}} (\mathbf{u}F) \cdot \mathbf{n}\, dA \approx \sum_f (\mathbf{u}F)_f\, A_f, \tag{3.52}$$

with the sum taken over all faces f of the control volume, and A_f representing the area of face f.

Diffusion Term:

$$\int_{\partial V_{i,j,k}} \Gamma \nabla F \cdot \mathbf{n}\, dA \approx \sum_f \Gamma_f \frac{F_{\text{adj}} - F_{i,j,k}}{\Delta d} A_f, \tag{3.53}$$

where F_{adj} is the scalar value in the cell adjacent to face f and Δd is the distance between cell centers.

Source Term:

$$\int_{V_{i,j,k}} S\, dV \approx S_{i,j,k}\, V_{i,j,k}. \tag{3.54}$$

Thus, the fully discretized equation for each control volume becomes:

$$\frac{F_{i,j,k}^{n+1} - F_{i,j,k}^{n}}{\Delta t} V_{i,j,k} + \sum_f (\mathbf{u}F)_f\, A_f - \sum_f \Gamma_f \frac{F_{\text{adj}} - F_{i,j,k}}{\Delta d} A_f = S_{i,j,k}\, V_{i,j,k}. \tag{3.55}$$

In practice, boundary conditions may specify either the value of F (Dirichlet) or its flux (Neumann) along the domain boundaries. For diffusion-dominated problems, central differencing is often preferred, whereas upwind or higher-order schemes may be more suitable for advection-dominated scenarios. The resulting system of algebraic equations is typically solved using iterative techniques such as Gauss-Seidel, conjugate gradient, or multigrid methods.

3.8.1 TWO-POINT FLUX APPROXIMATION

The Two-Point Flux Approximation (TPFA) method is one of the most straightforward and computationally efficient techniques for discretizing flow equations in porous media. It is particularly effective on structured grids where the grid orientation aligns with the principal permeability directions, which ensures that the underlying assumption of linear pressure variation between adjacent cells remains valid. In TPFA, the flux across the interface between two adjacent control volumes is approximated by considering only the two cell-centered pressures that share the face. The basic formulation is given by:

$$u_i \approx T_{ij}(p_2 - p_1), \tag{3.56}$$

where T_{ij} is the transmissibility coefficient that encapsulates the grid geometry, the permeability of the medium, and fluid properties such as viscosity. Here, p_1 and p_2 represent the pressures in the neighboring cells. An alternative, physically explicit expression for the flux is:

$$u_i \approx -\frac{k}{\mu}\frac{p_j - p_i}{d_{ij}}A_{ij},$$

with k denoting the permeability, μ the fluid viscosity, d_{ij} the distance between cell centers, and A_{ij} the area of the interface between the two cells.

The derivation of the transmissibility coefficient often involves averaging the permeability values at the interface—, commonly using a harmonic mean to account for variations in heterogeneity, and incorporating the geometric dimensions of the grid. This approach results in a locally conservative formulation, meaning that mass is conserved across each control volume, an essential property for reliable simulation of fluid flow.

In two-dimensional problems, the TPFA discretization typically leads to a 5-point stencil, as illustrated in Figure 3.2. This stencil connects the central cell to its four immediate neighbors (north, south, east, and west), forming a cross-shaped pattern that facilitates the efficient assembly and solution of the linear system. In three-dimensional applications, the stencil extends to a 7-point configuration, where the central cell interacts with its six face-adjacent neighbors along the primary coordinate directions.

Despite its advantages, TPFA has inherent limitations. Its accuracy depends critically on the alignment of the grid with the principal directions of the

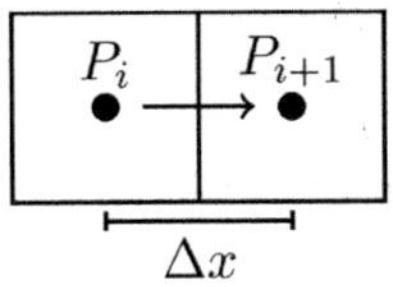

Figure 3.2 Illustration of a 5-point stencil in 2D TPFA, connecting a central cell with its four immediate neighbors.

permeability tensor. When grids are distorted or unstructured, or when the permeability is highly anisotropic (i.e., best described by a full tensor), the assumption of a linear pressure gradient between cell centers becomes less valid, potentially leading to significant errors. In such cases, the TPFA may not adequately capture the directional dependency of the flow, and more advanced techniques, such as the Multipoint Flux Approximation (MPFA)—might be required to improve accuracy.

Moreover, while TPFA is computationally attractive due to the simplicity of its stencil and the resulting sparse matrix structure, it is generally only first-order accurate in space. This may be insufficient for problems requiring high precision, particularly when small-scale heterogeneities play a critical role in the overall flow dynamics.

Algorithm 2 outlines the implementation of TPFA, including grid discretization, transmissibility computation, system assembly, pressure solution, and flux calculation, ensuring mass conservation and efficient numerical performance.

3.8.2 MULTI-POINT FLUX APPROXIMATION

The Multi-Point Flux Approximation (MPFA) method offers significant advantages, particularly when dealing with media whose anisotropy direction is not aligned with the coordinate system (see Chapter 6). In such cases, the permeability tensor contains non-zero off-diagonal elements, requiring specialized numerical techniques.

In two-dimensional applications, the MPFA method establishes an interaction region (indicated by the dashed line in Figure 3.3) centered on four adjacent cells. Within each interaction region, four fluxes must be calculated. Each individual flux is approximated using the general expression:

$$f \approx \sum_{i \in \Omega} T_i P_i \tag{3.57}$$

where T_i represents the transmissibility coefficient and P_i denotes the potential gradient (such as pressure gradient) at the center of cell i. The set Ω varies depending on the dimensionality of the problem being analyzed.

A fundamental principle in MPFA is local flux conservation, achieved by ensuring that inflow and outflow fluxes are equal. Referring to Figure 3.3, the

Algorithm 2 Two-Point Flux Approximation (TPFA) for Flow in Porous Media

Input: Grid size (n_x, n_y, n_z), permeability field k, viscosity μ, grid spacing $(\Delta x, \Delta y, \Delta z)$, boundary conditions p_B, source term q **Output:** Pressure field p at cell centers, flux field u at interfaces

Step 1: Discretize Computational Domain Generate structured grid and define control volume centers Compute face areas A_{ij} and cell center distances d_{ij}

Step 2: Compute Transmissibility Coefficients Compute harmonic mean permeability at cell interfaces Compute transmissibility $T_{ij} = \frac{k}{\mu}\frac{A_{ij}}{d_{ij}}$

Step 3: Assemble Discretized Flow Equations Construct coefficient matrix A using TPFA stencil (5-point in 2D, 7-point in 3D) Construct right-hand side vector b including source terms and boundary conditions

Step 4: Solve for Pressure Field Solve linear system $Ap = b$ using an iterative solver (e.g., Conjugate Gradient, Gauss-Seidel)

Step 5: Compute Fluxes at Cell Interfaces Compute flux using $u_i = T_{ij}(p_j - p_i)$ for each interface

Step 6: Post-Processing and Validation Ensure global mass conservation by verifying $\sum u = q$ Apply boundary conditions to adjust flux at domain boundaries

Return Pressure field p and flux field u

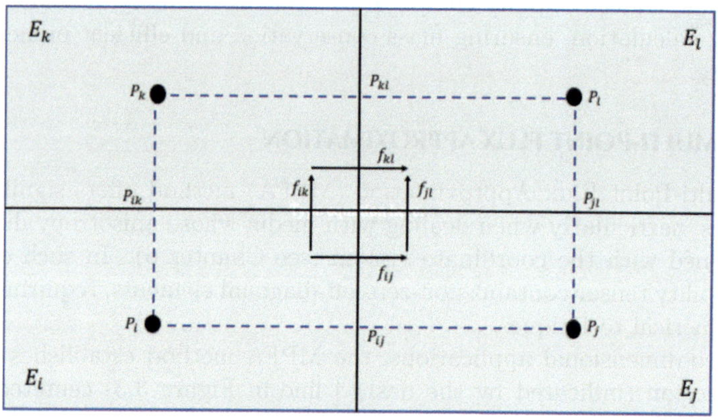

Figure 3.3 Numerical stencil used in the two-dimensional MPFA.

half flux f_{ij} is expressed as:

$$f_{ij} \approx -t_i^{xx}(p_{ij} - p_i) - t_i^{xy}(p_{ik} - p_i) \tag{3.58}$$

$$f_{ij} \approx -t_j^{xx}(p_j - p_{ij}) - t_j^{xy}(p_{jl} - p_j) \tag{3.59}$$

Similar expressions can be derived for the half fluxes f_{ik}, f_{jk}, and f_{kl}. Each flux computation involves two adjacent cells, as shown in Equations (3.58) and (3.59). From each interaction region, four systems of equations must be solved locally. The full flux calculation involves two interaction regions and consequently incorporates six pressure variables from neighboring cells.

The construction of the MPFA stencil presents considerable complexity. To address this challenge, we implement the experimenting field approach developed by Sun et al. [32], which systematically solves a set of local problems to automatically construct the global system (Salama et al. [33]). In this approach, matrix coefficients are determined by solving several local problems using predefined experimenting pressure fields. These fields are specifically designed to extract global matrix coefficients through local problem analysis. A study by Negara et al. [35] presented a 3D numerical simulation of subsurface flow in anisotropic porous media using the multipoint flux approximation (MPFA) method, making it suitable for realistic reservoir modeling scenarios.

In practice, the MPFA formulation introduces additional unknowns, such as interface pressures, to rigorously enforce local flux continuity. These intermediate variables can be systematically eliminated to reduce the system to a more compact form involving only cell-centered pressures and fluxes:

$$R\,u = C\,p, \tag{3.60}$$

where R is a matrix derived from the transmissibility coefficients, C is the coefficient matrix linking pressures and fluxes, u is the vector of fluxes, and p is the vector of cell-centered pressures.

Although MPFA is computationally more intensive than TPFA due to its larger stencils and increased algebraic complexity, its enhanced accuracy in modeling anisotropic permeability, distorted grids, and heterogeneous media makes it an invaluable tool for advanced reservoir simulations and groundwater flow modeling.

3.9 FINITE ELEMENT METHOD

The Finite Element Method (FEM) is a powerful and widely used numerical technique for solving differential equations, particularly in engineering and applied sciences. Originating in the mid-19th century with applications in solid mechanics [36], FEM has evolved through extensive theoretical and computational refinements [36, 30]. Today, it is a cornerstone of numerical analysis, addressing complex problems in structural mechanics, fluid flow, heat transfer, and electromagnetic field simulations. FEM transforms functions from an infinite-dimensional space into a finite-dimensional one, enabling numerical solutions by discretizing the domain into small elements (e.g., triangles). The dependent variables are approximated within each element using polynomial basis functions, typically linear or quadratic. A residual error is minimized by multiplying the governing PDE by a test function, leading to the weak form, which eliminates spatial derivatives and reduces the problem to algebraic equations (AEs) for steady-state cases or ordinary differential equations (ODEs) for transient cases. The Galerkin method is often used to solve this weak form.

Several FEM variants exist, including standard FEM, control volume FEM, discontinuous FEM, mixed FEM (MFEM), characteristic FEM, and adaptive FEM. This book focuses primarily on MFEM and control volume FEM due to their effectiveness for elliptic PDEs, though the standard FEM is introduced first to build a foundational understanding.

3.9.1 STANDARD FINITE ELEMENT METHOD

To illustrate FEM, we start with a one-dimensional Poisson equation, commonly used in porous media flow:

$$-p''(x) = f(x), \quad x \in (0,1) \tag{3.61}$$

with homogeneous boundary conditions:

$$p(0) = p(1) = 0 \tag{3.62}$$

We define the function space:

$$\Sigma_0(\Omega) = \{p \mid p \text{ is sufficiently smooth}, \, p|_{\partial\Omega} = 0\} \tag{3.63}$$

where "sufficiently smooth" means $p \in H^1(\Omega)$. Multiplying the equation by a test function $\phi \in \Sigma_0(\Omega)$ and integrating over Ω gives:

$$\int_0^1 \phi p'' dx = \int_0^1 \phi f dx \tag{3.64}$$

Using integration by parts and boundary conditions:

$$\int_0^1 \phi' p' dx = \int_0^1 \phi f dx \tag{3.65}$$

This is the weak form of the equation.

To discretize, we divide Ω into n subintervals $[x_{k-1}, x_k]$ with uniform spacing $\Delta x = \frac{1}{n}$. The piecewise linear approximation of $p(x)$ is:

$$\widetilde{p}(x) = \sum_{j=0}^{n} p_j \phi_j(x) \tag{3.66}$$

where ϕ_j are piecewise linear basis functions.

Substituting the approximations of p and ϕ into the weak form results in:

$$\sum_{i=0}^{n} p_i \int_0^1 \phi_i' \phi_j' dx = \int_0^1 f \phi_j dx, \quad \forall j \tag{3.67}$$

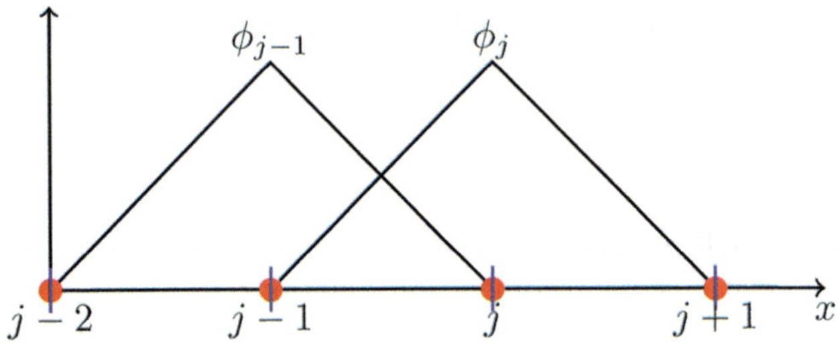

Figure 3.4 Finite element basis function.

which leads to the matrix equation:

$$M\mathbf{p} = \mathbf{f} \tag{3.68}$$

The element-wise matrix representation is:

$$M_k^e = \frac{1}{\Delta x}\begin{bmatrix} 1 & -1 \\ -1 & 1 \end{bmatrix}, \quad \forall k \tag{3.69}$$

For nonhomogeneous boundary conditions, modifications are made to $p(x)$, and for Neumann boundary conditions, additional terms are added to $\mathbf{f}$. The final solution is obtained by solving the algebraic system.

PROBLEM FORMULATION

Consider an incompressible, non-gravitational, single-phase flow within a polygonal or polyhedral Lipschitz domain $\Omega \subset \mathbb{R}^d$, where $d \in \{2, 3\}$. On this domain, we define the standard $L^2(\Omega)$ function space, with $L^2(\Omega) \equiv (L^2(\Omega))^d$ for vector-valued functions. The governing equations for the flow, based on Darcy's law, are expressed as:

$$\begin{aligned} K^{-1}\mathbf{u} &= -\nabla p && \text{in } \Omega, \\ \nabla \cdot \mathbf{u} &= q && \text{in } \Omega, \\ p &- p_B && \text{on } \partial\Omega, \end{aligned} \tag{3.70}$$

where p_B is the prescribed boundary pressure on $\partial\Omega$. Ω represents the domain of interest, while $\partial\Omega$ is its boundary. It is noted that the closure of Ω is denoted as $\overline{\Omega}$, such that $\Omega \cup \partial\Omega = \overline{\Omega}$.

The Finite Element Method offers several advantages for solving the above-described equations. FEM can handle complex geometries, including irregular boundaries, with high accuracy. FEM is well-suited for problems involving anisotropic materials, where the permeability tensor K varies spatially or directionally. The method can accommodate nonlinear source terms or boundary conditions with minimal modifications. The FEM framework is compatible with higher-dimensional problems, making it extensible to 3D domains.

REMARKS:

Mass Conservation: The governing equations of the system consist of two fundamental relationships; momentum equation, $K^{-1}\mathbf{u} = -\nabla p$, which describes the relationship between the pressure gradient ∇p and the Darcy velocity $\mathbf{u}$. The term K^{-1} acts as a proportionality factor that accounts for the fluid's viscosity and the medium's permeability. The continuity equation $\nabla \cdot \mathbf{u} = q$ ensures mass conservation, where the divergence of the velocity field $\mathbf{u}$ equals the volumetric source term q. The boundary condition $p = p_B$ ensures that the pressure field is prescribed along the boundary $\partial\Omega$. This setup accommodates Dirichlet boundary conditions, but more general boundary conditions, such as Neumann or mixed conditions, may also be imposed depending on the problem.

Discontinuities: The permeability tensor K is moved to the left-hand side of the momentum equation to ensure numerical stability and continuity. This rearrangement is particularly important when integrating ∇p by parts, as discontinuities in K (common in heterogeneous or layered porous media) can lead to errors in numerical solutions.

WEAK FORMULATION

The weak form of the governing equations (3.70) is given as:

$$\begin{aligned} &(K^{-1}\mathbf{u}, \mathbf{v}) = -\langle p_B, \mathbf{v}\cdot\mathbf{n}\rangle + (p, \nabla\cdot\mathbf{v}), && \forall \mathbf{v} \in L^2(\Omega) \times H(\text{div}, \Omega), \\ &(\nabla\cdot\mathbf{u}, w) = (q, w), && \forall w \in L^2(\Omega), \end{aligned} \tag{3.71}$$

where the inner product over the domain Ω is defined as:

$$(f, g)_\Omega = \int_\Omega f(x)g(x)\,dV, \quad \forall f, g : \Omega \to \mathbb{R}. \tag{3.72}$$

For $d = 1$, the inner product simplifies to:

$$(f, g)_\Omega = \int_\Omega f(x)g(x)\,dx, \quad \forall f, g : \Omega \to \mathbb{R}. \tag{3.73}$$

The inner product over the boundary $\partial\Omega$ is defined as:

$$\langle f, g\rangle_{\partial\Omega} = \int_{\partial\Omega} fg\,dS, \quad \forall f, g : \partial\Omega \to \mathbb{R}. \tag{3.74}$$

FUNCTION SPACES

The function space $L^2(\Omega)$ is defined as:

$$L^2(\Omega) = \{f : \Omega \to \mathbb{R} : \int_\Omega f^2\, dx < +\infty\}. \tag{3.75}$$

To ensure that the weak formulation (3.71) is well-posed, we must select appropriate function spaces for the variables. Specifically:

$u, \mathbf{v} \in L^2(\Omega)$ so that $(K^{-1}\mathbf{u}, \mathbf{v})$ is well-defined.
$p, \nabla \cdot \mathbf{v} \in L^2(\Omega)$ so that $(p, \nabla \cdot \mathbf{v})$ is well-defined.
$p, w \in L^2(\Omega)$ and $u, \mathbf{v} \in H(\text{div}, \Omega)$.

The space $H(\text{div}, \Omega)$ is defined as:

$$H(\text{div}, \Omega) = \{\mathbf{f} \in L^2(\Omega) : \nabla \cdot \mathbf{f} \in L^2(\Omega)\}. \tag{3.76}$$

It is a Hilbert space equipped with the norm:

$$\|\mathbf{f}\|^2_{H(\text{div},\Omega)} = \|\mathbf{f}\|^2 + \|\nabla \cdot \mathbf{f}\|^2, \tag{3.77}$$

where both the function $\mathbf{v}$ and its divergence $\nabla \cdot \mathbf{v}$ must be square-integrable over Ω.

In the case of a scalar function p, we select the space $L^2(\Omega)$. If $\mathbf{v} \in L^2(\Omega)$ and $\nabla \cdot \mathbf{v} \in L^2(\Omega)$, it follows that $\mathbf{v} \in H(\text{div}, \Omega)$. In the context of mixed finite elements, $H(\text{div}, \Omega)$ is the typical finite element space.

Thus, the solution we seek satisfies:

$$(p, \mathbf{u}) \in L^2(\Omega) \times H(\text{div}, \Omega), \tag{3.78}$$

where both p and $\mathbf{u}$ are sufficiently smooth.

For numerical discretization, we define two finite-dimensional spaces:

$$W_h \subset L^2(\Omega), \quad \dim(W_h) < +\infty, \tag{3.79}$$

$$V_h \subset H(\text{div}, \Omega), \quad \dim(V_h) < +\infty. \tag{3.80}$$

The condition $V_h \subset H(\text{div}, \Omega)$ imposes continuity requirements on the normal components of $\mathbf{u}$ across the inter-element boundaries $E \in E_h$ of the mesh K_h. In other words, the normal component $\mathbf{u} \cdot \mathbf{n}$ must remain continuous across inter-element boundaries.

Raviart-Thomas ($\mathbf{RT}_0$) Space

The RT_0 elements are designed specifically to approximate the space $H(\text{div}, \Omega)$. The Raviart–Thomas space provides a robust framework for solving mixed finite element problems where continuity of normal flux across element boundaries is required. The spaces W_h and V_h for the RT_0 formulation are defined as:

$$W_h = \{w \in L^2(\Omega) : w|_E \in P_0(E),\ E \in E_h\}, \tag{3.81}$$

$$V_h = \{\mathbf{v} \in H(\text{div}, \Omega) : \mathbf{v}|_E \in P_1(E),\ E \in E_h\}, \tag{3.82}$$

where w is a discontinuous, piecewise constant function on each element $E \in E_h$. $\mathbf{v}$ is a piecewise linear vector field on each element $E \in E_h$.

For $d = 2$, the basis functions for $\mathbf{v}|_E$ on a given element E can be expressed as:

$$\mathbf{v}|_E(x, y) = \begin{bmatrix} a_1 x + b_1 \\ a_2 y + b_2 \end{bmatrix}, \quad a_1, b_1, a_2, b_2 \in \mathbb{R}. \tag{3.83}$$

For $d = 3$, the basis functions are represented as:

$$\mathbf{v}|_E(x, y, z) = \begin{bmatrix} a_1 x + b_1 \\ a_2 y + b_2 \\ a_3 z + b_3 \end{bmatrix}, \quad a_1, b_1, a_2, b_2, a_3, b_3 \in \mathbb{R}. \tag{3.84}$$

These basis functions are continuous only in the normal direction across element boundaries. That is, for any element E with normal vector $\mathbf{n}$, the continuity condition is satisfied as:

$$\mathbf{v}|_E \cdot \mathbf{n} = \begin{cases} 1, & \text{if } i = j, \\ 0, & \text{if } i \neq j, \end{cases} \tag{3.85}$$

where $\mathbf{v}|_E = \{\mathbf{v}_i\}$ represents the basis functions, and $\mathbf{n} = \{\mathbf{n}_j\}$ are the normal vectors associated with sides i and j.

The solution $(p_h, \mathbf{u}_h) \in W_h \times V_h$ can be obtained by solving the following discretized weak formulations, which correspond to the weak formulation (3.71):

1. Momentum equation:

$$(K^{-1}\mathbf{u}_h, \mathbf{v})_\Omega = -\langle p_B, \mathbf{v} \cdot \mathbf{n} \rangle_{\partial\Omega} + (p_h, \nabla \cdot \mathbf{v})_\Omega, \quad \forall \mathbf{v} \in V_h. \tag{3.86}$$

2. Continuity equation:

$$(\nabla \cdot \mathbf{u}_h, w)_\Omega = (q, w)_\Omega, \quad \forall w \in W_h. \tag{3.87}$$

To ensure the numerical solution is well-posed, the following criteria must be satisfied:

Uniqueness: The system of equations must be solvable uniquely.
Convergence: The error $\|\mathbf{u}_h - \mathbf{u}\|$ between the numerical solution $\mathbf{u}_h$ and the exact solution $\mathbf{u}$ must be small.
Computational Efficiency: The solution process should be computationally efficient, especially for large-scale problems.

FINITE ELEMENT DISCRETIZATION

The representation of unknown scalar and vector fields using finite element basis functions is a fundamental aspect of the weighted residual formulation in finite element methods. In this formulation, different basis functions are used for the velocity components and pressure. The continuous vector field $\mathbf{u}$ is approximated as the discretized vector field ($\mathbf{u} \approx \mathbf{u}_h$), and the continuous scalar field p is approximated as the discretized scalar field ($p \approx p_h$). These approximations are expressed as:

$$\mathbf{u}_h = \sum_{k=1}^{d} \sum_{j=1}^{m} u_j^k n^k N_j, \tag{3.88}$$

$$p_h = \sum_{j=1}^{n} p_j^k L_j, \tag{3.89}$$

where $\{u_j^k\}$ is the unknown vector of velocity components:

$$\{u_j^k\} = [u_1^1, u_2^1, \ldots, u_m^1, u_1^2, u_2^2, \ldots, u_m^2, \ldots, u_1^d, u_2^d, \ldots, u_m^d], \tag{3.90}$$

$\{p^k\}$ is the unknown vector of pressure values:

$$\{p^k\} = [p_1, p_2, \ldots, p_n], \tag{3.91}$$

N_j is the jth basis function for velocity, L_j is the jth basis function for pressure, and $\{n^k\}$ are unit vectors in the Cartesian coordinates for d dimensions.

The Galerkin method consists of using N_j as the weighting functions for the velocity equations and L_j as the weighting functions for the pressure equations. Applying this method to the governing equations results in a system of $d{\cdot}m{+}n$ equations for $d{\cdot}m{+}n$ unknowns in d spatial dimensions. The pressure gradient term in the governing equations is integrated by parts, leading to the following discrete equations:

$$\sum_{j=1}^{m} A_{ij} u_j^k + \sum_{j=1}^{n} B_{ij}^k p_j = 0, \quad i = 1, 2, \ldots, m, \quad k = 1, 2, \ldots, d, \tag{3.92}$$

$$\sum_{k=1}^{d} \sum_{j=1}^{m} B_{ji}^k u_j^k = q_i, \quad i = 1, 2, \ldots, n. \tag{3.93}$$

Here $\mathbf{u}_h = \sum_{j=1}^{m} u_j^k N_j$ and $p_h = \sum_{j=1}^{n} p_j L_j$, $A_{ij} = \int_\Omega N_i N_j \, d\Omega$, $B_{ij}^k = -\int_\Omega \frac{\partial N_i}{\partial x_k} \, d\Omega$, and $q_i = \int_\Omega q L_i \, d\Omega$.

The above equations can be compactly expressed in matrix form as:

$$\begin{bmatrix} A & B^T \\ B & 0 \end{bmatrix} \begin{bmatrix} \mathbf{u}_h \\ p_h \end{bmatrix} = \begin{bmatrix} \mathbf{f} \\ \mathbf{q} \end{bmatrix}, \tag{3.94}$$

where A is an $m \times m$ matrix representing velocity contributions, and B is an $n \times m$ matrix coupling velocity and pressure.
If we multiply the first equation by A^{-1}, we obtain:

$$\mathbf{u}_h = A^{-1}\mathbf{f} - A^{-1}B^T p_h. \tag{3.95}$$

Substituting this result into the second equation yields:

$$-BA^{-1}B^T p_h = \mathbf{q} - BA^{-1}\mathbf{f}. \tag{3.96}$$

Here, $BA^{-1}B^T$ is known as the *pressure Schur complement.*

Remarks:

The matrix $M = \begin{bmatrix} A & B^T \\ B & 0 \end{bmatrix}$ is not invertible because the pressure Schur complement $BA^{-1}B^T$ is not invertible. If the velocity and pressure are approximated using the same type of finite elements, the pressure becomes singular, potentially causing oscillations.

Babuska-Brezzi Conditions:

To ensure the invertibility of M, the following conditions must be satisfied:

(a) The matrix A must be sufficiently invertible, i.e.,

$$\mathbf{u}_h^T A\mathbf{u}_h > 0, \quad \forall \mathbf{u}_h \in \text{Ker}(B). \tag{3.97}$$

(b) The pressure Schur complement $BA^{-1}B^T$ is invertible if:

$$\text{Ker}(B^T) = \{0\}. \tag{3.98}$$

As listed below, Algorithm 3 outlines the standard FEM implementation, detailing the discretization process, matrix assembly, system solution, and post-processing to ensure mass conservation and numerical accuracy.

3.10 MIXED FINITE ELEMENT METHOD

The Mixed Finite Element Method (MFEM) is a numerical approach particularly suited for addressing elliptic problems that characterize flow in porous media. While traditional finite element methods (FEM) often encounter instabilities when solving such problems [36, 30], the MFEM overcomes these challenges by introducing mixed formulations. These methods are locally stable, capable of higher-order approximations, and extend seamlessly to more complex scenarios.

The MFEM provides a framework to approximate two fundamental variables in a parabolic system: the scalar quantity (e.g., pressure) and its flux (e.g., velocity). These variables belong to distinct finite element spaces and

Algorithm 3 Finite Element Method (FEM) for Incompressible Flow

Input: Domain Ω, grid discretization (finite element mesh), permeability tensor K, viscosity μ, boundary conditions p_B, source term q
Output: Approximated velocity field $\mathbf{u}_h$, pressure field p_h
Step 1: Discretize the Computational Domain
Generate a finite element mesh for Ω with nodal coordinates
Define basis functions N_j for velocity and L_j for pressure
Step 2: Assemble System Matrices
Compute element-wise contributions to the global stiffness matrix A
Compute gradient coupling matrix B
Assemble the source term vector $\mathbf{q}$
Step 3: Apply Boundary Conditions
Modify system matrices to enforce Dirichlet boundary conditions on p
Adjust the right-hand side vector for Neumann boundary conditions
Step 4: Solve the Discrete System
Formulate the block system:

$$\begin{bmatrix} A & B^T \\ B & 0 \end{bmatrix} \begin{bmatrix} \mathbf{u}_h \\ p_h \end{bmatrix} = \begin{bmatrix} \mathbf{f} \\ \mathbf{q} \end{bmatrix}$$

Solve for velocity $\mathbf{u}_h$ and pressure p_h using a direct or iterative solver
Step 5: Compute Post-Processing Quantities
Compute derived quantities such as velocity gradients and fluxes
Validate mass conservation and numerical accuracy
Return Approximated velocity $\mathbf{u}_h$ and pressure p_h

are estimated simultaneously in the mixed formulation. This results in higher-order approximations for both variables.

One critical aspect of MFEM is the compatibility condition between the two finite element spaces. This compatibility ensures the stability of the numerical scheme by extending the concepts of consistency and convergence while imposing additional restrictions on the discretization.
Let $\Omega \subset \mathbb{R}^d$ $(d = 2, 3)$ be a polygonal and bounded domain. We adopt the standard notations for function spaces:

$L^p(\Omega)$: The Lebesgue space of functions $f : \Omega \to \mathbb{R}$ such that $\|f\|_{L^p(\Omega)} < \infty$, where $1 \leq p \leq \infty$.
$W^{k,p}(\Omega)$: The Sobolev space of functions with derivatives up to order k in $L^p(\Omega)$. In particular, $W^{k,2}(\Omega)$ is denoted by $H^k(\Omega)$.

The corresponding norms are denoted by $\|\cdot\|_{L^p(\Omega)}$, $\|\cdot\|_{W^{k,p}(\Omega)}$, and $\|\cdot\|_{H^k(\Omega)}$, respectively. For simplicity, $\|\cdot\|_{L^p(\Omega)}$ is also written as $\|\cdot\|$. The inner product on $L^2(\Omega)$ or $(L^2(\Omega))^d$ is denoted by $(\cdot,\cdot)$. For a vector-valued function

$\mathbf{v} \in (L^p(\Omega))^d$ and a matrix-valued function $\mathbf{A} \in (L^p(\Omega))^{d\times d}$, the norms are defined as:

$$\|\mathbf{v}\|_{(L^p(\Omega))^d} := \||\mathbf{v}|\|_{L^p(\Omega)}, \quad \|\mathbf{A}\|_{(L^p(\Omega))^{d\times d}} := \||\mathbf{A}|\|_{L^p(\Omega)}, \tag{3.99}$$

where $|\cdot|$ denotes the 2-norm of a vector in $\mathbb{R}^d$ or a matrix in $\mathbb{R}^{d\times d}$. For simplicity, the norms $\|\cdot\|_{L^2(\Omega)}$, $\|\cdot\|_{(L^2(\Omega))^d}$, and $\|\cdot\|_{(L^2(\Omega))^{d\times d}}$ are also written as $\|\cdot\|$ when dealing with scalar-, vector-, and matrix-valued functions, respectively.

Let $T > 0$ represent the final simulation time. For a given normed space X and $1 \le p \le \infty$, we define the space $L^p(0,T;X)$ as:

$$L^p(0,T;X) = \{\phi : \phi(t) \in X, \|\phi\|_X \in L^p(0,T)\}, \tag{3.100}$$

which is a normed linear space equipped with the norm:

$$\|\phi\|_{L^p(0,T;X)} := \|\|\phi\|_X\|_{L^p(0,T)}. \tag{3.101}$$

Here, $\|\phi\|_X$ represents the norm of ϕ in the space X, and the outer norm is taken over the time interval $(0,T)$.

The MFEM ensures numerical stability through the compatibility of the finite element spaces used for the scalar and flux variables. Additionally, it extends the numerical framework by ensuring local conservation properties. Providing higher-order accuracy for both primary variables. Supporting stable approximations for complex domains and heterogeneous media. By leveraging the norms and function spaces defined above, MFEM provides a mathematically rigorous foundation for solving flow problems in porous media.

3.10.1 MIXED WEAK FORMULATION

The mixed weak formulation for the wormhole problem can now be expressed as follows:
Continuity Equation:

$$(\nabla \cdot \mathbf{u}, \phi) = 0, \quad \forall \phi \in L^2(\Omega). \tag{3.102}$$

Momentum Equation:

$$(K^{-1}\mathbf{u}, \mathbf{v}) = (p, \nabla \cdot \mathbf{v}), \quad \forall \mathbf{v} \in V_0. \tag{3.103}$$

where K^{-1} represents the inverse hydraulic conductivity, $\phi \in L^2(\Omega)$ serves as the test function for the continuity equation and $\mathbf{v} \in V_0$ is the test function for the momentum equation, ensuring zero normal flux on $\partial\Omega$.

Equation (3.102) enforces the divergence-free condition for the velocity field $\mathbf{u}$ in the absence of sources or sinks within the domain. The space $H(\text{div};\Omega)$ ensures that both the velocity field and its divergence are square-integrable, making the weak form well-posed. The subspace $H_0(\text{div};\Omega)$ imposes natural boundary conditions on $\mathbf{v}$ by requiring $\mathbf{v} \cdot \mathbf{n} = 0$ on $\partial\Omega$.

3.10.2 MIXED FINITE ELEMENT FORMULATION

Let $J = (0, T]$, where $T > 0$ represents the simulation time interval. The mixed finite element formulation seeks the following time-dependent variables:

$p_h(\cdot, t), c_h(\cdot, t) \in W$: Scalar fields (e.g., pressure or concentration),
$\mathbf{u}_h(\cdot, t) \in X_0$: Velocity field,
$\mathbf{q}_h(\cdot, t) \in Y_0$: Flux field,

such that the following equations hold for all $t \in J$:
Continuity Equation:

$$(\nabla \cdot \mathbf{u}_h, \phi) = 0, \quad \forall \phi \in W. \tag{3.104}$$

Momentum Equation:

$$(K^{-1}\mathbf{u}_h, \mathbf{v}) = (p_h, \nabla \cdot \mathbf{v}), \quad \forall \mathbf{v} \in X_0. \tag{3.105}$$

Let $\mathcal{E}_h$ denote a partition of the domain Ω, composed of triangles or quadrilaterals if $d = 2$, and tetrahedra, prisms, or hexahedra if $d = 3$. For each element $E \in \mathcal{E}_h$, let h_E denote the diameter of E, and define the mesh size as:

$$h = \max_{E \in \mathcal{E}_h} h_E, \tag{3.106}$$

with the condition:

$$\frac{h}{h_E} \leq C, \tag{3.107}$$

where C is a constant independent of the mesh size. We assume that $\mathcal{E}_h$ is non-degenerate. This means each element E is convex, and there exists $\rho > 0$ such that every sub-triangle (if $d = 2$) or sub-tetrahedron (if $d = 3$) within E contains a ball of radius ρh_E in its interior.
The finite element approximation is defined on the following subspaces:

$X \subset V$ and $X_0 \subset V_0$: Subspaces for the velocity field, typically represented using the r-th order Raviart-Thomas space (RT_r).
$Y \subset H(\mathrm{div}; \Omega)$ and $Y_0 \subset H_0(\mathrm{div}; \Omega)$: Subspaces for divergence-conforming vector fields.
$W \subset L^2(\Omega)$: Subspace for scalar fields, such as pressure or concentration.

Here, $r \geq 0$ represents the order of the Raviart-Thomas space, and the subspaces are defined on the partition $\mathcal{E}_h$ of the domain.

REMARKS

The use of Raviart-Thomas spaces ensures that the velocity field $\mathbf{u}_h$ conforms to the divergence condition. This property is critical for accurately modeling flow in porous media.

The assumptions on the mesh, such as convexity and the existence of a sub-ball in each element, ensure that the numerical scheme is stable and convergent (Mesh Regularity).
The inclusion of time-dependent terms allows the formulation to capture transient phenomena, such as evolving pressure and concentration fields.

LADYZHENSKAYA-BABUSKA-BREZZI (LBB) CONDITION

The Ladyzhenskaya-Babuska-Brezzi (LBB) condition, also known as the inf-sup condition, is a mathematical stability criterion that ensures well-posedness for mixed finite element formulations. This condition plays a critical role in problems where multiple interdependent variables are approximated simultaneously, such as velocity and pressure in fluid dynamics.

Definition

Let $\mathbf{V}$ be the function space for velocity and P the function space for pressure. The LBB condition requires the existence of a constant $\beta > 0$ such that:

$$\inf_{q\in P} \sup_{\mathbf{v}\in \mathbf{V}} \frac{b(\mathbf{v}, q)}{\|\mathbf{v}\|_{\mathbf{V}} \|q\|_P} \geq \beta > 0, \tag{3.108}$$

where $b(\mathbf{v}, q)$ is the bilinear form representing the coupling between velocity and pressure (e.g., the divergence operator in fluid dynamics). $\|\cdot\|_{\mathbf{V}}$ and $\|\cdot\|_P$ are the norms for the velocity and pressure spaces, respectively. β is a constant dependent on the geometry of the domain and the choice of finite element spaces. The LBB condition ensures the stability of the mixed finite element formulation by guaranteeing compatibility between the velocity and pressure approximation spaces. For compressible flow problems, the velocity space $\mathbf{V}_h$ must be chosen to ensure compatibility with the divergence operator, which couples it with the pressure space P_h.

EXAMPLE: 1D FLOW IN POROUS MEDIA

For simplicity, let us consider a one-dimensional steady-state flow in porous media, governed by the system of equations:

$$\begin{cases} u' = f(x), \\ p' = u, \end{cases} \tag{3.109}$$

defined on the spatial domain $\Omega = (0, 1)$. Here, $u(x)$ represents the velocity-like variable, while $p(x)$ can be interpreted as a pressure or potential function. The source term $f(x)$ acts as an external forcing function, driving the system.

To derive the weak form, we multiply both equations by suitable test functions ϕ and ψ, chosen from appropriate function spaces, and integrate over the domain:

$$\int_0^1 u'\phi\,dx = \int_0^1 f\phi\,dx, \quad \int_0^1 p'\psi\,dx = \int_0^1 u\psi\,dx. \tag{3.110}$$

Applying integration by parts to shift derivatives onto the test functions while assuming suitable boundary conditions (e.g., homogeneous Dirichlet or Neumann conditions), we obtain:

$$\int_0^1 u\phi'\,dx = \int_0^1 f\phi\,dx, \quad \int_0^1 p\psi'\,dx = -\int_0^1 u\psi\,dx. \tag{3.111}$$

This weak formulation is advantageous as it allows the solution to belong to function spaces with fewer regularity constraints compared to the classical (strong) form.

To discretize the problem using the Finite Element Method (FEM), we approximate $u(x)$ and $p(x)$ using a set of basis functions:

$$\widetilde{u} = \sum_{j=0}^{n} u_j\psi_j, \quad \widetilde{p} = \sum_{j=0}^{n} p_j\phi_j, \tag{3.112}$$

where ψ_j and ϕ_j are basis functions defined on a finite element mesh, and u_j, p_j are the corresponding unknown coefficients to be determined.

Substituting these approximations into the weak formulation and enforcing the equations in a discrete sense leads to the following matrix system:

$$\begin{bmatrix} -A & 0 \\ B & A^T \end{bmatrix} \begin{bmatrix} \mathbf{u} \\ \mathbf{p} \end{bmatrix} = \begin{bmatrix} \mathbf{f} \\ 0 \end{bmatrix}, \tag{3.113}$$

where the system matrices are defined as:

$$A_{ij} = \int_0^1 \phi_i\psi_j'\,dx, \quad B_{ij} = \int_0^1 \psi_i\psi_j\,dx, \quad \mathbf{f}_i = \int_0^1 \phi_i f\,dx. \tag{3.114}$$

It is noteworthy that the block matrix structure reveals an important coupling between the unknown variables. The presence of A^T suggests a symmetry property often encountered in mixed finite element formulations. Moreover, the weak form naturally accommodates irregular geometries and heterogeneous material properties, making FEM suitable for complex porous media simulations. This formulation extends naturally to higher dimensions and is widely used in reservoir simulations, groundwater flow modeling, and subsurface transport phenomena. This formulation provides a rigorous foundation for numerical simulation of flow in porous media, highlighting the flexibility and robustness of MFEM in handling such problems.

Algorithm 4 Mixed Finite Element Method (MFEM) for Incompressible Flow

Input: Domain Ω, mesh $\mathcal{E}_h$, polynomial order r, permeability tensor K, boundary conditions p_B, simulation time interval J, time step Δt
Output: Approximated velocity field $\mathbf{u}_h$, flux field $\mathbf{q}_h$, and pressure field p_h
Step 1: Discretization of the Computational Domain
Partition Ω into a finite element mesh $\mathcal{E}_h$
Define basis functions from Raviart-Thomas space RT_r for velocity
Define basis functions from $L^2(\Omega)$ for pressure
Step 2: Formulate Mixed Weak Formulation
Define velocity space X_0, flux space Y_0, and scalar space W
Compute the weak form of the continuity and momentum equations
Step 3: Assemble System Matrices
Compute mass and stiffness matrices for mixed formulation
Construct discrete divergence operator
Assemble right-hand side vector including boundary conditions
Step 4: Time Discretization (if time-dependent)
Apply implicit time stepping using backward Euler or Crank-Nicholson schemes
Discretize time derivatives and update system matrices
Step 5: Solve the Mixed System
Solve coupled system:

$$\begin{bmatrix} M & B^T \\ B & 0 \end{bmatrix} \begin{bmatrix} \mathbf{u}_h \\ p_h \end{bmatrix} = \begin{bmatrix} \mathbf{f} \\ \mathbf{q} \end{bmatrix}$$

Use direct solvers (LU decomposition) or iterative solvers (GMRES, MINRES)
Step 6: Compute Fluxes and Post-Processing
Compute flux field $\mathbf{q}_h$ from velocity approximation
Ensure mass conservation and numerical stability
Return Approximated velocity $\mathbf{u}_h$, flux $\mathbf{q}_h$, and pressure p_h

In the following, Algorithm 5 outlines the implementation steps, including domain discretization, system assembly, time discretization (if applicable), and solving the coupled saddle-point problem to obtain velocity, flux, and pressure fields.

3.10.3 CHARACTERISTIC FINITE ELEMENT METHODS

Characteristic Finite Element Methods (CFEMs), also known as characteristic-Galerkin or characteristic-based schemes, are specialized approaches designed to handle transport-dominated problems, such as advection-diffusion

Algorithm 5

Algorithmic Steps

1. *Preprocessing*: Compute $\mathbf{q}(\mathbf{x}, t)$ from a flow model (e.g., a Darcy solver).
2. *Backward Characteristic Tracing*: For each time step, and for each node (or integration point) at time t, solve:

$$\frac{d\mathbf{X}(s)}{ds} = \frac{\mathbf{q}(\mathbf{X}(s), s)}{\phi}, \quad \mathbf{X}(t) = \mathbf{x}. \tag{3.115}$$

Integrate backward from t to t_0 to find $\mathbf{X}(t_0)$.
3. *Interpolation*: Interpolate $c(\mathbf{X}(t_0), t_0)$ from the known solution field at the previous time step.
4. *Assembly*: Form the global system of equations using the characteristic-based evaluation for the advection term, along with standard finite element integration for diffusion and reaction terms.
5. *Solution*: Solve the resulting linear or nonlinear system for $\mathbf{c}^t$.
6. *Repeat*: Advance to the next time step and repeat the process.

equations and hyperbolic conservation laws. Traditional finite element methods often struggle with stability and accuracy when simulating problems where convection effects dominate diffusion, leading to oscillations and non-physical solutions. CFEMs address this issue by incorporating characteristics—curves along which information travels in time, directly into the numerical formulation [37].

The key idea behind CFEMs is to trace fluid parcels or solution features along their trajectories (characteristics) and to use these trajectories when defining test and trial functions or when formulating the weak form of the governing equations. By doing so, the method inherently aligns the discretization with the physics of the problem. Consider a time-dependent advection-dominated PDE of the form:

$$\frac{\partial u}{\partial t} + \mathbf{v} \cdot \nabla u = f(\mathbf{x}, t), \tag{3.116}$$

where $\mathbf{v}$ is a velocity field and f is a source term. In a characteristic framework, one identifies characteristic curves $\mathbf{X}(t)$ defined by:

$$\frac{d\mathbf{X}(t)}{dt} = \mathbf{v}(\mathbf{X}(t), t), \tag{3.117}$$

along which the solution is constant or changes only according to the source term. By integrating the PDE along these characteristics, one can derive a formulation that reduces numerical diffusion and stabilizes the solution.

In the finite element context, CFEMs typically rewrite the variational form of the PDE using a semi-Lagrangian approach. Instead of integrating purely over fixed spatial elements at successive time steps, the method accounts for the movement of information along characteristics. This can be done by mapping back the solution at a new time level to its position at the previous time level via the characteristic trajectories. As a result, the variational problem involves integrals of the solution along these trajectories, leading to enhanced numerical stability and fewer artificial oscillations.

One challenge in implementing CFEMs is the additional complexity of computing and storing characteristic curves. Since each node or integration point may move backward in time along a characteristic to locate the corresponding departure position, efficient algorithms and data structures are required, especially for large-scale or multidimensional problems. Another consideration is that the integrals along characteristics can lead to non-symmetric and time-dependent system matrices, adding complexity to the solution procedure.

3.10.4 CFEMS FOR GAS TRANSPORT IN POROUS MEDIA

Characteristic Finite Element Methods (CFEMs) offer a robust and stable framework for simulating gas transport in porous media, particularly in advection-dominated regimes. Such problems can be described by an advection-diffusion-reaction equation governing the concentration $c(\mathbf{x}, t)$ of a gas (e.g., CO_2, CH_4) in a porous medium:

$$\phi \frac{\partial c}{\partial t} + \nabla \cdot (\mathbf{q}c) - \nabla \cdot (D\nabla c) = R(c, \mathbf{x}, t), \tag{3.118}$$

where ϕ is porosity, $\mathbf{q} = \mathbf{v}\rho$ is the Darcy flux (with $\mathbf{v}$ being the Darcy velocity and ρ the fluid density), D is the diffusion-dispersion coefficient (potentially tensor-valued), and $R(c, \mathbf{x}, t)$ is a source or reaction term.

Standard Galerkin finite element methods often produce oscillatory, nonphysical solutions when simulating highly advective flows. CFEMs mitigate these issues by incorporating characteristics—curves along which fluid parcels or solution features move directly into the numerical formulation [37, 39].

Characteristic Formulation

Consider a scenario dominated by advection. The governing equation reduces approximately to:

$$\phi \frac{\partial c}{\partial t} + \nabla \cdot (\mathbf{q}c) \approx 0. \tag{3.119}$$

The characteristic curves $\mathbf{X}(t)$ satisfy:

$$\frac{d\mathbf{X}}{dt} = \frac{\mathbf{q}(\mathbf{X}(t), t)}{\phi}. \tag{3.120}$$

Along these curves, the solution evolution is determined by integrating any source or reaction terms. In a purely advective setting, the concentration is nearly constant along a characteristic.

For the full advection-diffusion-reaction equation, CFEMs handle the advective portion via characteristics and correct for diffusion and reaction within a variational (weak) framework. This often involves a semi-Lagrangian approach, where one traces backward along characteristics from the current time level t to the previous time level t_0, thereby relating $c(\mathbf{x}, t)$ to $c(\mathbf{x}(t_0), t_0)$.

Weak Formulation and Discretization

Starting from the weak form of the PDE:

$$\int_\Omega \phi \frac{\partial c}{\partial t} w \, d\Omega + \int_\Omega \mathbf{q} c \cdot \nabla w \, d\Omega - \int_\Omega D(\nabla c \cdot \nabla w) \, d\Omega = \int_\Omega R(c, \mathbf{x}, t) w \, d\Omega, \quad (3.121)$$

where w is a suitable test function. To incorporate characteristics, one adopts a characteristic-based formulation for the advection term. Instead of evaluating $c(\mathbf{x}, t)$ directly at nodes, one traces each node's position backward along a characteristic to find its departure point at t_0:

$$c(\mathbf{x}, t) \approx c(\mathbf{X}(t_0), t_0), \quad (3.122)$$

where $\mathbf{X}(t_0)$ is the departure position of the characteristic passing through $\mathbf{x}$ at time t.

After discretizing the spatial domain using finite elements and introducing shape functions and nodal values, this backward characteristic tracing ensures that the system of equations incorporates upwind information naturally. The resulting discrete equations have improved stability and reduced numerical oscillations.

In porous media, properties such as porosity, permeability, and diffusivity may be heterogeneous and anisotropic. CFEM can accommodate these complexities by:

- Incorporating spatially varying $\phi(\mathbf{x})$ and $D(\mathbf{x})$ into the characteristic tracing and weak form integrals.
- Using advanced search techniques or spatial data structures to efficiently locate departure points in complex geometries.
- Coupling with nonlinear reaction terms, $R(c, \mathbf{x}, t)$, which may require iterative solvers or fixed-point methods.

3.11 SPECTRAL ELEMENT METHODS

Spectral element methods combine the geometric flexibility of finite element methods with the high-order accuracy of spectral methods. These methods are valuable for simulating flow in porous media, where high resolution and accuracy are required to capture complex flow patterns across multiple scales [40, 41]. The spectral element method is based on approximating solutions using high-order polynomial basis functions. Unlike traditional finite element methods that typically use linear or quadratic basis functions, spectral methods employ high-degree polynomials that provide exponential convergence rates for smooth solutions.

Let us consider the pressure equation for single-phase flow in porous media:

$$-\nabla \cdot (K(\mathbf{x})\nabla p) = q(\mathbf{x}) \quad \text{in } \Omega \tag{3.123}$$

where $K(\mathbf{x})$ is the permeability tensor, p is the pressure, and q is the source/sink term.

In the spectral element method, we divide the domain Ω into elements Ω_e and approximate the solution within each element using high-order polynomials. The key distinction from standard finite element methods lies in the choice and degree of these polynomials.

Let us start with a simple 1D example to illustrate the fundamental concepts. Consider steady-state flow in a one-dimensional heterogeneous porous medium with length L:

$$-\frac{d}{dx}\left(K(x)\frac{dp}{dx}\right) = q(x) \quad \text{for } x \in [0, L] \tag{3.124}$$

with boundary conditions:

$$p(0) = p_0, \quad p(L) = p_L \tag{3.125}$$

We divide the domain into E elements, each mapped to a reference element $\xi \in [-1, 1]$.

For an element e spanning $[x_e, x_{e+1}]$, the mapping from the reference coordinate ξ to the physical coordinate x is given by:

$$x(\xi) = \frac{x_{e+1} + x_e}{2} + \frac{x_{e+1} - x_e}{2}\xi \tag{3.126}$$

Within each element, we approximate the pressure using Lagrange interpolation polynomials based on Gauss-Lobatto-Legendre (GLL) points:

$$p(x) \approx \sum_{i=0}^{N} p_i^e \ell_i(\xi) \tag{3.127}$$

where $\ell_i(\xi)$ are the Lagrange polynomials of degree N based on GLL points $\{\xi_j\}_{j=0}^{N}$, and p_i^e are the pressure values at these points within element e.

For a polynomial degree N, the GLL points are the roots of:

$$(1 - \xi^2)L_N'(\xi) = 0 \tag{3.128}$$

where $L_N(\xi)$ is the Legendre polynomial of degree N. These points include the endpoints $\xi = \pm 1$ and are optimal for both interpolation and quadrature.

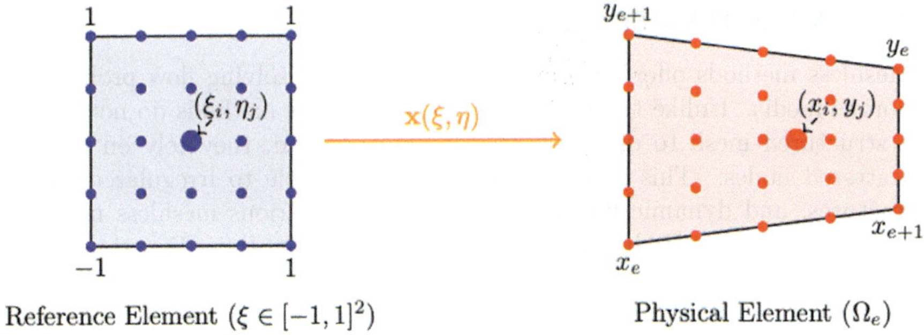

Figure 3.5 Mapping between the reference element $\xi \in [-1, 1]^2$ and a physical element Ω_e in the spectral element method. The transformation is defined using shape functions $N_i(\xi, \eta)$, which interpolates the physical coordinates x_i from the reference space.

The associated Lagrange polynomials are:

$$\ell_i(\xi) = \prod_{j=0, j \neq i}^{N} \frac{\xi - \xi_j}{\xi_i - \xi_j} \tag{3.129}$$

To efficiently approximate solutions within each element, we employ a transformation from a reference element to the physical domain. This transformation, shown in Figure 3.5, enables the use of high-order interpolation polynomials based on Gauss-Lobatto-Legendre (GLL) points. By mapping the computational domain in this way, we maintain spectral accuracy while accommodating complex geometries in porous media flow simulations.

The weak form of our equation, after integration by parts, is:

$$\int_{\Omega_e} K(x) \frac{dp}{dx} \frac{dv}{dx} dx = \int_{\Omega_e} q(x) v(x) dx + \text{boundary terms} \tag{3.130}$$

For each element, we compute the element stiffness matrix and load vector:

$$A_{ij}^e = \int_{x_e}^{x_{e+1}} K(x) \frac{d\ell_i}{dx} \frac{d\ell_j}{dx} dx \tag{3.131}$$

$$b_i^e = \int_{x_e}^{x_{e+1}} q(x) \ell_i(x) dx \tag{3.132}$$

The spectral element method offers several advantages for porous media flow applications. It provides high-order accuracy with exponential convergence rates for smooth solutions, requiring fewer degrees of freedom to achieve a given accuracy. Its geometric flexibility allows it to conform to complex domain geometries and align element boundaries with material interfaces.

3.12 MESHLESS METHODS

Meshless methods offer a promising alternative for solving flow problems in porous media. Unlike traditional methods, meshless methods do not require a structured mesh to discretize the domain; instead, they rely on a set of scattered nodes. This makes them highly adaptable to irregular domains, fractures, and dynamic boundaries. Among the various meshless methods, Smoothed Particle Hydrodynamics (SPH) [42], Radial Basis Function (RBF) methods [43], Meshless Local Petrov-Galerkin (MLPG) methods [44], and Element-Free Galerkin (EFG) methods [45] have been applied to flow simulations in porous media.

Applications of meshless methods in porous media include single-phase and multiphase flow, fractured systems, reactive transport, and subsurface remediation. These methods are particularly advantageous for complex problems, such as flow through heterogeneous and anisotropic media or domains with evolving interfaces [49]. However, challenges such as computational cost, accuracy, and the imposition of boundary conditions remain active areas of research. In the following, let us explore some common meshless methods applied to porous media.

3.12.1 SMOOTHED PARTICLE HYDRODYNAMICS (SPH)

Smoothed Particle Hydrodynamics (SPH) is a fully Lagrangian, particle-based numerical method originally developed for astrophysical applications [47, 48], and it has since been widely applied in various fields of fluid mechanics. Unlike traditional grid-based approaches, SPH does not rely on a fixed mesh; instead, it represents the fluid by a set of moving particles. Each particle carries its own physical properties, such as mass, velocity, and pressure, and moves according to the governing equations of fluid flow. This particle-based paradigm naturally accommodates complex boundaries, large deformations, and free-surface dynamics without requiring mesh reconstruction.

A central idea in SPH is the use of smoothing kernels to approximate field variables. These kernels provide a weighting function that determines how the properties of neighboring particles contribute to the value of a field at a given particle position. For a field variable $A(\mathbf{r})$ at particle position $\mathbf{r}$, the SPH approximation is:

$$A(\mathbf{r}) = \sum_j m_j \frac{A_j}{\rho_j} W(\mathbf{r} - \mathbf{r}_j, h), \tag{3.133}$$

where m_j and ρ_j are the mass and density of the j-th particle, A_j is the value of the field at the j-th particle, W is the smoothing kernel, and h is the smoothing length that determines the spatial extent of the kernel. Spatial derivatives can be computed by differentiating the kernel function:

$$\nabla A(\mathbf{r}) = \sum_j m_j \frac{A_j}{\rho_j} \nabla W(\mathbf{r} - \mathbf{r}_j, h). \tag{3.134}$$

This mesh-free formulation enables SPH to naturally handle complex scenarios like free surfaces, splashing, and fragmentation [42, 51]. Since particles move with the fluid flow, tracking evolving interfaces and fluid boundaries becomes straightforward, making SPH well-suited for simulations of multiphase flows, geophysical processes, and industrial mixing problems.

However, SPH can be computationally demanding, with a naive complexity of $O(n^2)$ due to the need to identify neighboring particles at each time step. To address this, efficient neighbor search algorithms, such as tree structures, linked lists, or grid-based methods, are commonly implemented to reduce the computational overhead. Another challenge involves imposing boundary conditions. Techniques such as ghost particles, kernel corrections, and modified kernel functions are often employed to accurately represent solid walls and interfaces.

Despite these challenges, SPH's flexibility and robustness have contributed to its growing popularity. It has been successfully applied to a wide range of problems, from astrophysical phenomena to engineering applications and environmental flows.

3.12.2 RADIAL BASIS FUNCTION (RBF) METHODS

Radial Basis Function (RBF) methods are a class of mesh-free approximation techniques used for interpolating functions, reconstructing surfaces, and solving partial differential equations (PDEs) without relying on structured grids. Unlike finite difference or finite element methods that require a predefined mesh or element connectivity, RBF methods work directly with scattered nodes placed throughout the computational domain. This flexibility makes RBF approaches well-suited for complex, irregular geometries and problems where adaptive node refinement is desirable.

In RBF methods, the unknown field variable $u(\mathbf{x})$ is represented as a weighted sum of radially symmetric functions centered at each node:

$$u(\mathbf{x}) = \sum_{j=1}^{N} c_j \phi(\|\mathbf{x} - \mathbf{x}_j\|), \tag{3.135}$$

where ϕ is the chosen radial basis function (e.g., Gaussian, multiquadric, inverse multiquadric, or polyharmonic spline) and c_j are unknown coefficients determined by enforcing boundary conditions and governing equations at the chosen collocation points [43]. This results in a system of equations that can be solved for the coefficients c_j. By selecting different forms for ϕ, one can adjust the smoothness, shape, and support of the basis functions.

One of the main advantages of RBF methods is their capability to handle scattered and arbitrarily distributed nodes with ease. This is particularly useful in complex domains where constructing a structured mesh would be challenging or time-consuming. Additionally, RBFs provide spectral-like accuracy for smooth functions and can be combined with local refinement strategies. Local RBF approaches, which consider only a subset of neighboring nodes for each interpolation, help reduce the computational complexity compared to global RBF methods and can improve the conditioning of the resulting linear systems [52].

However, global RBF methods often lead to fully dense linear systems of size $N \times N$, where N is the number of nodes. Such systems can be computationally expensive to solve, scaling roughly as $O(N^3)$, and can suffer from ill-conditioning as N grows or nodes become unevenly distributed. Various techniques, such as using local RBFs, employing hierarchical decomposition methods, or incorporating specialized preconditioners, have been developed to alleviate these issues and improve scalability and stability. When successfully implemented, RBF methods excel in problems involving irregular boundaries, complex topologies, and dynamic node distributions. They have found applications in fields such as fluid dynamics, geophysical modeling, image processing, and machine learning.

3.12.3 MESHLESS LOCAL PETROV-GALERKIN METHODS

The Meshless Local Petrov-Galerkin (MLPG) method is a mesh-free, Galerkin-based approach that avoids the need for a global background mesh or fixed elements. Instead, it relies on a set of scattered nodes distributed throughout the computational domain. By constructing local subdomains around each node, MLPG transforms the governing partial differential equations (PDEs) into a series of locally defined weak formulations. This approach builds upon the general framework of Petrov-Galerkin methods but applies them in a localized, element-free setting.

In MLPG, the governing equations are represented in a weak (integral) form over these small local subdomains. For an unknown field u, subject to a differential operator $\mathcal{L}$ (often representing spatial derivatives arising from diffusion, convection, or elasticity), and boundary conditions involving a boundary function g, the localized weak form can be written as:

$$\int_{\Omega_s} \delta u \, \mathcal{L}(u) \, d\Omega_s = \int_{\Gamma_s} \delta u \, g \, d\Gamma_s, \tag{3.136}$$

where Ω_s is a local subdomain associated with a chosen node, Γ_s is its boundary, δu is a test (weight) function, and $\mathcal{L}(u)$ represents the PDE in strong form [44]. Each local subdomain can be chosen to overlap partially with others, ensuring coverage of the entire domain while maintaining flexibility in node placement.

To approximate the unknown field u, MLPG commonly employs mesh-free interpolation schemes like Moving Least Squares (MLS) or Radial Basis Functions (RBFs) for constructing shape functions. MLS, for example, determines shape functions by minimizing a weighted least squares error within a local neighborhood, resulting in highly smooth and flexible approximations. Similarly, RBF-based shape functions can handle scattered nodes and complex geometries with relative ease.

One of the key advantages of MLPG is its suitability for complex problems, such as anisotropic and heterogeneous porous media, where material properties vary dramatically within the domain [50]. Because the method does not rely on predefined element connectivity, it can adapt easily to changes in node distribution, making it appealing for problems that involve evolving boundaries, moving interfaces, or complex internal structures.

However, a central challenge in MLPG lies in the computation of the integrals over arbitrarily shaped local subdomains. Unlike methods that rely on structured grids or elements with known integration rules, MLPG must contend with irregularly shaped subdomains formed by scattered nodes. Accurately performing numerical integration within these domains can be computationally expensive and intricate. Techniques for improving efficiency include using background cells for integration, applying adaptive quadrature rules, or implementing efficient node placement strategies.

Despite these computational considerations, MLPG's flexibility and mesh-free character have made it popular in various engineering and scientific applications, including geomechanics, fluid flow in porous media, and multiphysics problems.

3.12.4 ELEMENT-FREE GALERKIN (EFG) METHODS

The Element-Free Galerkin (EFG) method is a meshless approach that constructs shape functions without relying on predefined elements or element connectivity. Instead, it uses a set of scattered nodes placed throughout the computational domain, similar to other mesh-free methods. The cornerstone of EFG is the Moving Least Squares (MLS) approximation, which provides continuous and differentiable shape functions used to represent the unknown field variables.

In EFG, the goal is to approximate a field variable $u(\mathbf{x})$ using a weighted least squares fit to a polynomial basis $P(\mathbf{x})$. Consider a set of nodes $\mathbf{x}_1, \mathbf{x}_2, \ldots, \mathbf{x}_n$ scattered in the domain. The MLS shape functions are obtained by minimizing the following weighted error functional:

$$J = \sum_{i=1}^{n} w(\|\mathbf{x} - \mathbf{x}_i\|)[u(\mathbf{x}_i) - P(\mathbf{x}_i)]^2, \tag{3.137}$$

where $w(\|\mathbf{x} - \mathbf{x}_i\|)$ is a nonnegative weight function that decreases with the distance between the field point $\mathbf{x}$ and the nodal point $\mathbf{x}_i$ [45]. By minimizing

J with respect to the unknown coefficients in the polynomial basis $P(\mathbf{x})$, one obtains a set of MLS shape functions that interpolate the field variable $u(\mathbf{x})$.

A key advantage of EFG is its high accuracy and flexibility. The shape functions are constructed adaptively based on the chosen weight function and polynomial basis. This flexibility makes EFG highly suitable for complex geometries, moving boundaries, and problems where mesh generation and refinement would be difficult or overly time-consuming. Furthermore, the smoothness and continuity of MLS shape functions often result in very accurate solutions, especially for problems involving large deformations or complicated boundary shapes.

However, EFG also comes with certain challenges. Although it does not require elements for creating shape functions, EFG typically relies on a background integration grid or cell structure to perform numerical integration of the weak form of the governing equations. This background grid introduces some complexity and partial loss of the fully element-free nature of the method. Additionally, imposing essential boundary conditions (e.g., prescribed displacements or temperatures on boundaries) can be more difficult than in traditional finite element methods, as the shape functions in EFG do not necessarily satisfy the Kronecker delta property (i.e., they do not generally take a value of 1 at a node and 0 elsewhere).

Despite these challenges, EFG methods have been successfully applied to problems in solid mechanics, structural analysis, fluid-structure interaction, and other areas where mesh generation is nontrivial. The method's ability to handle irregular node distributions and complex domains is often a decisive factor in choosing EFG over classical element-based approaches.

CONCLUSION

The landscape of computational methods for transport phenomena in porous media is characterized by remarkable complexity and continuous innovation. This chapter has demonstrated that no single numerical approach is universally optimal; instead, the selection of an appropriate method depends on the specific problem characteristics, domain geometry, and desired computational efficiency.

Finite Difference, Finite Element, Mixed Finite Element, Spectral Element, and Meshless methods each offer unique advantages in modeling transport processes. Finite Difference methods provide computational simplicity, while Finite Element approaches excel in handling complex geometries. Mixed Finite Element methods offer enhanced stability for coupled flow problems, Spectral Element methods deliver high-order accuracy, and Meshless techniques provide unprecedented flexibility in handling irregular domains.

PROBLEMS

1. Consider a one-dimensional advection-diffusion equation in a porous medium:
$$\frac{\partial c}{\partial t} + u\frac{\partial c}{\partial x} - D\frac{\partial^2 c}{\partial x^2} = 0$$
Implement the forward-time central-space (FTCS) scheme with a uniform grid spacing. Analyze the stability conditions and discuss the impact of the Courant-Friedrichs-Lewy (CFL) number.
2. Develop a mixed finite element formulation for a steady-state groundwater flow problem with heterogeneous permeability. Implement the Raviart-Thomas element and discuss how the method ensures mass conservation.
3. For a simple diffusion problem, compare the convergence rates of a spectral element method using polynomials of different degrees. Investigate how the polynomial order affects the solution accuracy and computational efficiency.
4. Implement a Smoothed Particle Hydrodynamics (SPH) and a Radial Basis Function (RBF) method for simulating fluid transport in a heterogeneous porous medium. Compare their performance in terms of accuracy, computational cost, and ability to handle complex geometries.
5. Analyze the stability of different temporal discretization schemes (Forward Euler, Backward Euler, Crank-Nicolson) for a nonlinear reaction-diffusion equation in a porous medium. Determine the conditions under which each method provides stable numerical solutions.
6. Develop a numerical method to simulate fluid flow in a porous medium with a strongly anisotropic permeability tensor. Discuss how different discretization approaches handle directional variations in permeability.
7. Implement a characteristic-based finite element method for an advection-dominated transport problem. Compare its performance with standard Galerkin methods in terms of numerical oscillations and solution accuracy.
8. Design an adaptive time-stepping algorithm for a two-phase flow simulation in a heterogeneous porous medium. Develop criteria for dynamically adjusting the time step based on local flow dynamics and numerical stability.
9. Investigate the impact of different boundary condition treatments (Dirichlet, Neumann, Robin) on the numerical solution of a transport problem in a porous medium. Analyze how these conditions affect solution accuracy and stability.
10. Compare the computational efficiency of different numerical methods (Finite Difference, Finite Element, Meshless) for simulating transport in a complex, highly heterogeneous porous domain. Develop metrics for assessing computational performance beyond traditional accuracy measures.

DECLARATION

During the preparation of this work, the author used ChatGPT in order to improve the language and readability. After using this tool/service, the author reviewed and edited the content as needed and takes full responsibility for the content of the publication.

4 Single-Phase Flow in Porous Media

4.1 INTRODUCTION

Porous media flow plays a fundamental role in a wide range of scientific and engineering applications, including groundwater hydrology, petroleum reservoir simulation, environmental remediation, and subsurface heat storage. Single-phase flow, in particular, serves as a foundational concept in understanding more complex multiphase and reactive transport processes. It describes the movement of a single fluid, such as water, oil, or gas, through a porous solid matrix under the influence of pressure gradients and physical properties of the medium. This chapter provides an exploration of incompressible single-phase flow through porous media, combining analytical derivations, numerical modeling, and computational implementation. Beginning with the formulation of the governing equations based on Darcy's law and the principle of mass conservation, the chapter builds toward solving these equations under various boundary and initial conditions. Analytical solutions are first presented for simplified geometries, offering insights into pressure distribution and the influence of key physical parameters such as permeability, viscosity, and flow rate. These solutions serve as validation benchmarks for more sophisticated numerical approaches. Subsequently, numerical methods, including finite difference and finite element methods, are introduced to solve the pressure equation in higher-dimensional and more realistic settings. To bridge theory and application, this chapter also presents MATLAB® and Python implementations for simulating flow in two-dimensional domains, supported by graphical visualizations. Furthermore, advanced topics such as solute transport, desiccant-based water harvesting, and stability criteria are introduced to highlight the relevance of single-phase flow modeling in both environmental and industrial contexts.

The goal of this chapter is not only to provide mathematical rigor and computational proficiency but also to establish a clear conceptual framework that supports the reader in understanding and simulating flow processes in porous media. This foundation will serve as a stepping stone for subsequent chapters addressing multiphase, thermal, and other flow systems.

4.2 INCOMPRESSIBLE SINGLE-PHASE FLOW

The incompressible fluid flow through a porous medium is governed by the continuity equation, which ensures mass conservation under steady-state conditions:

DOI: 10.1201/9781032724973-4

$$\nabla \cdot \mathbf{u} = 0 \tag{4.1}$$

where $\mathbf{u}$ is the Darcy velocity (fluid velocity through the porous medium).

According to Darcy's law, the relationship between the fluid velocity $\mathbf{u}$ and the pressure gradient ∇P in a porous medium is given by:

$$\mathbf{u} = -\frac{K}{\mu}\nabla P \tag{4.2}$$

where K is the permeability of the medium (in m^2), μ is the dynamic viscosity of the fluid (in Pa · s), P is the pressure (in Pa).

By substituting Darcy's law into the continuity equation, we obtain the governing momentum equation for incompressible flow in a 2D porous medium:

$$\nabla \cdot \left(\frac{K}{\mu}\nabla P\right) = -\frac{q}{\mu} \tag{4.3}$$

where q is the volumetric flow rate per unit volume (in m^3/s). In regions where there are no sources or sinks (i.e., $q = 0$), the equation simplifies to:

$$\nabla \cdot \left(\frac{K}{\mu}\nabla P\right) = 0 \tag{4.4}$$

Assuming constant permeability and viscosity, this further simplifies to the Laplace equation in two dimensions $\nabla^2 P = 0$, with left boundary (Inlet, $x = 0$)), which is constant pressure boundary condition: $P = P_{\text{left}}$ (e.g., $P = 100{,}000\,\text{Pa}$). The right boundary (Outlet, $x = L_x$), which is a constant flux boundary condition: The flux is specified as $\frac{\partial P}{\partial x} = -\frac{q\mu}{K}$. The top and bottom boundaries (No-flow boundaries) are a no flux across the boundary, which implies $\frac{\partial P}{\partial y} = 0$.

4.3 ANALYTICAL SOLUTIONS

Analytical solutions are usually available for some standard ODEs or PDEs via methods handled by differential equations courses. These analytical solutions mainly serve as checks of numerical solutions or as verified simplified cases, such as around-well flow. To address the problem within a specific context, such as a one-dimensional domain, we simplify the governing equation to the following form:

$$\frac{\partial}{\partial x}\left(-\frac{k}{\mu}\frac{\partial P}{\partial x}\right) = q$$

Upon expanding the derivative, we obtain:

$$-\frac{k}{\mu}\frac{\partial^2 P}{\partial x^2} = q,$$

which is a second-order ordinary differential equation (ODE) in terms of P with x as the independent variable. To solve this ODE, it is requisite to carry out integration with respect to x twice, considering two boundary conditions to ascertain the constants of integration. Let us posit two generic boundary conditions for our solution: $P(0) = P_0$, representing the pressure at the origin where $x = 0$. $P(L) = P_L$, denoting the pressure at a point $x = L$ within the domain. The integration yields the general form of the solution:

$$P(x) = -\frac{1}{2}\frac{\mu q x^2}{k} + C_1 x + C_2$$

Applying our boundary conditions, we derive:

$$P_0 = C_2 \quad \text{(since } x = 0 \text{ nullifies the other terms)}$$

$$P_L = -\frac{1}{2}\frac{\mu q L^2}{k} + C_1 L + P_0$$

Solving for C_1, we get:

$$C_1 = \frac{P_L - P_0 + \frac{1}{2}\frac{\mu q L^2}{k}}{L}$$

Substituting back the constants C_1 and P_0 into our general solution, we finalize:

$$P(x) = -\frac{1}{2}\frac{\mu q}{k}x^2 + \left(\frac{P_L - P_0 + \frac{1}{2}\frac{\mu q L^2}{k}}{L}\right)x + P_0$$

This equation delineates the pressure profile $P(x)$ along the one-dimensional domain extending from 0 to L, considering the specified boundary pressures at $P(0) = P_0$ and $P(L) = P_L$.

To study the effect of each parameter on the pressure profile $P(x)$, we plot several curves, each varying one parameter while keeping the others constant. We create a series of plots for each parameter: μ (viscosity), q (flow rate), k (permeability), P_0 (initial pressure), and P_L (final pressure).

The quadratic nature of the equation suggests a parabolic pressure profile. The term $-\frac{1}{2}\frac{\mu q}{k}x^2$ indicates a pressure decrease with the square of the distance, characteristic of flow resistance due to viscosity. The term $\left(\frac{P_L - P_0 + \frac{1}{2}\frac{\mu q L^2}{k}}{L}\right)x$ represents a linear change in pressure, modified by viscosity and flow rate.

4.3.1 LINEAR FLOW

Darcy's law describes the flow of a fluid through a porous medium. Figure 4.1 illustrates fluid moving linearly through a porous medium from a higher-pressure region (P_1) to a lower-pressure region (P_2) across the distance L. The general form of Darcy's law for steady-state, single-phase flow is:

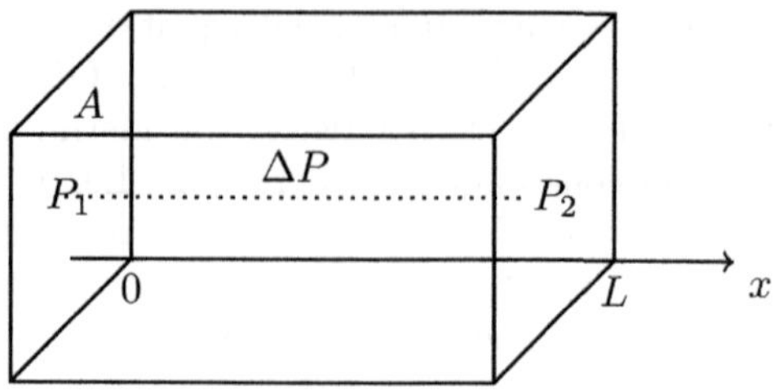

Figure 4.1 Schematic representation of linear flow in a porous medium.

$$\frac{q}{A} = -\frac{k}{\mu}\frac{dP}{dx} \tag{4.5}$$

For linear flow through a porous medium, the differential form of Darcy's law is:

$$dP = -\frac{q\mu}{Ak}dx \tag{4.6}$$

Integrating over the length of the medium from 0 to L:

$$\int_{P_1}^{P_2} dP = -\frac{q\mu}{Ak}\int_0^L dx \tag{4.7}$$

This results in:

$$P_1 - P_2 = \frac{q\mu L}{Ak} \tag{4.8}$$

Solving for the flow rate:

$$q = \frac{k}{\mu}\frac{A}{L}(P_1 - P_2) \tag{4.9}$$

4.3.2 RADIAL FLOW

For radial flow, the area through which the fluid flows is a function of the radius, r, and is given by $A = 2\pi rh$, where h is the thickness of the reservoir (see Figure 4.2). The differential form of Darcy's law is:

$$dP = -\frac{q\mu}{2\pi kh}\frac{dr}{r} \tag{4.10}$$

Integrating over the radius from r_w (well radius) to r_e (external boundary radius):

$$\int_{P_w}^{P_e} dP = -\frac{q\mu}{2\pi kh}\int_{r_w}^{r_e}\frac{dr}{r} \tag{4.11}$$

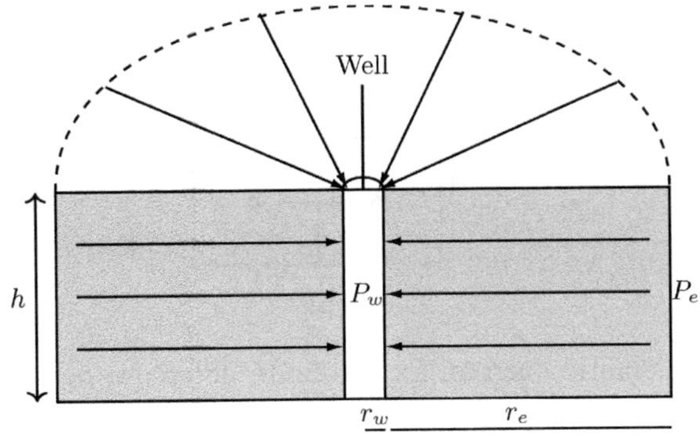

Figure 4.2 Schematic representation of radial flow toward a well. The fluid moves from an external boundary at pressure P_e to the well at pressure P_w.

This results in:

$$P_e - P_w = \frac{q\mu}{2\pi kh} \ln\left(\frac{r_e}{r_w}\right) \tag{4.12}$$

Solving for the flow rate:

$$q = \frac{2\pi kh}{\mu \ln\left(\frac{r_e}{r_w}\right)}(P_e - P_w) \tag{4.13}$$

4.3.3 FINITE DIFFERENCE METHOD

The Finite Difference Method (FDM) is used to discretize the continuous 2D Laplace equation. Figure 4.3 shows the grid points represent the locations where the pressure values P are computed, and the distance between grid points in the x-direction is Δx. The domain $[0, L_x] \times [0, L_y]$ is discretized into a grid of points with spacing Δx in the x-direction and Δy in the y-direction. Number of grid points: N_x in the x-direction, N_y in the y-direction. Grid points (x_i, y_j) where $x_i = i\Delta x$ and $y_j = j\Delta y$, for $i = 1, \ldots, N_x$ and $j = 1, \ldots, N_y$. The second-order central difference approximation for the Laplacian in 2D at a grid point (i, j) is:

$$\nabla^2 P \approx \frac{P_{i+1,j} - 2P_{i,j} + P_{i-1,j}}{\Delta x^2} + \frac{P_{i,j+1} - 2P_{i,j} + P_{i,j-1}}{\Delta y^2} \tag{4.14}$$

This leads to the following equation for each interior grid point:

$$-\left(\frac{1}{\Delta x^2} + \frac{1}{\Delta y^2}\right) P_{i,j} + \frac{P_{i+1,j}}{\Delta x^2} + \frac{P_{i-1,j}}{\Delta x^2} + \frac{P_{i,j+1}}{\Delta y^2} + \frac{P_{i,j-1}}{\Delta y^2} = 0 \tag{4.15}$$

The boundary conditions are stated as left boundary (constant pressure): $P_{1,j} = P_{\text{left}}$. The right boundary (constant flux): $\frac{P_{N_x,j} - P_{N_x-1,j}}{\Delta x} = -\frac{q\mu}{K}$. the

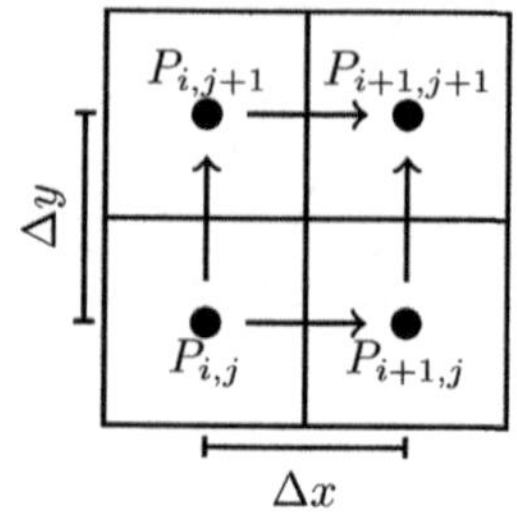

Figure 4.3 Schematic diagram for the finite difference discretization of the domain.

top/bottom boundaries (no-flow): $\frac{P_{i,N_y} - P_{i,N_y-1}}{\Delta y} = 0$. The discrete equations for each grid point form a system of linear equations $A\mathbf{P} = \mathbf{b}$, where A is a sparse matrix representing the finite difference coefficients. $\mathbf{P}$ is the vector of unknown pressures at each grid point. $\mathbf{b}$ is the right-hand side vector incorporating boundary conditions.

MATLAB® IMPLEMENTATION

This section focuses on the implementation of the pressure distribution in a 2D porous medium based on Darcy's law, which governs fluid flow through porous materials (see 4.2). When the source term is zero, Darcy's law reduces to the 2D Laplace equation, which is solved using the Finite Difference Method (FDM) under specified boundary conditions. Both MATLAB® and Python implementations are used to construct the resulting system of linear equations, solve for the pressure distribution, and visualize the results. The pressure distribution is illustrated through 2D contour and 3D surface plots, showing the variation in pressure across the domain from high to low values. These implementations offer efficient and clear methods for modeling and visualizing pressure distributions in porous media. Here is the MATLAB® code for solving the 2D pressure distribution problem using the finite difference method (FDM):

Listing 4.1: MATLAB® Code for Modeling Pressure Distribution

```
clc;
clear all;
close all;

% Parameters
Lx = 100;  % Length of the domain in x-direction [m]
Ly = 100;  % Length of the domain in y-direction [m]
Nx = 50;   % Number of grid points in x-direction
Ny = 50;   % Number of grid points in y-direction
K = 1e-12; % Permeability [m^2]
mu = 1e-3; % Fluid viscosity [Pa*s]
q = 1e-4;  % Flow rate [m^3/s]
```

```
% Grid setup
dx = Lx / (Nx-1);
dy = Ly / (Ny-1);
[x, y] = meshgrid(linspace(0, Lx, Nx), linspace(0, Ly, Ny));

% Construct the coefficient matrix
N = Nx * Ny;
A = sparse(N, N);

for i = 1:Ny
    for j = 1:Nx
        n = (i-1)*Nx + j;

        if i == 1 || i == Ny || j == 1 || j == Nx
            A(n, n) = 1;  % Boundary condition
        else
            A(n, n) = -2/dx^2 - 2/dy^2;
            A(n, n-1) = 1/dx^2;
            A(n, n+1) = 1/dx^2;
            A(n, n-Nx) = 1/dy^2;
            A(n, n+Nx) = 1/dy^2;
        end
    end
end

% Construct the right-hand side
b = zeros(N, 1);

% Left boundary: constant pressure
b(1:Nx:end) = 1e5;

% Right boundary: constant flux
b((Nx-1)*Ny+1:N) = -q*mu/(K*dy);

% Solve the system
P = A \ b;

% Reshape the solution to 2D
P = reshape(P, Nx, Ny)';

% Plot the results
figure;
surf(x, y, P);
xlabel('X␣[m]');
ylabel('Y␣[m]');
zlabel('Pressure␣[Pa]');
title('Pressure␣Distribution␣in␣2D␣Porous␣Medium');
colorbar;

figure;
contourf(x, y, P, 20);
xlabel('X␣[m]');
ylabel('Y␣[m]');
title('Pressure␣Contours␣in␣2D␣Porous␣Medium');
colorbar;
```

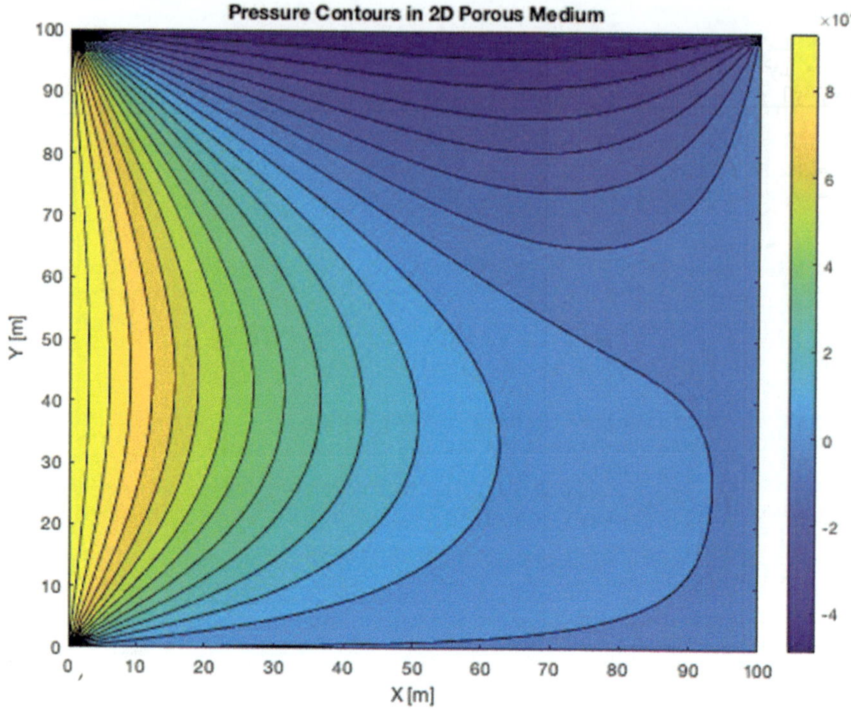

Figure 4.4 Pressure contours in the porous medium.

Note that the sparse matrix A represents the finite difference approximation of the Laplace equation. Interior grid points are filled based on the finite difference scheme. Boundary conditions are enforced by modifying the appropriate rows in A and $\mathbf{b}$. The right-hand side vector b contains values corresponding to boundary conditions. The pressure distribution is obtained by solving the linear system $\mathbf{P} = A^{-1}\mathbf{b}$, which is done using MATLAB®'s backslash operator (\). Reshape is the resulting vector $\mathbf{P}$ is reshaped into a 2D matrix for visualization.

Figure 4.4 shows that high-pressure regions are located on the left side, where the constant pressure boundary condition is applied, and pressure decreases toward the right boundary.

PYTHON IMPLEMENTATION

Below is the Python implementation for solving the 2D pressure distribution problem using the finite difference method (FDM):

Listing 4.2: Python Code for 2D Pressure Distribution

```
import numpy as np
import matplotlib.pyplot as plt
from scipy.sparse import diags, csr_matrix
from scipy.sparse.linalg import spsolve

# Parameters
Lx = 100  # Length of the domain in x-direction [m]
Ly = 100  # Length of the domain in y-direction [m]
Nx = 50   # Number of grid points in x-direction
Ny = 50   # Number of grid points in y-direction
K = 1e-12 # Permeability [m^2]
mu = 1e-3 # Fluid viscosity [Pa*s]
q = 1e-4  # Flow rate [m^3/s]

# Grid setup
dx = Lx / (Nx-1)
dy = Ly / (Ny-1)
x = np.linspace(0, Lx, Nx)
y = np.linspace(0, Ly, Ny)
X, Y = np.meshgrid(x, y)

# Construct the coefficient matrix
N = Nx * Ny
main_diag = -2/dx**2 - 2/dy**2
side_diag = 1/dx**2
up_down_diag = 1/dy**2

diagonals = [main_diag * np.ones(N),
             side_diag * np.ones(N-1),
             side_diag * np.ones(N-1),
             up_down_diag * np.ones(N-Nx),
             up_down_diag * np.ones(N-Nx)]

offsets = [0, -1, 1, -Nx, Nx]

A = diags(diagonals, offsets, shape=(N, N))

# Set boundary conditions in the matrix
for i in range(Ny):
    for j in range(Nx):
        n = i*Nx + j
        if i == 0 or i == Ny-1 or j == 0 or j == Nx-1:
            A[n, :] = 0
            A[n, n] = 1

A = csr_matrix(A)

# Construct the right-hand side
b = np.zeros(N)

# Set boundary conditions
# Left boundary: constant pressure
b[0:N:Nx] = 1e5
```

```
# Right boundary: constant flux
b[Nx-1:N:Nx] = -q*mu/(K*dy)

# Top and bottom boundaries: no flow
# (Already set to zero in b)

# Solve the system
P = spsolve(A, b)

# Reshape the solution to 2D
P = P.reshape((Ny, Nx))

# Plot the results
plt.figure(figsize=(12, 5))

# 3D surface plot
plt.subplot(121, projection='3d')
surf = plt.gca().plot_surface(X, Y, P, cmap='viridis')
plt.xlabel('X [m]')
plt.ylabel('Y [m]')
plt.gca().set_zlabel('Pressure [Pa]')
plt.title('Pressure Distribution in 2D Porous Medium')
plt.colorbar(surf)

# 2D contour plot
plt.subplot(122)
contour = plt.contourf(X, Y, P, levels=20, cmap='viridis')
plt.xlabel('X [m]')
plt.ylabel('Y [m]')
plt.title('Pressure Contours in 2D Porous Medium')
plt.colorbar(contour)

plt.tight_layout()
plt.show()
```

The domain dimensions L_x and L_y, number of grid points N_x and N_y, permeability K, viscosity μ, and flow rate q are defined at the beginning. The grid spacing is computed as $\Delta x = \frac{L_x}{N_x-1}$ and $\Delta y = \frac{L_y}{N_y-1}$. The grid is constructed using "numpy.linspace" to create a 1D array of grid points in both the x and y directions. These are combined into a 2D grid using "meshgrid", which is used for plotting. The finite difference method (FDM) discretizes the Laplace equation, creating a system of linear equations that is stored in matrix A. The diagonals of the matrix represent the relationships between neighboring grid points, and the boundary conditions are enforced by modifying the matrix at the boundaries. The vector b represents the boundary conditions. The left boundary is set to a constant pressure, and the right boundary is set to a constant flux. The top and bottom boundaries have no flow, which is already accounted for by initializing b to zero. The system $A\mathbf{P} = \mathbf{b}$ is solved using the sparse solver "spsolve" from the "scipy.sparse.linalg" library. The solution vector $\mathbf{P}$ is then reshaped into a 2D array for visualization. Two plots are generated, including a 3D surface plot showing the pressure distribution across the 2D domain. In addition to a 2D contour plot showing the pressure contours across the domain, with the color map representing pressure levels.

4.4 CCFD SOLUTIONS

Consider an incompressible single-phase flow in a two-dimensional domain $\Omega \subset \mathbb{R}^2$ with boundary $\partial\Omega$. Assuming constant density ρ and neglecting gravity effects, the governing partial differential equations (PDEs) are:

$$\mathbf{u} = -K\nabla p, \tag{4.16}$$
$$\nabla \cdot \mathbf{u} = q, \tag{4.17}$$
$$p = p_B, \tag{4.18}$$
$$\mathbf{u} \cdot \mathbf{n} = u_B. \tag{4.19}$$

where $K = \dfrac{k}{\mu}$ is the hydraulic conductivity tensor (with k being the permeability and μ the viscosity), p_B is the specified pressure on the Dirichlet boundary Γ_D, u_B is the specified normal velocity on the Neumann boundary Γ_N, and $\mathbf{n}$ is the outward unit normal vector on the boundary.

By substituting the expression for velocity from Darcy's law into the continuity equation, the pressure equation becomes:

$$-\nabla \cdot (K\nabla p) = q \quad \text{in } \Omega. \tag{4.20}$$

We discretize the rectangular domain into $n_x \times n_y$ cells, defining the mesh as:

$$\{(x_i, y_j) \mid i = 0, 1, \ldots, n_x;\ j = 0, 1, \ldots, n_y\}, \tag{4.21}$$

where $x_0 = 0 < x_1 < \cdots < x_{n_x} = L_x$ and $y_0 = 0 < y_1 < \cdots < y_{n_y} = L_y$. Each cell is centered at $(x_{i+\frac{1}{2}}, y_{j+\frac{1}{2}})$, where:

$$\begin{aligned} x_{i+\frac{1}{2}} &= \frac{x_i + x_{i+1}}{2}, \quad \Delta x_i = x_{i+1} - x_i, \quad i = 0, 1, \ldots, n_x - 1, \\ y_{j+\frac{1}{2}} &= \frac{y_j + y_{j+1}}{2}, \quad \Delta y_j = y_{j+1} - y_j, \quad j = 0, 1, \ldots, n_y - 1. \end{aligned} \tag{4.22}$$

PRESSURE EQUATION DISCRETIZATION

In the CCFD method, the pressure is approximated at the center of each cell:

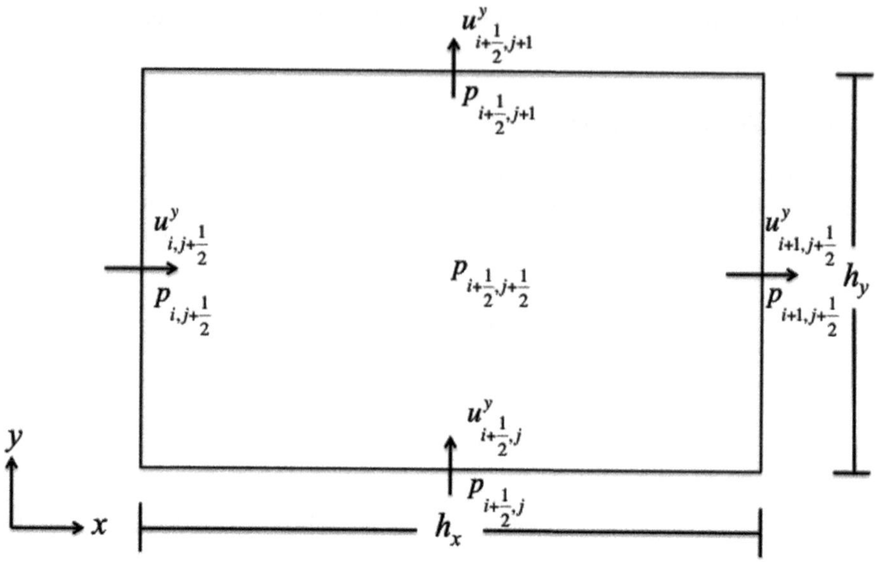

Figure 4.5 Generic CCFD rectangular cell.

$$p_{i+\frac{1}{2},j+\frac{1}{2}} = p(x_{i+\frac{1}{2}}, y_{j+\frac{1}{2}}). \tag{4.23}$$

The discretized form of the pressure equation (3.9) is derived by applying the divergence operator to the flux across each cell face. The discrete pressure equation becomes:

$$\frac{1}{\Delta x_i}\left[-K_{i,j+\frac{1}{2}}\frac{p_{i+\frac{1}{2},j+\frac{1}{2}}-p_{i-\frac{1}{2},j+\frac{1}{2}}}{x_{i+\frac{1}{2}}-x_{i-\frac{1}{2}}}+K_{i+1,j+\frac{1}{2}}\frac{p_{i+\frac{3}{2},j+\frac{1}{2}}-p_{i+\frac{1}{2},j+\frac{1}{2}}}{x_{i+\frac{3}{2}}-x_{i+\frac{1}{2}}}\right]$$
$$+\frac{1}{\Delta y_j}\left[-K_{i+\frac{1}{2},j}\frac{p_{i+\frac{1}{2},j+\frac{1}{2}}-p_{i+\frac{1}{2},j-\frac{1}{2}}}{y_{j+\frac{1}{2}}-y_{j-\frac{1}{2}}}+K_{i+\frac{1}{2},j+1}\frac{p_{i+\frac{1}{2},j+\frac{3}{2}}-p_{i+\frac{1}{2},j+\frac{1}{2}}}{y_{j+\frac{3}{2}}-y_{j+\frac{1}{2}}}\right]$$
$$=q^{n+1}_{i+\frac{1}{2},j+\frac{1}{2}}, \tag{4.24}$$

where K represents the hydraulic conductivity tensor at the cell faces, and q is the source term at the cell center.

DARCY'S LAW DISCRETIZATION

To accurately capture the permeability variations, the permeability tensor components at the cell faces are calculated using the weighted harmonic average. The x-component of the permeability at the face between cells i and $i+1$ is:

$$K^{xx}_{i,j+\frac{1}{2}} = \left(\frac{x_{i+1} - x_{i-1}}{\dfrac{x_{i+1} - x_i}{K^{xx}_{i+\frac{1}{2},j+\frac{1}{2}}} + \dfrac{x_i - x_{i-1}}{K^{xx}_{i-\frac{1}{2},j+\frac{1}{2}}}} \right), \tag{4.25}$$

and the y-component at the face between cells j and $j+1$ is:

$$K^{yy}_{i+\frac{1}{2},j} = \left(\frac{y_{j+1} - y_{j-1}}{\dfrac{y_{j+1} - y_j}{K^{yy}_{i+\frac{1}{2},j+\frac{1}{2}}} + \dfrac{y_j - y_{j-1}}{K^{yy}_{i+\frac{1}{2},j-\frac{1}{2}}}} \right). \tag{4.26}$$

Using these permeability values, the discrete form of Darcy's law for the x-component is:

$$u^x_{i,j+\frac{1}{2}} = -K^{xx}_{i,j+\frac{1}{2}} \frac{p_{i+\frac{1}{2},j+\frac{1}{2}} - p_{i-\frac{1}{2},j+\frac{1}{2}}}{x_{i+\frac{1}{2}} - x_{i-\frac{1}{2}}}, \quad i = 1, \ldots, n_x; \quad j = 0, \ldots, n_y - 1. \tag{4.27}$$

BOUNDARY CONDITIONS

At the boundaries, special considerations are made:

West Boundary ($i = 0$):

$$u^x_{0,j+\frac{1}{2}} = \begin{cases} -K^{xx}_{0,j+\frac{1}{2}} \dfrac{p_{\frac{1}{2},j+\frac{1}{2}} - p^W_{B,j+\frac{1}{2}}}{x_{\frac{1}{2}} - x_0}, & \text{if } j = 0, \ldots, n_I, \quad 0 < n_I < n_y, \\ 0, & \text{if } j = n_I + 1, \ldots, n_y, \end{cases} \tag{4.28}$$

East Boundary ($i = n_x$):

$$u^x_{n_x,j+\frac{1}{2}} = -K^{xx}_{n_x,j+\frac{1}{2}} \frac{p^E_{B,j+\frac{1}{2}} - p_{n_x-\frac{1}{2},j+\frac{1}{2}}}{x_{n_x} - x_{n_x-\frac{1}{2}}}, \quad j = 0, \ldots, n_y - 1. \tag{4.29}$$

Here, $p^W_{B,j+\frac{1}{2}}$ and $p^E_{B,j+\frac{1}{2}}$ denote the specified pressures at the west and east boundaries, respectively. The permeabilities at the boundaries are set equal to those of the nearest interior cells:

$$\begin{cases} K^{xx}_{0,j+\frac{1}{2}} = K^{xx}_{\frac{1}{2},j+\frac{1}{2}}, & j = 0,\dots,n_y-1, \\ K^{xx}_{n_x,j+\frac{1}{2}} = K^{xx}_{n_x-\frac{1}{2},j+\frac{1}{2}}, & j = 0,\dots,n_y-1. \end{cases} \tag{4.30}$$

Similarly, the discrete form of Darcy's law for the y-component is:

$$u^y_{i+\frac{1}{2},j} = -K^{yy}_{i+\frac{1}{2},j} \frac{p_{i+\frac{1}{2},j+\frac{1}{2}} - p_{i+\frac{1}{2},j-\frac{1}{2}}}{y_{j+\frac{1}{2}} - y_{j-\frac{1}{2}}}, \quad i = 0,\dots,n_x-1; \quad j = 1,\dots,n_y. \tag{4.31}$$

At the south ($j = 0$) and north ($j = n_y$) boundaries, the velocities are set to zero (no-flow boundary condition):

$$u^y_{i+\frac{1}{2},0} = 0, \quad u^y_{i+\frac{1}{2},n_y} = 0, \quad i = 0,1,\dots,n_x-1. \tag{4.32}$$

The treatment of boundary conditions depends on their types, which can be Dirichlet, Neumann, or mixed. We introduce an indicator function δ to distinguish between them. $\delta = 1$ represents the Dirichlet boundary condition (specified pressure). While $\delta = 0$ represents the Neumann boundary condition (specified flux or impermeable boundary). An example of the general boundary condition treatment at the west boundary is:

$$u^x_{0,j+\frac{1}{2}} = \begin{cases} -K^{xx}_{\frac{1}{2},j+\frac{1}{2}} \dfrac{p_{\frac{1}{2},j+\frac{1}{2}} - p_{0,j+\frac{1}{2}}}{x_{\frac{1}{2}} - x_0}, & \text{if } \delta^{\text{Diri}}_{0,j+\frac{1}{2}} = 0, \\ u_B^{0,j+\frac{1}{2}}, & \text{if } \delta^{\text{Diri}}_{0,j+\frac{1}{2}} = 1. \end{cases} \tag{4.33}$$

This approach allows for a unified treatment of boundary conditions by adjusting the indicator function δ accordingly. Similar formulations can be applied to the other boundaries, ensuring consistent and accurate implementation of the physical boundary conditions in the numerical model.

To numerically solve the governing pressure equation for incompressible single-phase flow in porous media using the Cell-Centered Finite Difference (CCFD) method. The solution procedure involves discretizing the computational domain, assembling the pressure equation, applying boundary conditions, solving the resulting linear system, and computing velocity components using Darcy's law. For a vectorized MATLAB® implementation, see Chapter 3.

4.5 MRST IMPLEMENTATION

The equation is discretized using the Two-Point Flux Approximation (TPFA), a finite volume method, which ensures the conservation of mass across control volumes. The grid is divided into control volumes (cells), and the flux across cell faces is computed. The pressure field is solved iteratively to balance the fluxes and satisfy boundary conditions. The flux between two cells i and j is computed as:

$$F_{ij} = -\frac{T_{ij}}{\mu}(p_j - p_i), \tag{4.34}$$

where T_{ij} is the transmissibility factor between cells i and j (depends on grid geometry and permeability), and p_j and p_i are pressures in cells j and i.

Boundary conditions are applied directly to modify the flux or pressure in specific grid cells. Examples include the Dirichlet boundary condition, which specifies fixed pressure values at boundaries (e.g., injection or production wells), and the Neumann boundary condition, which specifies flux values across boundaries (e.g., no-flow boundaries).

The solution involves matrix assembly, which is the discretized form expressed as a linear system:

$$Ap = b, \tag{4.35}$$

where A is the coefficient matrix derived from the transmissibility and grid structure, p is the pressure vector (unknowns to solve for), and b is the right-hand side vector containing source/sink terms and boundary conditions. MRST uses iterative linear solvers, such as GMRES or direct solvers for small-scale problems to solve the linear system efficiently. It includes post-processing, such as computing fluxes and velocities using the solved pressure field.

MRST STEPS:

Step 1: Adding MRST to the Path

Before using MRST functions, you need to add MRST to the MATLAB® path and initialize its modules:

```
addpath('path_to_mrst'); % Replace 'path_to_mrst' with the actual
    path to MRST
mrstModule add all; % Add all MRST modules
mrstModule add incomp; % Add incompressible flow module
```

This ensures MRST is accessible in MATLAB® and loads the required modules for incompressible flow simulation.

Step 2: Creating the Simulation Grid

Define the computational grid using the 'cartGrid' function:

```
nx = 20; % Number of cells in x-direction
ny = 20; % Number of cells in y-direction
nz = 5;  % Single layer in z-direction
physDims = [100, 100, 10]; % Physical dimensions of the reservoir
    (m)
G = cartGrid([nx, ny, nz], physDims); % Generate the grid
G = computeGeometry(G); % Compute cell geometry
```

This creates a Cartesian grid with specified dimensions and computes geometrical properties like centroids and face normals.

Step 3: Defining Rock Properties

Assign permeability and porosity to the reservoir rock:

```
rock = makeRock(G, 100 * milli * darcy, 0.2); % Permeability (mD)
    and porosity
```

Step 4: Defining Fluid Properties

Set the properties of the single-phase fluid:

```
fluid = initSingleFluid('mu', 1 * centi * poise, ... % Viscosity
    (Pa.s)
                        'rho', 1000 * kilogram / meter^3); %
                            Density (kg/m^3)
```

Step 5: Defining Boundary Conditions

Specify pressure at the inlet (left boundary) and outlet (right boundary):

```
bc = pside([], G, 'LEFT', 1 * barsa); % Set pressure on left
    boundary (inlet)
bc = pside(bc, G, 'RIGHT', 0); % Set pressure on right boundary (
    outlet)
```

Step 6: Initializing the Simulation State

Initialize the pressure state of the reservoir:

```
state = initResSol(G, 0); % Initialize with uniform pressure (0
    Pa)
```

Step 7: Solving the Flow Equation

Compute transmissibilities and solve the flow equation:

```
T = computeTrans(G, rock); % Compute transmissibilities
state = incompTPFA(state, G, T, fluid, 'bc', bc); % Solve for
    pressure
```

Step 8: Visualizing the Pressure Distribution

Plot the pressure distribution across the grid:

```
figure;
plotCellData(G, state.pressure);
colorbar;
title('Pressure␣Distribution␣(Pa)');
xlabel('X-direction');
ylabel('Y-direction');
view(3);
```

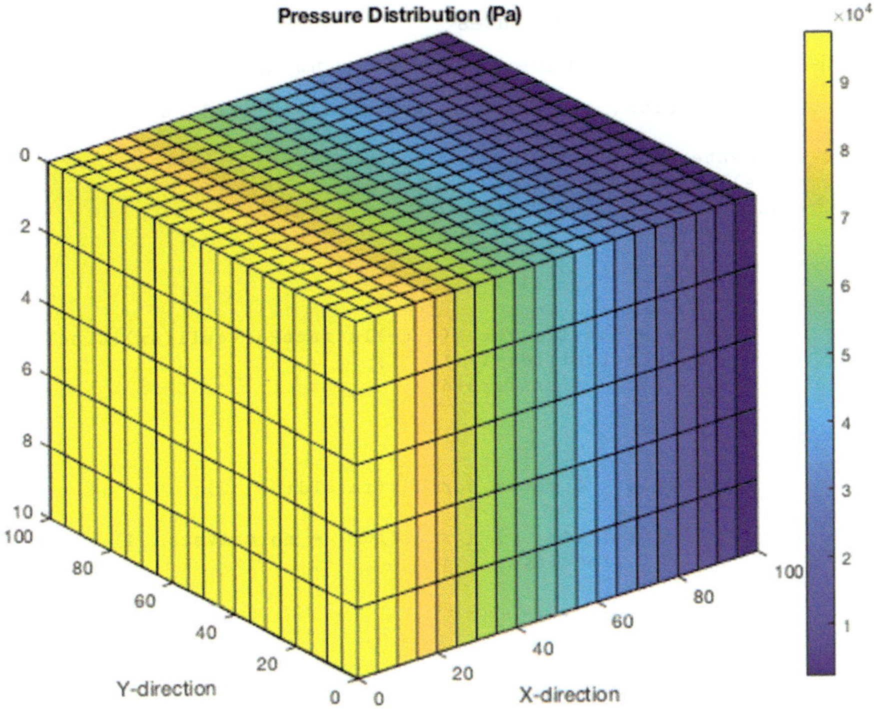

Figure 4.6 Pressure distribution (Pa) from the MRST simulation.

Step 9: Computing the Velocity Field

Compute the velocity field based on fluxes and face normals:

```
[cellFlux, cellCenters] = getCellFluxes(G, state);
```

Step 10: Visualizing the Flow Field

Plot the velocity vectors at cell centroids:

```
% Extract x and y components of fluxes
vx = cellFlux(:, 1); % x-component of velocity
vy = cellFlux(:, 2); % y-component of velocity

% Extract cell centroids
cx = cellCenters(:, 1);
cy = cellCenters(:, 2);

% Visualize the flow field
figure;
quiver(cx, cy, vx, vy, 'r'); % Red flow vectors
title('Flow␣Field');
xlabel('X-direction');
ylabel('Y-direction');
axis equal;
```

Assistant Function: `getCellFluxes`

This helper function computes velocity vectors for each grid cell:

```
function [cellFlux, cellCenters] = getCellFluxes(G, state)
    % Compute fluxes and map to cell centers
    cellCenters = G.cells.centroids; % Get cell centroids
    numCells = G.cells.num;
    cellFlux = zeros(numCells, 2); % Initialize fluxes (2D: x and
         y components)

    for cellIdx = 1:numCells
        % Find the faces associated with this cell
        faceIdx = G.cells.faces(G.cells.facePos(cellIdx):G.cells.
            facePos(cellIdx+1)-1, 1);
        flux = state.flux(faceIdx); % Extract flux for these
            faces

        % Approximate velocity by averaging flux over faces
        normalVecs = G.faces.normals(faceIdx, 1:2); % Normals (x,
             y)
        cellFlux(cellIdx, :) = sum(flux .* normalVecs, 1) /
            length(faceIdx); % Average velocity
    end
end
```

EXAMPLE: QUARTER FIVE-SPOT WITH TPFA IN MRST

We simulate single-phase incompressible flow in a homogeneous, isotropic porous medium using the Two-Point Flux Approximation (TPFA) method within the MATLAB® Reservoir Simulation Toolbox (MRST). The physical domain is a quarter of a classic five-spot pattern with an injection well at the bottom-left corner and a production well at the top-right corner. The injection rate is set to 1.84×10^{-6} m^3/s (equivalent to 100 STB/day), while the production rate is slightly lower (0.5×10^{-6} m^3/s), which results in net positive injection and avoids negative pressures. The domain size is 100×100 meters, the grid resolution is 50×50, permeability is 100 mD ($= 9.869 \times 10^{-14}$ m^2), porosity is 0.2, fluid viscosity is $\mu = 1 \times 10^{-3}$ Pa·s, and fluid density is $\rho = 1000$ kg/m^3. The code below sets up and solves the pressure field using TPFA:

```
clear; clc; close all;
mrstModule add incomp

% Define domain and grid
Lx = 100; Ly = 100;
Nx = 50; Ny = 50;
G = cartGrid([Nx, Ny], [Lx, Ly]);
G = computeGeometry(G);

% Rock properties
perm_mD = 100;
rock.perm = perm_mD * 9.869e-16 * ones(G.cells.num, 1);
```

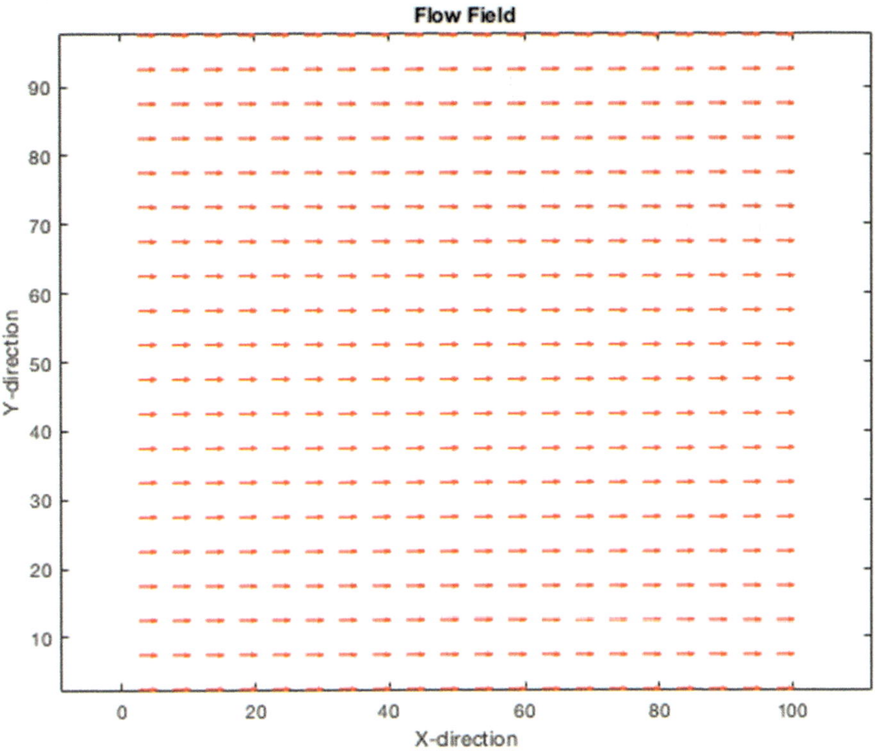

Figure 4.7 Flow field visualization showing velocity vectors.

```
rock.poro = 0.2 * ones(G.cells.num, 1);

% Fluid properties
mu = 1e-3;
rho = 1000;
fluid = initSimpleFluid('mu', [mu, mu], 'rho', [rho, rho]);

% Compute transmissibilities
T = computeTrans(G, rock);
[A, b] = getTPFAMatrix(G, T);

% Injection and production configuration
q = zeros(G.cells.num, 1);
q(1) = 1.84e-6;          % Injection at cell 1
q(end) = -0.5e-6;        % Production at last cell
b = b + q;

% Reference pressure at injector
refCell = 1;
A(refCell,:) = 0;
A(refCell,refCell) = 1;
b(refCell) = 1e7;        % 10 MPa
```

```
% Solve system
p = A \ b;
p_mpa = p / 1e6;        % Convert to MPa

% Plot pressure
figure;
plotCellData(G, p_mpa, 'EdgeColor', 'none');
axis equal tight;
colorbar;
title('Pressure␣Field␣(MPa)␣Quarter␣Five-Spot');
set(gca, 'XTick', [], 'YTick', []);

% Helper function
function [A, b] = getTPFAMatrix(G, T)
    N = G.cells.num;
    A = sparse(N, N);
    b = zeros(N, 1);
    for f = 1:G.faces.num
        c = G.faces.neighbors(f, :);
        T_f = T(f);
        if all(c > 0)
            A(c(1), c(1)) = A(c(1), c(1)) + T_f;
            A(c(1), c(2)) = A(c(1), c(2)) - T_f;
            A(c(2), c(1)) = A(c(2), c(1)) - T_f;
            A(c(2), c(2)) = A(c(2), c(2)) + T_f;
        end
    end
end
```

The simulation generates a pressure distribution, as illustrated in Figure 4.8. The simulation produces a smooth and realistic pressure field.

4.6 SOLUTE TRANSPORT IN POROUS MEDIA

The study of solute transport in porous media is crucial for understanding various environmental and industrial processes, such as groundwater contamination, enhanced oil recovery, and chemical transport in aquifers. These processes involve the movement of a solute through a porous matrix under the influence of advection, diffusion, and dispersion. This section introduces the governing equations that describe the flow and transport of a solute, using Darcy's law for flow in the advection-diffusion equation for solute transport. The fluid flow in the porous medium is modeled under the assumptions of incompressibility and low Reynolds number flow, which are characteristic of systems dominated by viscous and buoyancy forces. The continuity equation ensuring incompressibility is expressed as:

$$\nabla \cdot \mathbf{u} = 0, \tag{4.36}$$

where $\mathbf{u}$ is the Darcy velocity vector $[m/s]$, representing the volumetric flow rate per unit area in the porous medium. The momentum balance is governed

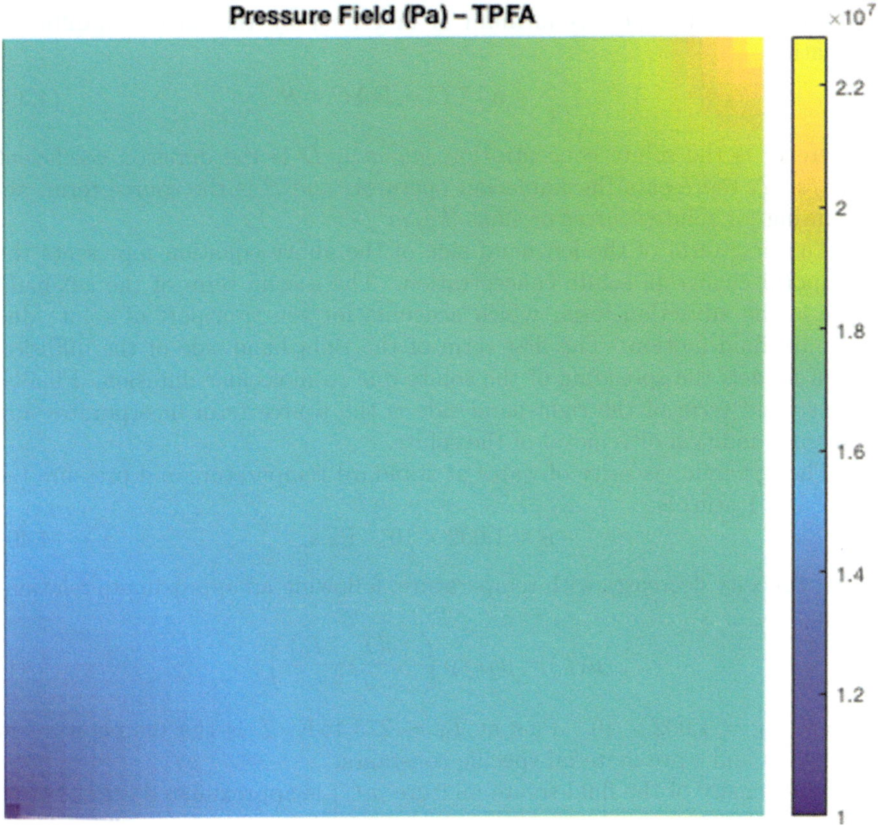

Figure 4.8 Pressure distribution obtained from the TPFA simulation.

by Darcy's law:

$$\mathbf{u} = -\frac{K}{\mu}\nabla p + \frac{K\,\rho(C)\mathbf{g}}{\mu}, \tag{4.37}$$

where K is the permeability of the porous medium $[m^2]$, μ is the dynamic viscosity of the fluid $[Pa \cdot s]$, ∇p is the pressure gradient driving the flow, $\rho(C)$ is the density of the fluid as a function of solute concentration C, and $\mathbf{g}$ is the gravitational acceleration vector $[m/s^2]$.

Substituting $\mathbf{u}$ from Darcy's law into the continuity equation results in a Poisson-type equation for pressure p:

$$\nabla \cdot \left(-\frac{K}{\mu}\nabla p + \frac{K\,\rho(C)\mathbf{g}}{\mu} \right) = 0. \tag{4.38}$$

This equation provides the relationship between the pressure field and the solute concentration through the density term $\rho(C)$. The transport of the solute in the porous medium is described by the advection-diffusion equation:

$$\frac{\partial C}{\partial t} + \mathbf{u} \cdot \nabla C = D \Delta C + S, \tag{4.39}$$

where C is the solute concentration $[kg/m^3]$, D is the diffusion coefficient $[m^2/s]$, Δ represents the Laplacian operator, and S is the source term, accounting for solute sources or sinks $[kg/m^3/s]$.

The first term of the left-hand side of the above equation represents the temporal change in solute concentration. The second term of the left-hand side is the advection term, which accounts for the transport of solute due to bulk fluid motion. The first term of the right-hand side of the diffusion term models the spreading of the solute due to molecular diffusion. Finally, the second term of the right-hand side is the source term incorporates any external addition or removal of the solute.

The dynamic viscosity of water at standard temperature and pressure (at $0°C$ and 1 atm) is:

$$\mu \approx 1.002 \times 10^{-3}\,\text{Pa.s.} \tag{4.40}$$

The viscosity decreases with temperature, following an approximate relationship:

$$\mu(T) = \mu_0 \exp\left(-\frac{b(T - T_0)}{T + c}\right), \tag{4.41}$$

where $\mu_0 = 1.002 \times 10^{-3}\,\text{Pa·s}$ at $T_0 = 273.15\,\text{K}$, T is the temperature in Kelvin, b and c are material-specific constants.

The density of the fluid-solute mixture $\rho(C)$ is approximated as:

$$\rho(C) = \rho_{\text{water}}(1 + \beta C), \tag{4.42}$$

where $\rho_{\text{water}} \approx 1000\,\text{kg/m}^3$ (density of pure water) and β is the relative density coefficient of the solute. For dilute solute concentrations, the density coefficient β is constant.

Initial values for p and C are required:

$$p(\mathbf{x}, 0) = p_0(\mathbf{x}), \quad C(\mathbf{x}, 0) = C_0(\mathbf{x}), \tag{4.43}$$

for $\mathbf{x} \in \Omega$.

Boundary conditions are imposed as follows to ensure a well-posed problem, which involves prescribed pressures, fluxes, and solute concentration. The Dirichlet boundary condition is applied as a fixed pressure is applied at a reference node:

$$p = p_{\text{in}}. \tag{4.44}$$

The Neumann boundary condition is applied at impermeable walls, where the velocity is zero:

$$\mathbf{u} \cdot \mathbf{n} = 0. \tag{4.45}$$

For concentration, the Dirichlet boundary condition (inflow concentration) is applied as a fixed solute concentration is specified at the inlet:

$$C = C_{\text{in}}. \tag{4.46}$$

Also, the Neumann boundary condition (Zero Normal Flux) is applied at impermeable walls and outflow boundaries:

$$\nabla C \cdot \mathbf{n} = 0. \tag{4.47}$$

4.6.1 FINITE ELEMENT APPROXIMATION

Let q be a test function. The weak form of the pressure equation is obtained by multiplying the Poisson equation by q and integrating over Ω:

$$\int_\Omega \left(-\frac{K}{\mu}\Delta p + \frac{K}{\mu}\nabla \cdot [\beta(C)\mathbf{g}] \right) q \, d\Omega = 0. \tag{4.48}$$

Integrating by parts the Δp term:

$$\int_\Omega \frac{K}{\mu}\nabla p \cdot \nabla q \, d\Omega = \int_\Omega \frac{K}{\mu}\beta(C)\mathbf{g} \cdot \nabla q \, d\Omega, \tag{4.49}$$

assuming suitable boundary conditions on p.

For the concentration, let w be a test function. After time discretization (e.g., backward Euler) from t^n to t^{n+1}, the weak form is:

$$\int_\Omega \frac{C^{n+1} - C^n}{\Delta t} w \, d\Omega + \int_\Omega (\mathbf{u}^{n+1} . \nabla C^{n+1}) w \, d\Omega + \int_\Omega D \nabla C^{n+1} . \nabla w \, d\Omega = \int_\Omega S w \, d\Omega. \tag{4.50}$$

Since $\mathbf{u}$ is given by Darcy's law, it is directly expressed in terms of p and C.

A stabilized finite element method can be employed to discretize the problem in space, and a backward Euler scheme is used in time. Appropriate stabilization techniques (e.g., SUPG) can be applied to the concentration equation.

Solving the pressure equation gives p, and subsequently, the velocity field $\mathbf{u}$ is obtained from Darcy's law. With $\mathbf{u}$ known, the concentration equation is solved for C.

Let $\Omega \subset \mathbb{R}^3$ be the computational domain. We consider the following Hilbert spaces, $H^1(\Omega)$ is the space of square-integrable functions with square-integrable first derivatives, i.e.,

$$H^1(\Omega) = \left\{ v \in L^2(\Omega) \mid \nabla v \in (L^2(\Omega))^3 \right\}. \tag{4.51}$$

We denote by $P_h \subset H^1(\Omega)$ the finite element subspace for the pressure approximation. Typically, P_h is constructed from continuous piecewise polynomial functions that vanish on the Dirichlet boundaries as required. $Q_h \subset H^1(\Omega)$ the finite element subspace for the concentration approximation. Similar to P_h, Q_h consists of continuous piecewise polynomial functions defined on the same mesh. By choosing P_h and Q_h appropriately (e.g., piecewise linear or higher-order polynomials), we ensure that these spaces are conforming subspaces of $H^1(\Omega)$, allowing for a straightforward application of the variational formulation.

4.6.1.1 Approximation Scheme

Let $\Omega \subset \mathbb{R}^3$ be our domain, and let $P_h \subset H^1(\Omega)$ and $Q_h \subset H^1(\Omega)$ be finite element approximation spaces for pressure p and concentration C, respectively. We use continuous finite elements (e.g., piecewise linear elements) for both p and C. The velocity $\mathbf{u}$ is not a primary unknown but is computed after obtaining p via:

$$\mathbf{u}^{n+1} = -\frac{K}{\mu}\nabla p^{n+1} + \frac{K\,\beta(C^n)}{\mu}\mathbf{g}. \tag{4.52}$$

We use a backward Euler scheme in time. Given C^n and p^n at time t^n, we proceed as follows:

Step 1:

The pressure equation in weak form is obtained by taking $q \in P_h$ as a test function:

$$\left(\frac{K}{\mu}\nabla p^{n+1}, \nabla q\right) = \left(\frac{K}{\mu}\beta(C^n)\mathbf{g}, \nabla q\right). \tag{4.53}$$

Solving this linear problem gives p^{n+1}.

Step 2:

Once p^{n+1} is found, compute:

$$\mathbf{u}^{n+1} = -\frac{K}{\mu}\nabla p^{n+1} + \frac{K\,\beta(C^n)}{\mu}\mathbf{g}. \tag{4.54}$$

Step 3:

The concentration equation in weak form uses a test function $\phi \in Q_h$:

$$\frac{1}{\Delta t}(C^{n+1} - C^n, \phi) + (\mathbf{u}^{n+1} \cdot \nabla C^{n+1}, \phi) + (D\nabla C^{n+1}, \nabla \phi) = (S, \phi). \tag{4.55}$$

This is a nonlinear problem due to the advective term, but since $\mathbf{u}^{n+1}$ is known at this stage, it becomes a linear or weakly nonlinear problem in C^{n+1}. Solve this system to obtain C^{n+1}.

Step 4:

Check convergence criteria; if met, proceed to the next time step.

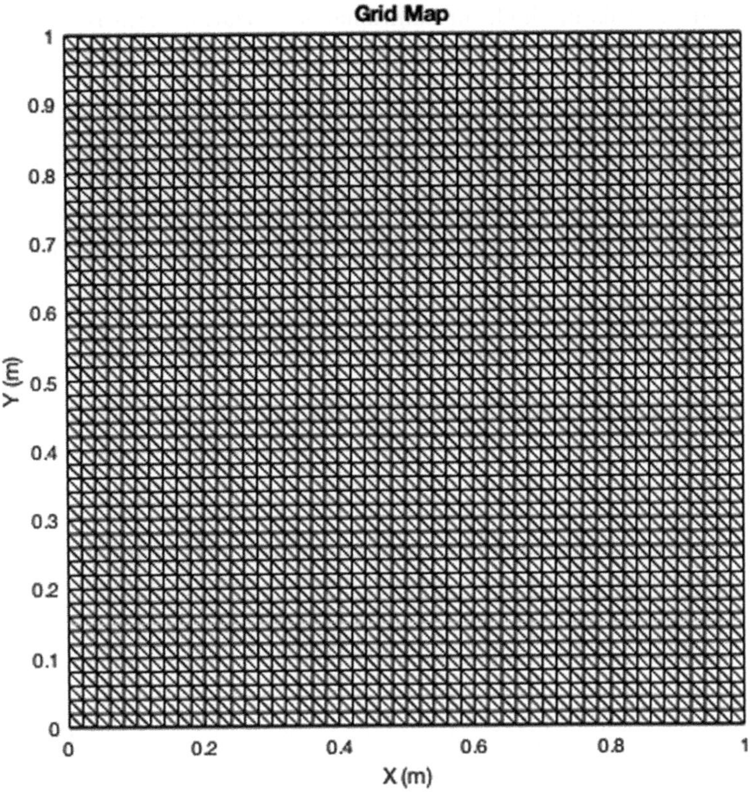

Figure 4.9 Grid map used for the finite element simulation.

For large-scale problems, efficient solvers such as BiCGSTAB(L) with suitable preconditioners can be employed. The number of elements, degrees of freedom, and chosen time step will affect the computational cost. Parallelization and efficient data structures can be used to manage large problem sizes.

The grid map shown in Figure 4.9 represents the discretization of the domain into triangular finite elements. This structured mesh is generated using a Delaunay triangulation algorithm with the domain as a square with dimensions $1\,\text{m} \times 1\,\text{m}$. The mesh consists of 50 elements along both the x- and y-axes, resulting in a fine discretization suitable for capturing the flow and transport phenomena. Triangular elements ensure flexibility in representing complex domains while maintaining computational efficiency.

4.6.2 STABILITY CONSIDERATIONS

To improve numerical stability in advection-dominated problems, the Streamline-Upwind Petrov-Galerkin (SUPG) stabilization method is employed. In this method, an additional stabilization term is added to the weak form of the concentration equation, aligned along the direction of the total velocity, which includes both Darcy and solute-induced density components. The stabilized weak form for the concentration equation is:

$$\begin{aligned}\int_\Omega \frac{C^{n+1} - C^n}{\Delta t} w \, d\Omega + \int_\Omega \mathbf{u}_{\text{total}} \cdot \nabla C^{n+1} \, w \, d\Omega \\ + \int_\Omega D \nabla C^{n+1} \cdot \nabla w \, d\Omega + \int_\Omega \tau_{\text{SUPG}} (\mathbf{u}_{\text{total}} \cdot \nabla w)(\mathbf{u}_{\text{total}} \cdot \nabla C^{n+1}) \, d\Omega \\ = \int_\Omega S w \, d\Omega.\end{aligned} \tag{4.56}$$

Here $\mathbf{u}_{\text{total}}$ is the total velocity, given by:

$$\mathbf{u}_{\text{total}} = -\frac{K}{\mu} \nabla p + \frac{K\beta(C)}{\mu} \mathbf{g}, \tag{4.57}$$

where the second term accounts for solute-induced effects due to variations in solute concentration. τ_{SUPG} is the stabilization parameter, computed as:

$$\tau_{\text{SUPG}} = \frac{h_{\text{element}}}{2\|\mathbf{u}_{\text{total}}\| + \epsilon}, \tag{4.58}$$

where h_{element} is the characteristic element size, $\|\mathbf{u}_{\text{total}}\|$ is the magnitude of the total velocity, and ϵ is a small positive parameter to avoid division by zero.

The SUPG stabilization is incorporated by modifying the advection matrix for each element. The stabilized advection term is:

$$C_e(i,j) = \int_\Omega (\mathbf{u}_{\text{total}} \cdot \nabla N_j) \, N_i \, d\Omega + \tau_{\text{SUPG}} \int_\Omega (\mathbf{u}_{\text{total}} \cdot \nabla N_i) \, (\mathbf{u}_{\text{total}} \cdot \nabla N_j) \, d\Omega, \tag{4.59}$$

where N_i and N_j are the shape functions for the i-th and j-th nodes in the element.

The solute-induced term $\frac{K\beta(C)}{\mu}\mathbf{g}$ is explicitly included in $\mathbf{u}_{\text{total}}$, ensuring that the stabilization reflects the dominant physics in regions where buoyancy effects are significant.

Incorporating buoyancy effects in the stabilization ensures accurate alignment of the stabilization term with the actual flow direction. It improved numerical stability in regions with strong solute-induced density advection. It ensures the reduction of spurious oscillations near steep concentration gradients. This modification is especially important in the simulation of solute dispersion, where solute-induced effects dominate due to the solute low density and high solute-induced coefficient.

4.6.3 BASIS FUNCTIONS

The finite element method (FEM) implementation in this code uses linear triangular basis functions to approximate the solution (e.g., pressure and concentration) within each triangular element.

For a triangular element with vertices $\mathbf{x}_1 = (x_1, y_1)$, $\mathbf{x}_2 = (x_2, y_2)$, and $\mathbf{x}_3 = (x_3, y_3)$, the basis functions ϕ_1, ϕ_2, and ϕ_3 are defined as:

$$\begin{aligned} \phi_1(x,y) &= \frac{a_1 + b_1 x + c_1 y}{2A}, \quad \phi_2(x,y) = \frac{a_2 + b_2 x + c_2 y}{2A}, \\ \phi_3(x,y) &= \frac{a_3 + b_3 x + c_3 y}{2A}, \end{aligned} \tag{4.60}$$

where A is the area of the triangular element, given by:

$$A = \frac{1}{2} \left| x_1(y_2 - y_3) + x_2(y_3 - y_1) + x_3(y_1 - y_2) \right|. \tag{4.61}$$

The coefficients a_i, b_i, and c_i are computed as:

$$a_1 = x_2 y_3 - x_3 y_2, \qquad b_1 = y_2 - y_3, \qquad c_1 = x_3 - x_2, \tag{4.62}$$

$$a_2 = x_3 y_1 - x_1 y_3, \qquad b_2 = y_3 - y_1, \qquad c_2 = x_1 - x_3, \tag{4.63}$$

$$a_3 = x_1 y_2 - x_2 y_1, \qquad b_3 = y_1 - y_2, \qquad c_3 = x_2 - x_1. \tag{4.64}$$

Each basis function $\phi_i(x, y)$ is linear in x and y. At a vertex $\mathbf{x}_j$, the basis function satisfies:

$$\phi_i(\mathbf{x}_j) = \begin{cases} 1 & \text{if } i = j, \\ 0 & \text{if } i \neq j. \end{cases} \tag{4.65}$$

Additionally, each basis function ϕ_i is nonzero only within the triangular element to which it belongs.

The stiffness matrix for the pressure equation is computed using the gradients of the basis functions:

$$K_{ij} = \int_T \nabla\phi_i \cdot \nabla\phi_j \, dA, \tag{4.66}$$

where $\nabla\phi_i$ is the gradient of the basis function ϕ_i.

The mass matrix for the concentration equation is computed as:

$$M_{ij} = \int_T \phi_i \phi_j \, dA. \tag{4.67}$$

The advection term uses the basis functions and velocity field:

$$C_{ij} = \int_T (\mathbf{u} \cdot \nabla\phi_i)\phi_j \, dA. \tag{4.68}$$

The advection matrix is modified using a stabilization parameter τ_{SUPG} and the basis functions' gradients:

$$C_{ij} \to C_{ij} + \tau_{\text{SUPG}} \int_T (\mathbf{u} \cdot \nabla\phi_i)(\mathbf{u} \cdot \nabla\phi_j)\, dA. \tag{4.69}$$

The linear basis functions have several advantages, including simplicity, local support, and first-order accuracy. Linear basis functions are straightforward to compute and implement. Basis functions are nonzero only within one element, making the global stiffness matrix sparse. Suitable for relatively simple problems or fine mesh.

4.6.4 STABILITY CONSIDERATIONS

The Peclet number is a dimensionless quantity that characterizes the relative importance of advection and diffusion in the transport equation. It is defined as:

$$\text{Peclet Number} = \frac{|\mathbf{u}|L}{D}, \tag{4.70}$$

where $|\mathbf{u}|$ is the magnitude of the velocity field, L is the characteristic length scale of the domain, and D is the diffusion coefficient.

For advection-dominated flows, the Peclet number is expected to be large. However, excessively high values may indicate extreme velocity magnitudes or negligible diffusion, which can lead to numerical instability.

The Courant-Friedrichs-Lewy (CFL) condition ensures stability for explicit time-stepping schemes. It relates the time step size (Δt) to the spatial resolution (Δx) and velocity field ($|\mathbf{u}|$):

$$\text{CFL} = \frac{|\mathbf{u}|\Delta t}{\Delta x}. \tag{4.71}$$

For stability, the CFL condition must be satisfied:

$$\text{CFL} \leq 1. \tag{4.72}$$

The following corrections were made to ensure numerical stability: The time step Δt is computed dynamically based on the CFL condition:

$$\Delta t \leq \frac{\Delta x}{|\mathbf{u}|_{\max}}, \tag{4.73}$$

where Δx is the grid spacing, and $|\mathbf{u}|_{\max}$ is the maximum velocity magnitude.

Large pressure gradients can lead to extreme velocities. The inlet pressure was reduced to 1×10^6 Pa, while the outlet pressure was adjusted using hydrostatic pressure:

$$p_{\text{outlet}} = 1 \times 10^5 + \rho g h, \tag{4.74}$$

where ρ is the fluid density, g is gravitational acceleration, and h is the height of the outlet node.

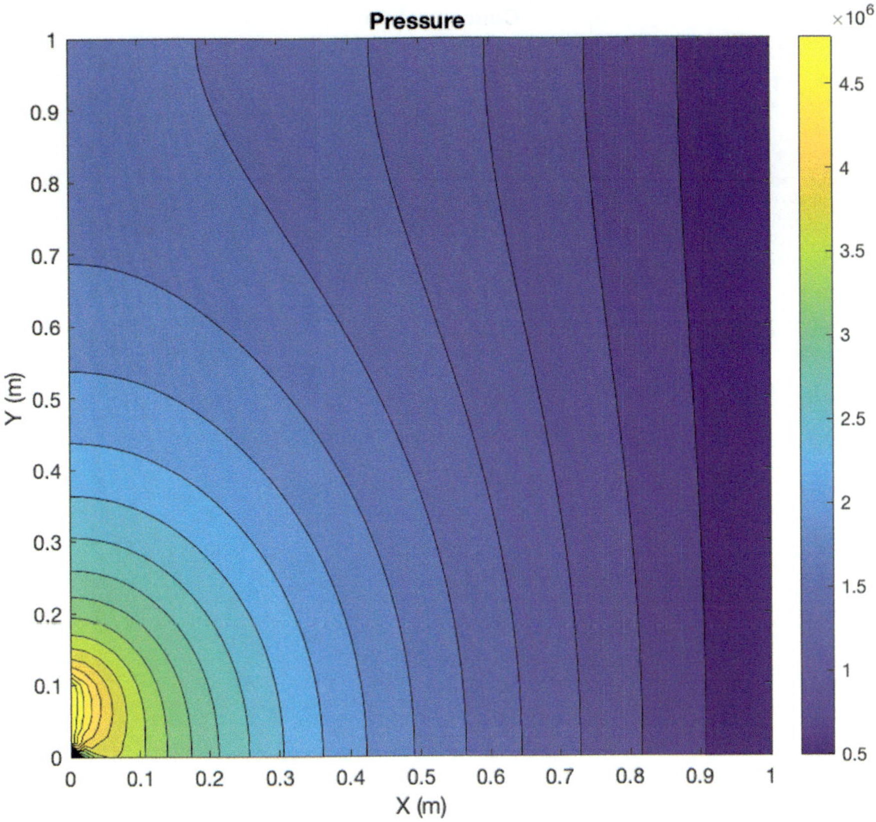

Figure 4.10 Pressure contours showing the pressure distribution in the domain.

Implementing the above numerical method, some results are discussed here. The pressure distribution in the domain is shown in Figure 4.10. The highest pressure is observed near the inlet, which gradually decreases toward the outlet. This distribution is consistent with Darcy's law, which predicts a linear pressure gradient for laminar flow in a porous medium.

The contour plot in Figure 4.11 shows the concentration distribution of the solute in the domain. The solute concentration is highest near the inlet and gradually decreases toward the outlet due to advection and diffusion. The concentration gradient near the inlet suggests a strong advective influence, while further into the domain, the distribution becomes smoother due to diffusive mixing.

Figure 4.12 shows the concentration profiles along the x-direction for different values of y. At $y = 0.1$, the concentration is highest and exhibits a gradual decrease with increasing x. Profiles at higher y values exhibit lower

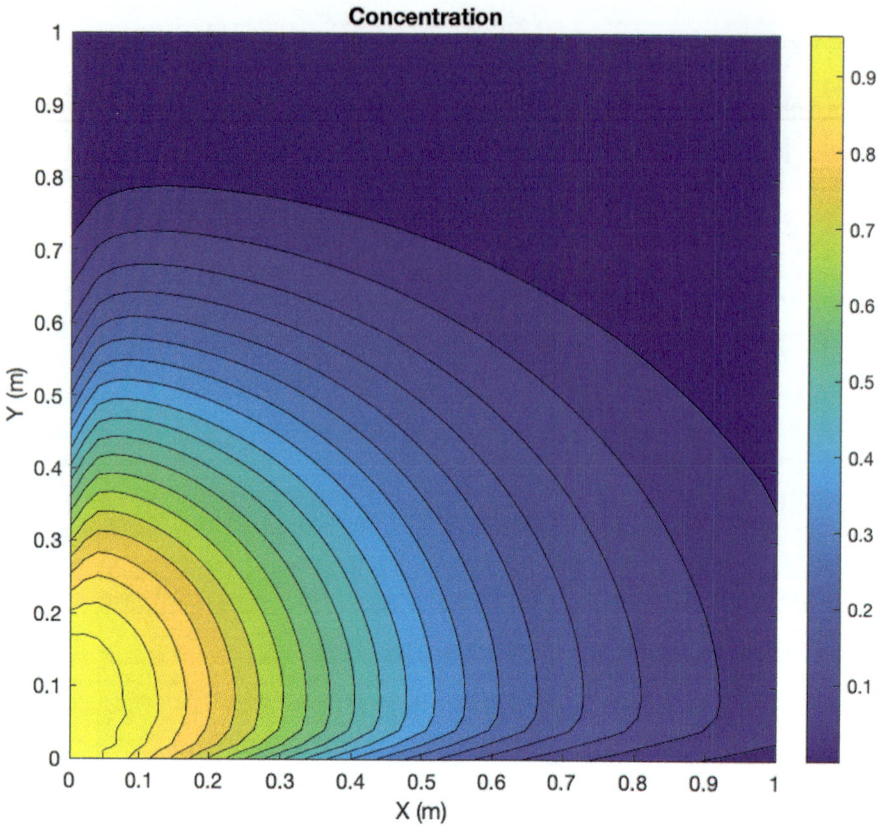

Figure 4.11 Concentration contours showing the distribution of solute in the domain.

overall concentrations, which indicates the dominance of advection near the lower boundary of the domain.

The concentration profiles along the y-direction for different values of x are shown in Figure 4.13. At $x = 0.1$, the concentration is nearly uniform due to proximity to the inlet. For larger x values, the profiles gradually decrease, indicating the diffusion-dominated region away from the inlet.

4.6.5 SOLUTE TRANSPORT IN ANISOTROPIC MEDIA

The transport of fluids and solutes through anisotropic porous media exhibits unique characteristics that distinguish it from transport in isotropic systems. In this section, we develop a comprehensive framework for modeling transport processes in anisotropic media, focusing on the governing equations and their physical interpretation.

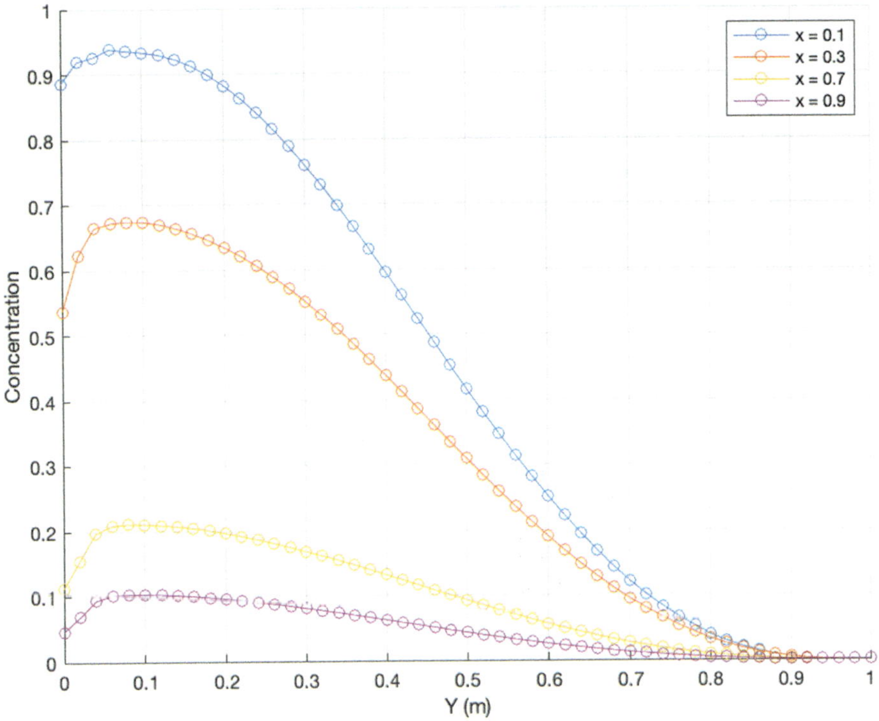

Figure 4.12 Concentration profiles along the x-direction for different values of y.

The fundamental equation governing transport phenomena in porous media is the advection-dispersion equation. For anisotropic media, this equation must be modified to account for directional dependencies in both flow and dispersion processes:

$$\phi\frac{\partial C}{\partial t} + \nabla \cdot (\mathbf{u}C - \phi\mathbf{D}\nabla C) = q_c \tag{4.75}$$

where ϕ represents porosity, C is concentration, $\mathbf{u}$ is the Darcy velocity vector, $\mathbf{D}$ is the hydrodynamic dispersion tensor, and q_c denotes source/sink terms.

The hydrodynamic dispersion tensor $\mathbf{D}$ in anisotropic media requires special attention, as it combines both mechanical dispersion and molecular diffusion effects. For anisotropic systems, this tensor takes the form:

$$\mathbf{D} = \alpha_T|\mathbf{u}|\mathbf{I} + (\alpha_L - \alpha_T)\frac{\mathbf{u}\mathbf{u}^T}{|\mathbf{u}|} + \mathbf{D}^* \tag{4.76}$$

where α_L and α_T are the longitudinal and transverse dispersivity values, respectively, $\mathbf{I}$ is the identity tensor, and $\mathbf{D}^*$ represents the effective molecular diffusion tensor in the porous medium.

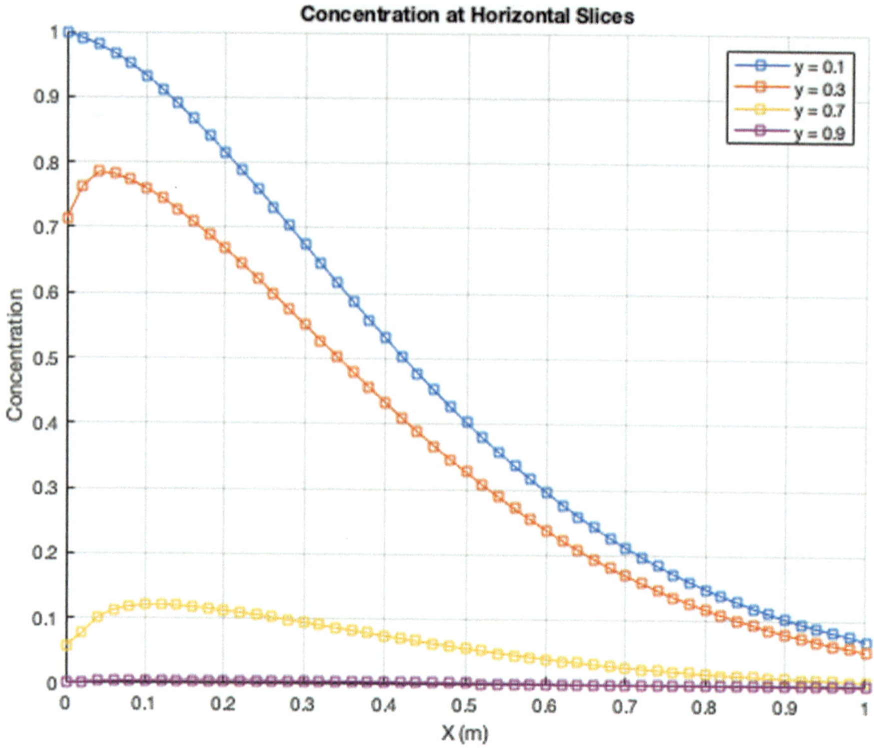

Figure 4.13 Concentration profiles along the y-direction for different values of x.

In anisotropic media, solute transport is primarily influenced by advection and dispersion processes. The bulk movement of solutes driven by the fluid flow, represented by the term $\nabla \cdot (\mathbf{u}C)$. The spreading of solutes due to mechanical dispersion and molecular diffusion. The dispersion tensor $\mathbf{D}(\mathbf{u})$ captures the anisotropic nature of this process.

The dispersion tensor $\mathbf{D}(\mathbf{u})$ characterizes the spreading of solutes and is symmetric and positive definite. In anisotropic media, the tensor depends on the directional components of the Darcy velocity $\mathbf{u} = [u_x, u_y, u_z]$. It is expressed as:

$$\mathbf{D} = \begin{bmatrix} D_{xx} & D_{xy} & D_{xz} \\ D_{yx} & D_{yy} & D_{yz} \\ D_{zx} & D_{zy} & D_{zz} \end{bmatrix}.$$

A widely used formulation for the coefficients of the dispersion tensor is given by:

$$D_{xx} = \frac{1}{u}\left(a_L u_x^2 + a_{TH} u_y^2 + a_{TV} u_z^2\right), \tag{4.77}$$

$$D_{yy} = \frac{1}{u}\left(a_{TH} u_x^2 + a_L u_y^2 + a_{TV} u_z^2\right), \tag{4.78}$$

$$D_{zz} = \frac{1}{u}\left(a_{TV}(u_x^2 + u_y^2 + u_z^2)\right), \tag{4.79}$$

$$D_{xy} = D_{yx} = \frac{1}{u}[a_L - a_{TH}]u_x u_y, \tag{4.80}$$

$$D_{xz} = D_{zx} = \frac{1}{u}[a_L - a_{TV}]u_x u_z, \tag{4.81}$$

$$D_{yz} = D_{zy} = \frac{1}{u}[a_L - a_{TV}]u_y u_z. \tag{4.82}$$

Here, a_L, a_{TH}, and a_{TV} are the longitudinal, transverse horizontal, and transverse vertical dispersion coefficients, respectively. An alternative formulation consistent with tensor transformation rules can written as:

$$D_{xx} = \left[a_L \frac{u_x^2}{u^2} + a_{TH}\frac{u_y^2}{u^2} + a_{TV}\frac{u_x^2 u_z^2}{u^4}\right]u, \tag{4.83}$$

$$D_{yy} = \left[a_L \frac{u_y^2}{u^2} + a_{TH}\frac{u_x^2}{u^2} + a_{TV}\frac{u_y^2 u_z^2}{u^4}\right]u, \tag{4.84}$$

$$D_{zz} = \left[a_L \frac{u_z^2}{u^2} + a_{TV}\frac{u_x^2 + u_y^2}{u^2}\right]u, \tag{4.85}$$

$$D_{xy} = D_{yx} = \left[a_L \frac{u_x u_y}{u^2} + a_{TH}\frac{u_x u_y}{u^2} + a_{TV}\frac{u_x u_y u_z}{u^4}\right]u, \tag{4.86}$$

$$D_{xz} = D_{zx} = [a_L - \alpha_{TV}]\frac{u_x u_z}{u^2}u, \tag{4.87}$$

$$D_{yz} = D_{zy} = [a_L - \alpha_{TV}]\frac{u_y u_z}{u^2}u. \tag{4.88}$$

The coefficients are further refined based on the flow direction:

$$\alpha_L = \alpha_1 + \alpha_2 - \alpha_3 \cos^2\theta, \tag{4.89}$$

$$\alpha_{TV} = \alpha_1 + \alpha_3(1 - \cos^2\theta), \tag{4.90}$$

$$\alpha_{TH} = \alpha_1, \tag{4.91}$$

where θ is the angle between the material axis and the flow direction.

Anisotropy significantly influences transport phenomena through several mechanisms, including preferential flow pathways, enhanced spreading and mixing, scale-dependent transport behavior, and breakthrough curve distortion. In anisotropic media, fluid naturally follows paths of least resistance, creating preferential flow channels aligned with the direction of maximum permeability. This directional bias leads to non-uniform transport patterns that cannot be observed in isotropic systems. The interaction between anisotropic permeability and the resulting velocity field creates complex flow patterns that enhance the spreading and mixing of solutes. This effect is particularly pronounced when the principal directions of permeability are not aligned with the mean flow direction. Anisotropy often varies with scale in geological formations, leading to scale-dependent transport behavior. Microscopic anisotropy at the pore scale combines with macroscopic anisotropy at the formation scale to create multi-scale transport phenomena. The breakthrough curves for solutes in anisotropic media exhibit characteristic asymmetry and tailing effects that differ significantly from the more symmetric profiles observed in isotropic systems.

For practical applications, it is often necessary to define effective transport parameters that capture the combined effects of anisotropy across multiple scales. These parameters include effective dispersion tensor, directional sorption coefficients, and effective reaction rates. A macroscopic parameter that incorporates the effects of both microscopic dispersion and flow field heterogeneity resulting from anisotropic permeability. In anisotropic media, sorption processes may also exhibit directional dependencies, requiring direction-specific retardation factors. Chemical and biological reactions in anisotropic media may proceed at different rates depending on the local flow conditions, which are influenced by the directional permeability. The accurate determination of these effective parameters typically requires a combination of laboratory experiments, field measurements, and numerical upscaling techniques that explicitly account for the anisotropic nature of the medium.

4.6.6 NUMERICAL METHODS

Simulating transport processes in anisotropic porous media presents significant computational challenges, particularly in accurately representing the full permeability tensor. Traditional numerical approaches often struggle with anisotropic properties, necessitating specialized techniques for reliable simulations. While researchers continue to develop more effective methods, several challenges remain in this area.

The conventional two-point flux finite difference method, widely used for isotropic systems, exhibits significant limitations when applied to anisotropic media. This method cannot adequately handle diagonal permeability tensors, leading to substantial errors in flow and transport predictions. More sophisticated approaches, including finite element and finite volume methods, possess the theoretical capability to incorporate anisotropy effects but typically involve complex implementation procedures that can be computationally intensive.

To overcome these limitations, the multi-point flux approximation (MPFA) has emerged as a promising alternative. This approach fundamentally differs from conventional methods by accounting for flow contributions from multiple neighboring points rather than relying on two-point approximations. In MPFA, flux components are influenced by additional nearby points, providing a more accurate representation of the complex flow patterns characteristic of anisotropic media.

Various researchers have proposed intuitive techniques to extend cell-centered finite-difference methods to accommodate permeability tensors. These efforts have led to the systematic formulation of finite-difference approaches using multi-point flux approximations. A critical requirement for effective MPFA implementation is the ability to account for cross-flow potential arising from off-diagonal elements in the permeability tensor [53].

The computational complexity of MPFA increases significantly with dimensionality. In two-dimensional domains, the flow vector at each cell face center is influenced by six neighboring pressure values. This complexity expands dramatically in three-dimensional fields, where eighteen adjacent pressure values affect the flux vector at each cell face center. Consequently, constructing the finite difference stencil for two-dimensional problems requires determining six coefficients (or eighteen in three-dimensional scenarios), representing a substantial computational burden.

Recent methodological advances by Sun et al. [34] and Salama et al. [32] have streamlined this process by developing techniques for automatic coefficient determination through the solution of local problems within each cell. This approach significantly facilitates the construction of the global system of equations by decomposing the complex problem into more manageable components. Solving numerous local problems proves substantially more efficient than directly constructing the intricate stencil required for the global equation system. This methodology has been successfully applied to diverse scenarios involving both single-phase and two-phase flows in porous media, with anisotropy as a key focus area [33, 35, 54].

The development of multi-point flux approaches has proceeded along two distinct pathways. The first approach utilizes control volume approximation, which involves bisecting each edge (in two dimensional cases) and assigning individual flux vectors to edge segments. This methodology has been refined by Aavatsmark et al. [55], Aavatsmark [56], and numerous other researchers. The second approach, pursued by Brezzi et al. [57], Wheeler and Yotov [58], and others, is grounded in mixed finite element formulations within BDM space. By applying specific quadrature algorithms for integral estimation, this approach generates MPFA stencils compatible with cell-centered finite difference methods, particularly when applied to structured grids.

Despite methodological advances, deriving the MPFA stencil analytically remains complex (see Chapter 3), particularly due to flux approximation and divergence calculations requiring multiple neighboring points. For two-dimensional systems, flux approximation involves interactions with neighboring cells and is generally represented by:

$$f \approx \sum_{i \in \Omega} t_i p_i \tag{4.92}$$

where t_i denotes transmissibility coefficients and p_i represents pressure gradients.

In practice, local flux continuity is enforced through specific systems of equations. For instance, the x-component flux across an interface can be explicitly expressed as:

$$v_x = a_1 p_1 + a_2 p_2 + a_3 p_3 + a_4 p_4 + a_5 p_5 + a_6 p_6 \tag{4.93}$$

The coefficients a_i encapsulate discretization parameters and permeability characteristics, derived using the numerical experimenting pressure approach. Specifically, for node (i, j), the local flux continuity equations take the form:

$$A[v_x^-, v_y^+, v_x^+, v_y^-]^T = B[p_1, p_2, p_3, p_4]^T \tag{4.94}$$

where matrices A and B are defined as:

$$[A]_{ij} = \frac{1}{4}\begin{bmatrix} \frac{k_{yy}^{E_1}}{detE_1}A^{E_1} + \frac{k_{yy}^{E_2}}{detE_2}A^{E_2} & -\frac{k_{xy}^{E_2}}{detE_2}A^{E_2} & 0 & -\frac{k_{xy}^{E_1}}{detE_1}A^{E_1} \\ -\frac{k_{xy}^{E_2}}{detE_2}A^{E_2} & \frac{k_{xx}^{E_2}}{detE_2}A^{E_2} + \frac{k_{xx}^{E_3}}{detE_3}A^{E_3} & -\frac{k_{xy}^{E_3}}{detE_3}A^{E_3} & 0 \\ 0 & -\frac{k_{xy}^{E_3}}{detE_3}A^{E_3} & \frac{k_{yy}^{E_3}}{detE_3}A^{E_3} + \frac{k_{yy}^{E_4}}{detE_4}A^{E_4} & -\frac{k_{xy}^{E_4}}{detE_4}A^{E_4} \\ -\frac{k_{xy}^{E_1}}{detE_1}A^{E_1} & 0 & -\frac{k_{xy}^{E_4}}{detE_4}A^{E_4} & \frac{k_{xx}^{E_1}}{detE_1}A^{E_1} + \frac{k_{xx}^{E_4}}{detE_4}A^{E_4} \end{bmatrix} \tag{4.95}$$

$$[B]_{ij} = \frac{1}{2}\begin{bmatrix} \Delta y_{E_1} & -\Delta y_{E_1} & 0 & 0 \\ 0 & \Delta x_{E_2} & -\Delta x_{E_2} & 0 \\ 0 & 0 & -\Delta y_{E_3} & \Delta y_{E_3} \\ \Delta x_{E_4} & 0 & 0 & -\Delta x_{E_4} \end{bmatrix} \tag{4.96}$$

Here, Δx_{E_i} and Δy_{E_i} denote horizontal and vertical lengths, respectively, and A^{E_i} signifies the area of cell E_i. The determinant "det" refers to the hydraulic conductivity tensor for each cell. This approach systematically constructs a global algebraic system for pressure determination:

$$C\mathbf{p} = \mathbf{b} \tag{4.97}$$

where C is the coefficient matrix, $\mathbf{p}$ the pressure vector, and $\mathbf{b}$ the vector incorporating source terms and boundary conditions.

4.7 ATMOSPHERIC WATER HARVESTING

Water scarcity is one of the most pressing challenges of the 21st century, driven by rapid population growth, urbanization, climate change, and inefficient water resource management. In arid and semi-arid regions, the crisis is exacerbated by minimal precipitation, rising temperatures, and overexploitation of limited freshwater resources [59]. Nearly two-thirds of the world's population experiences severe water shortages for at least one month per year [60]. The need for alternative water sources is thus urgent. Among the emerging solutions, atmospheric water harvesting (AWH) stands out as a viable method to supplement existing water supplies. Despite representing only 0.001% of the Earth's total water volume, the atmosphere contains approximately 12,900 km^3 of water, enough to significantly alleviate water shortages if efficiently utilized [60]. Technologies that tap into this overlooked resource can offer a decentralized and renewable solution to water scarcity, especially in remote or underserved communities.

Atmospheric water harvesting refers to the process of extracting water vapor from the air and converting it into liquid water for human use. The atmosphere contains water in the form of clouds, fog, and vapor. AWH technologies utilize various mechanisms such as condensation, adsorption, and membrane separation to capture and collect this moisture. Fog and dew collection methods rely on natural condensation processes and have been historically utilized by ancient civilizations. Recent developments have enhanced the efficiency of these systems, making them more suitable for large-scale and low-humidity environments. Modern AWH systems incorporate active and passive techniques, including refrigeration-based condensers and advanced sorbent materials like metal-organic frameworks (MOFs) and anhydrous salts [61, 62].

The concept of extracting water from the atmosphere is not new. Ancient societies employed techniques such as dew ponds and fog fences to collect moisture in arid environments. For example, dew ponds in England were shallow basins used for collecting dew for livestock, while fog fences constructed from mesh materials captured droplets from wind-driven fog in mountainous regions. Technological innovations in the 20th and 21st centuries, including refrigeration cycles, thermoelectric coolers, and solar-powered desorption systems, have significantly improved the efficiency and viability of AWH systems. Devices like the Warka Water towers and MOF-based harvesting systems have emerged as practical tools for delivering potable water in water-stressed areas.

Recent advancements in materials science have expanded the operational capabilities of AWH systems. MOFs, for example, can adsorb water even at relative humidity levels as low as 10–20% [63]. These materials, coupled with passive solar heating, enable water recovery in arid environments without requiring extensive external energy inputs. Other innovations include refrigeration-based atmospheric water generators (AWGs), fog collectors, and hygroscopic salts. AWGs are more effective in environments with moderate to high humidity but are limited by energy consumption and operational costs in

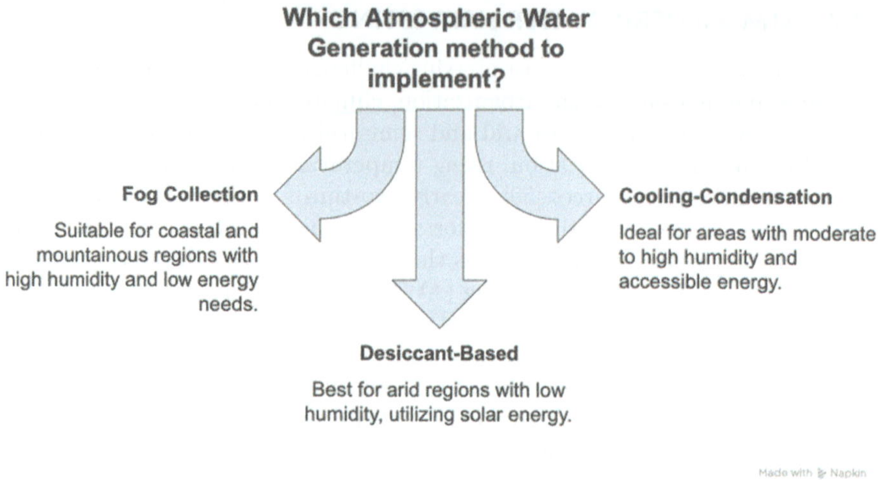

Figure 4.14 Classification of atmospheric water harvesting techniques.

drier climates. Passive fog collectors, on the other hand, are environmentally friendly and effective in regions where fog is frequent.

Feasibility studies from diverse global regions such as Oman, India, Chile, and the UAE highlight both the potential and limitations of AWH. These studies assess capital costs, maintenance needs, energy requirements, and long-term economic returns. While initial investment is typically high, the benefits may outweigh costs in areas where water is scarce and conventional alternatives are infeasible [318].

4.7.1 COMMON AWH TECHNIQUES

Atmospheric Water Generators (AWGs) are innovative devices designed to extract water from ambient air. These systems operate using a variety of physical and chemical principles, allowing them to function across a range of climatic conditions. Broadly, AWGs can be classified based on the mechanism of water extraction into two primary categories: fog water collection and water vapor condensation systems.

Figure 4.14 presents a schematic overview of the main categories of atmospheric water harvesting (AWH) techniques. These are generally divided into fog collection systems and water vapor collection systems.

FOG WATER COLLECTION

Fog water collection utilizes mesh nets or similar structures to capture microscopic water droplets suspended in the air. This passive method is particularly effective in locations characterized by persistent fog, such as coastal areas adjacent to cold ocean currents and mountainous regions.

As fog flows through the mesh net, droplets condense and coalesce on the fibers, eventually dripping into collection channels. Typically constructed from polypropylene or specialized polymers designed to maximize droplet adhesion and drainage. Successful large-scale implementations have been established in regions such as the Atacama Desert in Chile [65].

WATER VAPOR COLLECTION

Water vapor collection systems extract invisible moisture from the atmosphere and condense it into usable water. These systems can be categorized into several subtypes:

Passive Dew Collection: Relies on surfaces that radiatively cool below the ambient temperature during the night, allowing dew to form.
Active Condensation Systems: Utilize mechanical cooling (e.g., refrigeration or thermoelectric cooling) to lower air temperature below the dew point, triggering condensation.
Desiccant-Based Absorption: Employ hygroscopic materials, such as silica gel or salt-based compounds, to absorb atmospheric moisture, which is later released via heating.
Advanced Sorbent Materials: Leverage novel materials like metal-organic frameworks (MOFs) and hygroscopic nanomaterials to capture water vapor efficiently, even under low humidity conditions.

Each method has unique advantages depending on environmental conditions and energy availability. Continuous research efforts aim to enhance these technologies in terms of efficiency, cost-effectiveness, and scalability.

The future of atmospheric water harvesting (AWH) is driven by several promising technological advancements. One of the most significant trends is material innovation, particularly the development of high-capacity sorbent materials that require minimal energy for regeneration. These advanced materials enhance water capture efficiency, even in low-humidity environments. In parallel, the integration of renewable energy sources, such as solar and wind, plays a crucial role in reducing the carbon footprint of AWH systems, making them more sustainable and suitable for deployment in remote or off-grid areas. Additionally, the application of artificial intelligence (AI) and machine learning (ML) enables smart system optimization, allowing real-time environmental data to guide predictive control and improve performance. AWH systems are also becoming increasingly scalable, evolving from small, individual units to larger, community- or industrial-scale installations that can support agricultural and municipal water needs. Collectively, these advancements establish atmospheric water harvesting as a vital and adaptable solution to global water scarcity, particularly in regions most vulnerable to the impacts of climate change.

4.7.2 DESICCANT-BASED AWGS

Desiccant-based atmospheric water generators (AWGs) utilize hygroscopic materials that can absorb moisture from the air and release it upon heating. These systems are especially well-suited for arid regions with limited energy resources, as they are often powered by solar energy. The operating mechanism involves capturing water vapor under ambient conditions using desiccants, such as silica gel or advanced materials like metal-organic frameworks (MOFs). Once the desiccant is saturated, solar-induced heating triggers the desorption process, during which the collected moisture is released as vapor and subsequently condensed into liquid water. This approach offers several advantages, including effective operation in low-humidity environments, minimal energy requirements, and compatibility with off-grid settings. The practical viability of atmospheric water harvesting technologies is evidenced by a growing number of commercial AWG products available on the market. For instance, Watergen produces both mobile and stationary units tailored for community-level water generation, including deployment in disaster relief scenarios. EcoloBlue focuses on residential and office environments, offering AWGs that employ active condensation systems. Commercial innovations illustrate the expanding role of AWGs across diverse applications, from emergency water supply to environmentally conscious infrastructure development.

Atmospheric water generators can be broadly categorized into three main types, each with unique operational profiles and ideal use cases. Fog collection systems operate efficiently in high-humidity, fog-prone areas and require minimal energy input, making them well-suited for coastal and mountainous regions. Cooling-condensation systems, on the other hand, function best in moderate to high humidity conditions and rely on medium to high levels of energy input, typically from electrical sources, limiting their use to areas with reliable power infrastructure. Finally, desiccant-based systems offer versatility across a wide humidity range and are particularly advantageous in arid and remote areas due to their compatibility with solar energy and low operational demands. This comparative evaluation highlights the importance of selecting the appropriate AWG technology based on environmental conditions and resource availability.

4.7.3 THERMODYNAMICS ASPECTS

Water vapor is the most abundant greenhouse gas and a vital part of Earth's hydrological cycle. It forms from evaporation (liquid to vapor) and sublimation (solid to vapor), and returns to solid or liquid via condensation or deposition. In the atmosphere, water vapor impacts climate, weather, and serves as a resource through atmospheric water harvesting (AWH). Moist air properties such as Humidity Ratio, Enthalpy, Relative Humidity, and Dew Point Temperature are worth mentioning here [66].

The Humidity Ratio takes the form:

$$W = 0.622 \cdot \frac{P_w}{P - P_w} \tag{4.98}$$

The enthalpy can be written as:

$$h = 1.006T + W(2501 + 1.86T) \tag{4.99}$$

While the relative humidity, which is of great interest to our study, is expressed as:

$$\phi = \frac{P_w}{P_{ws}} \times 100\% \tag{4.100}$$

Finally, the dew point temperature has the form:

$$T_{dp} = \frac{243.04 \ln(P_w/6.1094)}{17.625 - \ln(P_w/6.1094)} \tag{4.101}$$

4.7.4 DESICCANT SORPTION

Desiccant-based water harvesting systems rely on sorption processes, where hygroscopic materials absorb water vapor from the air. The sorption rate, which quantifies the amount of water vapor captured per unit time and area, can be expressed as:

$$\mu = (\phi - \phi_c)P_{ws}f(T, A) \tag{4.102}$$

Here, μ is the water vapor sorption rate (e.g., kg/m^2·s), ϕ is the ambient relative humidity, ϕ_c is the critical relative humidity below which the sorbent becomes ineffective, P_{ws} is the saturation vapor pressure at temperature T, and $f(T, A)$ is a function representing the desiccant's temperature-dependent absorption efficiency over area A. This formulation reflects that sorption is more efficient at higher humidity and temperature levels above the material's threshold.

The $CaCl_2$ sorption model can be written as:

$$\mu = \left[0.016T^2 + (4.537\phi - 2.094)T - 5.833\phi + 2.665\right] \cdot A \tag{4.103}$$

This empirical equation models the sorption rate for calcium chloride ($CaCl_2$), one of the most widely used desiccants. It expresses μ as a quadratic function of temperature T and relative humidity ϕ, scaled by the surface area A of the exposed desiccant. The coefficients are derived from experimental fitting and reflect the temperature-humidity synergy in hygroscopic performance.

Understanding sorption thermodynamics is essential for evaluating the equilibrium uptake of water vapor on a solid sorbent under varying pressure and temperature conditions.

4.7.5 LANGMUIR ISOTHERM

The Langmuir isotherm models monolayer adsorption and assumes uniform adsorption sites with no interaction between adsorbed molecules [67, 68].

$$\sigma(p) = \sigma_{\max} \cdot \frac{bp}{1 + bp} \tag{4.104}$$

Here, $\sigma(p)$ is the equilibrium adsorption amount (kg/kg or mol/g), $\sigma_{\max}$ is the maximum capacity, p is the partial pressure of water vapor, and b is the Langmuir constant related to the sorbent–adsorbate affinity.

At low pressures ($bp \ll 1$), the Langmuir isotherm simplifies to a linear form resembling Henry's Law:

$$\sigma(p) \approx Hp \quad \text{where} \quad H = \sigma_{\max} b \tag{4.105}$$

This linear regime is particularly relevant for low-humidity conditions, such as desert environments, where adsorption remains efficient despite reduced vapor content [63].

The affinity constant b in the Langmuir model is highly temperature-dependent and follows an Arrhenius-type relationship:

$$\ln b = -\frac{\Delta H_{\text{ads}}}{RT} + C \tag{4.106}$$

where ΔH_{ads} is the enthalpy of adsorption (J/mol), R is the universal gas constant, and C is a material-specific constant. This expression implies that increasing temperature generally reduces sorption affinity, making thermodynamic control crucial for system optimization [67].

When multiple vapors are present (e.g., water and organic compounds), the selectivity of the sorbent toward water vapor becomes critical:

$$S = \frac{b_{\text{H}_2\text{O}}}{b_{\text{other}}} \tag{4.107}$$

This dimensionless selectivity ratio indicates the preference of the sorbent for water vapor over competing molecules. A higher S implies superior water capture in mixed vapor environments, which is especially beneficial for atmospheric water generation in polluted or humid air containing volatile compounds [63].

4.7.6 MATHEMATICAL MODELING

One promising solution involves atmospheric water harvesting using hygroscopic materials like copper chloride ($CuCl_2$) and magnesium sulfate ($MgSO_4$), which can capture moisture from humid air even at low relative humidity. Recent advancements in the field have shown the effectiveness of these materials in transforming from solid states to saturated solutions, making them suitable for water generation in diverse climatic conditions. However, to optimize the performance of these materials and systems, it is crucial to develop accurate mathematical models that describe the adsorption and desorption processes involved in water vapor capture. Analytical and numerical models have been increasingly employed to simulate these interactions and predict system performance under various environmental conditions, significantly reducing the need for expensive and time-consuming physical experiments. For instance, El-Amin [69] developed analytical solutions for harvesting atmospheric water using desiccants, providing key insights into system optimization. Other works, such as the studies by Kabbaj and El-Amin [70] on adsorption and desorption isotherms, and Almasarani et al. [71] on experimental investigations and modeling of atmospheric water generation, have further contributed to the field by refining predictive models and validating them through experiments. Additionally, Sibie et al. [72] focused on modeling water generation from air using anhydrous salts, which complements the ongoing research efforts in this domain. Kabbaj and El-Amin [73] present a comprehensive numerical framework that incorporates thermodynamic principles to model the water adsorption-desorption processes in porous desiccant materials for atmospheric water harvesting applications.

Now, let us develop a mathematical model for water vapor adsorption and desorption in copper chloride ($CuCl_2$), assuming the material is isotropic, maintains local thermodynamic equilibrium, and behaves as an ideal incompressible fluid. Radiative transfers, chemical reactions, and freezing are neglected, and gravity is ignored. The mass conservation equation is formulated as:

$$\frac{\partial w}{\partial t} = \nabla \cdot (\delta_{p,a} \nabla P_v + \delta_l \nabla P_c) \tag{4.108}$$

where $w = \rho_0 \theta$ is the total water content, ρ_0 is the density of $CuCl_2$. $\theta = \frac{m_l + m_v}{m_s}$ is mass water content, m_l, m_v, and m_s are the masses of liquid water, water vapor, and $CuCl_2$, respectively. $\delta_{p,a}$ is the vapor permeability of air, P_v is the partial pressure of water vapor, δ_l is the liquid water permeability, P_c is the capillary pressure, and ρ_l is the density of liquid water.

If the liquid water is in equilibrium with the ambient and the capillary condensation occurs, the capillarity relationship is articulated by Kelvin's law [70]:

$$P_c = \frac{RT\rho_l}{M_l} \ln(R_h) \tag{4.109}$$

where ρ_l [kg.m^{-3}] designs here the water density.

On the other hand, relative humidity R_h is defined as the ratio of the partial pressure of water vapor P_v to the saturation vapor pressure P_{sat}, i.e., $R_h = \frac{P_v}{P_{\text{sat}}}$. The saturation vapor pressure depends on temperature T, given by $P_{\text{sat}} = e^{23.5771 - \frac{4042.9}{T - 37.58}}$, valid for temperatures between 0°C and 80°C. The rate of change of water content with respect to relative humidity R_h is described by, $\delta_{R_h} = \frac{\partial w}{\partial R_h}$.

The heat transfer equation, incorporating both sensible and latent heat, is:

$$\rho_0(C_0 + \frac{w}{\rho_0}C_l)\frac{\partial T}{\partial t} = \frac{\partial}{\partial x}\left(\lambda\frac{\partial T}{\partial x} + l_v\delta_p\frac{\partial T}{\partial x}\right) \tag{4.110}$$

where C_0 and C_l are the specific heat capacities of $CuCl_2$ and water, respectively. l_v is the latent heat of vaporization, and g_v is the vapor mass flux, which describes phase change processes. λ is the thermal conductivity of the material.

The initial conditions for relative humidity R_h and temperature T are defined based on pre-determined starting values. For the boundary conditions, both R_h and T are considered, with R_h varying over time. To account for this variation, an exponential interpolation, $a\exp(-b)$, was applied to smooth the transition between data points, providing a more accurate representation of the system's behavior. This approach ensures that the model reflects realistic changes in relative humidity over time while maintaining consistency with the experimental data.

To solve the mathematical model, we use the pdepe solver in MATLAB®, which handles one-dimensional parabolic and elliptic partial differential equations (PDEs). The solver uses the Method of Lines to discretize the spatial domain, transforming the PDEs into a system of ordinary differential equations (ODEs) that are integrated over time using MATLAB®'s ODE solvers. The governing equations (4.108) and (4.110) are solved with appropriate initial and boundary conditions, allowing us to model the coupled heat and mass transfer processes in the desiccant material. This numerical setup provides a robust framework for simulating the adsorption and desorption dynamics in $CuCl_2$.

```
function AWG_generator
    generate_rh_effect_figure();
    generate_temperature_effect_figure();
end

function generate_rh_effect_figure()
    t = linspace(0, 60, 100);

    % Different relative humidity values
    Rh_values = [0.4, 0.5, 0.6, 0.7];

    % Initialize figure
    figure('Position', [100, 100, 800, 600]);
    hold on;

    % Colors for different RH values
    colors = {'b-', 'r-', 'g-', 'm-'};

    for i = 1:length(Rh_values)
        Rh = Rh_values(i);

        % Desorption model: exponential decay with RH-dependent
            rate
```

```
        % Higher RH leads to faster desorption
        k = 0.02 + 0.05 * Rh; % Desorption rate constant
        theta_0 = 1.0; % Initial normalized water content
        theta_eq = 0.1 + 0.3 * (1 - Rh); % Equilibrium water
            content

        % Exponential approach to equilibrium
        theta = theta_eq + (theta_0 - theta_eq) * exp(-k * t);

        plot(t, theta, colors{i}, 'LineWidth', 2, 'DisplayName',
            sprintf('R_h␣=␣%.1f', Rh));
    end

    % Formatting
    xlabel('Time␣[min]', 'FontSize', 12, 'FontWeight', 'bold');
    ylabel('Normalized␣Water␣Content,␣\theta␣[-]', 'FontSize',
        12, 'FontWeight', 'bold');
    title('Effect␣of␣Relative␣Humidity␣on␣Water␣Desorption', '
        FontSize', 14, 'FontWeight', 'bold');
    legend('Location', 'northeast', 'FontSize', 11);
    grid on;
    xlim([0, 60]);
    ylim([0, 1.1]);

    % Set axes properties
    ax = gca;
    ax.FontSize = 11;
    ax.LineWidth = 1.2;

    % Save figure
    print('fig_rh_effect_theta', '-dpng', '-r300');

    fprintf('Figure␣1␣saved␣as:␣fig_rh_effect_theta.png\n');
end

function generate_temperature_effect_figure()
    % Generate figure showing effect of temperature on theta vs
        time

    % Time vector (0 to 60 minutes)
    t = linspace(0, 60, 100);

    % Different initial temperatures (K)
    T0_values = [300, 315, 335, 355];

    % Initialize figure
    figure('Position', [200, 200, 800, 600]);
    hold on;

    % Colors and line styles for different temperatures
    colors = {'b-', 'r-', 'g-', 'm-'};

    for i = 1:length(T0_values)
        T0 = T0_values(i);

        % Temperature-dependent desorption model
```

```
        % Higher temperature leads to faster desorption (
            Arrhenius-type)
        T_ref = 300; % Reference temperature
        Ea_R = 2000; % Activation energy / gas constant

        % Rate constant with Arrhenius dependence
        k_ref = 0.03; % Reference rate constant
        k = k_ref * exp(Ea_R * (1/T_ref - 1/T0));

        % Initial and equilibrium water contents
        theta_0 = 1.0;
        theta_eq = 0.05 + 0.15 * exp(-0.01 * (T0 - T_ref)); %
            Lower equilibrium at higher T

% Exponential desorption
        theta = theta_eq + (theta_0 - theta_eq) * exp(-k * t);

        plot(t, theta, colors{i}, 'LineWidth', 2, 'DisplayName',
            sprintf('T_0␣=␣%d␣K', T0));
    end

% Formatting
    xlabel('Time␣[min]', 'FontSize', 12, 'FontWeight', 'bold');
    ylabel('Normalized␣Water␣Content,␣\theta␣[-]', 'FontSize',
        12, 'FontWeight', 'bold');
    title('Effect␣of␣Initial␣Temperature␣on␣Water␣Desorption', '
        FontSize', 14, 'FontWeight', 'bold');
    legend('Location', 'northeast', 'FontSize', 11);
    grid on;
    xlim([0, 60]);
    ylim([0, 1.1]);

% Set axes properties
    ax = gca;
    ax.FontSize = 11;
    ax.LineWidth = 1.2;

% Save figure
    print('fig_HAWtemp', '-dpng', '-r300');

    fprintf('Figure␣2␣saved␣as:␣fig_HAWtemp.png\n');
end

% Enhanced AWG solver with figure generation capability
function AWG_solver_with_plots
    % Enhanced version of the original AWG solver that generates
        plots

% Spatial and time discretization
    x = linspace(0, 1, 100);  % Space domain (0 to 1 meter)
    t = linspace(0, 3600, 100); % Time domain (0 to 1 hour)

% Solve the system using pdepe
    sol = pdepe(0, @AWG_pde, @AWG_ic, @AWG_bc, x, t);

% Extract the solution for water content (w) and temperature (T)
```

```
    w = sol(:,:,1);
    T = sol(:,:,2);

% Plot the results - Water content evolution
    figure('Position', [300, 300, 800, 600]);

% Plot water content at different spatial positions
    subplot(2,1,1);
    plot(t/60, w(:,1), 'b-', 'LineWidth', 2, 'DisplayName', 'x␣=␣
        0␣m');
    hold on;
    plot(t/60, w(:,25), 'r--', 'LineWidth', 2, 'DisplayName', 'x␣
        =␣0.25␣m');
    plot(t/60, w(:,50), 'g-.', 'LineWidth', 2, 'DisplayName', 'x␣
        =␣0.5␣m');
    plot(t/60, w(:,75), 'm:', 'LineWidth', 2, 'DisplayName', 'x␣=
        ␣0.75␣m');
    plot(t/60, w(:,100), 'k-', 'LineWidth', 2, 'DisplayName', 'x␣
        =␣1␣m');

    xlabel('Time␣[min]', 'FontSize', 12);
    ylabel('Water␣Content␣[kg/kg]', 'FontSize', 12);
    title('Water␣Content␣Evolution␣at␣Different␣Positions', '
        FontSize', 14);
    legend('Location', 'best');
    grid on;

% Plot temperature evolution
    subplot(2,1,2);
    plot(t/60, T(:,1), 'b-', 'LineWidth', 2, 'DisplayName', 'x␣=␣
        0␣m');
    hold on;
    plot(t/60, T(:,25), 'r--', 'LineWidth', 2, 'DisplayName', 'x␣
        =␣0.25␣m');
    plot(t/60, T(:,50), 'g-.', 'LineWidth', 2, 'DisplayName', 'x␣
        =␣0.5␣m');
    plot(t/60, T(:,75), 'm:', 'LineWidth', 2, 'DisplayName', 'x␣=
        ␣0.75␣m');
    plot(t/60, T(:,100), 'k-', 'LineWidth', 2, 'DisplayName', 'x␣
        =␣1␣m');

    xlabel('Time␣[min]', 'FontSize', 12);
    ylabel('Temperature␣[K]', 'FontSize', 12);
    title('Temperature␣Evolution␣at␣Different␣Positions', '
        FontSize', 14);
    legend('Location', 'best');
    grid on;

 % Save the PDE solution figure
    print('AWG_PDE_solution', '-dpng', '-r300');

    fprintf('PDE␣solution␣figure␣saved␣as:␣AWG_PDE_solution.png\n
        ');
end

% Define the PDE system (same as original)
```

```
function [c,f,s] = AWG_pde(x,t,u,DuDx)
    % Parameters (example values)
    delta_p_a = 1e-5; % Vapor permeability
    delta_l = 1e-5;   % Liquid permeability
    rho_0 = 1000;     % Density of CuCl2
    C_0 = 1000;       % Specific heat of CuCl2
    C_l = 4186;       % Specific heat of water
    lam = 0.6;        % Thermal conductivity
    l_v = 2257e3;     % Latent heat of vaporization

 % Water content equation
    c(1) = 1;
    f(1) = delta_p_a * DuDx(1) + delta_l * DuDx(1);
    s(1) = 0;

% Heat transfer equation
    c(2) = rho_0 * (C_0 + (u(1) / rho_0) * C_l);
    f(2) = lam * DuDx(2) + l_v * delta_p_a * DuDx(2);
    s(2) = 0;
end

% Initial conditions
function u0 = AWG_ic(x)
    u0 = [0.01; 300]; % Initial water content and temperature
end

% Boundary conditions
function [pl,ql,pr,qr] = AWG_bc(xl,ul,xr,ur,t)
    % Left boundary conditions
    pl = [ul(1) - 0.02; ul(2) - 300];
    ql = [0; 0];

% Right boundary conditions (Neumann condition)
    pr = [0; 0];
    qr = [1; 1];
end
```

As shown in Figure 4.15, the desorption behavior is significantly influenced by relative humidity. At higher R_h, water desorbs more rapidly due to a greater vapor pressure difference between the desiccant surface and the surrounding air. Higher humidity ($R_h = 0.7$) leads to faster water loss, while lower humidity ($R_h = 0.4$) results in more gradual desorption. This demonstrates that the desorption kinetics of hygroscopic materials like $CuCl_2$ are highly sensitive to ambient humidity. Conversely, lower humidity levels lead to slower water loss and sustained retention.

As an example of implementation, Figure 4.16 compares the normalized water content θ as a function of time for different initial temperatures $T_0 =$ 300, 315, 335, and 355 K. This behavior highlights the significant influence of thermal conditions on the moisture dynamics.

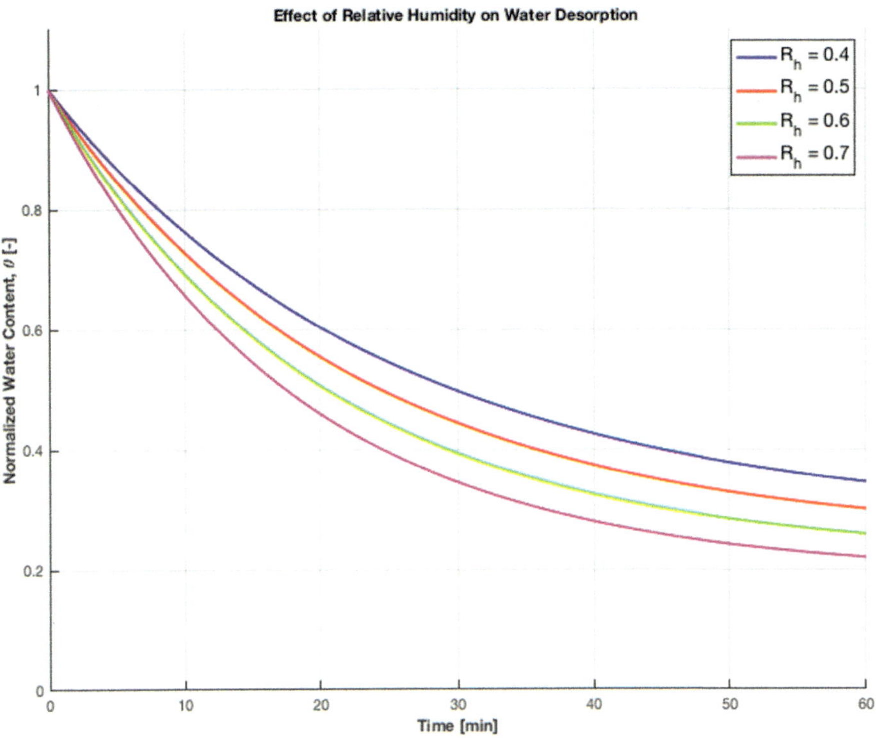

Figure 4.15 Effect of relative humidity R_h on the normalized water content θ over time.

CONCLUSION

This chapter has presented a detailed and structured treatment of single-phase flow in porous media, encompassing foundational theory, analytical solutions, and advanced numerical modeling techniques. Beginning with the mathematical formulation of flow governed by Darcy's law, we derived and solved canonical problems to build physical intuition and establish validation benchmarks. The transition from analytical to numerical methods, including finite difference, finite element, and MRST-based simulations, illustrates how real-world flow scenarios can be effectively modeled and visualized.

Applications such as solute transport and atmospheric water harvesting were examined to highlight the broader environmental and industrial relevance of porous media flow. Special emphasis was placed on integrating heat and mass transfer, stability analysis, and desiccant thermodynamics in these processes. The implementations provided in MATLAB® and Python not only support theoretical understanding but also serve as practical tools for engineers and researchers.

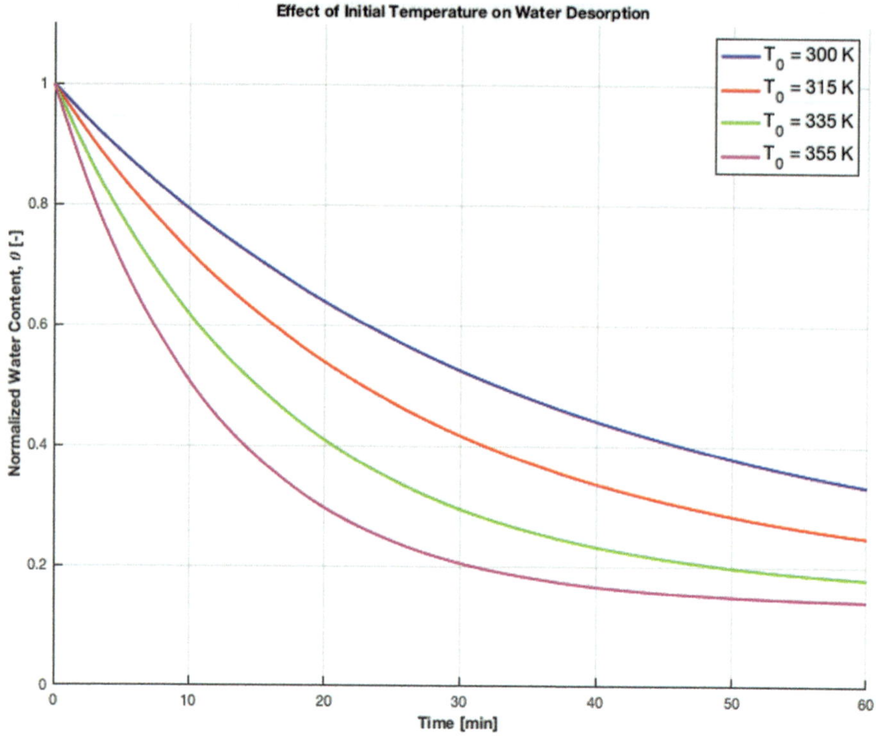

Figure 4.16 Normalized water content θ over time at different temperatures.

Overall, this chapter offers a foundational platform for advanced study and simulation of porous media systems. The techniques and models introduced here will be further implemented and extended in the subsequent chapters to address more complex multiphase and reactive transport phenomena.

REVIEW QUESTIONS

1. Describe how MATLAB®'s `pdepe` function is adapted to handle highly nonlinear PDEs in porous media, such as in two-phase flow simulations.
2. What are the primary differences between parabolic and elliptic PDEs, and how does MATLAB®'s `pdepe` function handle them?
3. Discuss how the transport of nanoparticles in porous media is modeled, particularly focusing on the interactions between nanoparticles and the solid matrix.
4. What is the significance of the adaptive time-stepping algorithm in simulations involving rapid changes in the solution, and how does it contribute to accuracy?

5. How is the water saturation equation formulated in two-phase flow models for enhanced oil recovery (EOR), and what factors influence the flow of water and oil?
6. Explain how atmospheric water harvesting using desiccant materials is modeled in MATLAB®, including the key factors that govern the adsorption and desorption processes.
7. What are the boundary and initial conditions typically required when solving transport problems in porous media using MATLAB's pdepe function?

PROBLEMS

Problem 1: Fluid is flowing through a fractured porous medium of length $L = 50$ units. The pressure at the inlet is constant, while at the outlet, the flow is zero:

$$p(0,t) = p_0, \quad \frac{\partial p}{\partial x}(L,t) = 0 \tag{4.111}$$

The fluid flow is governed by Darcy's law:

$$\frac{\partial p}{\partial t} = \frac{K}{\mu}\frac{\partial^2 p}{\partial x^2} \tag{4.112}$$

where K is the permeability of the medium, and μ is the fluid viscosity.

Problem 2: A contaminant is being adsorbed in a soil column of length $L = 15$ units. The contaminant concentration at the inlet is maintained at a constant value, while there is no contaminant at the outlet:

$$c(0,t) = c_0, \quad c(L,t) = 0 \tag{4.113}$$

The transport and adsorption of the contaminant are modeled by the following advection-diffusion equation with a reaction term:

$$\phi\frac{\partial c}{\partial t} + v\frac{\partial c}{\partial x} = \frac{\partial}{\partial x}\left(D\frac{\partial c}{\partial x}\right) - \lambda c \tag{4.114}$$

where v is the flow velocity, D is the dispersion coefficient, and λ is the adsorption rate constant.

Problem 3: Consider a water injection process for enhanced oil recovery in a porous reservoir. The water saturation in the porous medium is governed by the following equation:

$$\frac{\partial S_w}{\partial t} = \frac{\partial}{\partial x}\left(\lambda_w\frac{\partial p}{\partial x}\right) \tag{4.115}$$

where S_w is the water saturation, λ_w is the water mobility, and p is the pressure. The boundary conditions are:

$$S_w(0,t) = 1, \quad \frac{\partial S_w}{\partial x}(L,t) = 0 \tag{4.116}$$

Here, L is the length of the reservoir, and water is injected from the left boundary.

Problem 4: Nanoparticles are being injected into a porous medium of length $L = 100$ units. The nanoparticle concentration at the inlet is constant, while the outlet is no-flow:

$$c(0,t) = c_0, \quad \frac{\partial c}{\partial x}(L,t) = 0 \tag{4.117}$$

The transport of nanoparticles is modeled by the following advection-diffusion equation:

$$\phi \frac{\partial c}{\partial t} + \frac{\partial}{\partial x}(uc) = \frac{\partial}{\partial x}\left(\phi D \frac{\partial c}{\partial x}\right) \tag{4.118}$$

where u is the fluid velocity, D is the diffusion coefficient, and ϕ is the porosity.

Problem 5: Consider heat transfer with phase change in a porous medium during ice formation. The domain length is $L = 10$ units. The initial temperature is below freezing:

$$T(x,0) = T_0 \quad \text{where} \quad T_0 < T_f \tag{4.119}$$

The boundary condition is constant temperature at both ends:

$$T(0,t) = T_f, \quad T(L,t) = T_f \tag{4.120}$$

The heat transfer equation with phase change is:

$$\phi \rho C_p \frac{\partial T}{\partial t} + \rho L \frac{\partial f}{\partial t} = k \frac{\partial^2 T}{\partial x^2} \tag{4.121}$$

where f is the phase fraction, L is the latent heat, and T_f is the freezing temperature.

Problem 6: CO_2 is being injected into a deep saline aquifer, where it dissolves in the brine. The CO_2 concentration in the brine is governed by the following equation:

$$\phi \frac{\partial c}{\partial t} = D \frac{\partial^2 c}{\partial x^2} - R(c) \tag{4.122}$$

where $R(c)$ is a reaction term that accounts for the dissolution and precipitation of CO_2, and D is the diffusion coefficient. The boundary conditions are:

$$c(0,t) = c_0, \quad \frac{\partial c}{\partial x}(L,t) = 0 \tag{4.123}$$

where $L = 50$ units is the length of the aquifer.

Problem 7: Use MRST to simulate the pressure distribution in a Cartesian grid with dimensions $50 \times 50 \times 1$ and physical dimensions of $500\,\text{m} \times 500\,\text{m} \times 10\,\text{m}$. Assume the following properties: Permeability: 100 mD, Porosity: 0.2, Fluid viscosity: 1 cP, Fluid density: $1000\,\text{kg/m}^3$. Visualize the pressure distribution.

Problem 8: Modify the provided MRST simulation to compute and visualize the velocity vectors for a grid with 20 × 20 cells. Use Dirichlet boundary conditions with injection pressure of 5 bar at the left boundary, and outlet pressure of 0 bar at the right boundary.

Problem 9: For a 1D grid of 10 cells with uniform properties and a steady-state single-phase flow, analytically calculate the pressure drop across the grid using Darcy's law:

$$\mathbf{u} = -\frac{\mathbf{k}}{\mu}\nabla p, \tag{4.124}$$

Compare the results with MRST's numerical solution and highlight any differences, discussing potential reasons for discrepancies.

DECLARATION

During the preparation of this work, the author used ChatGPT in order to improve the language and readability. After using this tool/service, the author reviewed and edited the content as needed and takes full responsibility for the content of the publication.

5 Two-Phase Fluid Flow in Porous Media

5.1 INTRODUCTION

Multiphase flow in porous media underpins a wide range of subsurface engineering applications, such as petroleum recovery, CO_2 sequestration, and groundwater remediation—by describing how two immiscible fluids (e.g. water and oil, or gas and liquid) migrate through a porous rock matrix [74]. Extending Darcy's law to each phase requires introducing saturation-dependent relative permeabilities $k_{r,\alpha}(S)$, which capture the reduction in effective permeability due to the presence of another fluid, and capillary pressure $p_c(S)$, which embodies interfacial forces arising from pore geometry and fluid–fluid tension [75, 76, 77]. These constitutive functions give rise to hysteresis during drainage and imbibition, a hallmark of two-phase displacement in porous media [78].

A rigorous volume-averaging of the pore-scale mass and momentum equations yields macroscopic balance laws for each phase mass conservation coupled with a multiphase form of Darcy's law, forming the mathematical backbone of modern reservoir simulators [79, 80]. Over the past decades, empirical models (Corey, Brooks–Corey, Van Genuchten, Mualem, Stone) have been developed to represent wettability and pore-size effects [81, 82], while numerical schemes, from fully implicit solvers to IMPES and operator-splitting approaches, have matured to tackle the nonlinear coupling between pressure and saturation fields [86, 87].

More recently, the exigencies of CO_2 storage, hydrogen injection, and reactive transport have driven extensions to compositional, thermal, geomechanical, and geochemical frameworks [84, 83]. In this chapter, we review the fundamental definitions (porosity, saturation, capillary pressure, relative permeability), present the governing equations for immiscible two-phase flow, and survey classical constitutive models. We then examine numerical solution strategies, illustrate countercurrent imbibition in fractured systems, and introduce extensions to miscible and multicomponent flows, laying the groundwork for advanced multiphase simulations in porous media.

SATURATION

A phase is defined as a chemically homogeneous portion of a system under consideration that is separated from other portions by a definite physical boundary. In the case of a single-phase system, the void space of the porous medium is filled by a single fluid (e.g., water) or by several fluids completely miscible with each other (e.g., fresh water and saltwater).

DOI: 10.1201/9781032724973-5

In porous media with multiple flowing phases, saturation quantifies the volume fraction of a fluid phase within the pores, given by:

$$S_\alpha = \frac{V_\alpha}{V_{\text{pores}}} \tag{5.1}$$

The sum of phase saturations in a system must equal one.

$$\sum_\alpha S_\alpha = 1 \tag{5.2}$$

Specifically, in a two-phase system involving water and oil, the saturation of each phase quantifies how much of the pore space is occupied by that particular fluid. For a two-phase water–oil system, this constraint is expressed as:

$$S_{\text{water}} + S_{\text{oil}} = 1 \tag{5.3}$$

Porous media, such as reservoir rocks, consist of a solid framework (rock volume) and void spaces filled with fluids. As shown in Figure 5.7, the rock volume, while the pores are occupied by oil and water, illustrating the distribution of fluid saturations. This conceptual model is fundamental for understanding multiphase flow in petroleum reservoirs, where porosity and fluid saturations critically influence storage and transport properties.

CAPILLARY PRESSURE

Capillary pressure refers to the pressure differential across the interface of two immiscible fluids within a confined capillary space. This phenomenon is crucial in understanding fluid dynamics in porous media. Capillary pressure is defined as the curved interface between two immiscible fluids in a capillary

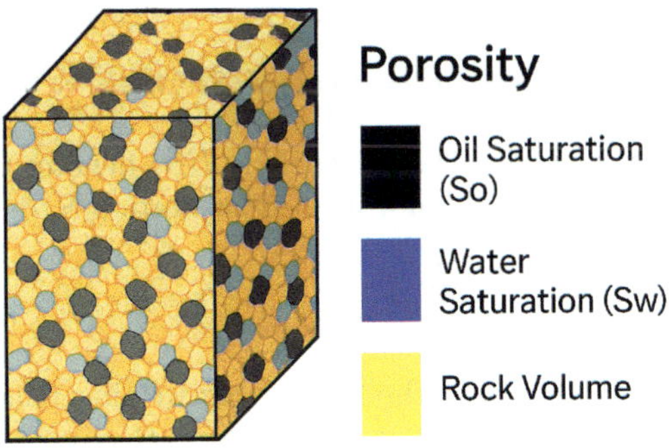

Figure 5.1 Schematic representation of porous rock volume showing oil saturation (So), water saturation (Sw), and rock matrix.

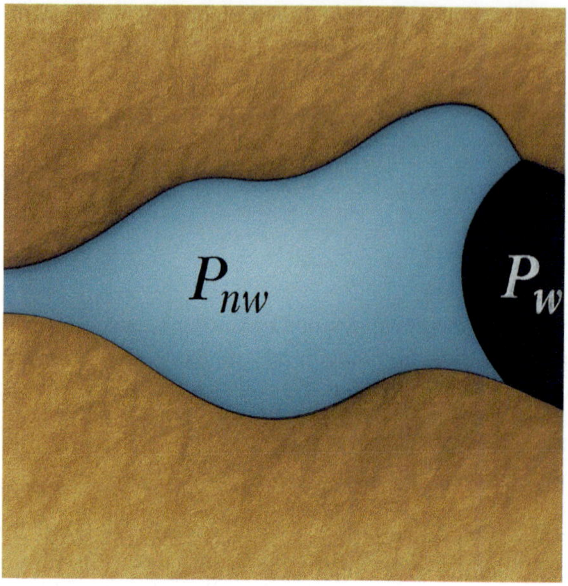

Figure 5.2 Illustration of capillary pressure in porous media (pore-scale geometry).

tube (or pore in the case of porous media, see Figure 5.2). This pressure differential is represented as P_c, given by:

$$P_c = P_{nw} - P_w \tag{5.4}$$

where P_{nw} represents the pressure of the non-wetting phase, and P_w represents the pressure of the wetting phase.

Capillary pressure in porous media depends on the geometry of the pores, fluid properties, and phase saturations. It can be mathematically represented as:

$$P_c = \gamma\sqrt{\frac{\phi}{K}}J(S) \tag{5.5}$$

In this equation, γ is the interfacial tension, $J(S)$ denotes the Leverett function, which is a function of the normalized saturation S, defined by:

$$S = \frac{S_w - S_{iw}}{1 - S_{nwr} - S_{iw}}, \quad 0 \leq S \leq 1 \tag{5.6}$$

where S_{iw} and S_{nwr} represent the irreducible wetting phase saturation and the residual non-wetting phase saturation, respectively.

The Leverett function, denoted as $J(S)$, typically spans a range defined by two distinct processes: drainage and imbibition, as illustrated in

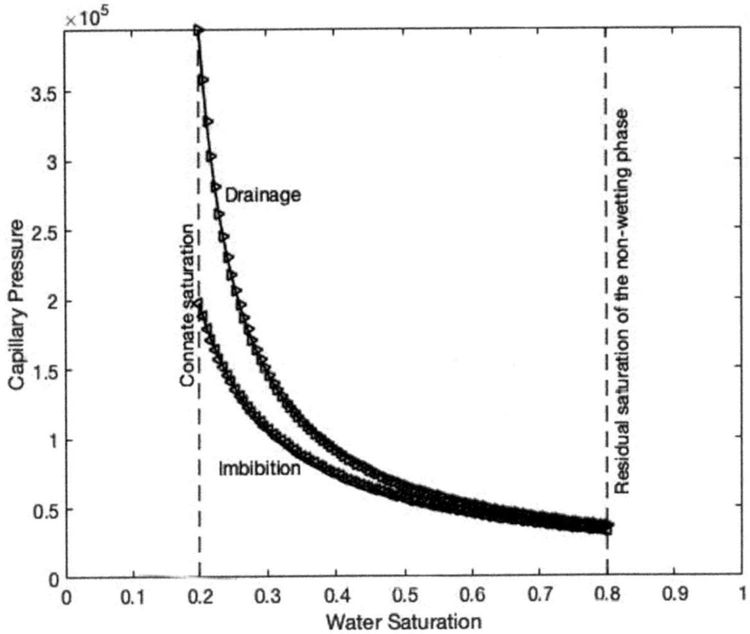

Figure 5.3 Capillary pressure as a function of water saturation illustrating the drainage and imbibition processes.

Figure 5.3 Various empirical models have been developed to characterize capillary pressure, with each model's applicability often tailored to specific types of applications in the field.

This function typically lies between two limits: *drainage* and *imbibition*, representing different phases of fluid displacement. In EOR processes such as water-nanoparticle injection, the goal is to alter the medium's wettability from oil-wet to water-wet, enhanced oil recovery. For mixed-wet systems, the capillary pressure correlation incorporates both drainage and imbibition processes, reflecting the wettability changes due to nanoparticles injection. One of the generalized capillary pressure correlations is given by [85]:

$$P_c = c_w \left(\frac{1 - S_{wr}}{S_w - S_{wr}} \right)^{a_w} + c_o \left(\frac{1 - S_{or}}{S_o - S_{or}} \right)^{a_o} \tag{5.7}$$

where c_w is the entry pressure for water during imbibition, c_o is the entry pressure for oil during drainage, S_{wr} and S_{or} are the residual saturations for water and oil, a_w and a_o are pore size distribution indices for water and oil.

RELATIVE PERMEABILITY

Relative permeability is a dimensionless measure of the permeability of porous media to one fluid phase, such as water or oil, in the presence of other fluid phases. It reflects how easily a fluid can flow through a porous material compared to its flow when the material is saturated with that fluid alone.

For the wetting phase, the relative permeability, denoted by k_{rw}, is described as a function of the normalized saturation S, given by the equation:

$$k_{rw} = k_{rw}^{0} S^{a} \tag{5.8}$$

For the non-wetting phase, the relative permeability, k_{rn}, is expressed as:

$$k_{rn} = k_{rn}^{0}(1 - S)^{b} \tag{5.9}$$

where k_{rw}^{0} and k_{rn}^{0} represent the end-point relative permeabilities of the wetting and non-wetting phases, respectively, corresponding to the saturation of $S = 1$ and $S = 0$ (see Figure 5.4).

Drainage curves represent the process where the non-wetting phase displaces the wetting phase, while imbibition curves depict the reverse, where the wetting phase displaces the non-wetting phase. These curves exhibit hysteresis effects due to factors such as phase trapping, irreversible changes in fluid configurations, and wettability influences. Residual saturations, marked on these curves, indicate the trapping efficiency of the phases, which is essential for understanding long-term storage or recovery dynamics.

Various models have been developed to represent this relationship for wetting and non-wetting phases:

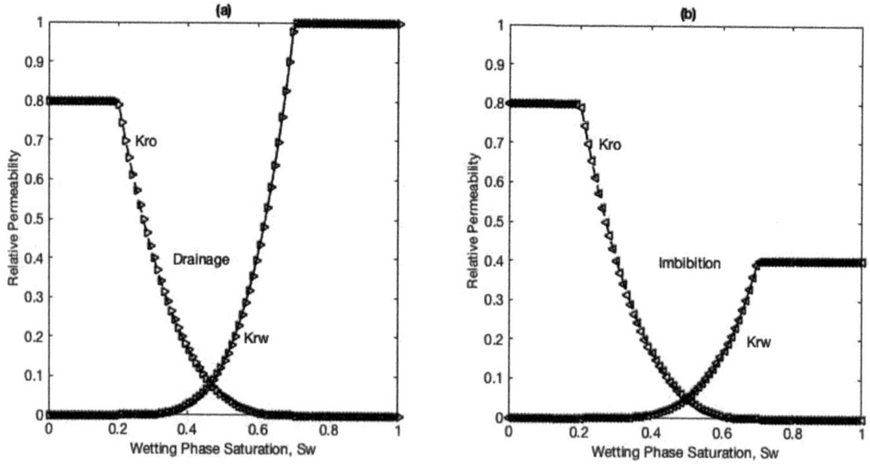

Figure 5.4 Relative permeabilities: (a) drainage and (b) imbibition.

Corey Model

$$k_{r,\alpha} = k_{r,\alpha}^{\text{end}} \left(\frac{S_\alpha - S_\alpha^{\text{res}}}{1 - S_{\text{w}}^{\text{res}} - S_{\text{nw}}^{\text{res}}} \right)^{n_\alpha} \tag{5.10}$$

This is a simple, power-law relationship suitable for homogeneous reservoirs or porous systems.

Van Genuchten-Mualem Model

$$k_{r,\text{w}} = S_{\text{e}}^{l}, \quad k_{r,\text{nw}} = (1 - S_{\text{e}})^2 (1 - S_{\text{e}}^{1/l}), \tag{5.11}$$

where S_{e} is the normalized saturation. This model provides smooth transitions and is ideal for heterogeneous porous media, particularly in systems where capillary effects are significant.

Brooks-Corey Model

$$k_{r,\text{w}} = S_{\text{e}}^{3+2/\lambda}, \quad k_{r,\text{nw}} = (1 - S_{\text{e}})^2 (1 - S_{\text{e}}^{1+2/\lambda}), \tag{5.12}$$

This model incorporates pore size distribution effects, making it suitable for capillary-dominated flows where pore geometry strongly influences flow behavior.

Hysteresis Models

$$k_{r,\alpha}^{\text{imb}} = k_{r,\alpha}^{\text{drain}} \left(1 - \beta \left(1 - \frac{S_\alpha^{\text{res}}}{S_\alpha^{\text{max}}} \right) \right), \tag{5.13}$$

Hysteresis models capture the differences between drainage (displacement of the wetting phase by the non-wetting phase) and imbibition (displacement of the non-wetting phase by the wetting phase). These models are essential for scenarios involving cyclic flow or trapping mechanisms.

Stone's Model

$$k_{r,\text{nw}} = k_{r,\text{nw}}^{\text{end}} \cdot (1 - S_{\text{w}}^{\text{eff}})^n, \tag{5.14}$$

Stone's model is widely used for systems with three-phase flow but can also be adapted to two-phase wetting and non-wetting systems.

WETTABILITY

Wettability is a critical property of porous media that dictates the interaction between the rock surface and immiscible fluid phases. It is defined as the tendency of one fluid to adhere to a solid surface in the presence of other immiscible fluids and is categorized as follows:

Wetting phase preference: The rock prefers the wetting phase, which spreads on its surface ($\theta < 90°$).

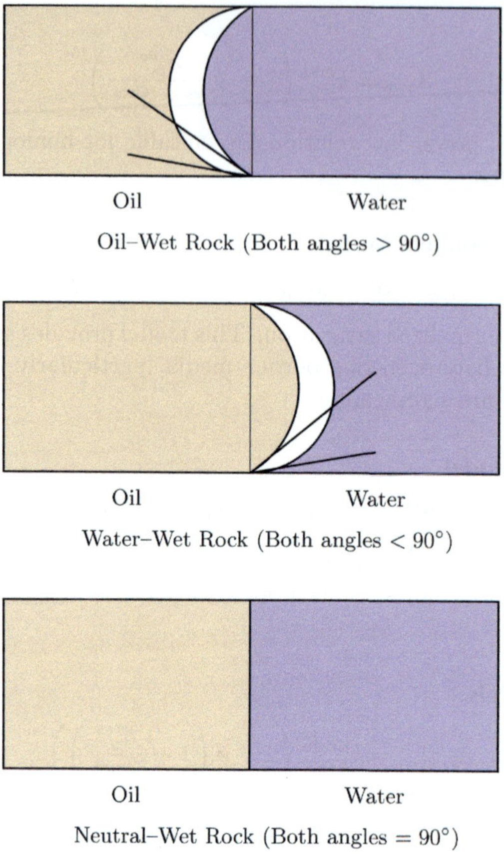

Figure 5.5 Illustration of wettability scenarios in a porous medium. The top draw represents an oil-wet rock, where both contact angles are greater than 90°. The middle drawing shows a water-wet rock, where both contact angles are less than 90°. The bottom diagram illustrates an intermediate–wet rock, where both contact angles equal 90°.

Non-wetting phase preference: The rock prefers the non-wetting phase, which coats the surface ($\theta > 90°$).
Intermediate-wet: The rock exhibits no strong preference for either phase ($\theta \approx 90°$).

Wettability influences fluid distribution, capillary trapping mechanisms, and relative permeability curves. It also affects the dynamics of fluid movement, with stronger wetting preferences enhancing capillary trapping of the non-wetting phase. Figure 5.5 visually demonstrates the concept of wettability in porous media, showcasing three distinct cases.

HYSTERESIS

Hysteresis in multiphase flow arises from differences in fluid configurations during drainage and imbibition processes. This behavior is influenced by:

Trapping of the non-wetting phase in pore spaces.
Irreversible changes in fluid distribution due to pore geometry.
Wettability effects at the fluid-solid interface.

Hysteresis enhances residual trapping, which is particularly critical for applications such as storage of non-wetting phases or enhanced recovery of wetting phases. Understanding these effects is essential for predicting flow behavior in porous media systems.

5.1.1 IMMISCIBLE TWO-PHASE FLOW

The two-phase flow model in porous media of two immiscible fluids, typically a wetting phase (e.g., water) and a non-wetting phase (e.g., oil or gas). This model is governed by several key equations that describe the conservation of mass, momentum, and the interactions between the two phases. The saturation equation ensures the conservation of mass for each fluid phase within the porous medium. It relates the rate of change of phase saturation to the divergence of the phase flux and any sources or sinks within the system:

$$\phi \frac{\partial(\rho_\alpha S_\alpha)}{\partial t} + \nabla \cdot (\rho_\alpha \mathbf{u}_\alpha) = \rho_\alpha q_\alpha, \quad \alpha = w, n \tag{5.15}$$

Here, ϕ is the porosity of the medium, ρ_α is the density of the phase α (where $\alpha = w$ for the wetting phase and $\alpha = n$ for the non-wetting phase), S_α is the saturation, $\mathbf{u}_\alpha$ is the phase velocity, and q_α represents external sources or sinks, such as injection or production wells. This equation states that the change in the mass of a fluid phase within a volume of porous medium over time is equal to the net flux of the fluid into or out of that volume, plus any sources or sinks

Darcy's law governs the flow of each fluid phase through the porous medium, driven by pressure gradients and gravitational forces. It describes the momentum balance for each phase:

$$\mathbf{u}_\alpha = -\frac{k_{r\alpha}}{\mu_\alpha} K \nabla p_\alpha - \rho_\alpha g z, \quad \alpha = w, n \tag{5.16}$$

In this equation, $k_{r\alpha}$ is the relative permeability of phase α, μ_α is the viscosity, K is the absolute permeability of the medium, ∇p_α is the pressure gradient, g is the acceleration due to gravity, and z is the vertical coordinate. This law highlights how the flow of each phase is influenced by the medium's permeability, the fluid's properties, and the pressure and gravitational fields.

Darcy's law states that the velocity of a fluid phase through a porous medium is proportional to the permeability and the pressure gradient, and inversely proportional to the viscosity. The gravitational term accounts for the buoyancy forces acting on the fluid.

The saturation constraint ensures that the sum of the saturations of the two phases at any point in the porous medium equals unity, reflecting the physical reality that the pore space is fully occupied by the fluids:

$$S_w + S_n = 1 \tag{5.17}$$

This constraint is fundamental to the two-phase flow model, as it couples the equations for the two phases.

Capillary pressure, p_c, arises due to the interfacial tension between the two immiscible fluids. It is defined as the difference between the pressures of the non-wetting and wetting phases:

$$p_c(S_w) = p_n - p_w \tag{5.18}$$

The capillary pressure is a function of the wetting phase saturation S_w and plays a critical role in determining the distribution of the two fluids within the porous medium.

Phase mobility, $\lambda_\alpha(S)$, represents the ease with which a particular phase can move through the porous medium. It is defined as the ratio of the phase's relative permeability to its viscosity:

$$\lambda_\alpha(S) = \frac{k_{r\alpha}}{\mu_\alpha}, \quad \alpha = w, n \tag{5.19}$$

Higher mobility indicates that the phase can flow more readily, which is influenced by the saturation-dependent relative permeability and the fluid's viscosity.

Total mobility, $\lambda_t(S)$, is the sum of the mobilities of both phases. It is an important quantity in multiphase flow as it affects the overall flow dynamics within the porous medium:

$$\lambda_t(S) = \lambda_w + \lambda_n \tag{5.20}$$

This total mobility is used to evaluate the combined influence of both phases on the flow properties.

Fractional flow, $f_\alpha(S)$, represents the fraction of the total flow that is attributable to each phase. It is defined as the ratio of the phase's mobility to the total mobility:

$$f_\alpha(S) = \frac{\lambda_\alpha}{\lambda_t}, \quad \alpha = w, n \tag{5.21}$$

The fractional flow is a key concept in understanding the distribution of flow between the two phases and is essential for predicting the behavior of two-phase flow in porous media, such as during oil recovery processes.

Together, these equations provide a comprehensive mathematical framework for modeling two-phase flow in porous media, capturing the essential physics that governs the interactions between the two immiscible fluids, their movement through the medium, and the impact of external forces such as pressure gradients and gravity.

The total velocity $\mathbf{u}_t = \mathbf{u}_w + \mathbf{u}_n$ is derived from Darcy's law, and by assuming incompressible fluids and applying mass conservation, the pressure equation is obtained. The saturation equation is then solved using the explicit approach, while the pressure equation is solved implicitly.

We begin by defining the total velocity $\mathbf{u}_t = \mathbf{u}_w + \mathbf{u}_n$ by utilizing Eq. (5.15), along with the algebraic constraints. This leads to the following expression:

$$\mathbf{u}_t = -K\left(\lambda_t(s)\nabla p - \lambda_w(s)\nabla p_c - (\lambda_w\rho_w + \lambda_n\rho_n)g\nabla z\right) \tag{5.22}$$

Here, $\mathbf{u}_t$ represents the combined velocity of both phases (wetting and non-wetting) within the porous medium. The first term, $\lambda_t(s)\nabla p$, captures the effect of the pressure gradient on the total mobility of the fluids. The second term, $\lambda_w(s)\nabla p_c$, accounts for the influence of the capillary pressure gradient. The last term, $(\lambda_w\rho_w + \lambda_n\rho_n)g\nabla z$, represents the impact of gravitational forces due to the density differences between the two fluids.

Assuming that the fluids are incompressible (i.e., the densities ρ_w and ρ_n are constant), we sum the two mass conservation equations given by Eq. (5.15), while recognizing that $s_w + s_n = 1$. This yields:

$$\nabla \cdot \mathbf{u}_t = q_t = q_w + q_n \tag{5.23}$$

This equation indicates that the divergence of the total velocity $\mathbf{u}_t$ is equal to the combined source terms for both the wetting and non-wetting phases.

Substituting Eq. (5.22) into Eq. (5.23), we derive the pressure equation as follows:

$$-\nabla \cdot (K\lambda_t(s)\nabla p) = q_t - \nabla \cdot \left(K\left(\lambda_w(s)\nabla p_c + (\lambda_w\rho_w + \lambda_n\rho_n)g\nabla z\right)\right) \tag{5.24}$$

This pressure equation encapsulates the combined effects of pressure gradients, capillary pressure, and gravitational forces on fluid flow within the porous medium. The left-hand side represents the diffusion of pressure, while the right-hand side includes contributions from source terms and the effects of capillary pressure and gravity.

5.2 COUNTERCURRENT IMBIBITION

In the context of enhanced oil recovery (EOR), particularly in fractured reservoirs, imbibition is a critical process by which oil is displaced by water. This occurs predominantly in water-wet fractured-matrix reservoirs. Imbibition can be classified into two types: cocurrent and countercurrent. In cocurrent

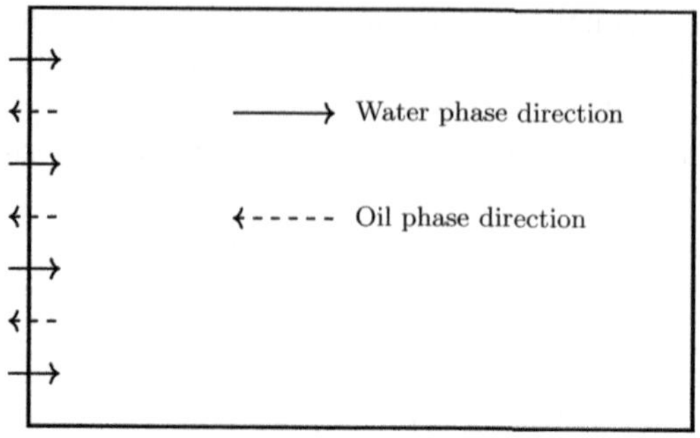

Figure 5.6 Schematic diagram of countercurrent imbibition.

imbibition, both water and oil flow in the same direction, with water displacing oil out of the matrix. In contrast, countercurrent imbibition occurs when water and oil flow in opposite directions, with water displacing oil into the fractures. In such cases, capillary forces dominate the process (see Figure 5.6).

In countercurrent imbibition, the sum of the velocities of the wetting (water) and non-wetting (oil) phases is zero:

$$u_t = u_w + u_o = 0 \tag{5.25}$$

Substituting this condition into the capillary pressure equation, we obtain:

$$\frac{\partial p_w}{\partial z} = -f_w \frac{\partial p_c}{\partial z} - \chi g \tag{5.26}$$

where $f_w = \frac{\lambda_w}{\lambda_t}$ and $f_o = \frac{\lambda_o}{\lambda_t}$ are the fractional flow functions of water and oil, $\lambda_w = \frac{k_{rw}}{\mu_w}$ and $\lambda_o = \frac{k_{ro}}{\mu_o}$ are the mobilities of the water and oil phases, $\lambda_t = \lambda_w + \lambda_o$ is the total mobility, $\chi = \frac{\rho_w \lambda_w + \rho_o \lambda_o}{\lambda_t}$, $\Delta\rho = \rho_w - \rho_o$ is the density difference between the water and oil. The velocity of the water phase can be expressed as:

$$u_w = K\lambda_w f_o \left(\frac{\partial p_c}{\partial z} - \Delta\rho g \right) \tag{5.27}$$

where K is the permeability, and the term $\Delta\rho g$ represents the gravitational forces. The saturation equation for the water phase then becomes:

$$\phi \frac{\partial S_w}{\partial t} + \frac{\partial}{\partial z} \left[K\lambda_w f_o \left(\frac{\partial p_c}{\partial z} - \Delta\rho g \right) \right] = 0 \tag{5.28}$$

This equation governs the transport of the water phase during the imbibition process, influenced by capillary pressure, fluid mobilities, and gravity.

The relative permeabilities of the water and oil phases are functions of the normalized saturation S. For the water and oil phases, the relative permeabilities can be expressed as:

$$k_{rw} = k_{rw}^0 S^a, \quad k_{ro} = k_{ro}^0 (1 - S)^b \tag{5.29}$$

where k_{rw}^0 and k_{ro}^0 are the relative permeabilities at the endpoints, S_{iw} is the irreducible water saturation, S_{ro} is the residual oil saturation, and a and b are positive constants representing the exponents for the water and oil phases.

MATLAB® Implementation

MATLAB's pdepe function is used to solve the nonlinear partial differential equations using the Galerkin method for spatial discretization and an adaptive time-stepping algorithm for time integration. The main function is defined as pdepe, which includes:

pdex4pde: Solves the equations.
pdex4ic: Defines the initial conditions.
pdex4bc: Defines the boundary conditions.

```
function OneD_TwoPhase_Imbibition
    clc;
    clear all;
    close all;

    % Coordinates Type: Cartesian
    m = 0;

    % Space and Time Discretization
    x = linspace(0, 0.2, 200);                                  %
        Spatial discretization over 0.2 m
    t = linspace(0, 200*24*3600, 200*24*60);                    % Time
         discretization over 200 days

    % Solving the system of PDEs using pdepe
    sol = pdepe(m, @pdex4pde, @pdex4ic, @pdex4bc, x, t);

    % Extract results: Water saturation (Sw)
    Sw  = sol(:,:,1);

    % Visualization
    % Plot for Water Saturation
    figure(1);
    plot(x, Sw(20*60,:), 'b', 'LineWidth', 2); hold on;
    plot(x, Sw(100*60,:), 'k', 'LineWidth', 2); hold on;
    plot(x, Sw(200*60,:), 'r', 'LineWidth', 2); hold off;
    legend('20␣hrs', '100␣hrs', '200␣hrs');
    title('Water␣Saturation␣Profile', 'FontSize', 14);
    xlabel('x␣[m]');
    ylabel('S_w');

end

% Solving the PDEs for two-phase water-oil flow
function [c, f, s] = pdex4pde(x, t, u, DuDx)

    % Physical Parameters
    phi0 = 0.3;          % Initial porosity
```

```
k0 = 0.02E-12;          % Absolute permeability
nn = 3.0;               % Permeability exponent
Sor = 0.001;            % Residual oil saturation
Swr = 0.001;            % Irreducible water saturation
a_w = 0.5;              % Capillary pressure exponent for water
    phase
a_o = 0.5;              % Capillary pressure exponent for oil
    phase
c_w = 500;              % Capillary pressure entry for water phase
c_o = -500;             % Capillary pressure entry for oil phase
krw0 = 1;               % Water phase relative permeability at
    endpoint
kro0 = 1;               % Oil phase relative permeability at
    endpoint
mu_w = 0.001;           % Water viscosity
mu_o = 0.001;           % Oil viscosity
a = 4;                  % Water relative permeability exponent
b = 4;                  % Oil relative permeability exponent
g = 9.80;               % Gravitational acceleration
rho_o = 660;            % Oil density
rho_w = 1000;           % Water density
drho = rho_w - rho_o; % Density difference between water and
    oil

% Porosity and permeability variation
phi = phi0;  % Assuming no change in porosity since
    nanoparticles are removed
k = k0;      % Constant permeability as nanoparticles are not
     involved

% Normalized Saturation
S_w = (u(1) - Swr) / (1 - Swr - Sor);

% Relative Permeability Variation
krw = krw0 * S_w .^ a;
kro = kro0 * (1 - S_w) .^ b;

% Mobility
lamw_p = krw / mu_w;
lamo_p = kro / mu_o;
lamt_p = lamw_p + lamo_p;
mobility = lamw_p .* lamo_p ./ lamt_p;

% Capillary Pressure and Derivative
p_c = c_w / ((u(1) - Swr) / (1 - Swr)) .^ a_w + c_o / ((1 - u
    (1) - Sor) / (1 - Sor)) .^ a_o;
dp_c_dx = DuDx(1) * (-a_w * c_w / (1 - Swr) / ((u(1) - Swr) /
     (1 - Swr)) .^ (1 + a_w) ...
                  + a_o * c_o / (1 - Sor) / ((1 - u(1) - Sor
                      ) / (1 - Sor)) .^ (1 + a_o));

% Water velocity
v_w = k .* mobility .* (drho * g + dp_c_dx);

% Coefficients for PDEs
c1 = phi;
```

```
    f1 = k .* mobility .* (drho * g - dp_c_dx);
    s1 = 0;

    % PDE system: water phase saturation only
    c = c1;
    f = f1;
    s = s1;
end

% Initial Conditions for water saturation
function u0 = pdex4ic(x)
    u0 = 0.01;     % Initial water saturation
end

% Boundary Conditions
function [pl, ql, pr, qr] = pdex4bc(xl, ul, xr, ur, t)
    pl = ul(1) - 0.99;  % Left boundary condition for water
        saturation
    ql = 0;
    pr = 0;             % Right boundary condition (zero flux)
    qr = 1;
end
```

Figure 5.7 shows the variation of saturation profiles over a distance x for three different times: after 1 day, 10 days, and 40 days. The results demonstrate that as time progresses, the saturation front moves further into the domain, which indicates the spread of fluid saturation over time. The initial

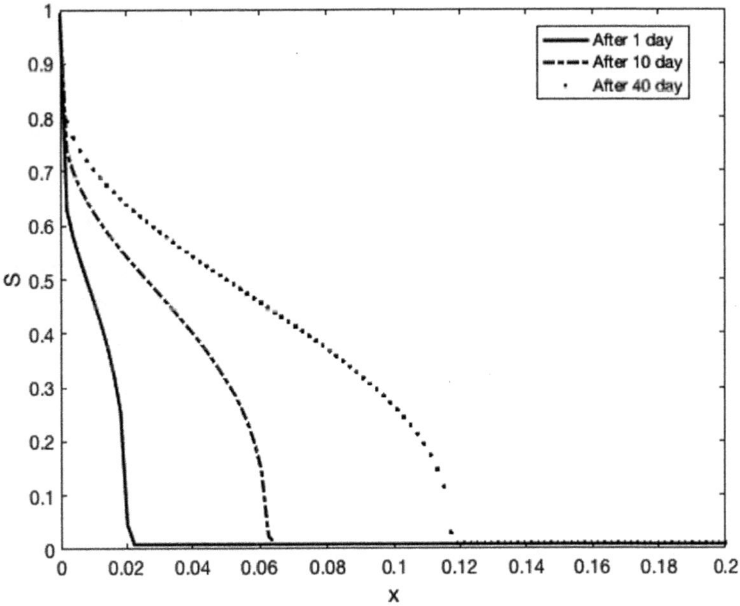

Figure 5.7 Saturation profiles after different time periods.

high saturation at the inlet boundary diminishes as the distance increases. The curves highlight how fluid displacement occurs in a porous medium, with the front becoming less steep as time increases.

5.2.1 COMPRESSIBLE IMMISCIBLE TWO-PHASE FLOW

The flow of compressible, immiscible fluids through a porous medium is governed by a set of equations that describe the conservation of mass for each fluid phase, the momentum of the fluids, and the interaction between the phases through capillary pressure and saturation. The main governing equations include the mass conservation equations, Darcy's law for each phase, and the capillary pressure-saturation relationship.

5.2.1.1 Fluid Compressibility

The compressibility of the fluid phases is represented by an equation of state (EOS), which relates the density of the fluid to its pressure. The fluid compressibility C_{f_α} for each phase is defined as:

$$C_{f_\alpha} = \frac{1}{\rho_\alpha}\left(\frac{\partial \rho_\alpha}{\partial P_\alpha}\right)_T, \qquad \alpha = n, w \tag{5.30}$$

This equation indicates how the density of a fluid phase changes with pressure. If the fluid compressibility is constant, the pressure-density relationship can be approximated as:

$$\rho_\alpha = \rho_\alpha^R e^{C_{f,\alpha}(P_\alpha - P_R)} \approx \rho_\alpha^R \left(1 + C_{f,\alpha}(P_\alpha - P_R)\right) \tag{5.31}$$

where ρ_α^R is the reference density at a reference pressure P_R. $C_{f,\alpha}$ is the compressibility of phase α. P_R is the reference pressure. This relationship is essential for modeling the behavior of compressible fluids under varying pressures.

5.2.1.2 Rock Compressibility

The compressibility of the porous medium itself is also considered, which accounts for changes in porosity as pressure changes:

$$C_R = \frac{1}{\phi}\left(\frac{\partial \phi}{\partial P}\right)_T \tag{5.32}$$

If the compressibility C_R is constant, the porosity as a function of pressure can be expressed as:

$$\phi = \phi_R e^{C_R(P - P_R)} \approx \phi_R \left(1 + C_R(P - P_R)\right) \tag{5.33}$$

where ϕ_R is the reference porosity at the reference pressure P_R. C_R is the compressibility of the porous medium. This equation helps in understanding how the pore space within the medium might expand or contract with changes in pressure.

5.2.1.3 Pressure Equation

The pressure equation for the fluid phases combines the effects of fluid and rock compressibility, expressed as:

$$\frac{\partial(\phi\rho_\alpha S_\alpha)}{\partial t} = \left(\phi\left(\frac{\partial\rho_\alpha}{\partial P}\right)_T + \rho_\alpha\left(\frac{\partial\phi}{\partial P}\right)_T\right)\frac{\partial P}{\partial t}S_\alpha + \phi\rho_\alpha\frac{\partial S_\alpha}{\partial t} \tag{5.34}$$

This equation can be rewritten as:

$$\rho_\alpha\phi\left(C_{f,\alpha} + C_R\right)\frac{\partial P}{\partial t}S_\alpha + \phi\rho_\alpha\frac{\partial S_\alpha}{\partial t} = \hat{q}_\alpha - \nabla\cdot(\rho_\alpha\mathbf{u}_\alpha) \tag{5.35}$$

Thus, the equation becomes:

$$\phi\left(C_{f,\alpha} + C_R\right)S_\alpha\frac{\partial P}{\partial t} + \phi\frac{\partial S_\alpha}{\partial t} = q_\alpha - \frac{1}{\rho_\alpha}\nabla\cdot(\rho_\alpha\mathbf{u}_\alpha) \tag{5.36}$$

When summing over all phases:

$$\phi\left(C_{f,w}S_w + C_{f,n}S_n + C_R\right)\frac{\partial P}{\partial t} + \sum_\alpha\frac{1}{\rho_\alpha}\nabla\cdot(\rho_\alpha\mathbf{u}_\alpha) = q_t \tag{5.37}$$

Therefore:

$$\begin{aligned}\phi C_t\frac{\partial P}{\partial t} - \frac{1}{\rho_w}\nabla\cdot(\rho_w\lambda_w K\nabla P_w) - \frac{1}{\rho_n}\nabla\cdot(\rho_n\lambda_n K\nabla P_n)\\ + \frac{1}{\rho_w}\nabla\cdot(\rho_w^2\lambda_w K\mathbf{g}) + \frac{1}{\rho_n}\nabla\cdot(\rho_n^2\lambda_n K\mathbf{g}) = q_t\end{aligned}$$

Alternatively:

$$\begin{aligned}\phi C_t\frac{\partial P}{\partial t} - \frac{1}{\rho_w}\nabla\cdot(\rho_w\lambda_w K\nabla P_w) - \frac{1}{\rho_n}\nabla\cdot(\rho_n\lambda_n K\nabla P_n) = q_t +\\ \frac{1}{\rho_n}\nabla\cdot(\rho_n\lambda_n K\nabla P_c) - \frac{1}{\rho_w}\nabla\cdot(\rho_w^2\lambda_w K\mathbf{g}) - \frac{1}{\rho_n}\nabla\cdot(\rho_n^2\lambda_n K\mathbf{g})\end{aligned}$$

where $C_t = C_{f,w}S_w + C_{f,n}S_n + C_R$ is the total compressibility. We define the velocity of the wetting phase u_w as:

$$u_w = f_w u_t + f_n\lambda_n k\left(\nabla P_c + (\rho_w - \rho_n)g\right) \tag{5.38}$$

where f_w and f_n are the fractional flow functions for the wetting and non-wetting phases, respectively. u_t is the total velocity of the fluid. λ_n is the mobility of the non-wetting phase. k is the absolute permeability of the porous medium. ∇P_c represents the gradient of the capillary pressure. $(\rho_n - \rho_w)g$

represents the gravitational effects due to the density difference between the non-wetting phase (ρ_n) and the wetting phase (ρ_w).

5.2.2 NUMERICAL METHODS FOR TWO-PHASE FLOW

Transport phenomena in porous media are governed by time-dependent partial differential equations (PDEs), particularly in the context of two-phase and multiphase flows. The time-dependent nature of these equations can lead to numerical instabilities and divergence if not handled properly. Thus, the discretization of the time-derivative term becomes crucial. This section discusses the convergence of these numerical methods against the Courant–Friedrichs–Lewy (CFL) condition and the introduction of a relaxation factor to stabilize the solution.

The time-dependent terms are not the only sources of instability; advection terms also require special treatment, such as the use of an upwind scheme. In the context of two-phase flow in porous media, the Implicit Pressure Explicit Saturation (IMPES) scheme is a widely used and efficient method. This method, however, necessitates the careful selection of time-step sizes for different variables, such as pressure and saturation, to achieve accurate results. Multiscale time-splitting and time-stepping techniques are often employed to address these challenges.

There are two primary strategies for time discretization in two-phase fluid flow models within porous media: the fully implicit scheme and the IMplicit-EXplicit (IMEX) scheme [86, 87]. The fully implicit scheme is unconditionally stable and preserves the inherent coupling of the two-phase flow model. On the other hand, the IMEX scheme, though conditionally stable, reduces the nonlinearity of the original equations. In the IMEX method, the Implicit Pressure Explicit Saturation (IMPES) approach is commonly applied, where the pressure equation is solved implicitly while the saturation is updated explicitly. However, due to its conditional stability, the IMPES method often requires smaller time steps, especially in highly heterogeneous porous media. Several modifications to the IMPES approach have been proposed to enhance its stability and efficiency, such as the linear approximation of capillary pressure in terms of saturation.

5.2.2.1 Implicit Pressure Explicit Saturation (IMPES)

The IMPES method, initially developed by Sheldon et al. [88] and later refined by Stone and Gardner [89], is a widely adopted technique in the petroleum industry for simulating two-phase flow in porous media. The key concept behind IMPES is the decoupling of the pressure and saturation equations, where the pressure is solved implicitly, and the saturation is updated explicitly.

The traditional IMPES approach, while efficient and memory-conservative, is conditionally stable, requiring small time steps for accurate saturation calculations. This method is particularly effective in scenarios where the pressure varies slowly compared to saturation, allowing for efficient computation.

To solve the saturation equation, we substitute the expression for u_w from Darcy's law into the wetting-phase saturation equation. This results in:

$$\phi \frac{\partial s_w}{\partial t} = q_w - \nabla \cdot \left\{ K f_w(s) \lambda_n(s) \left(\frac{\partial p_c}{\partial s} \nabla s + (\rho_o - \rho_w) g \nabla z \right) + f_w(s) u \right\} \tag{5.39}$$

This equation describes the temporal evolution of the wetting-phase saturation, incorporating the effects of capillary pressure, gravitational forces, and fluid motion within the porous medium.

To proceed with the numerical solution, we divide the total time interval $[0, T]$ into N_T steps, each with a length $\Delta t^n = t^{n+1} - t^n$, where $n+1$ denotes the current time step and n represents the previous one. We apply the backward Euler method to discretize the time derivatives in the saturation equations. The following two equations are solved simultaneously to compute the pressure:

$$-\nabla \cdot \left(\lambda_t(s_w^n) K \nabla p_w^{n+1} \right) - \nabla \cdot \left(\lambda_n(s_w^n) K \nabla p_c(s_w^n) \right) = q_t^n \tag{5.40}$$

$$\phi \frac{s_w^{n+1} - s_w^n}{\Delta t^n} - q_w^n = -\nabla \cdot \left(\lambda_t(s_w^n) K \nabla p_w^{n+1} \right) \tag{5.41}$$

The nonlinear capillary pressure function is linearized using a first-order approximation [90]:

$$p_c(S_w^{n+1}) \approx p_c(S_w^n) + \frac{dp_c}{dS_w}(S_w^n)(S_w^{n+1} - S_w^n) \tag{5.42}$$

Finally, the saturation is updated using the explicit equation:

$$\phi \frac{s_w^{n+1} - s_w^n}{\Delta t^n} + \nabla \cdot \left(f_w^n u_a^{n+1} \right) = q_w^{n+1} \tag{5.43}$$

This approach provides a structured method for solving the coupled pressure and saturation equations in a temporally discretized manner, ensuring stability and accuracy in the numerical solution.

5.2.2.2 Temporal Discretization

Temporal discretization in numerical simulations, especially in two-phase flow problems within porous media, is a critical aspect that requires careful consideration to maintain stability and accuracy. This section delves into the commonly employed numerical methods for temporal discretization, with a focus on the forward and backward Euler difference schemes. Additionally, we discuss the Courant–Friedrichs–Lewy (CFL) stability condition, the utility of a multiscale time-splitting scheme, and the role of the relaxation factor in conjunction with the CFL condition. Lastly, a stability analysis of the Implicit Pressure Explicit Saturation (IMPES) scheme is provided, highlighting its conditional stability and implications for computational efficiency.

5.2.2.3 CFL Stability Condition

The Courant–Friedrichs–Lewy (CFL) condition is a fundamental criterion for the stability of numerical schemes, particularly for explicit time-stepping methods. It dictates that the time step must be sufficiently small relative to the spatial discretization to ensure the stability of the solution.

In a two-dimensional domain, the CFL conditions are given by:

$$\mathrm{CFL}_x = \frac{u_x \Delta t}{\Delta x} \leq 1, \quad \mathrm{CFL}_y = \frac{u_y \Delta t}{\Delta y} \leq 1 \tag{5.44}$$

To maintain stability, an adaptive multi-stepping approach can be employed, where the time step is adjusted dynamically based on the CFL condition. If $\mathrm{CFL}_x > 1$ or $\mathrm{CFL}_y > 1$, the time step for saturation and concentration is halved, and the CFL conditions are recalculated. This process is repeated until the conditions $\mathrm{CFL}_x < 1$ and $\mathrm{CFL}_y < 1$ are satisfied, ensuring that the final adaptive time step is suitable for both saturation and concentration.

5.2.2.4 Multiscale Time Splitting

The multiscale time splitting method involves using different time step sizes for each time derivative term, depending on the physical processes they represent. For example, the time-step size for pressure can be larger than that for saturation, as pressure typically changes more slowly. The explicit-stepping CFL condition often requires a time step size smaller than $O(h^2)$, where h is the spatial mesh size, which can be overly restrictive in practical applications.

In this method, the pressure domain serves as the first level of time discretization. The total time interval $[0, T]$ is divided into N_p steps, with the time step length $\Delta t_i = t_{i+1} - t_i$. For the saturation, a smaller time step size is used, with each subinterval $[t_i, t_{i+1}]$ divided into N_{pf} sub-subintervals. If further refinement is needed, such as for concentration, additional discretization levels can be introduced.

This approach is particularly useful in complex systems, such as fractured porous media, where different physical processes occur at different rates, necessitating varying levels of temporal discretization.

5.2.2.5 Relaxation Factor

The CFL condition ensures stability for linear problems, but in highly nonlinear equations, additional stabilization may be required. A relaxation factor can be introduced at the advection terms to assist in satisfying the CFL condition.

The relaxation factor α_r is defined as [91]:

$$\alpha_r = \min\left(\frac{d_1}{\max(\text{CFL})}, d_2\right), \quad 0 \leq d_1, d_2 \leq 1 \tag{5.45}$$

This factor is multiplied by the convection terms to ensure that $\alpha_r \cdot \text{CFL} < 1$. While very small time steps can theoretically guarantee convergence, in practical computations, a larger time step size may be used, depending on the spatial resolution Δx. This is especially relevant in large domains, such as in reservoir simulations, where the mesh size and time step size can be adjusted accordingly.

5.2.3 MATLAB® IMPLEMENTATION

The numerical implementation of the two-phase flow equations relies on a cell-centered finite difference (CCFD) method, which is particularly suitable for problems with discontinuous coefficients such as those arising in porous media simulations. We employ this method on a nonuniform rectangular 2D mesh for both the advection and diffusion components of the system (see Chapter 3).

Now, we can build on what we have in Chapter 3 about MATLAB® coding of advection and diffusion operators. We start with the implementation of capillary pressure linearization is shown in the following code:

```
% Base capillary pressure from correlation
cap = -(log(So)) .* BCFun(permXX ./ md);

% Derivative of capillary pressure with respect to saturation
errcap = -sf * BCFun(permXX ./ md) ./ Se;

% Second derivative of capillary pressure
err2cap = BCFun(permXX ./ md) ./ (Se.^2);

% Linearize capillary pressure using first-order approximation
cap = cap + errcap .* (Swp - Sw);

% Convert to vector form for matrix operations
errcap = errcap(:);
derrcap = diag(sparse(ht * errcap(:)));
cap = cap(:);
```

The function BCFun defines the capillary pressure parameter based on permeability:

```
function bc = BCFun(perm)
    % BCFUN Calculates capillary pressure parameter based on
        permeability
    bar = 1.0;
    bc = bar * 50.0 ./ sqrt(perm);
end
```

For our simulations, we use a logarithmic capillary pressure function:

$$p_c(S_e) = -B_c \ln(S_e) \tag{5.46}$$

where $S_e = \frac{S_w - S_{rw}}{1 - S_{rw} - S_{rn}}$ is the effective saturation and B_c is a parameter dependent on permeability.

5.2.3.1 Solution Algorithm

The algorithm proceeds as follows:

1. Initialize saturation field
2. For each time step:
    a. Calculate phase mobilities and relative permeabilities
    b. Compute capillary pressure and its derivatives
    c. Assemble and solve the pressure equation
    d. Calculate phase velocities
    e. Solve the saturation equation explicitly
    f. Apply relaxation for stability if necessary
    g. Check convergence and adjust time step if needed
3. Repeat until final time

The core of the time-stepping loop in the non-iterative scheme is implemented as follows:

```
% Time stepping loop
for iT = 1:nTSteps
    % Calculate time step based on fractional derivative
        parameter alpha
    alph_Coef = alph ./ ((iT*ht+ht).^alph - (iT*ht).^alph);
    ht = 1 / alph_Coef;

    % Store Sw from previous time step
    Swp = reshape(Sw, [nx, ny]);

    % Calculate effective saturation
    Se = (Sw - Srw) ./ (ones(size(Sw)) - Srw - Srn);

    % Calculate relative permeabilities
    krw = (Se.^permType) ./ visw;
    krn = ((ones(size(Sw)) - Se).^permType) ./ visn;

    % Calculate total mobility and fractional flow
    lagt = krw + krn;     % Total mobility
    krt = krw ./ lagt;    % Water fractional flow

    % Calculate phase permeabilities
    pwpermXX = permXX .* lagt;
    pwpermYY = permYY .* lagt;

    % Capillary pressure equation permeabilities
    pcpermXX = permXX .* krn;
```

```
    pcpermYY = permYY .* krn;

    % Water flow equation permeabilities
    pspermXX = permXX .* krw;
    pspermYY = permYY .* krw;

    % Linearize capillary pressure
    cap = -(log(Se)) .* BCFun(permXX ./ md);
    errcap = -sf * BCFun(permXX ./ md) ./ Se;
    cap = cap + errcap .* (Swp - Sw);

    % Solve for water pressure
    totalA = pwmatA - pcmatA * psmatA;
    presW = totalA \ (pwrhs + pcrhs - pcmatA * cap);

    % Compute flux for saturation equation
    fluxX = para.difxxBar(:) .* ((para.adjW - para.adjE) * presW
        +
            para.adjBW * para.cBdryW - para.adjBE * para.cBdryE);

    % Update saturation using semi-implicit scheme
    Sw = Swp(:) + matM \ (ht * (rhs - matA * krt(:)));

    % Apply relaxation
    Swip = (1 - alp) * Swi + alp * Sw;
    Sw = Swip;
end
```

5.2.3.2 Numerical Example

As a test, we present a numerical example of two-phase flow simulation with capillary pressure effects in a 2D rectangular domain. The simulation mimics water injection into a reservoir initially containing mostly a non-wetting phase (e.g., oil). The domain dimensions is $L_x = 0.6$ m, $L_y = 0.4$ m, while the corresponding grid resolution is 30×20 cells. The simulation time is $T_{final} = 100$ and the number of time steps is 100. Rock and fluid properties are taken as follow; $\phi = 0.3$, $K = 100$ md (approximately 9.87×10^{-14} m^2), $\mu_w = 1.0$ cp (0.001 Pa·s), $\mu_n = 0.45$ cp (0.00045 Pa·s), $\rho_w = 1000$ kg/m^3, $\rho_n = 660$ kg/m^3, $S_{rw} = 0$, $S_{rn} = 10^{-4}$, $p_{c,max} = 50$ bar, $p_{c,min} = 5$ bar,

Initial water saturation is assumed to be $S_w^{init} = 10^{-4}$ nearly dry. At the west boundary, the water was injected at rate proportional to $0.2 \cdot L_x \cdot \phi$. At the east boundary, we assumed a fixed pressure (Dirichlet condition), while at north and South boundaries, there is no-flow (Neumann condition).

Unlike the above discussed relaxation factor, the code implements different relaxation factors (α) ranging from 0.9 to 1.1, with increments of 0.05. The relaxation factor affects the stability and convergence of the solution. This parametric study allows us to observe the sensitivity of the solution to this numerical parameter.

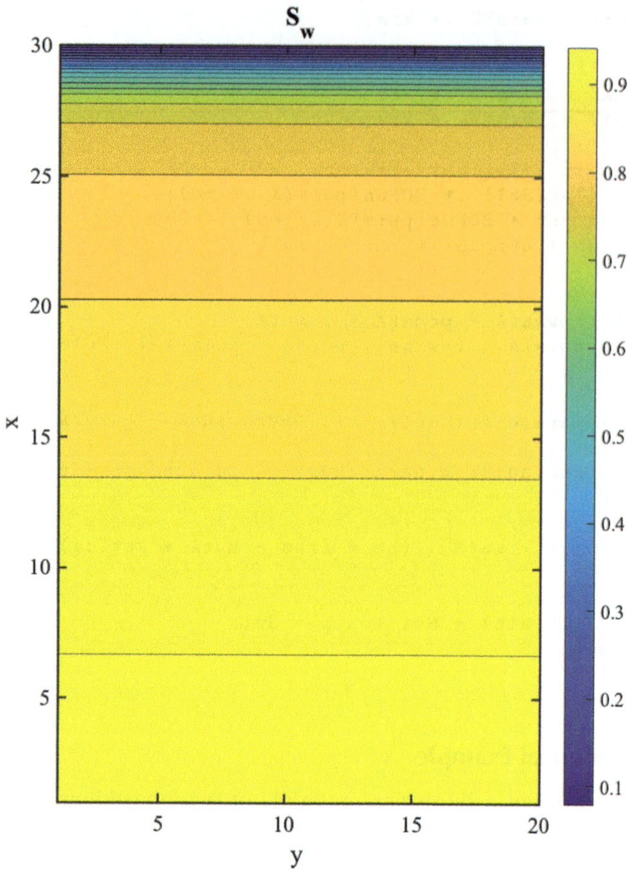

Figure 5.8 Water saturation contours at the end of simulation.

For each value of α, the saturation profiles are stored and compared. The relaxation factor is applied as:

$$S_w^{n+1} = (1 - \alpha)S_w^n + \alpha S_w^* \tag{5.47}$$

where S_w^* is the tentative solution before relaxation.

As an output for code run, Figure 5.8 shows the water saturation contours at the end of the simulation, and Figure 5.9 presents a comparison of water saturation profiles along a vertical cross-section (at $x = 10$) for different relaxation factor values.

Future improvements could focus on higher-order temporal discretization schemes, more sophisticated relaxation techniques, and extension to three-dimensional problems.

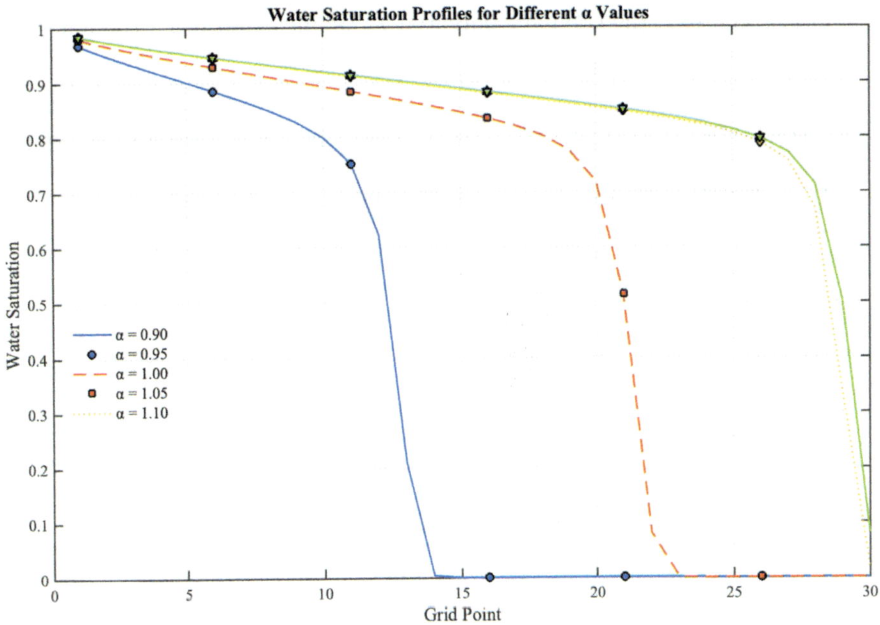

Figure 5.9 Water saturation profiles for different relaxation factor values.

5.2.4 ITERATIVE IMPES SCHEMES

In computational fluid dynamics, iterative approaches are crucial for solving complex, coupled, and highly nonlinear systems. Among these, the iterative Implicit Pressure Explicit Saturation (IMPES) scheme has emerged as a powerful tool, in the context of transport phenomena within porous media.

To discretize the total time interval $[0, T]$ into N_T steps, we define each time step as $\Delta t_n = t_{n+1} - t_n$, where $n+1$ refers to the current time step, and n denotes the previous one. Within each time step, we apply N_I iterations, labeling the current iteration as $k+1$ and the previous one as k.

For the discretization of time derivatives in the equations of saturation, the backward Euler method is employed. This method ensures stability and accuracy in the numerical solution. The pressure field, essential for fluid flow simulation, is computed by solving the following coupled equations:

$$-\nabla \cdot \left(\lambda_t(s_w^{n+1,k}) K \nabla P_w^{n+1,k+1}\right) - \nabla \cdot \left(\lambda_n(s_w^{n+1,k}) K \nabla P_c(s_w^{n+1,k+1})\right) = q^{t_{n+1}} \tag{5.48}$$

$$\phi \frac{s_w^{n+1,k+1} - s_w^n}{\Delta t_n} - q_w^{n+1} = -\nabla \cdot \left(\lambda_t(s_w^{n+1,k}) K \nabla P_w^{n+1,k+1}\right) \tag{5.49}$$

In this context, the capillary potential P_c is linearized using the following expression:

$$P_c(S_w^{n+1,k+1}) \approx P_c(S_w^{n+1,k}) + P_c'(S_w^{n+1,k})\left(S_w^{n+1,k+1} - S_w^{n+1,k}\right) \tag{5.50}$$

To update the saturation, an explicit equation is employed:

$$\phi\frac{s_w^{n+1,k+1} - s_w^n}{\Delta t_n} + \alpha r_1 \nabla\cdot\left(f_w^{n+1,k} u_a^{n+1,k+1}\right) = q_w^{n+1} \tag{5.51}$$

Applying the CCFD spatial scheme to discretize the system of equations (5.39) through (5.45), the discretized form of the pressure equation is given by:

$$A_a(s_w^{n+1,k})P_w^{n+1,k+1} + A_c(s_w^{n+1,k})P_d(s_w^{n+1,k+1}) = Q_a^{c,n+1} \tag{5.52}$$

Where the matrix-vector $P_d(s_w^{n+1,k+1})$ is defined as:

$$P_d(s_w^{n+1,k+1}) = P_c(s_w^{n+1,k}) + P_c'(s_w^{n+1,k})\left(s_w^{n+1,k+1} - s_w^{n+1,k}\right) \tag{5.53}$$

The spatial discretization of the saturation equation is similarly expressed as:

$$M\frac{s_w^{n+1,k+1} - s_w^n}{\Delta t_n} + A_w(s_w^{n+1,k})P_w^{n+1,k+1} = Q_w^{n+1} \tag{5.54}$$

In this equation, M represents a diagonal matrix that substitutes porosity, while the coupled matrix form of the pressure-saturation equation is represented as:

$$A_t(s_w^{n+1,k})P_w^{n+1,k+1} = Q_t(s_w^{n+1,k}) \tag{5.55}$$

where:

$$A_t(s_w^{n+1,k}) = A_a(s_w^{n+1,k}) - \Delta t_n A_c(s_w^{n+1,k})P_c'(s_w^{n+1,k})M^{-1}A_w(s_w^{n+1,k}) \tag{5.56}$$

and:

$$\begin{aligned} Q_t(s_w^{n+1,k}) &= Q_a^{c,n+1} - A_c(s_w^{n+1,k})P_c(s_w^{n+1,k}) + P_c'(s_w^{n+1,k})\left(s_w - s_w^{n+1,k}\right) \\ &\quad -\Delta t_n A_c(s_w^{n+1,k})P_c'(s_w^{n+1,k})M^{-1}Q_w^{n+1} \end{aligned} \tag{5.57}$$

The spatial discretization of equation (5.42) becomes:

$$M\frac{s_w^{n+1,k+1} - s_w^n}{\Delta t_n} + A_s(s_w^{n+1,k})P_w^{n+1,k+1} f_w(s_w^{n+1,k}) = Q_s^{n+1} \tag{5.58}$$

5.3 MISCIBLE TWO-PHASE FLOW

Miscible two-phase flow refers to the interaction of two fluid phases that mix at the molecular level without forming distinct physical interfaces. The governing dynamics are characterized by advection, diffusion, and phase-dependent transport properties. Unlike immiscible flow, where surface tension and interfacial dynamics dominate, miscible flow is driven primarily by molecular diffusion and convective transport. The phenomenon of miscible two-phase flow has significant applications across various industries, including petroleum engineering, chemical processing, and environmental management. For instance, these models are critical in enhanced oil recovery, where solvents or gases (e.g., CO_2) are injected into reservoirs to improve hydrocarbon extraction efficiency [92, 93]. They are equally important in carbon sequestration, simulating the injection of CO_2 into geological formations for long-term storage. Additionally, they are used in environmental engineering to model pollutant transport in subsurface environments and in chemical processing to optimize the mixing of reactants in industrial reactors. A more recent application is in underground hydrogen storage, where miscible two-phase flow models are employed to study and optimize the injection, withdrawal, and long-term stability of hydrogen stored in geological formations, ensuring its viability as a sustainable energy carrier.

Modeling miscible two-phase flow presents unique challenges due to the interplay of complex physical processes, including strong coupling between the phases through diffusion and convection, in addition to variability in fluid properties such as density and viscosity. Therefore, accurate numerical schemes are needed to resolve steep concentration gradients and maintain stability.

The governing equations for miscible two-phase flow in porous media are derived from conservation laws for mass and momentum and constitutive relations for diffusion and phase mobility.

The conservation of mass for the liquid phase is expressed as:

$$\frac{\partial G_w^w}{\partial t} + \nabla \cdot F_w^w = f_w, \tag{5.59}$$

where:

$$G_w^w = \phi S(q) \rho_w, \tag{5.60}$$

is the mass of the liquid phase per unit volume, and:

$$F_w^w = -K k_w(S(q)) \rho_w \nabla p_w + K k_w(S(q)) \rho_w g, \tag{5.61}$$

is the mass flux of the liquid phase governed by Darcy's Law, including the effect of gravity, K is the permeability of the porous medium, ϕ is the porosity of the medium (assumed constant), ρ_w is the density of the liquid phase, $k_w(S(q))$ is the relative mobility of the liquid phase (dependent on saturation), g is the gravitational acceleration vector, and $S(q)$ is the water saturation as a function of capillary pressure. The coupling between the liquid and gas phases is expressed through the capillary pressure:

$$q = p_w - p_g, \tag{5.62}$$

where p_w is the pressure of the liquid (water) phase, p_g is the pressure of the gas phase, and q is the capillary pressure, which depends on the liquid saturation $S(q)$.

The conservation of mass for the gas phase accounts for both free and dissolved gas:

$$\frac{\partial(G_w^g + G_g^g)}{\partial t} + \nabla \cdot (F_w^g + F_g^g) = f_g, \tag{5.63}$$

where

$$G_w^g = \phi S(q) \rho_w X(p_g), \tag{5.64}$$

is the mass of dissolved gas in the liquid phase,

$$G_g^g = \phi(1 - S(q))\rho_g, \tag{5.65}$$

is the mass of the gas phase per unit volume, and:

$$F_w^g = -K k_w(S(q))\rho_w X(p_g)\nabla p_w - D\rho_w S(q)\nabla X(p_g) + K k_w(S(q))\rho_w X(p_g) g, \tag{5.66}$$

is the mass flux of the dissolved gas in the liquid phase, while:

$$F_g^g = -K k_g(S(q))\rho_g \nabla p_g + K k_g(S(q))\rho_g g, \tag{5.67}$$

is the mass flux of the free gas phase, $X(p_g)$ is the mass fraction of the gaseous component dissolved in the liquid phase, $k_g(S(q))$ is the relative mobility of the gas phase, and D is the diffusivity coefficient for the dissolved gas in the liquid phase.

5.4 MULTICOMPONENT MISCIBLE TWO-PHASE FLOW

Multicomponent miscible two-phase flow in porous media involves the simultaneous transport and interaction of multiple chemical components within two fluid phases, typically liquid and gas. The modeling of such systems accounts for complex physical phenomena such as advection, diffusion, phase partitioning, and dynamic variations in fluid properties. Applications of multicomponent flow include enhanced oil recovery, carbon capture and storage, and pollutant transport in subsurface environments.

The governing equations for multicomponent miscible two-phase flow are derived from the principles of mass conservation for each component in both liquid and gas phases, as well as additional relations for capillary pressure and phase equilibrium.

Mass conservation for each component in the liquid phase can be given as:

$$\partial_t (G_{w,i}) + \nabla \cdot F_{w,i} = f_{w,i}, \tag{5.68}$$

where

$$G_{w,i} = \phi(x)S(q)\rho_w X_i, \tag{5.69}$$

represents the mass of component i in the liquid phase per unit volume,

$$F_{w,i} = -\phi\rho_w k_w(S(q))X_i\nabla p_w - D_{ij}\rho_w S(q)\nabla X_i, \tag{5.70}$$

is the flux of the component i in the liquid phase, $f_{w,i}$ is the source term for the component i in the liquid phase, and X_i is the mass fraction of the component i in the liquid phase. D_{ij} is the multicomponent diffusion coefficient tensor capturing interactions between components i and j.

The mass conservation for each component in the gas phase takes the form:

$$\partial_t\left(G_{g,i}\right) + \nabla\cdot F_{g,i} = f_{g,i}, \tag{5.71}$$

where

$$G_{g,i} = \phi(x)(1 - S(q))\rho_g Y_i \tag{5.72}$$

is the mass of the component i in the gas phase per unit volume

$$\Gamma_{g,i} = \quad \phi\rho_g k_g(S(q))Y_i\nabla p_g - D^g_{ij}\rho_g(1 - S(q))\nabla Y_i, \tag{5.73}$$

is the flux of the component i in the gas phase, and $f_{g,i}$ is the source term for component i in the gas phase.

The capillary pressure couples the liquid and gas phases:

$$q = p_w - p_g, \tag{5.74}$$

where p_w and p_g are the pressures in the liquid and gas phases, respectively.

The total mass conservation equation ensures the overall conservation of all components:

$$\sum_i \left(G_{w,i} + G_{g,i}\right) = \phi(x)\rho_{\text{total}}, \tag{5.75}$$

where ρ_{total} is the total density of the system.

Multicomponent diffusion in both phases is governed by the Stefan-Maxwell equations:

$$\sum_{j=1}^{N} \frac{X_j\mathbf{J}_i - X_i\mathbf{J}_j}{D_{ij}} = -\nabla P, \tag{5.76}$$

where $\mathbf{J}_i$ is the flux of component i, and D_{ij} is the diffusion coefficient for the pair of components i and j.

Components partition between the liquid and gas phases based on thermodynamic equilibrium relations:

$$X_i^{\text{liq}} = K_i Y_i^{\text{gas}}, \tag{5.77}$$

where K_i is the partition coefficient for component i.

The effective viscosity and density of each phase depend on the composition:

$$\mu_{\text{eff}} = \sum_{i=1}^{N} X_i \mu_i, \tag{5.78}$$

and

$$\rho_{\text{eff}} = \sum_{i=1}^{N} X_i \rho_i. \tag{5.79}$$

5.5 PARTIALLY MISCIBLE TWO-PHASE FLOW

Partially miscible two-phase flow occurs when two fluid phases (e.g., liquid and gas) interact in a porous medium with limited solubility. In such systems, dynamic mass transfer processes govern the exchange of components between phases, deviating from instantaneous equilibrium conditions such as those dictated by Henry's Law. Instead, the system is influenced by diffusion, convection, and kinetic mass transfer rates. This model captures the coupled dynamics of flow, phase interactions, and component transport within a porous medium. Applications include enhanced oil recovery, carbon sequestration, hydrogen storage, and groundwater contamination modeling. The governing equations describe mass conservation for components in each phase, Darcy's law for flow, and dynamic mass transfer between phases [94]. Mass conservation equations can be written as follows:

$$\frac{\partial(\phi \rho_w s_l)}{\partial t} + \nabla \cdot (\rho_w u_l + J_{w,l}) = 0, \tag{5.80}$$

$$\frac{\partial(\phi \rho_{h,l} s_l)}{\partial t} + \nabla \cdot (\rho_{h,l} u_l + J_{h,l}) = \Omega_{h,l}, \tag{5.81}$$

$$\frac{\partial(\phi \rho_{h,g} s_g)}{\partial t} + \nabla \cdot (\rho_{h,g} u_g + J_{h,g}) = \Omega_{h,g}, \tag{5.82}$$

$$\frac{\partial(\phi \rho_{w,g} s_g)}{\partial t} + \nabla \cdot (\rho_{w,g} u_g + J_{w,g}) = 0, \tag{5.83}$$

where ϕ is the porosity of the medium, s_α is the saturation of phase α (l for liquid, g for gas) with $s_l + s_g = 1$, $\rho_{\beta,\alpha}$ is the density of component β (w for water, h for hydrogen) in phase α, u_α is the Darcy velocity of phase α, $J_{\beta,\alpha}$ is the diffusive flux of component β in phase α, and $\Omega_{\beta,\alpha}$ is the mass transfer between components in phase α.

Darcy's law governs the velocity of each fluid phase:

$$u_\alpha = -K \frac{k_\alpha(s_\alpha)}{\mu_\alpha} (\nabla p_\alpha - \rho_\alpha g), \tag{5.84}$$

where K is the permeability tensor of the porous medium, is the $k_\alpha(s_\alpha)$: Relative permeability of phase α, μ_α is the viscosity of phase α, p_α is the pressure of phase α, and g is the gravitational acceleration.

The diffusive flux of components in each phase is modeled using Fick's law:

$$J_{\beta,\alpha} = -\rho_\alpha D_{\beta,\alpha} \nabla X_{\beta,\alpha}, \tag{5.85}$$

where $D_{\beta,\alpha}$ is the diffusion coefficient of component β in phase α, and $X_{\beta,\alpha}$ is the molar fraction of component β in phase α.

The pressures of the liquid and gas phases are coupled via the capillary pressure:

$$p_c(s_l) = p_g - p_l, \tag{5.86}$$

where $p_c(s_l)$ is a decreasing function of the liquid saturation s_l.

The mass transfer between phases is modeled dynamically as:

$$\Omega_{h,l} = -\sigma \rho_l (X_{h,l} - X_{\text{eq}}), \quad \Omega_{h,g} = -\Omega_{h,l}, \tag{5.87}$$

where σ is the inverse characteristic time for achieving equilibrium, and X_{eq} is the equilibrium concentration based on solubility.

5.6 CO_2 SEQUESTRATION

Carbon dioxide sequestration represents one of the most promising applications of multiphase flow theory in porous media, with significant implications for climate change mitigation [95, 96]. This process involves the injection of supercritical CO_2 into deep geological formations, particularly saline aquifers, for long-term storage. The effectiveness of this technology depends on accurately modeling and predicting the complex interaction between CO_2 and resident formation fluids.

The fundamental behavior of CO_2-brine systems can be described through a set of coupled partial differential equations that capture mass conservation and momentum transport [84]. For each phase (α = w for water/brine, CO_2 for carbon dioxide), the mass conservation equation is expressed as:

$$\frac{\partial}{\partial t}(\phi \rho_\alpha S_\alpha) + \nabla \cdot (\rho_\alpha \mathbf{u}_\alpha) = q_\alpha \tag{5.88}$$

where ϕ represents porosity (dimensionless), ρ_α is the density of phase α (kg/m^3), S_α denotes the saturation of phase α (dimensionless), $\mathbf{u}_\alpha$ is the Darcy velocity of phase α (m/s), and q_α represents source/sink terms (kg/m^3/s).

The fluid motion for each phase follows Darcy's law, extended to multiphase systems:

$$\mathbf{u}_\alpha = -\frac{k k_{r,\alpha}}{\mu_\alpha}(\nabla p_\alpha - \rho_\alpha g \nabla z) \tag{5.89}$$

where k is the absolute permeability tensor (m^2), $k_{r,\alpha}$ represents the relative permeability for phase α (dimensionless), μ_α is the viscosity of phase α (Pa·s),

p_α denotes the pressure of phase α (Pa), g is the gravitational acceleration (m/s^2), and z is the elevation (m).

These equations are constrained by the saturation condition:

$$S_{\mathrm{w}} + S_{\mathrm{CO_2}} = 1 \tag{5.90}$$

The interaction between phases is characterized by the capillary pressure relationship:

$$p_{\mathrm{c}} = p_{\mathrm{CO_2}} - p_{\mathrm{w}}, \tag{5.91}$$

which is typically modeled as a function of water saturation [76, 81]:

$$p_{\mathrm{c}}(S_{\mathrm{w}}) = p_{\mathrm{c,entry}} \cdot f(S_{\mathrm{w}}), \tag{5.92}$$

where $p_{\mathrm{c,entry}}$ is the entry capillary pressure (Pa), and $f(S_{\mathrm{w}})$ is a dimensionless function that depends on the specific capillary pressure model employed.

Relative permeability functions are critical components in modeling multiphase flow behavior during CO_2 sequestration. Several models have been developed to characterize relative permeability relationships [75, 81, 76]. The Corey model provides a straightforward power-law relationship applicable to relatively homogeneous reservoirs:

$$k_{r,\alpha} = k_{r,\alpha}^{\mathrm{end}} \left(\frac{S_\alpha - S_\alpha^{\mathrm{res}}}{1 - S_{\mathrm{w}}^{\mathrm{res}} - S_{\mathrm{CO_2}}^{\mathrm{res}}} \right)^{n_\alpha} \tag{5.93}$$

where $k_{r,\alpha}^{\mathrm{end}}$ is the endpoint relative permeability, S_α^{res} is the residual saturation of phase α, and n_α is the Corey exponent for phase α.

For more complex reservoir characteristics, the Van Genuchten-Mualem model offers sophisticated saturation-dependent relationships:

$$k_{r,\mathrm{w}} = S_{\mathrm{e}}^{l} \quad \text{and} \quad k_{r,\mathrm{CO_2}} = (1 - S_{\mathrm{e}})^2 (1 - S_{\mathrm{e}}^{1/l}) \tag{5.94}$$

where $S_{\mathrm{e}} = \frac{S_{\mathrm{w}} - S_{\mathrm{w}}^{\mathrm{res}}}{1 - S_{\mathrm{w}}^{\mathrm{res}} - S_{\mathrm{CO_2}}^{\mathrm{res}}}$ is the effective water saturation, and l is a pore-connectivity parameter.

The Brooks-Corey model accounts for pore size distribution effects:

$$k_{r,\mathrm{w}} = S_{\mathrm{e}}^{3+2/\lambda} \quad \text{and} \quad k_{r,\mathrm{CO_2}} = (1 - S_{\mathrm{e}})^2 (1 - S_{\mathrm{e}}^{1+2/\lambda}) \tag{5.95}$$

where λ characterizes the pore size distribution of the medium.

Hysteresis Effects

Hysteresis, the phenomenon where the relative permeability-saturation relationship follows different paths during drainage versus imbibition, plays a crucial role in CO_2 sequestration efficiency [78, 97]:

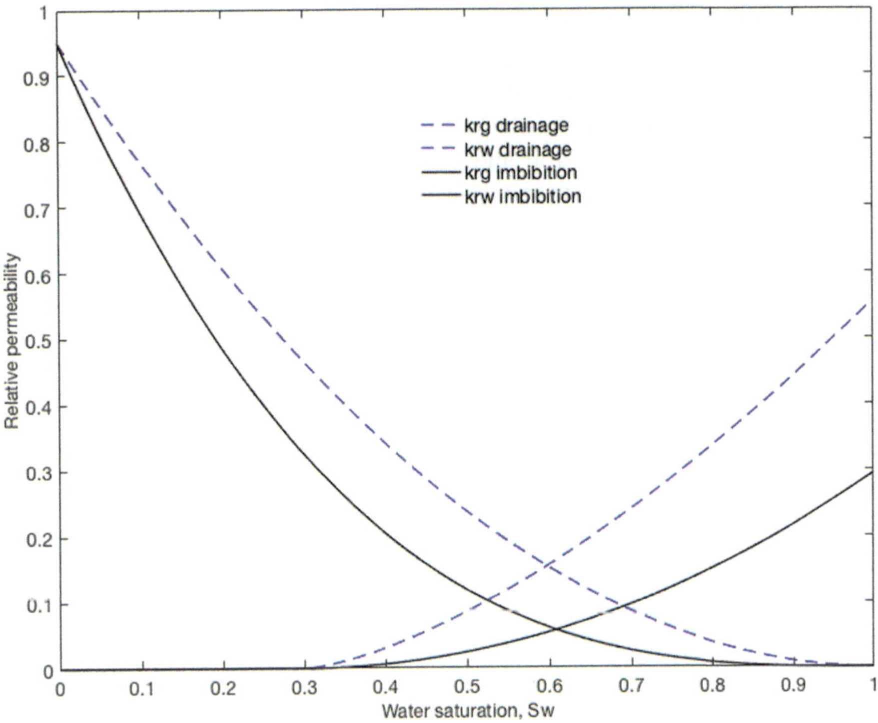

Figure 5.10 Relative permeability curves as a function of phase saturations, illustrating hysteresis effects between drainage and imbibition processes.

Hysteresis models capture this behavior through relationships such as:

$$k_{r,\alpha}^{\mathrm{imb}} = k_{r,\alpha}^{\mathrm{drain}}\left(1 - \beta(1 - S_{\alpha}^{\mathrm{res}}/S_{\alpha}^{\mathrm{max}})\right) \tag{5.96}$$

where $k_{r,\alpha}^{\mathrm{imb}}$ is the relative permeability during imbibition, $k_{r,\alpha}^{\mathrm{drain}}$ is the relative permeability during drainage, β is a hysteresis parameter, and $S_{\alpha}^{\mathrm{max}}$ is the maximum historical saturation of phase α.

The significance of hysteresis in CO_2 sequestration includes enhanced residual trapping, flow path asymmetry, and storage capacity implications. During imbibition, significantly more CO_2 becomes trapped as immobile bubbles compared to what would be predicted by drainage curves alone. The flow patterns during injection differ fundamentally from those during subsequent brine reimbibition. Hysteresis effects typically lead to higher storage capacity estimates compared to non-hysteretic models

In deep saline aquifers, initially water-wet conditions may evolve over time due to mineral dissolution, precipitation, and surface chemistry alterations induced by CO_2-brine interactions. This dynamic wettability affects capillary pressure-saturation relationships, relative permeability functions, residual trapping efficiency, and fluid distribution patterns within the pore network.

CO_2 Trapping Mechanisms

The long-term security of sequestered CO_2 depends on multiple trapping mechanisms [98, 99]. Initially, buoyant CO_2 accumulates beneath low-permeability caprocks and geological structures, creating physical barriers to upward migration (Structural and Stratigraphic Trapping). This represents the primary trapping mechanism during and immediately after injection. As the CO_2 plume migrates and water reimbibes into previously CO_2-saturated regions, capillary forces immobilize discrete CO_2 bubbles within pore spaces (Residual Trapping):

$$S_{\mathrm{CO_2,res}} = f(S_{\mathrm{CO_2,max}}, S_{\mathrm{w}}), \tag{5.97}$$

where $S_{\mathrm{CO_2,res}}$ is the residual saturation of CO_2, and $S_{\mathrm{CO_2,max}}$ is the maximum historical CO_2 saturation. A common relationship is:

$$S_{\mathrm{CO_2,res}} = S_{\mathrm{CO_2,max}}(1 - S_{\mathrm{w}}) \tag{5.98}$$

Residual trapping significantly enhances storage security by transforming mobile CO_2 into disconnected bubbles that cannot migrate under typical reservoir pressure gradients.

Over longer time periods, CO_2 dissolves into the formation brine (Solubility Trapping) according to Henry's law [100]:

$$C_{\mathrm{CO_2}}^{\mathrm{dissolved}} = H p_{\mathrm{CO_2}} \tag{5.99}$$

where $C_{\mathrm{CO_2}}^{\mathrm{dissolved}}$ is the concentration of dissolved CO_2, H is Henry's constant, $p_{\mathrm{CO_2}}$ is the partial pressure of CO_2.

This dissolution process increases brine density, potentially triggering convective mixing that accelerates the dissolution rate. Dissolved CO_2 no longer exists as a separate phase, eliminating the buoyancy-driven migration that poses leakage risks.

The ultimate fate of sequestered CO_2 involves geochemical reactions with formation minerals to form stable carbonate precipitates (Mineral Trapping). Geochemical reactions form stable carbonates [101].

$$\mathrm{CO_2 + Ca^{2+} + H_2O \rightarrow CaCO_3 + 2H^+} \tag{5.100}$$

These reactions modify porosity and permeability according to Xu et al. [102]:

$$\frac{\partial \phi}{\partial t} = \sum_m V_m \frac{\partial C_m}{\partial t} \tag{5.101}$$

where V_m is the molar volume of mineral m and C_m is its concentration. Although this process operates over centuries to millennia, it represents the most secure form of CO_2 storage, effectively removing carbon from the mobile fluid phases entirely.

Accurate simulation of CO_2 sequestration processes requires addressing several complex modeling challenges [103, 83]. One critical aspect is scale integration, which involves reconciling detailed pore-scale physics with larger-scale reservoir simulations to capture the essential multiscale behavior of CO_2 migration and trapping. Additionally, compositional effects must be considered, accounting for the mutual solubility of CO_2 and brine and the resulting changes in fluid density and viscosity. Geomechanical coupling is another important factor, as pressure changes induced by CO_2 injection can alter the stress field and, consequently, reservoir permeability. Furthermore, thermal effects arise when colder injected CO_2 interacts with the typically warmer reservoir formations, necessitating the modeling of non-isothermal flow and heat transfer. Finally, geochemical reactions such as mineral dissolution and precipitation significantly influence long-term storage integrity by modifying the reservoir's porosity and permeability. The evolution of a CO_2 storage system over time thus typically requires adaptive modeling approaches that emphasize different physical processes at different stages of the storage timeline.

5.6.1 MISCIBLE CO_2-BRINE SYSTEMS

Unlike immiscible flow models, miscible two-phase flow accounts for the significant mutual solubility between CO_2 and brine. This interphase mass transfer fundamentally alters flow dynamics and storage mechanisms. The mathematical framework requires a compositional approach that tracks component concentrations across phases [103, 104].

Component Conservation Equations

For a system with components i (where i = H_2O, CO_2, and potentially dissolved salts) distributed between phases α (where α = w for aqueous/brine phase, c for CO_2-rich phase), the component mass conservation equations are [105, 106, 107]:

$$\frac{\partial}{\partial t}\left(\phi \sum_{\alpha} \rho_\alpha S_\alpha \omega_i^\alpha\right) + \nabla \cdot \left(\sum_{\alpha} \rho_\alpha \omega_i^\alpha \mathbf{u}_\alpha\right) + \nabla \cdot \mathbf{J}_i = q_i \tag{5.102}$$

where:

ω_i^α is the mass fraction of component i in phase α (dimensionless)
$\mathbf{J}_i$ is the diffusive/dispersive flux of component i (kg/m²/s)
q_i is the source/sink term for component i (kg/m³/s)

Phase Velocities with Composition Dependence

Each phase follows a modified Darcy equation where fluid properties depend on composition [108, 106]:

$$\mathbf{u}_\alpha = -\frac{kk_{r,\alpha}(\mathbf{S},\omega)}{\mu_\alpha(\omega)}(\nabla p_\alpha - \rho_\alpha(\omega)g\nabla z) \tag{5.103}$$

where ω represents the vector of all mass fractions, and fluid properties (viscosity μ_α and density ρ_α) are functions of composition.

Diffusive/Dispersive Fluxes

The non-advective transport of components is given by [109, 110]:

$$\mathbf{J}_i = -\phi\sum_\alpha \rho_\alpha S_\alpha \mathbf{D}_{i,\alpha}\nabla\omega_i^\alpha, \tag{5.104}$$

where $\mathbf{D}_{i,\alpha}$ is the combined diffusion-dispersion tensor for component i in phase α [111]:

$$\mathbf{D}_{i,\alpha} = D_{i,\alpha}^{mol}\mathbf{I} + \alpha_T|\mathbf{u}_\alpha|\mathbf{I} + (\alpha_L - \alpha_T)\frac{\mathbf{u}_\alpha \otimes \mathbf{u}_\alpha}{|\mathbf{u}_\alpha|} \tag{5.105}$$

where $D_{i,\alpha}^{mol}$ as the molecular diffusion coefficient (m^2/s), α_L and α_T as the longitudinal and transverse dispersivities (m), $\mathbf{I}$ as the identity tensor, and $\otimes$ denoting the tensor product

Phase Equilibrium Relationships

Miscible systems require thermodynamic equilibrium constraints. For CO_2-brine systems, these are commonly represented using [112, 113]:

$$\omega_{\mathrm{CO_2}}^{\mathrm{w}} = K_{\mathrm{CO_2}}(T,p,\mathcal{I})\cdot f_{\mathrm{CO_2}}(p,T) \tag{5.106}$$

$$\omega_{\mathrm{H_2O}}^{\mathrm{c}} = K_{\mathrm{H_2O}}(T,p)\cdot f_{\mathrm{H_2O}}(p,T) \tag{5.107}$$

where K_i is the equilibrium partition coefficient for component i, f_i is the fugacity of component i, $\mathcal{I}$ is the ionic strength of the brine, T is temperature, and p is pressure.

For CO_2 dissolution in brine, the equilibrium relationship can often be approximated with a pressure, temperature, and salinity-dependent Henry's law [114, 115]:

$$\omega_{\mathrm{CO_2}}^{\mathrm{w}} = H_{\mathrm{CO_2}}(T,p,X_{salt})\cdot p_{\mathrm{CO_2}} \tag{5.108}$$

where X_{salt} is the salinity of the brine, and $H_{\mathrm{CO_2}}$ is the modified Henry's constant.

Equation of State Coupling

Density and viscosity of each phase vary with composition, pressure, and temperature according to appropriate equations of state [116, 117, 118]:

$$\rho_\alpha = \rho_\alpha(p, T, \omega_\alpha) \tag{5.109}$$

$$\mu_\alpha = \mu_\alpha(p, T, \omega_\alpha) \tag{5.110}$$

For the CO_2-rich phase, the Peng-Robinson or Span-Wagner equation of state is commonly employed [116], while brine properties often use empirical correlations that account for dissolved CO_2 and salt content [119].

Compositional Constraints

The system is constrained by mass balance requirements [104]:

$$\sum_i \omega_i^\alpha = 1 \quad \forall \alpha \tag{5.111}$$

And the saturation constraint remains:

$$\sum_\alpha S_\alpha = 1 \tag{5.112}$$

5.6.1.1 Phase Behavior and Solubility Effects

For miscible CO_2-brine systems, phase behavior is characterized by [112, 113]:

$$\mathrm{CO_2\ (g/sc)} \leftrightarrows \mathrm{CO_2\ (aq)} \tag{5.113}$$

$$\mathrm{H_2O\ (l)} \leftrightarrows \mathrm{H_2O\ (v)} \tag{5.114}$$

The mutual solubility between phases varies significantly with pressure, temperature, and salinity. At reservoir conditions typical for CO_2 sequestration (7–30 MPa, 30–90°C) [115, 108]:

CO_2 solubility in brine: approximately 2–6% by mass
H_2O solubility in the CO_2-rich phase: approximately 0.3–2% by mass

5.6.1.2 Modified Relative Permeability for Miscible Systems

In miscible systems, relative permeability functions are modified to account for composition-dependent interfacial tension [85, 120]:

$$k_{r,\alpha}(S_\alpha, \omega) = k^0_{r,\alpha}(S_\alpha) \cdot F(\mathrm{IFT}(\omega)) \tag{5.115}$$

where $k^0_{r,\alpha}$ is the reference relative permeability function and $F(\mathrm{IFT})$ is a scaling function that depends on the interfacial tension between phases.

As CO_2 dissolves into brine, the interfacial tension gradually decreases, which affects the relative permeability curves. The scaling function often takes forms such as [128, 129]:

$$F(\mathrm{IFT}) = \left(\frac{\mathrm{IFT}_0}{\mathrm{IFT}}\right)^n \tag{5.116}$$

where IFT_0 is a reference interfacial tension and n is an empirical exponent typically between 0.2 and 0.5.

Density-Driven Convection

A critical phenomenon in miscible CO_2-brine systems is density-driven convection. As CO_2 dissolves into brine, the solution density increases by approximately 0.1-1% [100, 110, 109]:

$$\rho_{\mathrm{w}}(\omega^{\mathrm{w}}_{\mathrm{CO_2}}) \approx \rho^0_{\mathrm{w}} + \gamma \cdot \omega^{\mathrm{w}}_{\mathrm{CO_2}} \tag{5.117}$$

where γ is typically between 10 and 30 kg/m^3 per percent of dissolved CO_2 by mass.

This small but significant density increase creates gravitational instabilities that lead to convective mixing, dramatically accelerating the dissolution process. The onset time for convection can be estimated by [110, 121]:

$$t_c \approx \frac{40\pi^2 \mu^2_{\mathrm{w}} D^{\mathrm{mol}}_{\mathrm{CO_2}}}{k^2 g^2 \Delta\rho^2} \tag{5.118}$$

where $\Delta\rho$ is the maximum density difference between CO_2-saturated and CO_2-free brine.

Enhanced Solubility Trapping

The dissolution of CO_2 in the aqueous phase represents a more secure trapping mechanism than structural or residual trapping, as dissolved CO_2 no longer forms a separate buoyant phase [96, 122]. The rate of dissolution can be described by [123]:

$$\frac{\partial M^{\mathrm{dissolved}}_{\mathrm{CO_2}}}{\partial t} = A_{\mathrm{contact}} \cdot J_{\mathrm{CO_2}} \tag{5.119}$$

where $M^{\mathrm{dissolved}}_{\mathrm{CO_2}}$ is the mass of dissolved CO_2, A_{contact} is the contact area between phases, and $J_{\mathrm{CO_2}}$ is the CO_2 mass flux across the interface.

In the convective regime, this flux can be approximated by [124]:

$$J_{\mathrm{CO_2}} \approx 0.017 \cdot \sqrt{\frac{k\Delta\rho g}{\mu_{\mathrm{w}}}} \cdot (\omega^{\mathrm{w}}_{\mathrm{CO_2,sat}} - \omega^{\mathrm{w}}_{\mathrm{CO_2,avg}}) \tag{5.120}$$

where $\omega^{\mathrm{w}}_{\mathrm{CO_2,sat}}$ is the saturation concentration of CO_2 in brine and $\omega^{\mathrm{w}}_{\mathrm{CO_2,avg}}$ is the average concentration in the domain.

Geochemical Reactivity in Miscible Systems

When CO_2 dissolves in brine, it forms carbonic acid [125, 126]:

$$CO_2 + H_2O \rightarrow H_2CO_3 \rightarrow H^+ + HCO_3^- \tag{5.121}$$

This acidification triggers mineral dissolution reactions, particularly with carbonate and aluminosilicate minerals [127]:

$$CaCO_3 + H^+ \rightarrow Ca^{2+} + HCO_3^- \tag{5.122}$$

$$NaAlSi_3O_8 + H^+ + H_2O \rightarrow Na^+ + Al(OH)_3 + H_4SiO_4 \tag{5.123}$$

CONCLUSION

Two-phase flow in porous rocks involves many interacting processes, from tiny pore-scale forces up to field-scale movement, that all need to be captured to make reliable predictions. Capillary pressure and relative permeability describe how fluids push against each other and the rock, while hysteresis (different paths for drainage and filling) and wettability (which fluid the rock "likes" more) control how much fluid stays trapped. Numerically, methods like IMPES, IMEX, or fully implicit schemes must balance accuracy and stability by choosing proper time steps (CFL) and splitting pressure from saturation calculations. Our countercurrent imbibition example shows how capillary forces and gravity work together to push oil out of rock fractures, and the MATLAB® code gives one way to implement these ideas. When fluids can mix (miscible flow) or contain multiple components (like CO_2 plus brine), additional effects, dissolution, density-driven convection, and rock–fluid chemistry, come into play. In future work, adding rock deformation, non-Darcy flow behavior, and even machine-learning upscaling will help us handle ever more complex reservoir problems.

REVIEW QUESTIONS

1. Define the saturation S_α of a fluid phase in porous media and explain why $\sum_\alpha S_\alpha = 1$.
2. Derive the expression for capillary pressure P_c in terms of non-wetting and wetting phase pressures.
3. Compare the Corey, Van Genuchten–Mualem, and Brooks–Corey relative permeability models, highlighting their key assumptions and applicability.
4. Explain the concept of hysteresis in relative permeability and describe one mathematical model that accounts for it.
5. What is the Leverett function $J(S)$? How does it relate capillary pressure to pore geometry and fluid properties?
6. Distinguish between immiscible, miscible, compressible, and incompressible two-phase flows, giving one application example of each.

7. Outline the steps of the IMPES (Implicit Pressure, Explicit Saturation) scheme and discuss its stability constraints.
8. State the Courant–Friedrichs–Lewy (CFL) condition for an explicit scheme in two dimensions and explain its physical significance.
9. Describe how countercurrent imbibition differs from cocurrent imbibition and why it is important in fractured reservoir EOR.
10. Summarize the main challenges in modeling CO_2–brine miscible flow and the role of compositional transport equations.

PROBLEMS

1. Given porosity ϕ, permeability K, interfacial tension γ, and a Leverett function $J(S) = S^{-0.5}(1-S)^{-2}$, derive the capillary pressure curve $P_c(S)$ and plot it over $S \in [0.1, 0.9]$.
2. For a Corey model with $k_{r,w}^{\text{end}} = 1$, $S_{wr} = 0.2$, $S_{or} = 0.05$, and exponent $n_w = 3$, compute and plot $k_{r,w}(S)$ versus S. Determine $k_{r,w}$ at $S = 0.5$.
3. Derive the fractional flow $f_w(S)$ given $\lambda_w(S) = S^2$, $\lambda_n(S) = (1-S)^2$. Find the saturation S^* where $f_w(S^*) = 0.5$.
4. On a uniform grid $\Delta x = \Delta y = 0.01$ m, with maximum velocity $u_{\max} = 10^{-4}$ m/s, calculate the maximum allowable explicit time step Δt to satisfy CFL ≤ 1.
5. Starting from $u_w + u_o = 0$ and Darcy's law, derive the one-dimensional capillary-driven imbibition ODE for $S_w(z)$. Solve it analytically for a simple linear capillary pressure law $P_c(S) = P_e(1-S)$.
6. Write the finite-difference stencil for the coupled pressure equation $\nabla \cdot (K\lambda_t \nabla p) = q_t$ on a 1D grid of spacing Δx, and show how to incorporate boundary Dirichlet conditions.
7. Implement the saturation update step

$$S_w^{n+1} = S_w^n - \frac{\Delta t}{\phi\,\Delta x}\left(F_{i+1/2} - F_{i-1/2}\right)$$

in pseudocode, where $F_{i+1/2} = f_w(S_i)\,u_t$ is the explicit flux.
8. Using the estimate

$$t_c \approx \frac{40\pi^2\mu^2 D}{k^2 g^2 \Delta\rho^2},$$

compute the onset time t_c for brine with $\mu = 0.001$ Pa.s, $D = 10^{-9}$ m²/s, $k = 10^{-13}$ mm², $\Delta\rho = 5$ kg/mm³.

DECLARATION

During the preparation of this work, the author used ChatGPT in order to improve the language and readability. After using this tool/service, the author reviewed and edited the content as needed and takes full responsibility for the content of the publication.

6 Transport in Fractured Porous Media

6.1 FRACTURED POROUS MEDIA

This section provides a comprehensive overview of the modeling techniques used to simulate fluid flow and transport phenomena in fractured porous media. These environments are common in natural reservoirs and other geological formations where a network of fractures intersects the porous rock matrix, creating complex pathways for fluid movement (see Figure 6.1). The approaches covered here lay the groundwork for the numerical solutions and computational implementations that follow.

Fractured porous media are prevalent in many natural systems, including hydrocarbon reservoirs and aquifers. These media are characterized by a matrix of porous rock intersected by a network of fractures, which significantly influence fluid flow and transport processes. For example, in hydrocarbon reservoirs, the fractures can either be natural or artificially induced, as in the case of hydraulic fracturing (fracking), to enhance gas recovery from shale formations. The principal aspects of flow in fractured porous media, including the development of conceptual models and the description of fracture characteristics in both static and dynamic systems, are introduced. Fractured media primarily consists of matrix blocks and fractures. Consequently, two distinct length scales are critical; the fracture thickness, which is approximately 10^{-4} meters, and the average distance between fracture planes, corresponding to the size of the matrix blocks, typically ranging from 0.1 to 1 meter. Various conceptual models are employed to describe fluid flow in fractured porous media [130], namely the single-continuum model, dual-continuum model, discrete network model, and hybrid fracture model. These fundamental approaches can be further subdivided into more specific models.

Flow in fractured media is modeled using four main approaches. The **Single-Continuum Model** treats the fractured medium as one continuous domain, and includes the *Single-Porosity Model* (represents rock matrix as a single porous medium) and the *Single-Permeability Model* (uses uniform permeability characteristics). The **Dual Continuum Model** represents the medium as two interacting continua (matrix and fractures). This approach includes the *Dual-Porosity Model* (uses different porosity values for matrix and fractures), the *Dual Permeability Model* (applies different permeability values), and the *Dual-Porosity Dual-Permeability Model* (combines both approaches). The **Discrete Fracture Model** explicitly models individual fractures, with methods such as the *Unstructured Discrete Fracture Model* (uses unstructured meshes for complex networks). Finally, the **Hybrid Model**

DOI: 10.1201/9781032724973-6

Figure 6.1 Natural fractured porous media.

combines continuum and discrete approaches, featuring the *Embedded Discrete Fracture Model* (integrates explicit fractures within a continuum matrix).

6.2 DUAL-CONTINUUM APPROACHES

The dual-continuum approach is a widely used method to model fluid flow in fractured porous media. This technique considers the fracture network and the porous matrix as two overlapping continua, each with its distinct properties of porosity and permeability [131, 132, 396]. The key challenge in these models is accurately capturing the fluid exchange between the matrix and the fractures, which is crucial for predicting flow behavior. In dual-porosity models, the fractures are treated as high-permeability pathways, while the matrix blocks are seen as storage zones with high porosity but low permeability. The interaction between these two continua is modeled through transfer functions representing the mass flux across the fracture-matrix interface. This interaction is significant in applications such as enhanced oil recovery, where the movement of fluids between the matrix and fractures can significantly impact the efficiency of the process. The dual-permeability model extends this concept by allowing for fluid flow within both the matrix and the fractures, providing a more detailed representation of the system. This model is particularly useful when the matrix itself has sufficient permeability to allow for

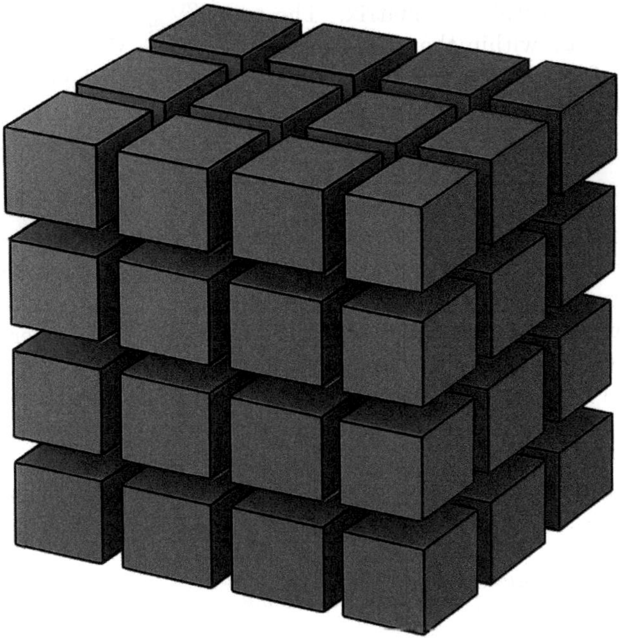

Figure 6.2 Idealized periodic model of fractured porous media.

internal flow, which can influence the overall fluid dynamics in the system.

Barenblatt et al. [131] introduced the dual-continuum approach, and Warren and Root developed an idealized dual-porosity model for flow in fractured porous media. These models assume that fluid potential differences control fluid transfer between the two continua. Warren and Root [132] pioneered an idealized model, drawing from dual continua concepts, to explore fluid flow in fractured porous media. This approach applies the Dual-Porosity model when the system exhibits two different porosity levels.

There are two main approaches to dual continua modeling: dual-porosity and dual-porosity dual-permeability. The dual-porosity model is utilized when the matrix blocks are disconnected, meaning that fluid flow is restricted to occur only from the matrix to the fractures. However, in cases where this assumption does not hold, the dual-porosity dual-permeability (DPDP) model is more appropriate. In this more complex scenario, fluid flow is considered to occur both within the matrix and between the matrix and fractures. The flow in each matrix block is described by the following equation:

$$\frac{\partial(\phi_m \rho_m)}{\partial t} + \nabla \cdot (\rho_m u_m) = -q_{mf} \tag{6.1}$$

This equation expresses the conservation of mass within the matrix block, where ϕ_m is the porosity of the matrix, ρ_m is the fluid density, and u_m is the fluid velocity within the matrix. The term $\frac{\partial(\phi_m \rho_m)}{\partial t}$ represents the rate of change of mass within the matrix over time, while $\nabla \cdot (\rho_m u_m)$ describes the net flux of mass out of the matrix block. In contrast, the flow within the fractures is governed by:

$$\frac{\partial(\phi_f \rho_f)}{\partial t} + \nabla \cdot (\rho_f u_f) = q_{mf} + q \tag{6.2}$$

In this equation, ϕ_f, ρ_f, and u_f denote the porosity, fluid density, and fluid velocity within the fractures, respectively. The term q_{mf} represents the mass transfer from the matrix to the fractures, often referred to as the matrix-fracture transfer term. The additional term q accounts for any external sources or sinks of fluid within the fracture network, which could be due to injection or production wells, or natural sources/sinks within the system.

To better understand the two-phase flow in fractured porous media under the dual-porosity dual-permeability framework, let's consider the flow equations for both wetting and non-wetting phases. For the matrix blocks, these are expressed as:

$$\frac{\partial(\phi_m \rho_{w,m} S_{w,m})}{\partial t} + \nabla \cdot (\rho_{w,m} u_{w,m}) = -q_{w,mf} \tag{6.3}$$

$$\frac{\partial(\phi_m \rho_{nw,m} S_{nw,m})}{\partial t} + \nabla \cdot (\rho_{nw,m} u_{nw,m}) = -q_{nw,mf} \tag{6.4}$$

Similarly, the equations for the wetting and non-wetting phases within the fractures are:

$$\frac{\partial(\phi_f \rho_{w,f} S_{w,f})}{\partial t} + \nabla \cdot (\rho_{w,f} u_{w,f}) = q_{w,mf} + q \tag{6.5}$$

$$\frac{\partial(\phi_f \rho_{nw,f} S_{nw,f})}{\partial t} + \nabla \cdot (\rho_{nw,f} u_{nw,f}) = q_{nw,mf} + q \tag{6.6}$$

It is essential to highlight that the external source or sink term q appears solely in the equations describing the flow within the fractures. This distinction underscores the complexity of modeling fluid dynamics in fractured reservoirs, where interactions between the matrix and fractures play a crucial role.

6.3 BOUNDARY CONDITIONS APPROACH

In another context of fluid flow within fractured porous media, the boundary conditions approach has been introduced. This approach, as developed by researchers like Barenblatt [131] and Pirson [134], focuses on specifying boundary conditions on the surfaces of matrix blocks to determine the rate of fluid exchange. It is particularly useful in cases where the physical conditions at the interface between the matrix and fractures are well understood. The flow within each matrix block can be represented by the following equation:

$$\frac{\partial(\phi_m \rho_m)}{\partial t} + \nabla \cdot (\rho_m u_m) = 0 \tag{6.7}$$

On the other hand, the fluid flow within the fractures, which play a significant role in determining the overall permeability and fluid transport characteristics of the system, is governed by:

$$\frac{\partial(\phi_f \rho_f)}{\partial t} + \nabla \cdot (\rho_f u_f) = q_{mf} + q \tag{6.8}$$

The matrix-fracture transfer term q_{mf} is critical in dual-porosity models, as it quantifies the exchange of fluid between the matrix and fractures. Two primary methodologies have been developed to characterize this term:

Consider a general boundary condition on the surface of each ith matrix block Ω_i, where the matrix-fracture transfer term q_{mf} is defined. The total mass of fluid leaving the ith matrix block of volume Ω_i per unit time is given by the surface integral:

$$\int_{\partial\Omega_i} \rho_m u \cdot n \, dl \tag{6.9}$$

Here, $\partial\Omega_i$ represents the boundary of the matrix block Ω_i, and n is the outward-pointing unit normal vector on this boundary. The dot product $u \cdot n$ captures the component of the fluid velocity that is normal to the boundary, which is the effective velocity of the fluid exiting the matrix block.

By applying the divergence theorem, which relates the flux of a vector field through a surface to the divergence of the field within the volume, and combining it with the mass conservation principle, we obtain:

$$\int_{\partial\Omega_i} \rho_m u \cdot n \, dl = \int_{\Omega_i} \nabla \cdot (\rho_m u) \, dx = \int_{\partial\Omega_i} \frac{\partial(\phi_m \rho_m)}{\partial t} \, dx \tag{6.10}$$

This equation implies that the total fluid mass leaving the matrix block over time is equal to the rate of change of fluid mass within the block, ensuring mass conservation.

The matrix-fracture transfer term q_{mf} can thus be expressed as:

$$q_{mf} = -\sum_i \chi_i(x) \frac{1}{|\Omega_a|^2} \int \frac{\partial(\phi_m \rho_m)}{\partial t} \, dx \tag{6.11}$$

In this expression, $\chi_i(x)$ is a characteristic function defined as:

$$\chi_i(x) = \begin{cases} 1 & \text{if } x \in \Omega_i \\ 0 & \text{otherwise} \end{cases} \tag{6.12}$$

The function $\chi_i(x)$ essentially acts as a switch, turning on the contribution from the ith matrix block only when the point x lies within that block. The factor $\frac{1}{|\Omega_a|^2}$ normalizes the contribution across the matrix blocks, ensuring that the total transfer term is appropriately scaled.

6.4 SHAPE-FACTOR APPROACH

The shape-factor approach simplifies the complex geometry of fractured media by using shape factors and transmissibility terms.

It is also known as the Warren-Root approach, the matrix-fracture transfer term q_{mf} is proportional to the matrix shape factor. This method provides a simplified yet effective way to model the exchange of fluids between the matrix and the fractures. For a matrix block m containing fluid and surrounded by another fluid within the fractures f, the pressure gradient across the block can be described as follows:

$$\Delta p_f = 0, \quad \Delta p_m = |\rho_f - \rho_m| g \tag{6.13}$$

Here, Δp_f is the pressure difference across the fractures, which is assumed to be negligible ($\Delta p_f = 0$), and Δp_m is the pressure difference across the matrix block, driven by the density difference between the fluids ρ_f and ρ_m and the gravitational constant g.

The matrix-fracture transfer term q_{mf} can be expressed as:

$$q_{mf} = \frac{T_m}{\mu} \left(p_m - p_{m,j} + L_c \Delta p_m \right) \tag{6.14}$$

where:

$$T_m = k\sigma \left(\frac{1}{l_x^2} + \frac{1}{l_y^2} + \frac{1}{l_z^2} \right) \tag{6.15}$$

or, in a more general form:

$$T_m = \sigma \left(k_{11} \frac{1}{l_x^2} + k_{22} \frac{1}{l_y^2} + k_{33} \frac{1}{l_z^2} \right), \quad k = \text{diag}(k_{11}, k_{22}, k_{33}) \tag{6.16}$$

In these equations, T_m represents the transmissibility of the matrix block, a measure of how easily fluid can move through the matrix. μ is the fluid viscosity. p_m and $p_{m,j}$ are the pressures within the matrix block and at its boundary, respectively. L_c is the characteristic length of the matrix block. σ is the shape factor, which encapsulates the geometric properties of the matrix block. l_x, l_y, and l_z are the dimensions of the matrix block along the x-, y-, and z-axes, respectively. - k represents the permeability tensor, where k_{11}, k_{22}, and k_{33} are the principal permeabilities along the corresponding axes.

6.5 DISCRETE FRACTURE MODEL (DFM)

The Discrete Fracture Model (DFM) [135, 136] is a powerful approach used to model fluid flow in fractured porous media. This model simplifies the complexity of fractures by treating them as lower-dimensional entities within the domain. Specifically, DFM reduces the geometric complexity of fractures by using $(n-1)$-dimensional grid cells within an n-dimensional domain. For example, in a 3D domain, fractures are represented as 2D grid cell interfaces, while in a 2D domain, fractures are simplified to 1D grid cell interfaces.

6.5.1 DFM FOR SINGLE-PHASE FLOW

To explicitly represent fractures in a fractured porous medium, the DFM is employed, in which fracture grid cells are treated as interfaces between matrix grid cells, where the matrix grid cells are n-dimensional and the fracture grid cells are $(n-1)$-dimensional. The domain is decomposed into the matrix domain Ω_m and the fracture domain Ω_f. The pressure equation in the matrix domain is given by:

$$-\nabla\cdot(K_m\nabla P_m)=q_m \tag{6.17}$$

Assuming that the pressure along the fracture width is constant, the pressure equation in the fracture becomes:

$$-\nabla\cdot(K_f\nabla P_f)=q_f+Q_f \tag{6.18}$$

The interface condition between the matrix and fracture is defined as:

$$P_m=P_f \quad \text{on} \quad \partial\Omega_m\cap\partial\Omega_f \tag{6.19}$$

Here, the subscript m refers to the matrix domain, and f refers to the fracture domain. Q_f represents the mass transfer across the matrix-fracture interface.

Consider two adjacent matrix grid cells K and K' with areas $|K|$ and $|K'|$, respectively, and a boundary $b\in\partial K\cap\partial K'$ where ξ is the flux across it. Let n_b be the unit normal vector pointing from K to K'. The Euclidean distance between the central points of cells K and K' is denoted by $d_{K,K'}$, while the distance from the center of cell K to boundary b is denoted by $d_{K,b}$. When b is on the domain boundary, the pressure is given by Dirichlet boundary conditions if $b\in\Gamma^D$; otherwise, the flux is determined by Neumann conditions.

The algebraic discretized equation can be written as:

$$\sum_{b\in\partial K}\xi_{a,m,b}=q_{t,m,K}|K| \tag{6.20}$$

If $b \in \partial K \cap \partial K'$ and $b \notin \Omega_f$, the fluxes are given by:

$$\xi_{a,m,b} = -\frac{|b|\chi_{t,b}\left(P_{w,m,K'} - P_{w,m,K}\right)}{d_{K,K'}} \tag{6.21}$$

where $\chi_{t,b}$ is defined as:

$$\chi_{t,b} = \frac{d_{K,K'}\lambda_{t,m,K}\lambda_{t,m,K'}K_{m,K}K_{m,K'}}{d_{K,b}\lambda_{t,m,K'}K_{m,K'} + d_{K',b}\lambda_{t,m,K}K_{m,K}} \tag{6.22}$$

On the other hand, if $b \in \Omega_f \cap \partial K \cap \partial K'$ and b is a grid cell of the fracture system, the flux is given by:

$$\xi_{a,m,b} \equiv \xi_{a,m,b,K} = -\frac{|b|\chi_{t,mf,b}\left(P_{w,f,b} - P_{w,m,K}\right)}{d_{K,b} + \frac{\epsilon}{2}} \tag{6.23}$$

where $\chi_{t,mf,b}$ is given by:

$$\chi_{t,mf,b} = \frac{\left(d_{K,b} + \frac{\epsilon}{2}\right)\lambda_{t,f,K}\lambda_{t,m,K}K_{f,K}K_{m,K}}{\frac{\epsilon}{2}\lambda_{t,m,K}K_{m,K} + d_{K,b}\lambda_{t,f,K}K_{f,K}} \tag{6.24}$$

where ϵ is the fracture width.

The above processes for deriving Eqs. (6.20) to (6.24) can be applied to a reduced-dimensional fracture system. Let b be a grid cell within the fracture network. The CCFD (Cell-Centered Finite Difference) scheme applied to Eq. (6.18) yields:

$$\sum_{\gamma \in \partial b} \xi_{a,f,\gamma} = q_{t,f,b}|b| + Q_{t,f,b}|b| \tag{6.25}$$

where γ is the face of grid cell b in the fracture network, and ξ is the flux across the boundary γ of the fracture grid cell b. The matrix-fracture transfer is treated as a source term in the fracture system:

$$Q_{t,f,b} = -\frac{\xi_{a,m,b,K} + \xi_{a,m,b,K'}}{\epsilon} \tag{6.26}$$

Now, consider the interface connecting multiple fractures. Let Λ_γ be the set composed of the fracture grid cells joined by γ. The discretization of the mass conservation equation is given by:

$$\sum_{b \in \Lambda_\gamma} \xi_{a,f,\gamma,e} = 0 \tag{6.27}$$

where $\xi_{a,f,\gamma,b,b} = \xi_{a,f,\gamma}$ is evaluated at the grid cell interface γ.

Finally, by implementing spatial discretization in Eq. (6.18), we obtain:

$$\sum_{\gamma \in \partial e} \xi_{a,f,\gamma} = q_{t,f,e}|e| + Q_{t,f,e}|e| \tag{6.28}$$

The matrix-fracture transfer is given by:

$$Q_{t,f,b} = -\frac{\xi_{a,m,e,K} + \xi_{a,m,b,K'}}{\epsilon} \tag{6.29}$$

where:

$$\xi_{a,m,b,K} \equiv \xi_{a,m,b} = -\frac{|e|\chi_{t,mf,b}\left(P_{w,f,b} - P_{w,m,K}\right)}{d_{K,b} + \frac{\epsilon}{2}} \tag{6.30}$$

with:

$$\chi_{t,mf,b} = \frac{\left(d_{K,b} + \frac{\epsilon}{2}\right)\lambda_{t,f,K}\lambda_{t,m,K}K_{f,K}K_{m,K}}{\frac{\epsilon}{2}\lambda_{t,m,K}K_{m,K} + d_{K,b}\lambda_{t,f,K}K_{f,K}} \tag{6.31}$$

6.5.2 DFM FOR TWO-PHASE FLOW

Now, let us address the DFM modeling for two-phase flow in porous media. The model geometry configuration is similar to the single-phase case but incorporates the appropriate equations and variables.

The pressure equation in the matrix domain Ω_m is:

$$-\nabla \cdot (\lambda_{t,m}K_m \nabla P_{w,m}) - \nabla \cdot (\lambda_{n,m}K_m \nabla P_{c,m}) = q_{t,m} \tag{6.32}$$

The pressure and saturation across the fracture width are assumed to be constant. Integrating across the fracture gives:

$$-\nabla \cdot (\lambda_{t,f}K_f \nabla P_{w,f}) - \nabla \cdot (\lambda_{n,f}K_f \nabla P_{c,f}) = q_{t,f} + Q_{t,mf} \tag{6.33}$$

With the interface conditions:

$$P_{w,m} = P_{w,f}, \quad P_{c,m} = P_{c,f} \quad \text{on } \partial\Omega_m \cap \Omega_f \tag{6.34}$$

Here, $Q_{t,mf}$ represents the mass transfer between the matrix and the fracture.

The saturation equation in the matrix domain Ω_m is:

$$\phi_m \frac{\partial S_{w,m}}{\partial t} - \nabla \cdot (\lambda_{w,m}K_m \nabla P_{w,m}) = q_{w,m} \tag{6.35}$$

For the fracture, it becomes:

$$\phi_f \frac{\partial S_{w,f}}{\partial t} - \nabla \cdot (\lambda_{w,f}K_f \nabla P_{w,f}) = q_{w,f} + Q_{w,mf} \tag{6.36}$$

where:

$$Q_{w,mf} = (\lambda_{w,f}K_f \nabla P_{w,f}) \cdot \mathbf{n} \tag{6.37}$$

with $\mathbf{n}$ as the unit normal from the matrix to the fracture.

To update saturation in the matrix:

$$\phi_m \frac{\partial S_{w,m}}{\partial t} + \nabla \cdot (f_{w,m} u_{a,m}) = q_{w,m} \tag{6.38}$$

And in the fracture:

$$\phi_f \frac{\partial S_{w,f}}{\partial t} + \nabla \cdot (f_{w,f} u_{a,f}) = q_{w,f} + \hat{Q}_{w,mf} \tag{6.39}$$

With:

$$\hat{Q}_{w,mf} = (f_{w,f} u_{a,f}) \cdot \mathbf{n} \tag{6.40}$$

6.5.3 MULTI-SCALE TIME-SPLITTING (MSTS)

MSTS for Pressure

To solve the highly nonlinear governing equations more efficiently, we apply a multiscale time-splitting (MSTS) approach. This technique assigns different time step sizes to different equations based on their physical characteristics. For instance, pressure equations use larger time steps than saturation and concentration equations. Also, fracture pressure equations use smaller steps than matrix pressure equations due to their faster dynamics. Saturation and concentration are discretized using the backward Euler method.

Divide the global time interval $[0, T]$ into $N_{p,m}$ steps:

$$0 = t^0 < t^1 < \cdots < t^{N_{p,m}} = T, \quad \Delta t^i = t^{i+1} - t^i \tag{6.41}$$

Each matrix time step $(t^i, t^{i+1}]$ is split into $N_{p,f}$ fracture sub-steps:

$$(t^i, t^{i+1}] = \bigcup_{j=0}^{N_{p,f}-1} (t^{i,j}, t^{i,j+1}], \quad \Delta t^{i,j} = t^{i,j+1} - t^{i,j} \tag{6.42}$$

with $t^{i,0} = t^i$, $t^{i,N_{p,f}} = t^{i+1}$.

MSTS for Saturation

Let the fracture time step $(t^{i,j}, t^{i,j+1}]$ be divided into $N_{s,m}$ sub-steps:

$$(t^{i,j}, t^{i,j+1}] = \bigcup_{k=0}^{N_{s,m}-1} (t^{i,j,k}, t^{i,j,k+1}], \quad t^{i,j,0} = t^{i,j}, \quad t^{i,j,N_{s,m}} = t^{i,j+1} \tag{6.43}$$

Then divide each matrix sub-step $(t^{i,j,k}, t^{i,j,k+1}]$ into $N_{s,f}$ fracture sub-steps:

$$(t^{i,j,k}, t^{i,j,k+1}] = \bigcup_{l=0}^{N_{s,f}-1} (t^{i,j,k,l}, t^{i,j,k,l+1}], \quad t^{i,j,k,0} = t^{i,j,k}, \quad t^{i,j,k,N_{s,f}} = t^{i,j,k+1} \tag{6.44}$$

The saturation equations in the matrix and fracture are solved explicitly:

$$\phi_m \frac{S_{w,m}^{i,j,k+1} - S_{w,m}^{i,j,k}}{\Delta t^{i,j,k}} + \nabla \cdot (f_{w,m}^{i,j,k} u_{a,m}^{i,j+1}) = q_{w,m}^{i,j,k} \tag{6.45}$$

$$\phi_f \frac{S_{w,f}^{i,j,k,l+1} - S_{w,f}^{i,j,k,l}}{\Delta t^{i,j,k,l}} + \nabla \cdot (f_{w,f}^{i,j,k,l} u_{a,f}^{i,j+1}) = q_{w,f}^{i,j,k,l} + Q_{w,f}^{i,j,k} \tag{6.46}$$

where $\Delta t^{i,j,k} = t^{i,j,k+1} - t^{i,j,k}$ and $\Delta t^{i,j,k,l} = t^{i,j,k,l+1} - t^{i,j,k,l}$.

6.5.4 HYBRID EMBEDDED FRACTURE MODEL (HEFM)

The Embedded Fracture Model (EFM) uses a hierarchical approach, where large fractures are treated as lower-dimensional objects and smaller-scale fractures are homogenized [137, 138]. The interaction between fractures and the surrounding rock matrix is modeled as a source term.

The Hybrid Embedded Fracture Model (HEFM) extends this by combining various modeling strategies: the dual-porosity/dual-permeability (DPDP) framework, the lower-dimensional Discrete Fracture Model (DFM), and the hierarchical fracture treatment of EFM. This hybrid approach enables accurate simulation of flow and transport in fractured porous media [139].

To model fracture behavior more precisely at the grid scale, HEFM incorporates two analytic local fracture functions [140, 142]. These functions allow each fracture segment to have a unique flow exchange rate based on its individual properties, such as length, aperture, and permeability.

Under isothermal conditions, Darcy's law and conservation of mass give the pressure equation:

$$-\nabla \cdot (\lambda \nabla P) = Q \quad \text{on } \Omega \subset \mathbb{R}^n \tag{6.47}$$

where P is pressure, $\lambda = \frac{K}{\mu}$ is mobility (with K as permeability and μ as viscosity), and Q is the source or sink term.

In fractured porous media, the scale and permeability of fractures differ from the surrounding matrix. For instance, fracture permeability k_f is related to aperture d by:

$$k_f = \frac{d^2}{12} \tag{6.48}$$

To capture these differences, HEFM solves two coupled equations: one for the matrix and one for the fracture:

$$-\nabla \cdot (\lambda^m \nabla P^m) = q_m + \alpha(P^f - P^m) \quad \text{on } \Omega^m \subset \mathbb{R}^n \tag{6.49}$$

$$-\nabla \cdot (\lambda^f \nabla P^f) = q_f + \alpha(P^m - P^f) \quad \text{on } \Omega^f \subset \mathbb{R}^{n-1} \tag{6.50}$$

Superscripts m and f denote matrix and fracture, respectively. The fracture is computed in one dimension less than the matrix, consistent with the DFM approach [138].

The terms $\alpha(P^f - P^m)$ and $\alpha(P^m - P^f)$ model fluid exchange between the matrix and fracture, similar to the exchange terms used in DPDP formulations [143] and classical well models [144].

6.5.4.1 Matrix-Fracture Flux Exchange

For a matrix volume Ω intersected by a fracture interface area A, the total flux between fracture and matrix is:

$$\int_{\Omega} \alpha(P^f - P^m)\, dV = \int_{A} \alpha(P^m - P^f)\, dA \tag{6.51}$$

This flux is normalized by the pressure difference and the matrix mobility λ^m. The exchange coefficient α is defined as:

$$\alpha = \alpha^* \frac{|L_f|\lambda_m}{h} \tag{6.52}$$

where $\lambda_m = \frac{K_m}{\mu}$, h is the mesh size (assumed equal in x and y directions), α^* is a dimensionless, grid-dependent coefficient derived from local fracture characteristics, and $|L_f|$ is the fracture length in each grid cell.

This concept extends to 3D matrix - 2D fracture cases with additional degrees of freedom. Here, we focus on the 2D/1D case. Analytic forms for α^*, proposed by Amir and Sun [140], allow fine-scale capture of fracture effects, including variable exchange rates based on pressure, fracture length, aperture, and other local properties.

6.6 LOW-PERMEABILITY RESERVOIRS

Gas transport in low-permeability reservoirs, particularly shale gas systems, has gained substantial attention due to its critical role in unconventional hydrocarbon extraction. Shale formations are characterized by their ultra-low permeability, organic-rich content, and complex pore structures, which pose significant challenges to effective gas recovery. This review highlights the key aspects of gas transport in low-permeability reservoirs, emphasizing shale gas systems.

Tight and shale formations consist of rocks with pores that are either extremely small or poorly connected, hindering the easy flow of oil and natural gas through them. Despite these challenges, hydrocarbons such as crude oil, natural gas, and natural gas liquids can exist in significant quantities within these formations. These hydrocarbons formed over millions of years through the burial and transformation of organic material, such as plants and microorganisms, under increasing heat and pressure. While some hydrocarbons

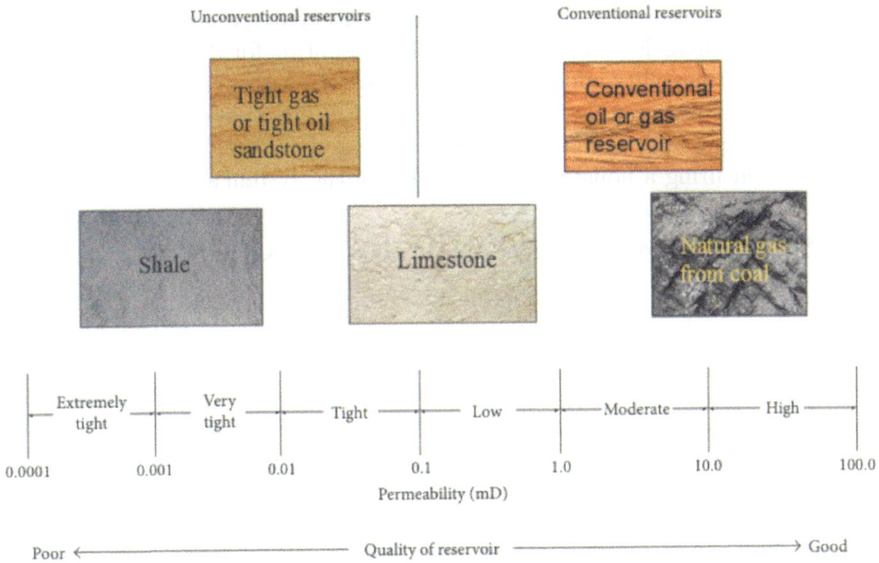

Figure 6.3 Classification of geologic formations based on permeability, from tight unconventional reservoirs to more permeable conventional reservoirs [141].

migrated to adjacent, more porous and permeable rock layers, making extraction relatively straightforward, much of the hydrocarbons remained trapped in the tighter, low-permeability formations.

The classification of petroleum reservoirs into conventional and unconventional reservoirs primarily depends on the permeability of the formations and the technology required for extraction. Conventional reservoirs, characterized by relatively high permeability, enable the easy transport of oil and gas, making conventional extraction methods effective. In contrast, unconventional reservoirs have much lower permeability, rendering traditional extraction methods ineffective. Specialized techniques such as hydraulic fracturing or horizontal drilling are often required to recover hydrocarbons from these formations.

Figure 6.3 illustrates the classification of geologic formations based on their permeability. It spans a spectrum from extremely tight shale formations to more permeable conventional reservoirs, highlighting the relationship between permeability and reservoir quality.

At the nanoscale, Knudsen diffusion becomes the dominant transport mechanism, especially when pore sizes are comparable to the mean free path of gas molecules. This phenomenon is typically incorporated into comprehensive models that couple diffusion with Darcy's flow to account for multi-scale transport processes. To further refine these predictions, Langmuir-type adsorption models are integrated with transport equations, enabling a more accurate representation of gas release and recovery dynamics. Additionally, geomechanical models play a crucial role in accounting for the effects of stress changes on permeability and porosity, which are key factors in unconventional reservoirs. For instance, in tight formations, the propagation of fractures is modeled using stress-strain relationships that are intricately linked with flow equations, ensuring a holistic understanding of the system's behavior.

6.7 SIMPLIFIED MODEL AND ANALYTICAL SOLUTION

The analytical solution is a valuable tool for validating numerical schemes and calculating the pressure distribution near a wellbore. In this section, we provide an analytical solution using the power-series method for the apparent permeability gas transport equation [145]. The development of the model is based on several assumptions and simplifications to ensure tractability while capturing the essential physics of gas flow in shale reservoirs. We consider single-phase gas flow within the reservoir, assuming that the gas behaves as an ideal gas. Additionally, the flow is considered isotropic, meaning the permeability tensor simplifies to $\mathbf{K} = k\mathbf{I}$, where k is a scalar permeability and $\mathbf{I}$ is the identity tensor. Furthermore, gas sorption and desorption effects are neglected. The mass balance equation for gas flow in a fractured porous medium is given by:

$$\frac{\partial M}{\partial t} + \nabla \cdot (\rho \mathbf{u}) = 0, \tag{6.53}$$

where $M = \varphi\rho$ is the mass accumulation, ρ is the gas density, φ is the porosity, and $\mathbf{u}$ is the velocity vector. Using the ideal gas assumption, the density is expressed as $\rho = \gamma p$, where $\gamma = \frac{M_g}{RT}$ is the proportionality constant.

Substituting $\rho = \gamma p$ and $\mathbf{u}$ from Darcy's law with the Klinkenberg correction for apparent permeability, the velocity is:

$$\mathbf{u} = -\frac{k}{\mu}\left(1 + \frac{b}{p}\right)\nabla p, \tag{6.54}$$

where b is the Klinkenberg correction factor. Substituting these expressions into the mass balance equation, we obtain the governing equation:

$$\gamma\varphi\frac{\partial p}{\partial t} - \nabla \cdot \left(\frac{\gamma k}{\mu}p\left(1 + \frac{b}{p}\right)\nabla p\right) = 0. \tag{6.55}$$

We consider a reservoir with infinite horizontal extent and axisymmetric properties around an isolated production well. Assuming a 1D radial cylindrical coordinate system and ignoring adsorption and source terms, the governing equation becomes:

$$\varphi\frac{\partial p}{\partial t} = \frac{1}{r}\frac{\partial}{\partial r}\left(\frac{rk_0}{\mu}(b + p)\frac{\partial p}{\partial r}\right), \tag{6.56}$$

where φ is the porosity, k_0 is the permeability, μ is the gas viscosity, and b is the Klinkenberg correction factor.

To simplify the equation, we apply the Boltzmann transformation:

$$y = \frac{r^2}{4t\chi}, \quad \text{where} \quad \chi = \frac{k_0}{\varphi\mu}. \tag{6.57}$$

Using this transformation, the governing equation reduces to:

$$(y+b+p)\frac{\mathrm{d}p}{\mathrm{d}y} + y\left(\frac{\mathrm{d}p}{\mathrm{d}y}\right)^2 + (b+p)y\frac{\mathrm{d}^2p}{\mathrm{d}y^2} = 0. \tag{6.58}$$

We assume the solution $p(y)$ can be expressed as a power series:

$$p(y) = \sum_{n=0}^{\infty} a_n y^n, \tag{6.59}$$

where $\{a_n\}$ are the coefficients to be determined. Substituting this series into the transformed equation and using series manipulation, we calculate the terms for $\frac{dp}{dy}$, $\frac{d^2p}{dy^2}$, and their products.

The product of two series $\sum_{i=0}^{\infty} a_i y^i$ and $\sum_{j=0}^{\infty} b_j y^j$ can be expressed as:

$$\sum_{i=0}^{\infty} a_i y^i \cdot \sum_{j=0}^{\infty} b_j y^j = \sum_{k=0}^{\infty} c_k y^k, \quad c_k = \sum_{l=0}^{k} a_l b_{k-l}. \tag{6.60}$$

Substituting back into the transformed equation, the series terms give a recurrence relation:

$$a_{n+1} = \frac{-1}{(b+a_0)(n+1)^2}\{na_n + \ldots\}, \quad n > 2, \tag{6.61}$$

where $\{a_n\}$ depend on boundary conditions. Boundary conditions are applied to determine the coefficients. At $y = \infty$, $p(L) = p_0$. At $y \to 0$, $p - p_w$ (Dirichlet) or $\frac{\mathrm{d}p}{\mathrm{d}y} = \frac{Q\mu}{2\pi kH}$ (Neumann). For example, with Dirichlet conditions, $p_w = a_0$. For Neumann conditions:

$$\frac{\mathrm{d}p}{\mathrm{d}y} \approx 3a_3y^2 + \ldots, \tag{6.62}$$

and the coefficient a_3 is:

$$a_3 = \frac{\frac{Q\mu}{2\pi kH}}{3y_0^2\left(1 - \frac{1+a_0}{4(b+a_0)}y_0\right)}. \tag{6.63}$$

For more details of the model derivation and analytical solution, refer to El-Amin [145].

6.8 ADSORBED GAS AND LANGMUIR ISOTHERM

In shale reservoirs, gas exists as free gas within the pore spaces and adsorbed gas on the surface of the matrix. The adsorbed gas mass per unit volume is described by the Langmuir isotherm:

$$q_a = \frac{\rho_s M_w}{V_{\text{std}}} \left(\frac{V_L p_m}{p_L + p_m} \right), \tag{6.64}$$

where q_a is the adsorbed gas mass per unit volume of shale [kg/m^3], ρ_s is the shale density [kg/m^3], M_w is the molecular weight of the gas [kg/mol], V_{std} is the molar volume under standard conditions [m^3/mol], V_L is the Langmuir volume [m^3/kg], and p_m and p_L are the matrix and Langmuir pressures, respectively, both measured in [Pa].

The total gas accumulation M in shale reservoirs includes contributions from both free gas and adsorbed gas. For free gas, the mass accumulation per unit volume is:

$$M_{\text{free}} = \varphi \rho, \tag{6.65}$$

where φ is the porosity (dimensionless) and ρ is the free gas density [kg/m^3].

The adsorbed gas contribution is:

$$M_{\text{adsorbed}} = (1 - \varphi) q_a, \tag{6.66}$$

where q_a is defined in Equation (1).

Thus, the total gas accumulation is:

$$M = M_{\text{free}} + M_{\text{adsorbed}} = \varphi \rho + (1 - \varphi) q_a. \tag{6.67}$$

The governing equation for gas transport in a porous medium is based on the principle of mass conservation:

$$\frac{\partial M}{\partial t} + \nabla \cdot (\rho \mathbf{u}) = q, \tag{6.68}$$

where M is the total gas mass per unit volume [kg/m^3], $\mathbf{u}$ is the velocity vector [m/s], and q is a source or sink term [kg/(m$^3 \cdot$ s)].

Substituting M from Equation (5) into Equation (6), we have:

$$\frac{\partial}{\partial t} (\varphi \rho + (1 - \varphi) q_a) + \nabla \cdot (\rho \mathbf{u}) = q. \tag{6.69}$$

For low-permeability systems, the apparent permeability k_m incorporates the Klinkenberg effect:

$$k_m = k_{m0} \left(1 + \frac{b_m}{p_m} \right), \tag{6.70}$$

where k_m is the apparent permeability [m^2], k_{m0} is the intrinsic permeability [m^2], b_m is the Klinkenberg parameter [Pa], and p_m is the matrix pressure [Pa].

The gas flux is given by Darcy's law:

$$\mathbf{u} = -\frac{k_m}{\mu}\nabla p, \tag{6.71}$$

where μ is the gas viscosity [Pa · s].

Substituting k_m from Equation (8) and $\mathbf{u}$ from Equation (9) into Equation (7), we obtain the final governing equation for gas transport:

$$\varphi\frac{\partial \rho}{\partial t} + \frac{\partial}{\partial t}\left[(1-\varphi)q_a\right] - \nabla\cdot\left(\frac{\rho k_m}{\mu}\nabla p\right) = q. \tag{6.72}$$

Here, q_a is explicitly defined in terms of p_m using the Langmuir isotherm. This equation governs the behavior of gas in shale reservoirs.

6.9 DUAL-CONTINUUM MODEL

The dual-continuum model has long been a cornerstone for studying flow behavior in fractured porous media. First introduced by Warren and Root [132], this idealized framework effectively described fluid flow in systems with dual porosities, one associated with the matrix and the other with fractures. Over time, this approach has been extensively adopted and refined to address the complexities of fractured reservoirs.

Bustin et al. [146] utilized a standard dual-continuum model to investigate the impact of permeability on gas flow, while Ozkan et al. [147] applied a dual-porosity model in which the matrix was represented as uniform spherical blocks. This model incorporated diffusive Darcy flow in the matrix and stress-dependent permeability in fractures but excluded sorption and desorption processes. Similarly, Moridis et al. [148] used the dual-continuum model to explore multiple mechanisms within kerogen, a critical component of shale reservoirs.

Wu and Fakcharoenphol [149] proposed a generalized dual-continuum methodology that formed the basis for advanced reservoir simulators. Although their model did not consider adsorption and desorption, it provided a robust framework for simulating fluid and heat flow in fractured unconventional reservoirs. Guo et al. [150] extended these efforts by developing a mathematical model integrating key mechanisms governing flow in tight shale gas formations. Recently, El Amin et al. [151] presented an analytical solution using the power-series method for the apparent permeability Klinkenberg model. Significant contributions to the theoretical and mathematical foundations of dual-porosity models have also been made by Arbogast et al. [152] and Showalter et al. [153].

The DPDP model is a mathematical framework that describes flow in fractured porous media. The model is governed by two coupled partial differential equations (PDEs), representing the pressures in the matrix (p_m) and fractures (p_f). These equations account for the transfer of gas between the two domains and are defined as follows [151]:

The matrix pressure equation is given by:

$$F_1(p_m)\frac{\partial p_m}{\partial t} - \nabla \cdot \left(\frac{\gamma k_{m0}}{\mu_g}\left(1+\frac{b_m}{p_m}\right)p_m \nabla p_m\right) = -S, \tag{6.73}$$

where $F_1(p_m)$ is a scaling function incorporating the compressibility of the porous matrix, gas properties, and interactions between the gas and the solid matrix. The parameters are: γ, the correction factor for gas deviation from ideal gas behavior; k_{m0}, the intrinsic permeability of the matrix [m^2]; μ_g, the gas viscosity [Pa · s]; b_m, the Klinkenberg parameter [Pa]; and $p_m \nabla p_m$, which drives the flow as a function of the pressure gradient. The term $-S$ represents the net transfer of gas from the matrix to fractures or other sinks.

The fracture pressure equation is:

$$F_2(p_f)\frac{\partial p_f}{\partial t} - \nabla \cdot \left(\frac{\gamma k_{f0}}{\mu_g}\left(1+\frac{b_f}{p_f}\right)p_f \nabla p_f\right) = S - q_p, \tag{6.74}$$

where $F_2(p_f)$ accounts for the compressibility and properties of the fracture network. The additional terms include k_{f0}, the intrinsic permeability of fractures [m^2]; b_f, the Klinkenberg parameter for fractures [Pa]; and q_p, the gas production or withdrawal rate from the fracture network [kg/(m^3 · s)]. The exchange term S represents the transfer of gas between the matrix and fractures. The scaling functions $F_1(p_m)$ and $F_2(p_f)$ introduce the effects of gas compressibility and the interplay between fluid and solid phases. For the matrix scaling function $F_1(p_m)$, we have [151]:

$$F_1(p_m) = \frac{\gamma}{Z}\left[\varphi_m + \varphi_m' p_m\right] + (1-\varphi_m)\frac{\rho_s M_w V_L p_L}{V_{\text{std}}\left(p_L+p_m\right)^2} - \frac{\rho_s M_w V_L \varphi_m'}{V_{\text{std}}\left(p_L+p_m\right)}, \tag{6.75}$$

where φ_m is the matrix porosity, φ_m' is the change in matrix porosity with pressure [Pa^{-1}], ρ_s is the solid matrix density [kg/m^3], M_w is the molecular weight of the gas [kg/mol], V_L is the Langmuir volume constant [m^3/kg], p_L is the Langmuir pressure constant [Pa], and V_{std} is the standard molar gas volume [m^3/mol]. For the fracture scaling function $F_2(p_f)$, we have:

$$F_2(p_f) = \frac{\gamma}{Z}\left(\varphi_f + \varphi_f' p_f\right), \tag{6.76}$$

where φ_f is the fracture porosity and φ_f' is the change in fracture porosity with pressure [Pa^{-1}].

The above equations are coupled through a transfer term that accounts for mass exchange, represented by a shape factor. Warren and Root [132] defined the shape factor for cubic matrix blocks as:

$$\sigma = \frac{2n(n+2)}{l^2}, \tag{6.77}$$

where n is the set of normal fractures, and l is a characteristic length defined by the geometry of the matrix block.

The gas transfer between the matrix and fractures is expressed as:

$$S = \frac{k_m \xi \sigma (p_m - p_f)}{\mu_g}, \tag{6.78}$$

where ξ is the correction factor for gas deviation from ideal gas behavior, σ is the shape factor for matrix-fracture interaction, and μ_g is the gas viscosity.

6.9.1 STRESS-DEPENDENT POROSITY AND PERMEABILITY

The concept of effective stress significantly influences both porosity and permeability in fractured systems. Biot and Willis [154] generalized this concept by expressing porosity as [151]:

$$\varphi_\beta = \varphi_{r,\beta} + (\varphi_{0,\beta} - \varphi_{r,\beta})\, e^{-a\sigma'_\beta}, \tag{6.79}$$

where $\varphi_{0,\beta}$ is the initial porosity, $\varphi_{r,\beta}$ is the reduced porosity, a is the stress sensitivity coefficient, and $\beta = m, f$ denotes matrix (m) and fractures (f). The effective stress is defined as:

$$\sigma'_\beta = \sigma_\beta - \alpha_\beta p_\beta, \tag{6.80}$$

where σ_β is the total stress, α_β is the Biot's effective stress coefficient, and p_β is the fluid pressure in the corresponding domain (p_m for the matrix and p_f for fractures). The exponential term $e^{-a\sigma'_\beta}$ in Equation (6) requires that the argument $-a\sigma'_\beta$ is dimensionless to ensure dimensional consistency. Here, the units of a must be the reciprocal of the units of σ'_β, ensuring that their product $a\sigma'_\beta$ is dimensionless. This critical requirement is essential for the correctness of the mathematical model.

The permeability of shale is strongly influenced by stress, as changes in effective stress can compress or expand the pore structure, altering fluid flow pathways. This relationship is often modeled as:

$$k = k_0 e^{-c\sigma}, \tag{6.81}$$

where k is the permeability under effective stress [m^2], k_0 is the initial permeability at zero effective stress [m^2], c is the stress sensitivity coefficient [Pa^{-1}], σ is the effective stress [Pa].

The exponential function requires that the argument $-c\sigma$ be dimensionless to ensure dimensional consistency. To satisfy this requirement, c must have units of [Pa^{-1}], making the product $c\sigma$ dimensionless. This ensures that the permeability model is physically and mathematically valid.

6.9.2 INITIAL AND BOUNDARY CONDITIONS

We assume that the entire computational domain for both the matrix blocks and fractures starts with the same initial pressure, expressed as:

$$p_m(x, y, t = 0) = p_f(x, y, t = 0) = p_0, \tag{6.82}$$

where p_0 represents the uniform initial pressure across the domain.
For the DPDP matrix continuum, Neumann boundary conditions are applied, which can be expressed as:

$$F_{m,n} = 0 \quad \left(\frac{\partial p_m}{\partial n} = 0 \right), \tag{6.83}$$

indicating no flow across the domain boundaries in the matrix continuum.
In the DPDP fracture continuum, Neumann boundary conditions are also applied for all boundaries except at the production well, which is treated as a Dirichlet boundary condition. These conditions are expressed as:

$$F_{f,n} = 0 \quad \left(\frac{\partial p_f}{\partial n} = 0 \right), \tag{6.84}$$

and

$$p_f(x, y, t) = p_w, \tag{6.85}$$

where p_w denotes the pressure at the production well.

To define the sink term, we adopt the model for gas production from a vertical well. The production rate is given by:

$$q_p = \frac{k_{f0}\theta\gamma}{Z\mu_g \left[\ln\left(\frac{r_e}{r_w}\right) + S + D_q \right]} p_f \left(p_{fe} - p_w \right), \tag{6.86}$$

where $\theta = 2\pi$ when the production well is placed at the center, and $\theta = \pi/2$ when it is located at the corner (Peaceman, 1983), p_w is the bottom hole pressure, p_{fe} is the average pressure in the fractures around the well, r_w is the well radius, and r_e is the drainage radius, which can be calculated as:

$$r_e = \begin{cases} 0.14\sqrt{2\left[(\Delta x)^2 + (\Delta y)^2\right]} & \text{if } k_x = k_y, \\ 0.28\dfrac{\sqrt{\left(\frac{k_y}{k_x}\right)^{\frac{1}{2}}\Delta x^2 + \left(\frac{k_x}{k_y}\right)^{\frac{1}{2}}\Delta y^2}}{\left(\frac{k_y}{k_x}\right)^{\frac{1}{4}} + \left(\frac{k_x}{k_y}\right)^{\frac{1}{4}}} & \text{if } k_x \neq k_y, \end{cases} \tag{6.87}$$

where Δx and Δy are the grid lengths in the x- and y-directions.

In the dual-porosity dual-permeability (DPDP) model, the source/sink term for fractures is expressed as $Q_f = S - q_p$, whereas for the matrix blocks, it is given by $Q_m = S$.

6.9.3 NUMERICAL SOLUTION

To solve the differential systems (6.73)-(6.74) numerically, we adopt the cell-centered finite difference (CCFD) method (see chapter 4 of this book). In the CCFD scheme, velocity is defined at cell edges, while pressure and permeability are defined at the cell centers. Permeability values are computed using harmonic averaging to ensure numerical stability and accuracy. The discretized equation for cell i becomes:

$$\frac{F_\alpha}{\Delta t} p_{\alpha,i}^{(k+1)} + \frac{1}{x_i}\left\{ \gamma \left[-\frac{2k_\alpha}{\mu_g}\left(1+\frac{b_\alpha}{p_i^{(k)}}\right) p_i^{(k)} \frac{p_{\alpha,i+1}^{(k+1)} - p_{\alpha,i}^{(k+1)}}{\Delta x_i + \Delta x_{i+1}} \right] - \left[-\frac{2k_\alpha}{\mu_g}\left(1+\frac{b_\alpha}{p_i^{(k)}}\right) p_i^{(k)} \frac{p_{\alpha,i}^{(k+1)} - p_{\alpha,i-1}^{(k+1)}}{\Delta x_i + \Delta x_{i-1}} \right] \right\} = Q_p^{(k)} + \frac{F_\alpha}{\Delta t} p_i^{(k)}. \tag{6.88}$$

Boundary conditions are critical to numerical discretization. If Dirichlet boundary conditions are applied at the west boundary $(i-1)$, the discretized equation becomes:

$$F_\alpha \frac{p_{\alpha,i-1}^{(k+1)} - p_{\alpha,i-1}^{(k)}}{\Delta t} + \frac{1}{\Delta x_{i-1}}\left\{ \gamma \left[-\frac{2k_\alpha}{\mu_g}\left(1+\frac{b_\alpha}{p_{i-1}^{(k)}}\right) p_{i-1}^{(k)} \frac{p_{\alpha,i}^{(k+1)} - p_{\alpha,i-1}^{(k+1)}}{\Delta x_{i-1} + \Delta x_i} \right] - \left[-\frac{2k_\alpha}{\mu_g}\left(1+\frac{b_\alpha}{p_{i-1}^{(k)}}\right) p_{i-1}^{(k)} \frac{p_{\alpha,i-1}^{(k+1)} - p_\alpha^{\mathrm{WB}}}{\Delta x_{i-1}} \right] \right\} = Q_p^{(k)}, \tag{6.89}$$

where p_α^{WB} is the pressure at the west Dirichlet boundary.
For the east boundary $(i+1)$, the discretized equation becomes:

$$F_\alpha \frac{p_{\alpha,i+1}^{(k+1)} - p_{\alpha,i+1}^{(k)}}{\Delta t} + \frac{1}{\Delta x_{i+1}}\left\{ \gamma \left[-\frac{2k_\alpha}{\mu_g}\left(1+\frac{b_\alpha}{p_{i+1}^{(k)}}\right) p_{i+1}^{(k)} \frac{p_\alpha^{\mathrm{EB}} - p_{\alpha,i+1}^{(k+1)}}{\Delta x_{i+1}} \right] - \left[-\frac{2k_\alpha}{\mu_g}\left(1+\frac{b_\alpha}{p_{i+1}^{(k)}}\right) p_{i+1}^{(k)} \frac{p_{\alpha,i+1}^{(k+1)} - p_{\alpha,i}^{(k+1)}}{\Delta x_{i+1} + \Delta x_i} \right] \right\} = Q_p^{(k)}, \tag{6.90}$$

where p_α^{EB} is the pressure at the east Dirichlet boundary.
A similar approach is applied for the y-direction discretization.
The computational domain consists of a 2D reservoir with a production well at the center, as shown in Figure 6.4. Due to symmetry, only the bottom-left quarter of the domain is simulated.
The DPDP model assumes uniform porosity and permeability throughout the domain.

The numerical simulations were conducted using the DPDP model to evaluate gas transport in fractured porous media. The reservoir domain was discretized into a uniform 2D grid with the CCFD method to ensure local conservation. Neumann boundary conditions were applied for non-production boundaries, while the production well was treated with a Dirichlet boundary condition. The initial pressure for both the matrix and fracture domains was set uniformly to $p_0 = 6.0 \times 10^6$ Pa. The physical properties used in the simulation include a porosity of $\varphi_m = 0.1$ for the matrix and $\varphi_f = 0.3$ for the fractures, with an initial permeability of $k_m = 1.0 \times 10^{-15}\,\mathrm{m}^2$ for the matrix

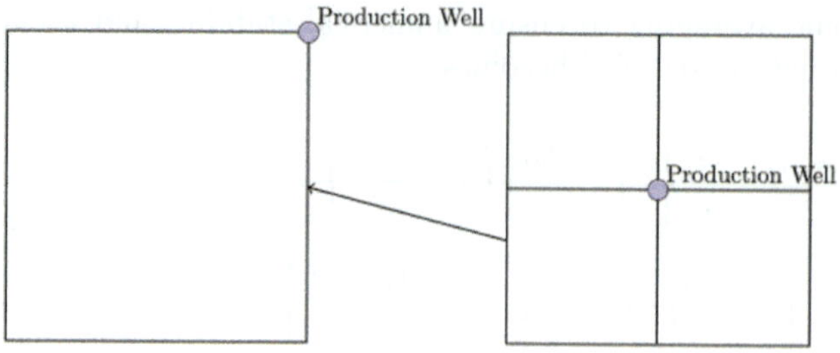

Figure 6.4 2D reservoir domain with a production well at the center. Symmetry allows modeling only one-quarter of the domain.

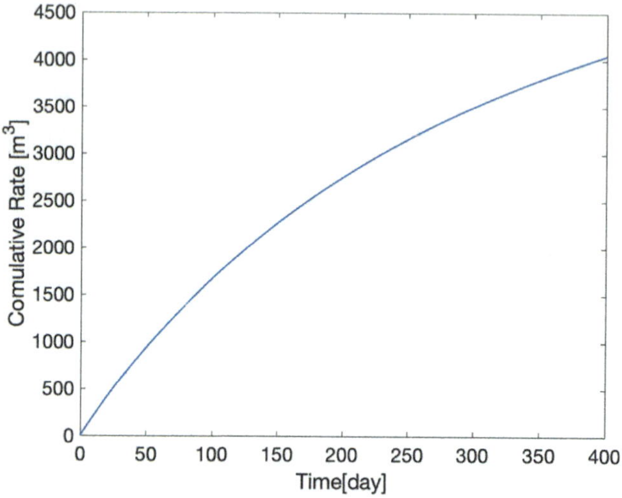

Figure 6.5 Cumulative gas production rate over time. The curve illustrates how the production accumulates in a 2D reservoir simulation. The steady increase demonstrates effective reservoir management under the applied model assumptions.

and $k_f = 1.0 \times 10^{-12}\,\mathrm{m}^2$ for the fractures. The Klinkenberg parameter b_m was set to 1.0×10^6 Pa, while gas viscosity was fixed at $\mu_g = 1.0 \times 10^{-5}\,\mathrm{Pa \cdot s}$. The production well maintained a bottom-hole pressure of $p_w = 1.0 \times 10^5$ Pa. These parameters were chosen to emulate realistic reservoir conditions and provide insight into the gas flow behavior within the dual-porosity framework.

The plot in Figure 6.5 shows the cumulative gas production rate over time. It highlights the reservoir's performance, indicating a gradual increase in production, which aligns with the natural depletion of pressure in the domain. This figure is essential for assessing the sustainability of production over the modeled time period.

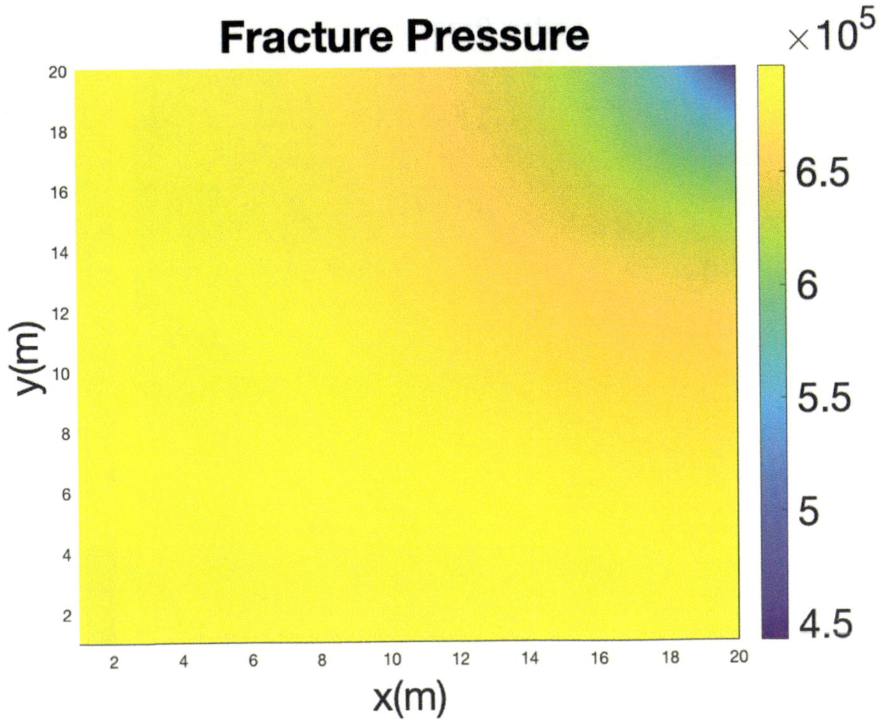

Figure 6.6 Fracture pressure distribution in the reservoir. Higher pressure values are concentrated near the outer boundaries, while the pressure depletes near the production well due to continuous withdrawal.

Figure 6.6 displays the pressure distribution within the fractures of the reservoir. The gradient from higher pressure in the boundary regions to lower pressure near the production well is consistent with expectations for a reservoir under production. This distribution is critical for understanding the pressure-driven flow mechanisms in the fractures.

Figure 6.7 illustrates the matrix pressure distribution. Unlike the fracture system, the pressure depletion in the matrix is more gradual, consistent with its lower permeability. This figure underscores the dual-continuum nature of the model, highlighting the interplay between the fracture and matrix systems.

6.10 MATLAB® CODE FOR DPDP PROBLEM

DESCRIPTION

```
clear all; close all
model.numDims = 2;
days =  0.1;
```

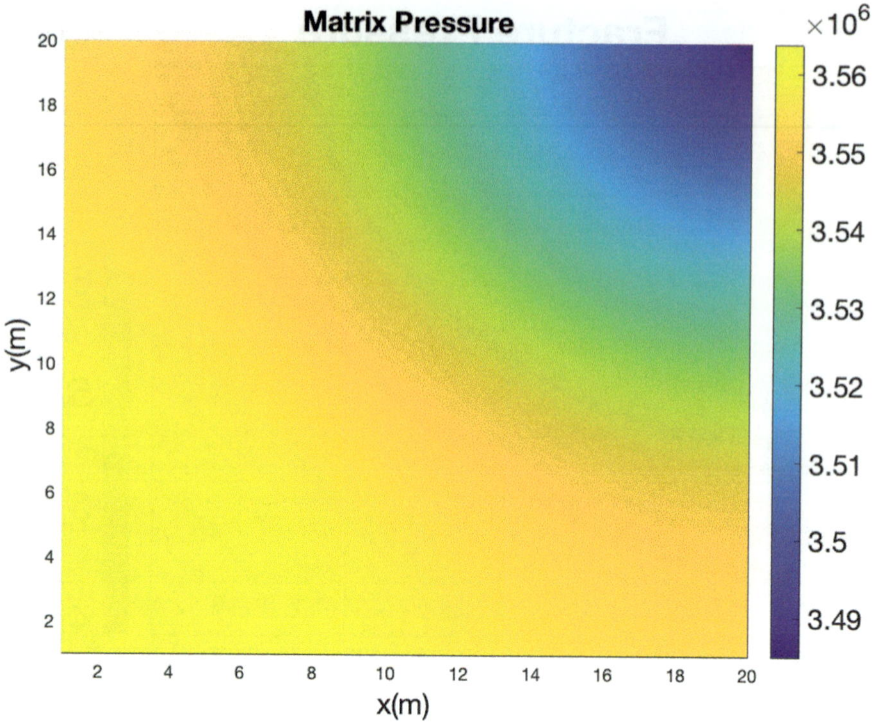

Figure 6.7 Matrix pressure distribution in the reservoir. The gradient is less steep compared to the fracture pressure, reflecting the lower permeability and slower depletion in the matrix.

```
nt = 4000;
model.xs = 0:1:20;
model.ys = 0:1:20;
md = 10^-15;                % Conversion to m^2
Kmi = 1.0E-04 * md;         % Matrix permeability (m^2)
Kfi = 10 * md;              % Fracture permeability (m^2)
Phi_m = 0.05;               % Matrix porosity
Phi_f = 0.001;              % Fracture porosity
model.Pw = 4.*10^5;         % Bottom-hole pressure (Pa)
timeEnd = days * 24 * 3600;
nx = numel(model.xs) - 1;
ny = numel(model.ys) - 1;
model.nx = nx;
model.ny = ny;
model.nt = nt;
model.delx = differ(model.xs); % Grid spacing in x
model.dely = differ(model.ys); % Grid spacing in y
model.deltaX = repmat(model.delx, ny, 1);
model.deltaY = repmat(model.dely, nx, 1)';
model.Spaces = model.deltaX .* model.deltaY;
```

```
model.ts = linspace(0, timeEnd, nt+1);
Mg = 0.016;                  % Mole weight of CH4 (kg/mol)
R = 8.314;                   % Gas constant (Pa*m^3/(mol*K))
Z = 1;                       % Gas compressibility factor
Temp = 343.15;               % Temperature (K)
model.gama = Mg / (Z * R * Temp);
model.Mu_g = 1.02E-05;  % Initial gas viscosity (Pa*s)
```

This section initializes the simulation environment for a quarter five-spot problem, where Neumann boundary conditions (B.C.) are applied to injection wells, and Dirichlet B.C. are used for production wells. Gravity is ignored for simplicity. The simulation duration is set to 0.1 days, and the number of time steps (`nt`) is 4000. The computational grid is defined with x-coordinates (`model.xs`) and y-coordinates (`model.ys`) ranging from 0 to 20. `md` is a scaling factor for permeability in m^2. Matrix and fracture permeabilities are initialized as `Kmi` and `Kfi`, respectively. Porosities (`Phi_m` and `Phi_f`) for matrix and fractures are set. `model.Pw` defines the bottom-hole pressure, and the total simulation time (`timeEnd`) is calculated in seconds. `nx` and `ny` represent the number of grid cells in the x and y directions. `model.delx` and `model.dely` store grid spacings in x and y, while `model.deltaX` and `model.deltaY` extend these spacings over the grid for computational purposes. The total simulation time is divided into evenly spaced time steps (`model.ts`). The gas properties (molecular weight, compressibility factor, etc.) and physical constants are initialized. `model.gama` is calculated based on these parameters.

PERMEABILITY AND POROSITY INITIALIZATION

```
KxxM = ones(nx, ny) * Kmi;
model.KxxM = KxxM;
model.KyyM = KxxM;

KxxF = ones(nx, ny) * Kfi;
model.KxxF = KxxF;
model.KyyF = KxxF;

poroM = ones(nx, ny) * Phi_m;
model.poroM = poroM;

poroF = ones(nx, ny) * Phi_f;
model.poroF = poroF;
```

The permeability (KxxM, KxxF, etc.) and porosity (`poroM`, `poroF`) for the matrix and fractures are initialized as constant values across all grid cells.

BOUNDARY AND INITIAL CONDITIONS

```
model.isDiriXF = zeros(2, ny);
model.isDiriXF(end, end) = 1;
model.isDiriYF = zeros(nx, 2);
model.isDiriYF(end, end) = 1;
```

```
model.PBdryXF = zeros(2, ny);
model.PBdryXF(end, end) = model.Pw;
model.PBdryYF = zeros(nx, 2);
model.PBdryYF(end, end) = model.Pw;

model.PInit = ones(nx, ny) * 10.4e6;
```

Boundary conditions (isDiriXF, PBdryXF, etc.) are set to indicate Dirichlet-type boundaries for specific edges. The initial pressure is set uniformly across the grid to 10.4 MPa.

SIMULATION EXECUTION

```
model = RST_fracture_SinglePhaseFlow(model);
MRST_plot(model);
```

The function RST_fracture_SinglePhaseFlow is called to perform the simulation. Results are visualized using MRST_plot.

VISUALIZATION

```
[X, Y] = meshgrid(model.xs(2:end), model.ys(2:end));

figure(1)
subplot(1, 2, 1)
plot(model.Time / 10, model.ProductionRateF1, 'LineWidth', 2);
xlabel('Time␣[day]', 'fontsize', 14)
ylabel('Production␣Rate␣[m^3/day]', 'fontsize', 14)

subplot(1, 2, 2)
plot(model.Time / 10, model.CumulativeProductionRateF1, '
    LineWidth', 2);
xlabel('Time␣[day]', 'fontsize', 14)
ylabel('Cumulative␣Production␣[m^3]', 'fontsize', 14)
```

The production rate and cumulative production are plotted against time, illustrating the simulation results.

ASSOCIATED MATLAB® FUNCTIONS

This document contains MATLAB® functions used to simulate a quarter five-spot problem. The functions are modular, focusing on specific tasks such as boundary condition handling, transmissibility calculations, and solving pressure and velocity fields.
This function computes the derivatives for a property across grid cells.

```
function del_PropD = differ(nd, ndo, PropD, deld)
    del_PropD = PropD(2:end) - PropD(1:end-1);
    del_PropD = [del_PropD; 0];
    del_PropD = reshape(del_PropD, nd + 1, ndo);
    del_PropD = del_PropD(1:end-1, :);
```

```
    for i = 1:nd
        del_PropD(i, :) = del_PropD(i, :) / deld(i);
    end
end
```

This function computes harmonic derivatives for a property.

```
function del_Kdd_Harm = differ_Harmonic_Avg(nd, ndo, Kdd_Harm,
    deld)
    del_Kdd_Harm = Kdd_Harm(2:end) - Kdd_Harm(1:end-1);
    del_Kdd_Harm = [del_Kdd_Harm; 0];
    del_Kdd_Harm = reshape(del_Kdd_Harm, nd + 1, ndo);
    del_Kdd_Harm = del_Kdd_Harm(1:end-1, :);
    for i = 1:nd
        del_Kdd_Harm(i, :) = del_Kdd_Harm(i, :) / deld(i);
    end
end
```

This function calculates harmonic averages for transmissibility across grid cells.

```
function [Kdd_Harm] = Harmonic_Avg(nd, ndo, deld, Kdd_vec)
    k = 0;  % Counter for layers
    Kdd_Harm = zeros((nd + 1) * ndo, 1);
    for i = 1:(nd + 1) * ndo
        j = i - k * (nd + 1);
        if (i == k * (nd + 1) + 1)  % At West or South Boundary
            Kdd_Harm(i) = Kdd_vec(i - k);
        elseif (mod(i, nd + 1) == 0)  % At East or North Boundary
            Kdd_Harm(i) = Kdd_vec(i - 1 - k);
            k = k + 1;
        else
            Kdd_Harm(i) = (deld(j - 1) + deld(j)) / ...
                          (deld(j - 1) / Kdd_vec(i - 1 - k) + ...
                           deld(j) / Kdd_vec(i - k));
        end
    end
end
```

This function applies North South boundary conditions to the main diagonal of a matrix.

```
function [Main_diag_Y, RHS] = Main_diag_SN_BC(nx, ny, del, isDiri
    , PBdry, UBdry, CondY, Main_diag_Y, RHS, gama, p_vector,
    bf_vector)
    k = 0;
    for i = 1:nx
        if (isDiri(i, 1) == 1)
            Main_diag_Y(i) = Main_diag_Y(i) + ...
                             ((2 * gama) / del(1)^2) * ...
                             (-CondY(i + k * ny)) * ...
                             (p_vector(i + k * ny) + bf_vector(i
                                 + k * ny));
            RHS(i) = RHS(i) + ...
                     ((2 * gama) / del(1)^2) * ...
                     (-PBdry(i, 1)) * ...
```

```
                    (-CondY(i + k * ny)) * ...
                    (p_vector(i + k * ny) + bf_vector(i + k * ny
                        ));
            k = k + 1;
        else
            RHS(i) = RHS(i) + gama * UBdry(i, 1) / del(1);
            k = k + 1;
        end
        if (isDiri(i, 2) == 1)
            Main_diag_Y((ny - 1) * nx + i) = Main_diag_Y((ny - 1)
                * nx + i) + ...
                                        ((2 * gama) / del(end
                                            )^2) * ...
                                        (-CondY(i * ny + i))
                                            * ...
                                        (p_vector((ny - 1) *
                                            nx + i) +
                                            bf_vector((ny -
                                            1) * nx + i));
            RHS((ny - 1) * nx + i) = RHS((ny - 1) * nx + i) - ...
                                ((2 * gama) / del(end)^2) *
                                    ...
                                PBdry(i, 2) * ...
                                (-CondY(i * ny + i)) * ...
                                (p_vector((ny - 1) * nx + i)
                                    + bf_vector((ny - 1) *
                                    nx + i));
        else
            RHS((ny - 1) * nx + i) = RHS((ny - 1) * nx + i) -
                gama * UBdry(i, 2) / del(end);
        end
    end
end
```

This is the main function to solve the single-phase fracture flow problem.

```
function model = RST_fracture_SinglePhaseFlow(model)
    % Preparing variables
    model.KyyM = model.KyyM';
    model.UBdryXM(1, :) = -model.UBdryXM(1, :);
    model.UBdryYM(:, 1) = -model.UBdryYM(:, 1);

end
```

The following function computes the side diagonals for North and South directions.

```
function [Side_diag_Y] = Side_diag_SN(nx, ny, Coeff_Y, CondY,
    Pressure, bf)
    Side_diag_Y = zeros(ny - 1, nx);
    CondY_trim = reshape(CondY, ny + 1, nx);
    CondY_trim = CondY_trim(2:end-1, :);
    for i = 1:size(CondY_trim, 1)
        Side_diag_Y(i, :) = -Coeff_Y(i) .* CondY_trim(i, :);
    end
    Side_diag_Y = Side_diag_Y .* (Pressure(1:ny-1, :) + bf(1:ny
        -1, :));
```

```
    Side_diag_Y = Side_diag_Y';
    Side_diag_Y = reshape(Side_diag_Y, nx * (ny - 1), 1);
end
```

The following function computes the side diagonals for East and West directions.

```
function [Side_diag_X] = Side_diag_WE(nx, ny, Coeff_X, CondX,
    Pressure, bf)
    Side_diag_X = zeros(nx, ny);
    CondX_trim = reshape(CondX, nx + 1, ny);
    CondX_trim = CondX_trim(2:end-1, :);
    for i = 1:size(CondX_trim, 1)
        Side_diag_X(i, :) = -Coeff_X(i) .* CondX_trim(i, :);
    end
    Side_diag_X = Side_diag_X .* (Pressure + bf);
    Side_diag_X = reshape(Side_diag_X, nx * ny, 1);
    Side_diag_X = Side_diag_X(1:end-1);
end
```

The following function performs upwinding in the X direction.

```
function [C_upwind, ymeth_upwind, yprop_upwind] = Upwind_X(C, yG1
    , yG2, UG)
    Q = UG >= 0;
    C_upwind = Q .* C(:, 1:end-1) + (1 - Q) .* C(:, 2:end);
    ymeth_upwind = Q .* yG1(:, 1:end-1) + (1 - Q) .* yG1(:, 2:end
        );
    yprop_upwind = Q .* yG2(:, 1:end-1) + (1 - Q) .* yG2(:, 2:end
        );
end
```

The following function performs upwinding in the Y direction.

```
function [C_upwind, yG1_upwind, yG2_upwind] = Upwind_Y(C, yG1,
    yG2, UG)
    Q = UG >= 0;
    C_upwind = Q .* C(1:end-1, :) + (1 - Q) .* C(2:end, :);
    yG1_upwind = Q .* yG1(1:end-1, :) + (1 - Q) .* yG1(2:end, :);
    yG2_upwind = Q .* yG2(1:end-1, :) + (1 - Q) .* yG2(2:end, :);
end
```

The following function rearranges vectors for processing

```
function [C, yG1, yG2] = Vector_Rearrange(nd, ndo, C, yG1, yG2)
    C = reshape(C, (nd + 1) * ndo, 1);
    yG1 = reshape(yG1, (nd + 1) * ndo, 1);
    yG2 = reshape(yG2, (nd + 1) * ndo, 1);
end
```

The following function computes velocity components for interior cells.

```
function UD_Int = Vel_Interior(nd, ndo, P, CondD, sumdeld, model)
    k = 0;
    CondD_Trim = reshape(CondD, nd + 1, ndo);
    CondD_Trim = CondD_Trim(2:end-1, :);
```

```
    CondD_Trim = CondD_Trim';
    UD_Int = zeros(ndo, nd - 1);
    for i = 1:ndo
        UD_Int(i, :) = -CondD_Trim(i, :) .* ...
            (2 * (model.gamma / model.Mu_g) * ...
            (P(i, 2:end) - P(i, 1:end-1)) ./ sumdeld);
        k = k + 1;
    end
end
```

The following function computes velocity properties, including averages and magnitudes.

```
function [Ux_avg, Uy_avg, U_Mod] = Vel_Prop(nx, ny, Ux, Uy)
    Ux_avg = zeros(ny, nx);
    Uy_avg = zeros(ny, nx);
    for i = 1:nx
        Ux_avg(:, i) = (Ux(:, i) + Ux(:, i + 1)) / 2;
    end
    for i = 1:ny
        Uy_avg(i, :) = (Uy(i, :) + Uy(i + 1, :)) / 2;
    end
    U_Mod = sqrt(Ux_avg.^2 + Uy_avg.^2);
end
```

The following function adds boundary effects to the X velocity component.

```
function UX = VelX_BC_Eff(nx, ny, UX, P, PBdry, CondX, del,
    isDiri, grav, DensiX, model)
    for i = 1:ny
        if (isDiri(1, i) == 1)
            UX(i, 1) = -CondX(i) * ...
                (2 * (model.gamma / model.Mu_g) * ...
                (P(i, 1) - PBdry(1, i)) / del(1) - grav * DensiX(
                    i));
        end
        if (isDiri(2, i) == 1)
            UX(i, end) = -CondX(i + i * nx) * ...
                (2 * (model.gamma / model.Mu_g) * ...
                (PBdry(2, i) - P(i, end)) / del(end) - grav *
                    DensiX(i + i * nx));
        end
    end
end
```

The following function adds boundary effects to the Y velocity component.

```
function UY = VelY_BC_Eff(nx, ny, UY, P, PBdry, CondY, del,
    isDiri, grav, DensiY, model)
    k = 0;
    for i = 1:nx
        if (isDiri(i, 1) == 1)
            UY(1, i) = -CondY(i + k * ny) * ...
                (2 * (model.gamma / model.Mu_g) * ...
                (P(1, i) - PBdry(i, 1)) / del(1) - grav * DensiY(
                    i + k * ny));
```

```
            end
            if (isDiri(i, 2) == 1)
                UY(end, i) = -CondY(i + i * ny) * ...
                    (2 * (model.gamma / model.Mu_g) * ...
                    (PBdry(i, 2) - P(end, i)) / del(end) - grav *
                        DensiY(i + i * ny));
            end
            k = k + 1;
        end
end
```

The following function calculates the production rate Q_p based on pressure and reservoir parameters.

```
function Qp = Compute_Qp(Pressure, Kxx, deltaX, deltaY, gama, MuG
    , theta, rw, SS, Dq, Pw)
    re = 0.14 * sqrt(2 * (deltaX^2 + deltaY^2));
    PressureAvg = mean(mean(Pressure));
    Qp1 = -((Kxx .* gama .* Pressure) / MuG) .* ...
        (theta ./ (log(re / rw) + SS + Dq)) .* (PressureAvg -
            Pw);
    Qp = mean(mean(Qp1));
end
```

6.10.1 MIXED FINITE ELEMENT APPROXIMATION

In this section, we present the mixed finite element approximation for a dual-porosity model applied to matrix and fracture domains [155]. Let Ω_m and Ω_f represent the matrix and fracture regions, respectively, within a polygonal or polyhedral Lipschitz domain $\Omega \subset \mathbb{R}^d$, where $d \in \{1, 2, 3\}$. The function spaces $L^2(\Omega)$ and $\mathbb{L}^2(\Omega)$ are defined as follows:

$$\mathbb{L}^2(\Omega) \equiv \left(L^2(\Omega)\right)^d,$$

where $L^2(\Omega)$ is the standard space of square-integrable functions. The boundaries of the matrix and fracture domains are denoted as $\partial\Omega_m = \Gamma_D^m \cup \Gamma_N^m$ and $\partial\Omega_f = \Gamma_D^f \cup \Gamma_N^f$, respectively. The dual-porosity model governing the fluid flow in the matrix and fracture domains can be expressed as follows. In the matrix domain Ω_m, we have:

$$f_1(p_m)\frac{\partial p_m}{\partial t} + \nabla \cdot \mathbf{u}_m = -S(p_m, p_f), \quad \text{in } \Omega_m \times (0, T), \tag{6.91}$$

$$\mathbf{D}_m(p_m)^{-1}\mathbf{u}_m = -\nabla p_m, \quad \text{in } \Omega_m \times (0, T), \tag{6.92}$$

while in the fracture domain Ω_f, we have:

$$f_2(p_f)\frac{\partial p_f}{\partial t} + \nabla \cdot \mathbf{u}_f = S(p_m, p_f) - Q(p_f), \quad \text{in } \Omega_f \times (0, T), \tag{6.93}$$

$$\mathbf{D}_f(p_f)^{-1}\mathbf{u}_f = -\nabla p_f, \quad \text{in } \Omega_f \times (0,T). \tag{6.94}$$

Here, $f_1(p_m)$ and $f_2(p_f)$ are storage terms associated with the matrix and fracture domains, $S(p_m, p_f)$ is the exchange term between the matrix and fracture, and $Q(p_f)$ is the sink/source term in the fracture.

The permeability tensors $\mathbf{D}_m(p_m)$ and $\mathbf{D}_f(p_f)$ are defined as:

$$\mathbf{D}_m(p_m) = \frac{\rho(p_m)}{\mu} k_{0,m}\left(1 + \frac{b_m}{p_m}\right), \tag{6.95}$$

$$\mathbf{D}_f(p_f) = \frac{\rho(p_f)}{\mu} k_{0,f}\left(1 + \frac{b_f}{p_f}\right), \tag{6.96}$$

where $\rho(p)$ is the density, μ is the viscosity, $k_{0,m}$ and $k_{0,f}$ are the intrinsic permeabilities of the matrix and fracture, and b_m, b_f are compressibility factors.

To ensure continuity and smoothness in the numerical solution, the weak form of the governing equations is constructed. By integrating equations (6.91), (6.94) by parts and using test functions $\varphi \in W_h$ and $\omega \in V_h$, the weak formulation is stated as:

$$\left(f_1(p_m)\frac{\partial p_m}{\partial t}, \varphi\right) + (\nabla \cdot \mathbf{u}_m, \varphi) + (S(p_m, p_f), \varphi) = 0, \tag{6.97}$$

$$\left(\mathbf{D}_m(p_m)^{-1}\mathbf{u}_m, \omega\right) = (p_m, \nabla \cdot \omega), \tag{6.98}$$

$$\left(f_2(p_f)\frac{\partial p_f}{\partial t}, \varphi\right) + (\nabla \cdot \mathbf{u}_f, \varphi) - (S(p_m, p_f), \varphi) = -\left(Q(p_f), \varphi\right), \tag{6.99}$$

$$\left(\mathbf{D}_f(p_f)^{-1}\mathbf{u}_f, \omega\right) = (p_f, \nabla \cdot \omega) - \langle p_w, \omega\rangle_{\Gamma_f^D}. \tag{6.100}$$

To discretize the weak form, let $V_h \subset H(\text{div}, \Omega)$ and $W_h \subset L^2(\Omega)$ be the r-th order Raviart-Thomas space (RT_r) on a partition $\mathcal{T}_h$. The discrete problem is then stated as follows: find $p_m^h, p_f^h \in W_h$ and $\mathbf{u}_m^h, \mathbf{u}_f^h \in V_h$ such that:

$$\left(f_1(p_m^h)\frac{\partial p_m^h}{\partial t}, \varphi\right) + (\nabla \cdot \mathbf{u}_m^h, \varphi) + \left(S(p_m^h, p_f^h), \varphi\right) = 0, \tag{6.101}$$

$$\left(\mathbf{D}_m(p_m^h)^{-1}\mathbf{u}_m^h, \omega\right) = \left(p_m^h, \nabla \cdot \omega\right), \tag{6.102}$$

$$\left(f_2(p_f^h)\frac{\partial p_f^h}{\partial t}, \varphi\right) + \left(\nabla \cdot \mathbf{u}_f^h, \varphi\right) - \left(S(p_m^h, p_f^h), \varphi\right) = -\left(Q(p_f^h), \varphi\right), \tag{6.103}$$

$$\left(\mathbf{D}_f(p_f^h)^{-1}\mathbf{u}_f^h, \omega\right) = \left(p_f^h, \nabla \cdot \omega\right) - \langle p_w, \omega\rangle_{\Gamma_f^D}. \tag{6.104}$$

For more details and stability analysis can be found in El-Amin et al. [155, 156, 157].

6.11 STOCHASTIC MODEL

To extend the deterministic model into a stochastic framework, we first introduce the deterministic model, which is subsequently modified to incorporate stochastic variations [158]. The stochastic model is essential for quantifying deviations in critical parameters due to inherent uncertainties and environmental fluctuations. This section presents a stochastic analysis of the Klinkenberg factor, as described in Eq. (6.55), by considering the factor b as a random parameter dependent on two random variables, α and β, expressed as:

$$b = \alpha k_0^{-\beta}. \tag{6.105}$$

The random variables α and β have their mean and standard deviations estimated based on empirical data. The advantage of this stochastic approach lies in its ability to evaluate the impact of uncertainties in the Klinkenberg factor b on pressure predictions. The randomness introduced into b renders the system stochastic, necessitating the evaluation of statistical properties such as mean and standard deviation. Polynomial Chaos Expansion (PCE) is employed to analyze the random system due to its computational efficiency and accuracy compared to other stochastic techniques.

Using PCE, we expand β as:

$$\beta = \sum_{i=0}^{M} \beta_i \psi_i, \tag{6.106}$$

where $M+1$ is the number of expansion terms, and ψ_i are orthogonal Hermite functions of the random variables. For $k_0^{-\beta}$, we have:

$$k_0^{-\beta} = e^{-\beta \ln k_0} = \sum_{m=0}^{\infty} \frac{(-1)^m (\ln k_0)^m \beta^m}{m!}. \tag{6.107}$$

Substituting the PCE expansion of β into $k_0^{-\beta}$, we get:

$$k_0^{-\beta} = \sum_{m=0}^{\infty} \frac{(-1)^m (\ln k_0)^m \left(\sum_{i=0}^{M} \beta_i \psi_i\right)^m}{m!}. \tag{6.108}$$

For a single random variable with Gaussian variations, this reduces to:

$$\left(\sum_{i=0}^{M} \beta_i \psi_i\right)^m \approx \beta_0^m \left(1 + m\frac{\beta_1}{\beta_0}\psi_1\right), \tag{6.109}$$

where $\frac{\beta_1}{\beta_0}$ is assumed small, allowing the approximation to hold. Consequently:

$$k_0^{-\beta} = k_0^{-\beta_0} \left(1 - \beta_1 \ln k_0 \psi_1\right). \tag{6.110}$$

Similarly, α is expanded using PCE as:

$$\alpha = \sum_{i=0}^{M} \alpha_i \psi_i = \alpha_0 + \alpha_1 \psi_2. \tag{6.111}$$

Combining these expansions, b becomes:

$$b = \alpha k_0^{-\beta} = \alpha_0 k_0^{-\beta_0} - \alpha_0 k_0^{-\beta_0} \beta_1 \ln k_0 \psi_1 + \alpha_1 k_0^{-\beta_0} \psi_2. \tag{6.112}$$

Neglecting higher-order terms, we have:

$$b = b_0 + b_1 \psi_1 + b_2 \psi_2, \tag{6.113}$$

where:

$$b_0 = \alpha_0 k_0^{-\beta_0}, \quad b_1 = -\alpha_0 k_0^{-\beta_0} \beta_1 \ln k_0, \quad b_2 = \alpha_1 k_0^{-\beta_0}.$$

The permeability k is adjusted using the stochastic b:

$$k = k_0 \left(1 + \frac{b}{p}\right) = k_0 \left(1 + \frac{b_0 + b_1 \psi_1 + b_2 \psi_2}{p_0}\right), \tag{6.114}$$

assuming the mean pressure p_0. The pressure perturbations p_1 and p_2 account for randomness in b and external factors, offering a realistic representation compared to deterministic assumptions. Using these expansions in the governing equation for apparent permeability:

$$\begin{aligned} &\gamma\varphi \frac{\partial}{\partial t} (p_0 + p_1 \psi_1 + p_2 \psi_2) - \nabla \\ &\cdot \left(\frac{\gamma}{\mu} (p_0 + p_1 \psi_1 + p_2 \psi_2) k_0 \left(1 + \frac{b_0 + b_1 \psi_1 + b_2 \psi_2}{p_0}\right)\right. \\ &\left. \nabla (p_0 + p_1 \psi_1 + p_2 \psi_2)\right) = 0. \end{aligned} \tag{6.115}$$

Applying the expectation operator and neglecting second-order terms:

$$\gamma\varphi \frac{\partial p_0}{\partial t} - \nabla \cdot \left(\frac{\gamma k_0}{\mu} p_0 \left(1 + \frac{b_0}{p_0}\right) \nabla p_0\right) = 0, \tag{6.116a}$$

$$\gamma\varphi \frac{\partial p_1}{\partial t} - \nabla \cdot \left(\frac{\gamma k_0}{\mu} \left[p_0 \left(1 + \frac{b_0}{p_0}\right) \nabla p_1 + b_1 \nabla p_0 + p_1 \left(1 + \frac{b_0}{p_0}\right) \nabla p_0\right]\right) = 0, \tag{6.116b}$$

$$\gamma\varphi \frac{\partial p_2}{\partial t} - \nabla \cdot \left(\frac{\gamma k_0}{\mu} \left[p_0 \left(1 + \frac{b_0}{p_0}\right) \nabla p_2 + b_2 \nabla p_0 + p_2 \left(1 + \frac{b_0}{p_0}\right) \nabla p_0\right]\right) = 0. \tag{6.116c}$$

6.11.1 NUMERICAL SOLUTION

The following algorithm summarizes the steps for analyzing the stochastic model:

1. Estimate α and β from experimental data. Assume standard deviations proportional to mean values, e.g., $\alpha_1 = 0.05\alpha_0$.
2. Compute b_0, b_1, and b_2 using stochastic expansions.
3. Solve Eq. (6.116a) to compute p_0 (mean pressure).
4. Solve Eq. (6.116b) to compute p_1 (pressure perturbations using p_0).
5. Solve Eq. (6.116c) to compute p_2 (pressure perturbations using p_0).
6. Evaluate the variance or standard deviation of pressure:

$$\text{Var}(p) = p_1^2 + p_2^2, \quad \text{SD}(p) = \sqrt{p_1^2 + p_2^2}.$$

This stochastic framework provides a robust methodology to quantify the impact of uncertainties and variations in the model parameters on predictions, enabling realistic assessments of permeability and pressure [158].

The following figures illustrate the results of the stochastic model solved numerically using the pdepe MATLAB® function, which was introduced in Chapter 2 of this book. These results highlight the impact of stochastic variations in the Klinkenberg factor and demonstrate the deviation in pressure and standard deviation due to randomness.

Figure 6.8 shows the variation of the mean pressure p_0 with distance x for different values of the stochastic parameter α_0. The results indicate that as

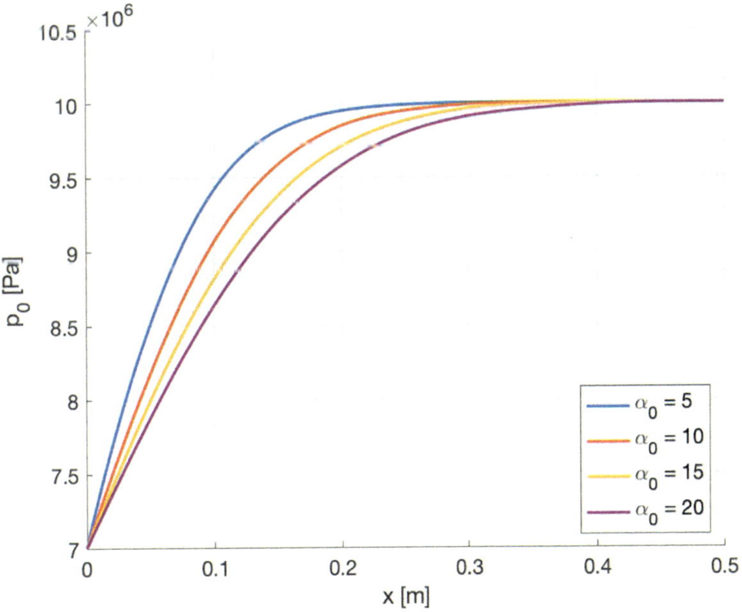

Figure 6.8 Variation of p_0 with x for different α_0 values.

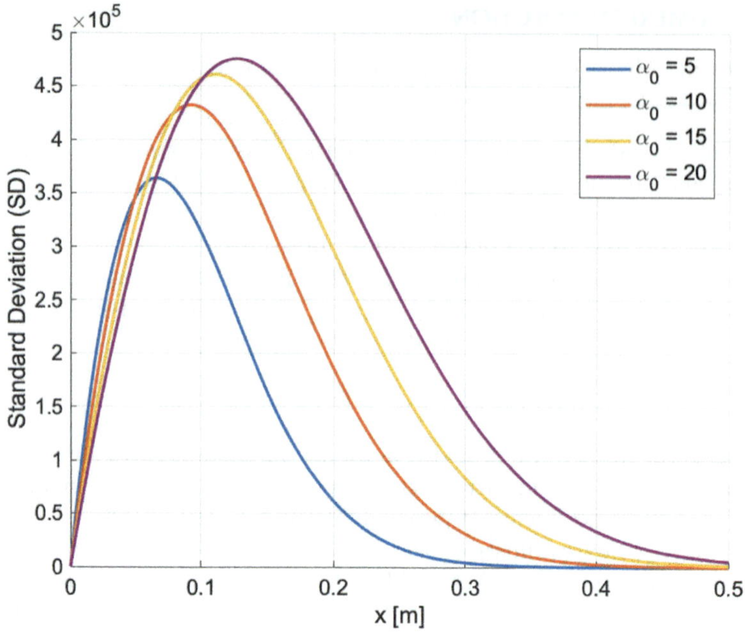

Figure 6.9 Variation of standard deviation (SD) with x for different α_0 values.

α_0 increases, the mean pressure p_0 increases and approaches a stable value for higher distances. The results reflect the influence of stochastic variations in the Klinkenberg factor on the pressure distribution.

Figure 6.9 shows the variation of the standard deviation (SD) of the pressure with distance x for different α_0 values. The peak SD occurs at a certain distance (around $x \approx 0.15$) and decreases significantly as x increases. This behavior demonstrates the regions of maximum sensitivity to stochastic variations and provides insight into where the model is most affected by uncertainties in the input parameters.

The numerical results presented in Figures 6.8 and 6.9 serve as examples of how the stochastic model accounts for randomness in the parameters. The use of the `pepde` MATLAB® function enabled efficient computation of these results, showing the capability of the stochastic approach to evaluate both the mean behavior and the associated uncertainties.

CONCLUSION

Fractured porous media pose significant modeling challenges due to their inherent multi-scale and multi-mechanism nature. This chapter addressed these complexities by introducing a suite of analytical, numerical, and hybrid approaches tailored to the characteristics of fractured reservoirs. Dual-continuum and discrete fracture models enable detailed simulation of matrix–fracture interactions, while stress-dependent and stochastic formulations enhance predictive accuracy under variable field conditions. The Langmuir isotherm captures adsorption effects critical for shale gas, and the integration of MATLAB® simulations supports practical implementation and validation. Together, these models form a robust toolkit for researchers and engineers tackling fluid and gas transport in fractured porous systems.

REVIEW QUESTIONS

1. Derive the governing equation for gas transport in a shale reservoir, incorporating the Klinkenberg correction for apparent permeability.
2. Explain the primary challenges of gas transport in low-permeability reservoirs such as shale formations.
3. What are the key differences between conventional and unconventional petroleum reservoirs?
4. Define the Klinkenberg effect and explain how it affects gas transport in porous media.
5. Describe the dual-porosity dual-permeability (DPDP) model and its significance in modeling fractured reservoirs.
6. What role does the Langmuir isotherm play in describing gas adsorption in shale reservoirs?
7. Explain the impact of stress-dependent porosity and permeability on gas flow in fractured systems.
8. How does the Boltzmann transformation simplify the governing equation for gas flow in a radial cylindrical coordinate system?

PROBLEMS

1. Derive a numerical solution for the pressure distribution $p(r,t)$ in a radial shale reservoir using the governing equation:
$$\varphi \frac{\partial p}{\partial t} = \frac{1}{r}\frac{\partial}{\partial r}\left(\frac{r k_0}{\mu}(b+p)\frac{\partial p}{\partial r}\right),$$
with the following parameters $\varphi = 0.15$, $k_0 = 1.0 \times 10^{-15}\,\text{m}^2$, $\mu = 1.0 \times 10^{-5}\,\text{Pa} \cdot \text{s}$, $b = 1.0 \times 10^{6}\,\text{Pa}$, $p(r,0) = 5.0\,\text{MPa}$, Boundary conditions: $p(0,t) = 5.0\,\text{MPa}$ (Dirichlet), $\frac{\partial p}{\partial r}(r \to \infty, t) = 0$. Implement this solution in MATLAB® using the `pdepe` function and plot $p(r,t)$ at $t = 0.1, 1, 10$ days.
2. Using the dual-porosity dual-permeability (DPDP) model equations for the matrix and fracture pressures, develop a finite difference numerical solution for p_m and p_f on a 2D reservoir domain (see Figure 6.4). Use Neumann boundary conditions and simulate for a production well at $p_w = 1.0\,\text{MPa}$. Evaluate how fracture permeability (k_{f0}) impacts the production rate over 30 days.

3. Consider a stochastic model for the Klinkenberg factor $b = \alpha k_0^{-\beta}$, where α and β are random variables. Using Polynomial Chaos Expansion (PCE), solve the stochastic governing equation:

$$\gamma\varphi\frac{\partial p}{\partial t} - \nabla\cdot\left(\frac{\gamma k_0}{\mu}p\left(1+\frac{b}{p}\right)\nabla p\right) = 0,$$

with the following parameters $\varphi = 0.1$, $k_0 = 1.0\times10^{-15}\,\text{m}^2$, $\mu = 1.0\times10^{-5}\,\text{Pa}\cdot\text{s}$, α= $1.0\times10^{6}\pm5\%$, and β= $0.5\pm2\%$, $p_0 = 5.0\,\text{MPa}$, Dirichlet boundary condition at $x = 0$, and Neumann at $x = L$.
Solve numerically using MATLAB® or Python and analyze how the randomness in b affects the mean and standard deviation of $p(x,t)$ over time.
4. Implement the following governing equation for gas flow in fractured porous media with stress-dependent porosity and permeability:

$$\frac{\partial}{\partial t}\left(\varphi(p)\rho + (1-\varphi(p))q_a(p)\right) - \nabla\cdot\left(\frac{\rho k_m(p)}{\mu}\nabla p\right) = q,$$

where:

$$\varphi(p) = \varphi_0 e^{-a\sigma'}, \quad k_m(p) = k_{m0}e^{-c\sigma}.$$

Use the following parameters $\varphi_0 = 0.2$, $k_{m0} = 1.0\times10^{-15}\,\text{m}^2$, $a = 0.005\,\text{Pa}^{-1}$, $c = 0.002\,\text{Pa}^{-1}$, $\sigma = 10\,\text{MPa}$, Dirichlet boundary condition: $p(0,t) = 6.0\,\text{MPa}$, Neumann: $\frac{\partial p}{\partial x} = 0$ at $x = L$.
Simulate this in MATLAB® using finite difference methods and plot $\varphi(p)$ and $k_m(p)$ along x at $t = 1, 10, 50$ days.
5. Using the Langmuir isotherm:

$$q_a = \frac{\rho_s M_w}{V_{\text{std}}}\left(\frac{V_L p_m}{p_L + p_m}\right),$$

and the governing equation for gas transport in shale:

$$\varphi\frac{\partial\rho}{\partial t} + \frac{\partial}{\partial t}\left[(1-\varphi)q_a\right] - \nabla\cdot\left(\frac{\rho k_m}{\mu}\nabla p\right) = q,$$

numerically solve for $p(x,t)$ and $q_a(x,t)$ using MATLAB®. Use the following parameters $\varphi = 0.15$, $\rho_s = 2500\,\text{kg/m}^3$, $M_w = 0.016\,\text{kg/mol}$, $V_{\text{std}} = 0.0224\,\text{m}^3/\text{mol}$, $V_L = 0.03\,\text{m}^3/\text{kg}$, $p_L = 2.0\,\text{MPa}$, $p_m = p(x,t)$, $k_m = 1.0\times10^{-15}\,\text{m}^2$, $\mu = 1.0\times10^{-5}\,\text{Pa}\cdot\text{s}$.
Evaluate the impact of varying p_L on the adsorption and plot $q_a(x,t)$ for different p_L values.
6. Simulate gas production from a vertical well in a 2D reservoir using the cumulative production rate formula:

$$q_p = \frac{k_{f0}\theta\gamma}{Z\mu_g\left[\ln\left(\frac{r_e}{r_w}\right) + S + D_q\right]}p_f\left(p_{fe} - p_w\right).$$

Assume $k_{f0} = 1.0 \times 10^{-12}\,\text{m}^2$, $\theta = \pi/2$, $r_w = 0.1\,\text{m}$, $\gamma = 0.002\,\text{Pa}^{-1}$, $Z = 1$, $\mu_g = 1.0 \times 10^{-5}\,\text{Pa}\cdot\text{s}$, $p_{fe} = 6.0\,\text{MPa}$, $p_w = 1.0\,\text{MPa}$, $r_e = 50\,\text{m}$.
Plot cumulative production over 30 days and discuss the sensitivity of q_p to p_{fe}.

DECLARATION

During the preparation of this work, the author used ChatGPT in order to improve the language and readability. After using this tool/service, the author reviewed and edited the content as needed and takes full responsibility for the content of the publication.

7 Fractional Modeling of Fluid Flow in Porous Media

7.1 INTRODUCTION

Fractional calculus generalizes the concepts of differentiation and integration by extending their orders from integer to real or complex numbers, thereby permitting the exploration of fractional-order dynamics. Rooted in the pioneering thoughts of Leibniz and Euler in the late 17th and early 18th centuries, fractional calculus has undergone substantial theoretical and methodological developments, especially in recent decades. Its relevance arises from the unique ability to represent memory effects, non-local interactions, and hereditary properties that classical integer-order calculus fails to adequately capture. These attributes make fractional calculus exceptionally suited to modeling phenomena characterized by complexity and irregularities, including anomalous diffusion processes, viscoelastic materials, turbulence, and fractal media.

Fractional calculus has found profound applications across diverse scientific disciplines such as fluid dynamics, geophysics, biology, and engineering, significantly enhancing the accuracy and scope of traditional modeling frameworks [159, 160, 168, 177, 178, 187, 192, 193, 194, 195]. Particularly, fractional-order differential equations have proven instrumental in accurately describing anomalous transport phenomena in porous media, such as groundwater contaminant migration modeled via fractional advection-dispersion equations. Moreover, fractional calculus facilitates the representation of complex flow behaviors of viscoelastic or non-Newtonian fluids, which are critical in industrial applications such as polymer processing and enhanced oil recovery techniques. It further contributes significantly to understanding heat and mass transfer processes in media with fractal geometries, frequently encountered in natural and engineered porous structures.

In the field of fluid flow in porous media, fractional calculus aids in modeling fluid dynamics within unconventional flows, where traditional models may not sufficiently capture fluid behavior due to inherent heterogeneity and anomalous flow characteristics. Incorporating fractional-order models into simulation and analysis thus provides enhanced predictive capabilities, facilitating improved media characterization, optimized resource recovery, and more effective management strategies.

Traditional modeling approaches, typically based on integer-order calculus, have successfully addressed a wide range of physical phenomena by assuming localized interactions and standard diffusive processes. Such models are governed by classic partial differential equations, including Darcy's law, which

DOI: 10.1201/9781032724973-7

assumes normal diffusion, homogeneity, and memoryless characteristics. However, many real-world phenomena exhibit complexities such as memory effects, spatial heterogeneity, and anomalous diffusion that cannot be accurately captured by these classical approaches.

Fractional modeling, utilizing fractional-order derivatives, provides a more comprehensive framework capable of incorporating non-local and memory-dependent processes. Unlike integer-order derivatives, fractional derivatives inherently account for the history of the state variables, capturing delayed responses and long-range spatial interactions within the media. This results in significantly enhanced modeling accuracy for complex systems, such as heterogeneous reservoirs, viscoelastic fluid flow, and anomalous diffusion in geological and biological materials.

Anomalous diffusion refers to transport processes that deviate from the classical linear diffusion described by Fick's law, often characterized by non-linear, subdiffusive, or superdiffusive behavior. Such deviations commonly occur in porous media where the structure is heterogeneous, fractured, or contains irregular geometries. In these contexts, solutes or fluids may spread at rates that differ significantly from those predicted by classical models.

Fractional-order derivatives are particularly effective in modeling anomalous diffusion because they inherently represent non-local phenomena, where the flux at a given location depends not only on local gradients but also on spatially distant interactions. Fractional diffusion equations, such as the fractional advection-dispersion equation, better capture complex temporal dynamics and spatial correlations. Thus, fractional calculus provides essential tools to more precisely describe and predict transport behavior in environments such as subsurface aquifers, biological tissues, and fractured reservoirs.

Traditional modeling approaches, typically based on integer-order calculus, have successfully addressed a wide range of physical phenomena by assuming localized interactions and standard diffusive processes. Such models are governed by classic partial differential equations, including Darcy's law, Fourier's law, and Fick's law, which assume normal diffusion, homogeneity, and memoryless characteristics [79]. However, many real-world phenomena exhibit complexities such as memory effects, spatial heterogeneity, and anomalous diffusion that cannot be accurately captured by these classical approaches [189, 190].

Fractional modeling, utilizing fractional-order derivatives, provides a more comprehensive framework capable of incorporating non-local and memory-dependent processes [182]. Unlike integer-order derivatives, fractional derivatives inherently account for the history of the state variables, capturing delayed responses and long-range spatial interactions within the media. This results in significantly enhanced modeling accuracy for complex systems, such as heterogeneous reservoirs, viscoelastic fluid flow, and anomalous diffusion in geological and biological materials [188, 165].

Anomalous diffusion refers to transport processes that deviate from the classical linear diffusion described by Fick's law, often characterized by nonlinear, subdiffusive, or superdiffusive behavior [189, 190]. Such deviations commonly occur in porous media where the structure is heterogeneous, fractured, or contains irregular geometries. In these contexts, solutes or fluids may spread at rates that differ significantly from those predicted by classical models.

Fractional-order derivatives are particularly effective in modeling anomalous diffusion because they inherently represent non-local phenomena, where the flux at a given location depends not only on local gradients but also on spatially distant interactions [200, 197]. Fractional diffusion equations, such as the fractional advection-dispersion equation, better capture complex temporal dynamics and spatial correlations [162, 163]. Thus, fractional calculus provides essential tools to more precisely describe and predict transport behavior in environments such as subsurface aquifers, biological tissues, and fractured reservoirs [164, 170]. Recent comprehensive developments and practical applications of fractional modeling in fluid flow and transport phenomena can be found in [175].

7.2 FRACTIONAL DERIVATIVES

Fractional operators generalize classical derivatives and integrals to non-integer orders, enabling a broader analytical approach to modeling complex phenomena that exhibit memory, history dependence, and spatial-temporal nonlocality. Classical derivatives define instantaneous rates of change through limits of incremental ratios:

$$\frac{df}{dx} = \lim_{h \to 0} \frac{f(x+h) - f(x)}{h}, \tag{7.1}$$

with the higher-order derivatives described as:

$$D^n f(x) = \frac{d^n f}{dx^n}, \quad n \in \mathbb{N}. \tag{7.2}$$

Fractional calculus extends these definitions to non-integer (fractional) orders $\alpha \in \mathbb{R}^+$. This extension allows modeling of phenomena beyond the classical scope, capturing memory effects and hereditary properties effectively. Fractional integration, initially formalized by Cauchy, leverages the gamma function for generalization:

$$I^\alpha f(x) = \frac{1}{\Gamma(\alpha)} \int_0^x (x-u)^{\alpha-1} f(u)\, du, \tag{7.3}$$

where $\Gamma(\alpha)$ is the gamma function, providing a meaningful generalization of repeated integrals.

7.2.1 COMMON FRACTIONAL DERIVATIVE

Two prominent fractional derivatives definitions utilized in modeling are:

Riemann-Liouville Fractional Derivative: A foundational formulation, defined as:

$$D_x^\alpha f(x) = \frac{d}{dx} I^{1-\alpha} f(x), \quad 0 < \alpha < 1, \tag{7.4}$$

emphasizing integration followed by differentiation, thus inherently embedding past state history within the derivative.

Caputo Fractional Derivative: Frequently employed due to its suitability for physical and engineering applications, defined as:

$$D_x^\alpha f(x) = I^{1-\alpha} \frac{d}{dx} f(x), \quad 0 < \alpha < 1, \tag{7.5}$$

characterized by differentiation followed by integration, facilitating the incorporation of traditional initial conditions, hence enhancing physical interpretability.

Both definitions illustrate the non-local nature of fractional derivatives, making them ideal for modeling phenomena influenced by historical states.

7.2.2 CONFORMABLE FRACTIONAL DERIVATIVE

Here, we present another approach known as the *conformable fractional derivative* (CFD), introduced by Khalil et al. [180]. This operator was developed to address fractional initial value problems of the form:

$$D_t^\alpha u(t) = f(t, u(t)), \quad u(t_0) = u_0, \quad a \leq t \leq b. \tag{7.6}$$

The CFD is defined as:

$${}_0^C T_t^\alpha(f)(t) = \lim_{\varepsilon \to 0} \frac{f(t + \varepsilon t^{1-\alpha}) - f(t)}{\varepsilon}, \tag{7.7}$$

which exists if and only if the classical derivative $\frac{d}{dt} f(t)$ exists. In such cases, the CFD simplifies to:

$${}_0^C T_t^\alpha(f)(t) = t^{1-\alpha} \frac{d}{dt} f(t). \tag{7.8}$$

To numerically approximate solutions, a finite difference scheme can be applied:

$$u_{k+1} - u_k = \frac{1}{\alpha} \left[(t_{k+1})^\alpha - (t_k)^\alpha \right] f(t_k, u_k), \tag{7.9}$$

which generalizes the classical Euler method (recovered when $\alpha = 1$). $h = \frac{b-a}{N}$ is the step size.

The conformable derivative ${}_0^C T_\tau^\alpha$ satisfies several important properties [180, 169], assuming u and v are α-differentiable at $\tau \in [0, \infty)$ and $\alpha \in (0, 1]$:

Linearity: ${}_0^C T_\tau^\alpha(C_1 u + C_2 v) = C_1 {}_0^C T_\tau^\alpha(u) + C_2 {}_0^C T_\tau^\alpha(v)$
Product Rule: ${}_0^C T_\tau^\alpha(uv) = v\,{}_0^C T_\tau^\alpha(u) + u\,{}_0^C T_\tau^\alpha(v)$
Quotient Rule: ${}_0^C T_\tau^\alpha\left(\frac{u}{v}\right) = \frac{1}{v^2}\left(v\,{}_0^C T_\tau^\alpha(u) - u\,{}_0^C T_\tau^\alpha(v)\right)$
Power Rule: ${}_0^C T_\tau^\alpha(\tau^m) = m\tau^{m-\alpha}$
Constant Rule: ${}_0^C T_\tau^\alpha(C) = 0$
Reduction to classical derivative: If $u(\tau)$ is differentiable, then ${}_0^C T_\tau^\alpha(u(\tau)) = \tau^{1-\alpha}\frac{du(\tau)}{d\tau}$

7.2.3 IMPORTANT SPECIAL FUNCTIONS

Fractional calculus heavily relies on specialized mathematical functions to represent and solve fractional-order differential equations effectively:

Gamma Function: Serving as a fractional factorial, the gamma function is central to fractional calculus and is defined by:

$$\Gamma(\alpha) = \int_0^\infty e^{-t} t^{\alpha-1} dt, \quad \mathrm{Re}(\alpha) > 0, \tag{7.10}$$

where its recursive property, $\Gamma(\alpha+1) = \alpha\Gamma(\alpha)$, simplifies the computation and application of fractional operators.
Beta Function: Closely related to the gamma function, defined as:

$$B(x,y) = \int_0^1 t^{x-1}(1-t)^{y-1} dt = \frac{\Gamma(x)\Gamma(y)}{\Gamma(x+y)}, \tag{7.11}$$

providing an essential tool for fractional integrals and derivatives involving multiple fractional parameters.
Mittag-Leffler Function: A pivotal function in fractional calculus, particularly for fractional-order differential equations solutions, defined by:

$$E_{\alpha,\beta}(z) = \sum_{n=0}^{\infty} \frac{z^n}{\Gamma(\alpha n + \beta)}, \tag{7.12}$$

generalizing the exponential function, making it instrumental for solving fractional differential equations and modeling anomalous diffusion.

Fractional operators have emerged as powerful tools in modeling and simulating fluid dynamics within porous media. They effectively capture complex phenomena, including memory effects, anomalous diffusion, non-local interactions, and viscoelastic behaviors, surpassing classical integer-order models in accuracy and predictive power. In engineering and environmental sciences, fractional modeling contributes significantly to improved reservoir management, groundwater contaminant transport prediction, and analysis of heat and mass transfer in structurally complex media. Thus, fractional calculus provides robust frameworks that enhance decision-making and operational strategies in practical applications such as petroleum extraction, hydrological forecasting, and material science.

7.3 FRACTIONAL MODELS OF SINGLE-PHASE FLOW

Recent developments in fractional calculus have improved these models by accounting for non-local and anomalous transport behaviors, which classical models overlook. Fractional derivatives introduce memory and scale effects, capturing flow in complex geometries and time-dependent behaviors. This leads to a generalized Darcy's law:

$$\mathbf{u} = -\eta \nabla^{\beta} p, \tag{7.13}$$

where $\mathbf{u} = (u_x, u_y, u_z)$ is the fluid velocity vector, directly linked to the pressure gradient; the constant $\eta = \frac{K}{\mu}$ captures the medium-fluid interaction; and ∇^{β} is a fractional gradient of order β. This operator introduces non-local pressure dependencies, making the model better suited to heterogeneous or anisotropic media.

To incorporate time-dependent changes like decreasing permeability due to clogging or chemical reactions, fractional time derivatives are used. Caputo's model introduces a memory effect:

$$\mathbf{u} = -\eta(t) \left(\frac{\partial^{\alpha}}{\partial t^{\alpha}} \nabla p \right), \tag{7.14}$$

where $0 \leq \alpha \leq 1$ quantifies the memory intensity. Obembe et al. [191] extended this by allowing the fractional order to vary over time:

$$\mathbf{u} = -\xi_0 \eta \left(\frac{\partial^{\alpha(t)}}{\partial t^{\alpha(t)}} \nabla p \right), \tag{7.15}$$

with ξ_0 as a transmissibility conversion factor, enabling greater flexibility as flow conditions evolve.

Spatial memory is modeled using fractional spatial derivatives:

$$\mathbf{u} = -\eta \nabla^{\alpha} p, \quad 0 \leq \alpha \leq 1, \tag{7.16}$$

with:

$$\nabla^{\alpha} = \left(\frac{\partial^{\alpha}}{\partial x^{\alpha}}, \frac{\partial^{\alpha}}{\partial y^{\alpha}}, \frac{\partial^{\alpha}}{\partial z^{\alpha}} \right). \tag{7.17}$$

Chang et al. [167] showed this model aligns with experimental data. For example, a fractional order $\alpha = 0.45$ fits well with permeability $k_1 = 2.36 \times 10^{-15}\,\mathrm{m}^{1.45}$, while $\alpha = 0.76$ matches $k_2 = 1.58 \times 10^{-16}\,\mathrm{m}^{1.76}$. For further reading, especially Chapters 3 and 10 of our book [175].

7.4 FRACTIONAL MULTIPHASE CONTINUITY EQUATION

We extend the mass conservation equation for multiphase flow in porous media using the Caputo fractional derivative. For a given phase γ, the fractional continuity equation is expressed as [174]:

$$\begin{aligned}
&\frac{\partial^{\alpha}}{\partial t^{\alpha}}(\phi \Delta V S_{\gamma}\rho_{\gamma}) + \Delta V \frac{t^{1-\alpha}\,\Gamma(2-\beta_x)}{x^{1-\beta_x}\,\Gamma(2-\alpha)}\left(\frac{\partial^{\beta_x}}{\partial x^{\beta_x}}(\rho_{\gamma}u_{\gamma})\right) + \\
&\Delta V \frac{t^{1-\alpha}\,\Gamma(2-\beta_y)}{y^{1-\beta_y}\,\Gamma(2-\alpha)}\left(\frac{\partial^{\beta_y}}{\partial y^{\beta_y}}(\rho_{\gamma}v_{\gamma})\right) + \\
&\Delta V \frac{t^{1-\alpha}\,\Gamma(2-\beta_z)}{z^{1-\beta_z}\,\Gamma(2-\alpha)}\left(\frac{\partial^{\beta_z}}{\partial z^{\beta_z}}(\rho_{\gamma}w_{\gamma})\right) = 0.
\end{aligned} \tag{7.18}$$

This equation holds for $n-1 < \alpha, \beta_x, \beta_y, \beta_z < n$, where n is a positive integer. Here, S_{γ} is the saturation of phase γ, defined as the fraction of pore volume occupied by that phase.

We now analyze the accumulation term in Eq. (7.18). Assuming the control volume ΔV, density ρ_{γ}, and porosity ϕ vary with pressure P, and $\alpha = 1$:

$$\begin{aligned}
\frac{\partial}{\partial t}(\phi \Delta V S_{\gamma}\rho_{\gamma}) =& \phi \Delta V \rho_{\gamma}\frac{\partial S_{\gamma}}{\partial t} + \\
&\left(\phi S_{\gamma}\rho_{\gamma}\frac{\partial \Delta V}{\partial P} + \Delta V S_{\gamma}\rho_{\gamma}\frac{\partial \phi}{\partial P} + \phi \Delta V S_{\gamma}\frac{\partial \rho_{\gamma}}{\partial P}\right)\frac{\partial P}{\partial t}
\end{aligned} \tag{7.19}$$

Total volume includes pore and solid volumes:

$$\Delta V = \Delta V_p + \Delta V_s \tag{7.20}$$

$$\Delta V_s = (1-\phi)\Delta V, \quad \Delta V_p = \phi \Delta V \tag{7.21}$$

$$\frac{\partial \Delta V}{\partial p} = \frac{\partial \Delta V_p}{\partial p} + \frac{\partial \Delta V_s}{\partial p} \tag{7.22}$$

Bulk, solid, and pore compressibilities are:

$$c_b = -\frac{1}{\Delta V}\frac{\partial \Delta V}{\partial p}, \quad c_s = -\frac{1}{\Delta V_s}\frac{\partial \Delta V_s}{\partial p}, \quad c_p = -\frac{1}{\Delta V_p}\frac{\partial \Delta V_p}{\partial p} \tag{7.23}$$

Then:

$$c_b = (1-\phi)c_s + \phi c_p \tag{7.24}$$

Porosity change with pressure is:

$$c_s = \frac{1}{1-\phi}\frac{\partial \phi}{\partial p} \tag{7.25}$$

Fluid compressibility is:

$$c_{f\gamma} = \frac{1}{\rho_{\gamma}}\frac{\partial \rho_{\gamma}}{\partial p} \tag{7.26}$$

Substituting into the accumulation term gives:

$$\begin{aligned}\frac{\partial}{\partial t}\left(S_\gamma \Delta V \phi \rho_\gamma\right) =& \Delta V \phi \rho_\gamma \frac{\partial S_\gamma}{\partial t} + \\ & S_\gamma \rho_\gamma \Delta V \left[(1-\phi)^2 c_s - \phi^2 c_p + \phi c_{f\gamma}\right] \frac{\partial P}{\partial t}\end{aligned} \tag{7.27}$$

Plugging Eq. (7.27) into Eq. (7.18) gives the spatial fractional continuity equation:

$$\begin{aligned}&\phi \rho_\gamma \frac{\partial S_\gamma}{\partial t} + S_\gamma \rho_\gamma \left[(1-\phi)^2 c_s - \phi^2 c_p + \phi c_{f\gamma}\right] \frac{\partial P}{\partial t} + \\ &\frac{\Gamma(2-\beta_x)}{x^{1-\beta_x}} \frac{\partial^{\beta_x}}{\partial x^{\beta_x}}\left(\rho_\gamma u_\gamma\right) + \frac{\Gamma(2-\beta_y)}{y^{1-\beta_y}} \frac{\partial^{\beta_y}}{\partial y^{\beta_y}}\left(\rho_\gamma v_\gamma\right) + \\ &\frac{\Gamma(2-\beta_z)}{z^{1-\beta_z}} \frac{\partial^{\beta_z}}{\partial z^{\beta_z}}\left(\rho_\gamma w_\gamma\right) = 0.\end{aligned} \tag{7.28}$$

Assuming incompressible fluids and negligible rock compressibility ($c_{f\gamma} = c_s = c_p = 0$):

$$\begin{aligned}&\phi \rho_\gamma \frac{\partial S_\gamma}{\partial t} + \frac{\Gamma(2-\beta_x)}{x^{1-\beta_x}} \frac{\partial^{\beta_x}}{\partial x^{\beta_x}}\left(\rho_\gamma u_\gamma\right) + \frac{\Gamma(2-\beta_y)}{y^{1-\beta_y}} \frac{\partial^{\beta_y}}{\partial y^{\beta_y}}\left(\rho_\gamma v_\gamma\right) + \\ &\frac{\Gamma(2-\beta_z)}{z^{1-\beta_z}} \frac{\partial^{\beta_z}}{\partial z^{\beta_z}}\left(\rho_\gamma w_\gamma\right) = 0.\end{aligned} \tag{7.29}$$

7.5 FRACTIONAL MULTIPHASE DARCY'S LAW

A commonly used form of fractional Darcy's law for multiphase flow with spatial memory is:

$$\mathbf{u}_\gamma = -\frac{K k_{r\gamma}}{\mu_\gamma}\left(\nabla^\alpha P_\gamma + \rho_\gamma g \nabla z\right), \tag{7.30}$$

where:

$$\mathbf{u}_\gamma = \left(u_\gamma, v_\gamma, w_\gamma\right), \quad \nabla^\alpha = \left(\frac{\partial^\alpha}{\partial x^\alpha}, \frac{\partial^\alpha}{\partial y^\alpha}, \frac{\partial^\alpha}{\partial z^\alpha}\right) \tag{7.31}$$

Here, $0 \leq \alpha \leq 1$, and g, z, K, μ_γ, P_γ, $k_{r\gamma}$, and ρ_γ have their usual meanings.

In time-fractional form, Darcy's law incorporates memory effects with time-varying transmissibility:

$$\mathbf{u}_\gamma = -\eta(t)_\gamma \frac{\partial^\alpha}{\partial t^\alpha}\left(\nabla P_\gamma + \rho_\gamma g \nabla z\right). \tag{7.32}$$

The mass balance for wetting (w) and non-wetting (n) phases becomes:

$$\phi\frac{\partial^\alpha S_w}{\partial t^\alpha}+\frac{t^{1-\alpha}}{\Gamma(2-\alpha)}\nabla\cdot\mathbf{u}_w=0 \tag{7.33}$$

$$\phi\frac{\partial^\alpha S_n}{\partial t^\alpha}+\frac{t^{1-\alpha}}{\Gamma(2-\alpha)}\nabla\cdot\mathbf{u}_n=0 \tag{7.34}$$

With saturation constraint:

$$S_w+S_n=1 \tag{7.35}$$

Time-fractional velocities are:

$$u_w=-\zeta_w(t)\frac{\Gamma(2-\beta)}{t^{1-\beta}}\frac{\partial^\beta}{\partial t^\beta}\left(\frac{\partial P_w}{\partial x}\right) \tag{7.36}$$

$$u_n=-\zeta_n(t)\frac{\Gamma(2-\beta)}{t^{1-\beta}}\frac{\partial^\beta}{\partial t^\beta}\left(\frac{\partial P_n}{\partial x}\right) \tag{7.37}$$

7.6 ANALYTICAL SOLUTION FOR FRACTIONAL GAS EQUATION

Power-series methods are a well-established approach for solving various types of differential equations, including fractional differential equations (FDEs). By expanding the unknown function as a series and matching coefficients, one can obtain analytical expressions that elucidate the behavior of fractional-order systems. Nonetheless, the success of this method hinges on the specific form of the equation and the imposed boundary conditions.

We now examine an example of single-phase gas flow within a laterally infinite reservoir exhibiting radial symmetry around a single well (Figure 7.1). The ideal gas law introduces the parameter $\gamma=\frac{M}{RT}$, where M is the molar concentration $\left[\mathrm{kg/Pa\,m^3}\right]$, $R=8.314\,\mathrm{Pa\,m^3/(mol\,K)}$ is the gas constant, and T[K] is the temperature. The other relevant quantities are porosity φ, gas pressure p, time t, gas viscosity μ, permeability k, and gas density ρ.

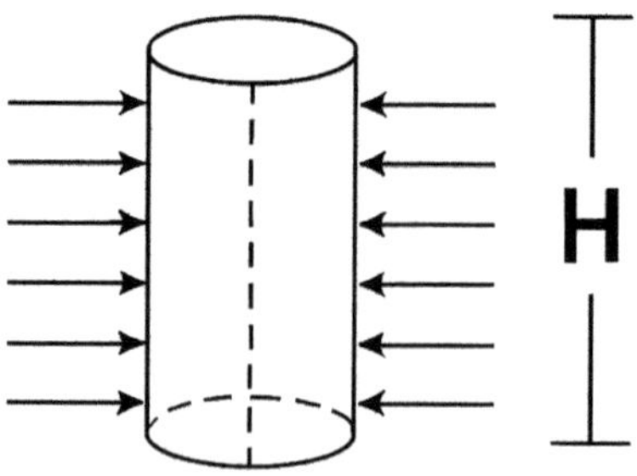

Figure 7.1 Schematic illustration of radial flow in the vicinity of a well.

Using a one-dimensional radial cylindrical coordinate system, the fractional governing equation can be cast as [145, 173]:

$$\frac{1}{\chi}\frac{\partial p}{\partial t} = \frac{\partial^{\alpha+1} p}{\partial r^{\alpha+1}} + \frac{1}{r}\frac{\partial^{\alpha} p}{\partial r^{\alpha}}, \tag{7.38}$$

where $\frac{1}{\chi} = \frac{k\,p_0}{\varphi\,\mu}$ and $0 < \alpha < 1$.

An initial condition at $t = 0$ is:

$$p(r, 0) = p_0, \quad 0 \le r < \infty, \tag{7.39}$$

where p_0 is a constant. The Dirichlet-Neumann boundary conditions are:

$$p(r,t) = p_w \quad \text{as} \quad r \to 0, \quad t > 0, \text{ and } r\,\frac{\partial p}{\partial r} = \frac{Q\,\mu}{2\,\pi\,k\,H'}, \text{ as } r \to \infty, \quad t > 0. \tag{7.40}$$

The Caputo model is appropriate for this analytical formulation and is given by

$$\frac{\partial^{\alpha} p(x)}{\partial r^{\alpha}} = \frac{1}{\Gamma(1-\alpha)} \int_{x_0}^{x} (x-\zeta)^{-\alpha}\, \frac{\partial p(x,\zeta)}{\partial \zeta}\, d\zeta, \tag{7.41}$$

where $\Gamma(\cdot)$ is the Gamma function.

El-Amin [145] introduced a generalized transformation:

$$\eta = \frac{r^{\alpha+1}}{2\,\Gamma(\alpha+2)\,t\,\chi}, \quad t > 0, \tag{7.42}$$

which simplifies to the classic Boltzmann variable when $\alpha = 1$:

$$\eta = \frac{r^2}{4\,t\,\chi}, \quad t > 0. \tag{7.43}$$

Inserting Eq. (7.42) into Eq. (8.21) and rearranging leads to:

$$\eta\frac{\partial p}{\partial \eta} + \frac{\partial^{\alpha} p}{\partial \eta^{\alpha}} + \frac{\alpha+1}{2}\,\eta\,\frac{\partial^{\alpha+1} p}{\partial \eta^{\alpha+1}} = 0, \tag{7.44}$$

with corresponding boundary conditions:

$$p = p_w, \quad \text{at} \quad \eta = 0, \tag{7.45}$$

and

$$\frac{\partial p}{\partial \eta} = \frac{Q\,\mu}{2\,\pi\,k\,H'}, \quad \text{as} \quad \eta = \infty. \tag{7.46}$$

To seek a power-series solution, suppose $p(\eta)$ is continuous and can be expanded as:

$$p(\eta) = \sum_{n=0}^{\infty} a_n\,\eta^{\alpha n}. \tag{7.47}$$

Then,

$$\eta \frac{dp}{d\eta} = \sum_{n=0}^{\infty} \alpha n \, a_n \, \eta^{\alpha n}. \tag{7.48}$$

Using the Caputo definition for a derivative of order α, we have:

$$\frac{d^{\alpha} p}{d\eta^{\alpha}} = \sum_{n=1}^{\infty} \frac{a_n \, \Gamma(\alpha n + 1)}{\Gamma(\alpha n - \alpha + 1)} \eta^{\alpha n - \alpha}. \tag{7.49}$$

Thus,

$$\eta \frac{d^{\alpha+1} p}{d\eta^{\alpha+1}} = \sum_{n=1}^{\infty} \frac{a_n \, \Gamma(\alpha n + 1)}{\Gamma(\alpha n - \alpha)} \eta^{\alpha n - \alpha}. \tag{7.50}$$

Substituting Eqs. (7.47) and (8.31) into Eq. (7.44) produces:

$$\sum_{n=1}^{\infty} \alpha n \, a_n \, \eta^{\alpha n} + \sum_{n=1}^{\infty} \frac{a_n \, \Gamma(\alpha n + 1)}{\Gamma(\alpha(n-1)+1)} \eta^{\alpha(n-1)} + \left(\frac{\alpha+1}{2}\right) \sum_{n=1}^{\infty} \frac{a_n \, \Gamma(\alpha n + 1)}{\Gamma(\alpha(n-1))} \eta^{\alpha(n-1)} = 0, \tag{7.51}$$

which can be rearranged as

$$\sum_{m=1}^{\infty} \alpha m \, a_m \, \eta^{\alpha m} + \sum_{n=1}^{\infty} \left(\frac{a_n \, \Gamma(\alpha n + 1)}{\Gamma(\alpha(n-1))} \left(\frac{1}{\alpha(n-1)} - \frac{\alpha+1}{2} \right) \right) \eta^{\alpha(n-1)} = 0. \tag{7.52}$$

From here, we note that the first sum has terms $\{\alpha, 2\alpha, 3\alpha, \dots\}$, whereas the second sum has powers $\{0, \alpha, 2\alpha, 3\alpha, \dots\}$. This offset implies $a_1 = 0$. Further manipulation yields the relationship

$$\alpha m \, a_m + \left(\frac{\Gamma(\alpha(m+1)+1)}{\Gamma(\alpha m)} \left(\frac{1}{\alpha m} - \frac{\alpha+1}{2} \right) \right) a_{m+1} = 0, \tag{7.53}$$

leading to the recurrence relation

$$a_{m+1} = - \frac{\alpha m}{\frac{\Gamma(\alpha(m+1)+1)}{\Gamma(\alpha m)} \left(\frac{1}{\alpha m} - \frac{\alpha+1}{2} \right)} a_m, \quad m \geq 1. \tag{7.54}$$

To determine a_0 and a_2, one applies the boundary conditions (Eqs. (7.45) and (7.46)) to the series solution. In particular, at $\eta \to 0$,

$$\frac{dp}{d\eta} = \alpha \sum_{n=2}^{\infty} n \, a_n \, \eta^{\alpha n - 1}, \tag{7.55}$$

since $a_1 = 0$. Letting $\eta \to 0$ in a dimensional sense (e.g., at $\eta = L$ for some characteristic length), one obtains

$$\frac{Q \, \mu}{2 \, \pi \, k \, H} = \alpha \left(2 \, a_2 \, L^{2\alpha-1} + 3 \, a_3 \, L^{3\alpha-1} + \dots \right). \tag{7.56}$$

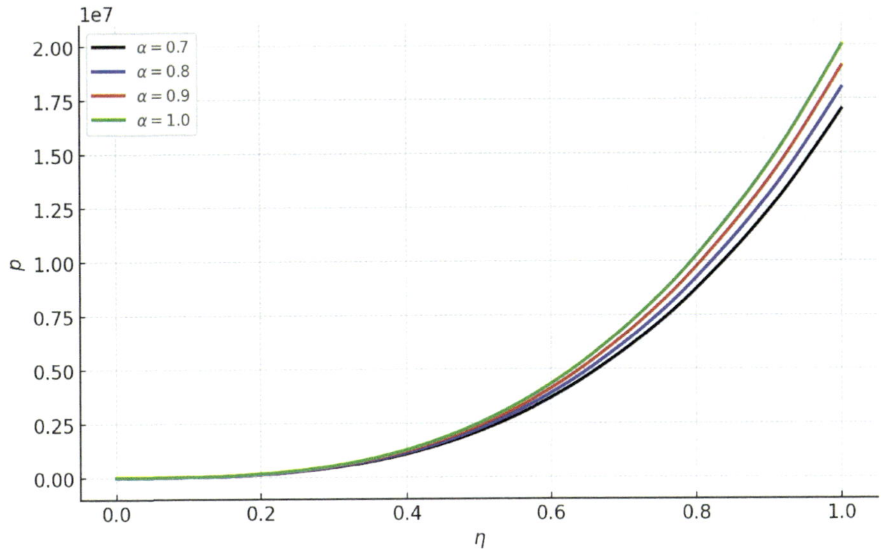

Figure 7.2 Impact of the fractional-order parameter in the case of Dirichlet-Neumann boundary conditions.

Ultimately, one can isolate a_2:

$$a_2 = \frac{\frac{Q\,\mu}{2\,\pi\,k\,H}}{\alpha\,L^{2\alpha-1}\left(2 + \sum_{j=1}^{\infty}(j+2)\,\beta_j\,L^{j\alpha}\right)}. \tag{7.57}$$

Figure 7.2 shows the pressure p as a function of the universal variable η. These profiles demonstrate that the effect of the fractional-order parameter α becomes more pronounced with increasing η.

7.6.1 MATLAB® CODE

To evaluate the analytical solution derived via the power-series method for the fractional-order gas flow equation, we implement the recurrence relation in MATLAB®. The algorithm computes the pressure distribution as a function of the similarity variable η and the fractional order α.

Listing 7.1: MATLAB® code to evaluate the analytical solution for fractional gas flow

```
clear; close all;

% Parameters
N = 500; M = 100;
p0 = 1e7; pw = 1e4;
Q = 0.1; H = 1000;
k = 1.2e-10; mu = 0.01;
h = 0.01; L = h * N;
```

```
FL = Q*mu / (4*pi*k*H);

% Fractional orders
alphas = 0.2:0.2:1.0;
eta = 0:h:L;
P = zeros(length(alphas), length(eta));

for i = 1:length(alphas)
    alpha = alphas(i);
    a = zeros(1, M); bet = zeros(1, M); sum_terms = 0;
    bet(1) = 2;

    for j = 1:M-3
        numer = (j+1) * alpha;
        denom = (gamma((j+2)*alpha + 1) / gamma((j+1)*alpha)) *
            (1/((j+1)*alpha) - (alpha+1)/2);
        bet(j+1) = -numer / denom * bet(j);
        sum_terms = sum_terms + bet(j+1) * (j+2) * L^(alpha * j);
    end

    a(1) = p0;
    a(2) = FL / (alpha * L^(2*alpha - 1) * (2 + sum_terms));

    for m = 2:M
        numer = alpha * m;
        denom = (gamma(alpha*(m+1)+1) / gamma(alpha*m)) * (1/(
            alpha*m) - (alpha+1)/2);
        a(m+1) = -numer / denom * a(m);
    end

    for j = 1:length(eta)
        eta_val = eta(j);
        p_val = 0;
        for m = 1:M
            p_val = p_val + a(m) * eta_val^(alpha * (m-1));
        end
        P(i, j) = p_val;
    end
end

% Plotting
figure;
hold on;
colors = lines(length(alphas));
for i = 1:length(alphas)
    plot(eta, P(i, :), 'LineWidth', 2, 'DisplayName', ['\alpha = 
        ', num2str(alphas(i))]);
end
xlabel('\eta', 'Interpreter', 'latex', 'FontSize', 12);
ylabel('Pressure $p(\eta)$ [Pa]', 'Interpreter', 'latex', '
    FontSize', 12);
title('Analytical Pressure Profiles for Different Fractional 
    Orders', 'Interpreter', 'latex');
legend('show', 'Location', 'northeast');
grid on;
```

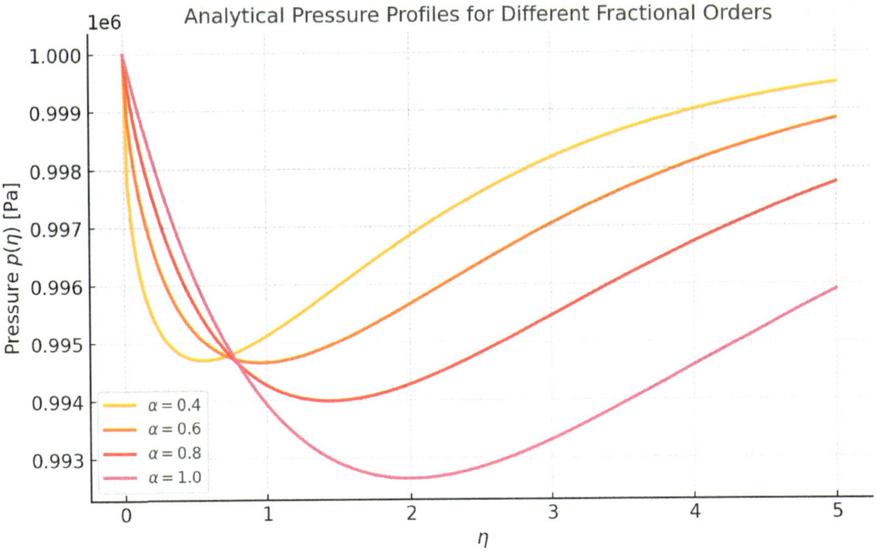

Figure 7.3 Pressure distribution $p(\eta)$ for various fractional orders α.

The MATLAB® code computes the analytical solution for a range of fractional orders $\alpha \in [0.2, 1.0]$ by initializing physical and numerical parameters and then defining the Neumann boundary condition. The term is translated into a coefficient $FL = \frac{Q\mu}{4\pi kH}$ for use in calculating the series coefficients. The pressure is expanded in a power series, and a recurrence relation is used to compute a_n for $n > 2$. The second coefficient a_2 is obtained from the Neumann boundary condition, and the coefficients β_j are computed iteratively. For each value of η, the pressure is calculated using the series expansion truncated to M terms. The results are plotted for various values of α to show the effect of the fractional order on pressure evolution. Figure 7.3 shows the pressure profiles $p(\eta)$ for different values of the fractional order α. The results highlight the sensitivity of the solution to α. As α increases toward 1, the diffusion becomes faster, and the pressure drop near the well becomes more pronounced.

7.7 NUMERICAL METHODS FOR FRACTIONAL MODELS

Fractional flow models arise in porous media, viscoelastic fluids, and anomalous transport processes. These models involve partial differential equations with fractional derivatives, representing memory and non-local effects. Numerical methods must account for these unique features while ensuring accuracy, stability, and efficiency. In this section, we detail the primary numerical techniques used to solve fractional flow models, including finite difference, finite element, and spectral methods, along with practical implementation considerations in MATLAB® and Python.

7.7.1 ENHANCED FINITE DIFFERENCE SCHEMES

Fractional finite difference methods extend classical numerical schemes to solve differential equations involving fractional derivatives, operators that generalize traditional integer-order differentiation. Unlike classical derivatives, fractional derivatives capture non-local effects by accounting for the complete history of a process. This intrinsic memory property makes them especially valuable in modeling anomalous diffusion, viscoelastic behavior, and other phenomena where past states influence current dynamics.

Two popular formulations in fractional calculus are the Riemann–Liouville and Caputo derivatives. The choice between these definitions is typically guided by the physical context, the specific treatment of initial and boundary conditions, and the desired interpretability. In any finite difference framework, the computational domain is discretized into a grid. However, due to the non-local nature of fractional derivatives, the approximation at each grid point incorporates contributions from many, if not all, preceding time levels rather than just immediate neighbors. One common discretization technique is the Grunwald–Letnikov method, which generalizes the classical finite difference approach by expressing the fractional derivative as a weighted sum of previous function values with weights determined by binomial coefficients and the order of differentiation.

The challenges in stability and convergence analyses are accentuated in the fractional context because each time step is influenced by a broad history of the solution. This complexity necessitates refined numerical techniques to ensure that the computed solution faithfully approximates the true behavior as the grid is refined. The increased computational cost and memory demands are also critical factors that drive ongoing research into more efficient algorithms and parallel computing strategies.

Recent developments in this area are focusing on improving algorithmic efficiency, extending the methods to non-linear and multi-term fractional differential equations, and rigorously establishing the theoretical foundations regarding stability and convergence. These advancements are enabling the application of fractional finite difference methods to a wide range of real-world problems where classical models fall short.

7.7.2 FRACTIONAL MODEL OF FLOW IN POROUS MEDIA

Combining Darcy's law with the continuity equation for single-phase flow in porous media leads to a diffusion-like equation for pressure. In classical models, Darcy's law is expressed as

$$q = -\frac{k}{\mu}\nabla p, \tag{7.58}$$

where q is the Darcy flux, k is the permeability, μ is the fluid viscosity, and p is the pressure. Coupled with the mass conservation (continuity) equation, and under the assumption of constant porosity ϕ and total compressibility c_t, the resulting pressure diffusion equation in one dimension is

$$\phi c_t \frac{\partial p}{\partial t} = \nabla \cdot \left(\frac{k}{\mu} \nabla p \right). \tag{7.59}$$

For the one-dimensional case, the above equation takes the form

$$\frac{\partial p}{\partial t} = \frac{k}{\phi \mu c_t} \frac{\partial^2 p}{\partial x^2}. \tag{7.60}$$

In many porous media, however, transport processes exhibit memory effects that are not adequately described by classical models. To account for such anomalous behavior, we extend the pressure diffusion equation to its fractional counterpart by replacing the first-order time derivative with a fractional derivative of order α. The resulting fractional diffusion equation becomes

$$\frac{\partial^\alpha p}{\partial t^\alpha} = D \frac{\partial^2 p}{\partial x^2}, \tag{7.61}$$

where

$$D = \frac{k}{\phi \mu c_t} \tag{7.62}$$

is the effective diffusivity, and the operator $\frac{\partial^\alpha}{\partial t^\alpha}$ is understood in the sense of either Caputo or Riemann–Liouville, depending on the physical context and boundary conditions.

7.7.3 GRUNWALD–LETNIKOV AND FDM APPROXIMATION

To numerically solve equation (7.61), the time-fractional derivative is approximated using the Grunwald–Letnikov scheme. This yields a weighted sum that captures the history-dependent nature of the flow:

$$\frac{\partial^\alpha p}{\partial t^\alpha} \approx \frac{1}{(\Delta t)^\alpha} \left[p(t) - \frac{\Gamma(\alpha)}{\Gamma(\alpha - 1)} p(t - \Delta t) + \sum_{m=2}^{N_t} (-1)^m \binom{\alpha}{m} p(t - m\Delta t) \right], \tag{7.63}$$

with the generalized binomial coefficient defined as

$$\binom{\alpha}{m} = \frac{\Gamma(\alpha + 1)}{\Gamma(m + 1)\,\Gamma(\alpha - m + 1)}. \tag{7.64}$$

Let p_j^k denote the pressure at spatial node j and time level k. Substituting the approximation (7.69) into the fractional pressure diffusion equation (7.61), and discretizing the spatial second derivative by a standard central difference, we obtain

$$\frac{1}{(\Delta t)^\alpha}\left[p_j^{k+1} - \frac{\Gamma(\alpha)}{\Gamma(\alpha-1)}p_j^k + \sum_{m=2}^{k}(-1)^m\binom{\alpha}{m}p_j^{k-m+1}\right] = D\,\frac{p_{j+1}^k - 2p_j^k + p_{j-1}^k}{(\Delta x)^2}. \tag{7.65}$$

Rearranging to solve for p_j^{k+1} gives the recurrence relation

$$\begin{aligned} p_j^{k+1} &= \left(\frac{\Gamma(\alpha)}{\Gamma(\alpha-1)} - (\Delta t)^\alpha\frac{2D}{(\Delta x)^2}\right)p_j^k + (\Delta t)^\alpha D\,\frac{p_{j+1}^k + p_{j-1}^k}{(\Delta x)^2} \\ &\quad - \sum_{m=2}^{k}(-1)^m\binom{\alpha}{m}p_j^{k-m+1}. \end{aligned} \tag{7.66}$$

This scheme captures the long-term memory effects inherent in anomalous transport phenomena within porous media, providing a more realistic model for pressure propagation when compared to classical approaches.

In this fractional framework, each new time step p_j^{k+1} is computed by not only considering the immediate spatial neighbors but also incorporating contributions from all previous time steps. This reflects the physical reality in many porous media where past states influence the current flow behavior. Applications of this model include enhanced recovery processes, contaminant transport, and scenarios where heterogeneities and anomalous diffusion are significant.

MATLAB® CODE

Here is the MATLAB® code for simulating the pressure of flow in porous media using a time-fractional diffusion-like equation. The simulation is carried out for multiple fractional orders α.

LOOP OVER FRACTIONAL ORDERS

The outer loop in the code iterates over different values of α:

```
for i = 1:length(alp_vals)
    alp = alp_vals(i);
    ...
```

Each α defines a different degree of memory effect. The case $\alpha = 1$ corresponds to classical diffusion, while $\alpha < 1$ introduces memory via fractional derivatives.

CLASSICAL CASE: $\alpha = 1$

When $\alpha = 1$, the code uses the explicit finite difference scheme:

$$p_j^{k+1} = p_j^k + \frac{D\Delta t}{\Delta x^2}\left(p_{j+1}^k - 2p_j^k + p_{j-1}^k\right)$$

This is implemented as:

```
p(j, k+1) = p(j,k) + dt * D * (p(j+1,k) - 2*p(j,k) + p(j-1,k)) /
    dx^2;
```

It solves the classical diffusion (or heat) equation and is subject to the stability condition:

$$\frac{D\Delta t}{\Delta x^2} \leq 0.5$$

FRACTIONAL CASE: $\alpha < 1$

When $\alpha < 1$, the fractional derivative is approximated using the Grünwald–Letnikov scheme. The fractional time derivative is expressed as a weighted sum over past time steps:

$$\frac{d^\alpha p(t)}{dt^\alpha} \approx \frac{1}{\Gamma(2-\alpha)} \sum_{m=0}^{k} w_m \, p(t - m\Delta t)$$

The weights w_m are computed recursively:

```
w(1) = 1;
for m = 1:memory_length
    w(m+1) = (1 - (alp+1)/m) * w(m);
end
```

The update equation becomes:

$$p_j^{k+1} = p_j^k - \frac{\Delta t^\alpha}{\Gamma(2-\alpha)} \sum_{m=0}^{M} w_m \, p_j^{k-m} + \frac{D\Delta t^\alpha}{\Delta x^2}(p_{j+1}^k - 2p_j^k + p_{j-1}^k)$$

In the code, this is written as:

```
memorySum = w(1) * p(j,k);
for m = 1:m_limit
    memorySum = memorySum + w(m+1) * p(j, k-m+1);
end

p(j, k+1) = p(j,k) - dt^alp/gamma(2-alp) * memorySum ...
            + coef * (p(j+1,k) - 2*p(j,k) + p(j-1,k));
```

Figure 7.4 shows the evolution of the pressure distribution along the 100 m spatial domain at two different time instances for fractional orders $\alpha = 0.90$, 0.95, and 0.99. Figure 7.5 illustrates the normalized pressure change over time at three monitoring locations for varying fractional orders α. The normalized change is computed as:

$$\text{Normalized Pressure Change}(t) = \frac{p(x,t) - p_{\text{boundary}}}{\max_t[p(x,t) - p_{\text{boundary}}]} \tag{7.67}$$

7.7.4 STABILITY CONDITION

In our discretization of the fractional pressure diffusion equation

$$\frac{\partial^\alpha p}{\partial t^\alpha} = D\frac{\partial^2 p}{\partial x^2}, \tag{7.68}$$

the time-fractional derivative is approximated by the Grünwald–Letnikov formula:

$$\frac{\partial^\alpha p}{\partial t^\alpha} \approx \frac{1}{(\Delta t)^\alpha}\left[p(t) - \frac{\Gamma(\alpha)}{\Gamma(\alpha-1)}p(t-\Delta t) + \sum_{m=2}^{N_t}(-1)^m\binom{\alpha}{m}p(t-m\Delta t)\right], \tag{7.69}$$

with the generalized binomial coefficient defined as

$$\binom{\alpha}{m} = \frac{\Gamma(\alpha+1)}{\Gamma(m+1)\,\Gamma(\alpha-m+1)}. \tag{7.70}$$

Discretizing the spatial second derivative with a central difference leads to the recurrence relation

$$\begin{aligned} p_j^{k+1} &= \left(\frac{\Gamma(\alpha)}{\Gamma(\alpha-1)} - (\Delta t)^\alpha\frac{2D}{(\Delta x)^2}\right)p_j^k + (\Delta t)^\alpha D\frac{p_{j+1}^k + p_{j-1}^k}{(\Delta x)^2} \\ &\quad - \sum_{m=2}^{k}(-1)^m\binom{\alpha}{m}p_j^{k-m+1}. \end{aligned} \tag{7.71}$$

VON NEUMANN STABILITY ANALYSIS

To assess the stability of the scheme, one typically performs a von Neumann analysis. Assume a solution of the form

$$p_j^k = \xi^k e^{ij\theta}, \tag{7.72}$$

where ξ is the amplification factor and θ is the wave number. Substituting this ansatz into the recurrence relation (and neglecting the memory terms for

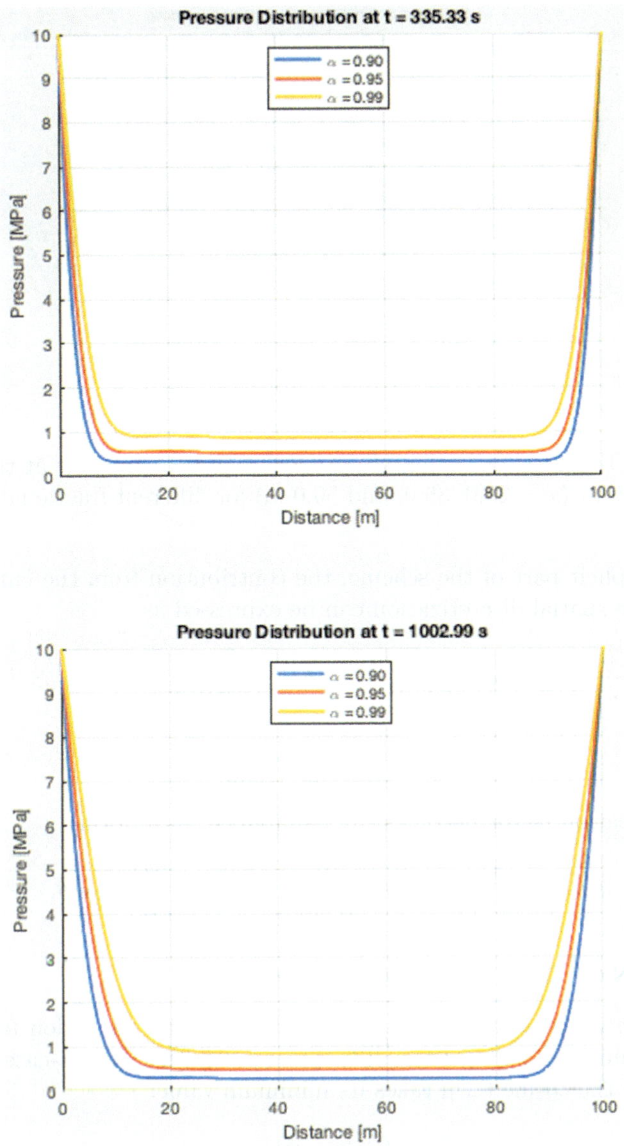

Figure 7.4 Pressure distribution along the spatial domain at two time instances ($t = 335.33$ s and $t = 1002.99$ s) for different fractional orders α.

the moment, as they introduce convolutional effects but do not change the form of the local amplification condition) yields a relation for ξ that depends on the coefficients of the current and neighboring nodes.

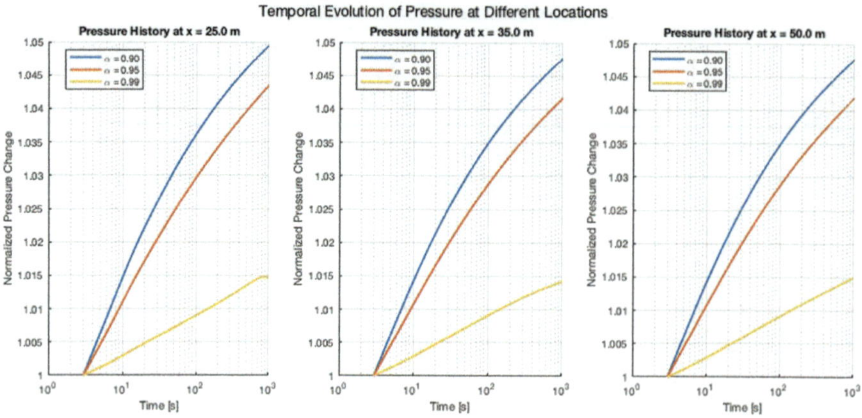

Figure 7.5 Temporal evolution of normalized pressure change at three monitoring locations ($x = 25.0$, 35.0, and 50.0 m) for different fractional orders α.

In the explicit part of the scheme, the contribution from the current time level and the spatial discretization can be expressed as

$$\left(\frac{\Gamma(\alpha)}{\Gamma(\alpha-1)} - (\Delta t)^{\alpha}\frac{2D}{(\Delta x)^2}\right) + (\Delta t)^{\alpha}\frac{D}{(\Delta x)^2}\left(e^{i\theta} + e^{-i\theta}\right). \tag{7.73}$$

Since

$$e^{i\theta} + e^{-i\theta} = 2\cos\theta. \tag{7.74}$$

The amplification factor becomes (in a simplified, local form)

$$\xi \approx \frac{\Gamma(\alpha)}{\Gamma(\alpha-1)} - (\Delta t)^{\alpha}\frac{2D}{(\Delta x)^2} + (\Delta t)^{\alpha}\frac{2D}{(\Delta x)^2}\cos\theta. \tag{7.75}$$

DERIVATION OF THE STABILITY CONDITION

For stability, we require that the magnitude of the amplification factor does not exceed unity, i.e., $|\xi| \leq 1$ for all θ. The most critical (worst-case) scenario occurs when the cosine term takes its minimum value:

$$\cos\theta = -1. \tag{7.76}$$

Substituting this into the expression for ξ gives

$$\xi_{\min} = \frac{\Gamma(\alpha)}{\Gamma(\alpha-1)} - (\Delta t)^{\alpha}\frac{2D}{(\Delta x)^2} - (\Delta t)^{\alpha}\frac{2D}{(\Delta x)^2}. \tag{7.77}$$

Thus, the condition for stability becomes

$$\left|\frac{\Gamma(\alpha)}{\Gamma(\alpha-1)} - 2(\Delta t)^{\alpha}\frac{2D}{(\Delta x)^2}\right| \leq \frac{\Gamma(\alpha)}{\Gamma(\alpha-1)}. \tag{7.78}$$

After simplifying (and noting that the coefficients are positive), this condition is essentially equivalent to requiring that

$$(\Delta t)^{\alpha}\frac{2D}{(\Delta x)^2} \quad \text{remains sufficiently small compared to} \quad \frac{\Gamma(\alpha)}{\Gamma(\alpha-1)}. \tag{7.79}$$

In the classical limit, when $\alpha = 1$, we have

$$\frac{\Gamma(1)}{\Gamma(0)} \tag{7.80}$$

which is not defined; however, the classical scheme is separately governed by

$$\frac{D\,\Delta t}{\Delta x^2} \leq 0.5. \tag{7.81}$$

For the fractional case ($0 < \alpha < 1$), the effective stability parameter is

$$r_{\text{frac}} = \frac{D\,(\Delta t)^{\alpha}}{(\Delta x)^2}, \tag{7.82}$$

and ensuring stability requires that r_{frac} is kept small relative to the coefficient $\frac{\Gamma(\alpha)}{\Gamma(\alpha-1)}$.

7.8 CONFORMABLE FRACTIONAL GAS EQUATION

This section presents the governing equation for natural gas transport in porous media using the conformable fractional derivative. The main transport equation is:

$$T_{\tau}^{\alpha}p = \nabla\cdot(\mathrm{D}\nabla p) + Q, \tag{7.83}$$

where p is the gas pressure, $T_{\tau}^{\alpha}p$ is the conformable fractional time derivative, and Q represents a source or sink term.

The diffusivity coefficient D is defined as:

$$\mathrm{D} = \frac{\rho_g}{\mu_g}Kp\left(1+\frac{b}{p}\right), \tag{7.84}$$

with

$$b = \sqrt{\frac{8\pi RT}{M_w}}\left(\frac{1}{r_w}\right)\left(\frac{2}{\alpha_m} - 0.995\right)\mu_g,$$

where T is the temperature, R is the universal gas constant, μ_g is the gas viscosity, r_w is the well radius, α_m is the Biot coefficient, and M_w is the molecular weight of the gas.

The relationship between permeability and porosity is modeled by:

$$K = K_0\exp\left(c\left(\frac{\phi}{\phi_0} - 1\right)\right), \tag{7.85}$$

where K_0 and ϕ_0 are the initial permeability and porosity, respectively, and c is a fitting coefficient.

Porosity variation with effective stress is described as:

$$\phi = \phi_r + (\phi_0 - \phi_r)\exp(-a\sigma'_d), \tag{7.86}$$

where ϕ_r is the reference porosity, a is a fitting parameter, and σ'_d is the effective stress.

Gas density follows the ideal gas law:

$$\rho_g = \frac{M_w}{RT}p. \tag{7.87}$$

Initial pressure condition:

$$p(x, y, 0) = p_0. \tag{7.88}$$

Dirichlet boundary conditions:

$$\begin{aligned} p(0, y) &= p_W, \quad p(L_x, y) = p_E, \\ p(x, 0) &= p_S, \quad p(x, L_y) = p_N. \end{aligned} \tag{7.89}$$

Neumann (zero-flux) boundary conditions:

$$\begin{aligned} \left.\frac{\partial p}{\partial x}\right|_{x=0} &= 0, \quad \left.\frac{\partial p}{\partial x}\right|_{x=L_x} = 0, \\ \left.\frac{\partial p}{\partial y}\right|_{y=0} &= 0, \quad \left.\frac{\partial p}{\partial y}\right|_{y=L_y} = 0. \end{aligned} \tag{7.90}$$

The conformable derivative is approximated by:

$$T_\tau^\alpha p \approx \alpha \frac{p_j^{n+1} - p_j^n}{(t_{n+1})^\alpha - (t_n)^\alpha}. \tag{7.91}$$

To solve numerically, we apply a modified conformable Euler method combined with Crank-Nicolson:

$$p_{i,j}^{n+1} - p_{i,j}^n = \frac{1}{\alpha}\left[(t_{n+1})^\alpha - (t_n)^\alpha\right]\frac{1}{2}\left(\nabla \cdot (diff\nabla p^{n+1}) + \nabla \cdot (diff\nabla p^n)\right). \tag{7.92}$$

Spatial discretization using central differences:

$$\begin{aligned} \nabla \cdot (\mathrm{D}\nabla p) \approx &\frac{D_{i+1/2,j}(p_{i+1,j} - p_{i,j}) - D_{i-1/2,j}(p_{i,j} - p_{i-1,j})}{\Delta x^2} \\ &+ \frac{D_{i,j+1/2}(p_{i,j+1} - p_{i,j}) - D_{i,j-1/2}(p_{i,j} - p_{i,j-1})}{\Delta y^2} \end{aligned} \tag{7.93}$$

Harmonic averages for diffusion coefficients:

$$\begin{aligned} D_{i+1/2,j} &= \frac{2D_{i,j}D_{i+1,j}}{D_{i,j} + D_{i+1,j}}, \quad D_{i-1/2,j} = \frac{2D_{i,j}D_{i-1,j}}{D_{i,j} + D_{i-1,j}}, \\ D_{i,j+1/2} &= \frac{2D_{i,j}D_{i,j+1}}{D_{i,j} + D_{i,j+1}}, \quad D_{i,j-1/2} = \frac{2D_{i,j}D_{i,j-1}}{D_{i,j} + D_{i,j-1}}. \end{aligned} \tag{7.94}$$

7.8.1 MATLAB® CODE FOR CONFORMABLE FRACTIONAL GAS EQUATION

PARAMETERS SETUP:

```
Lx = 100; Ly = 100;
Nx = 50; Ny = 50;
dx = Lx/Nx; dy = Ly/Ny;
T_final = 1;
Nt = 100;
dt = T_final/Nt;

alphas = [0.6, 0.8, 1.0];

p0 = 10e6;
T = 350;
R = 8.314;
mu_g = 1.8e-5;
Mw = 16e-3;
rw = 0.1;
alpha_m = 0.8;
K0 = 1e-15;
phi0 = 0.2;
phi_r = 0.05;
a = 5e-8;
c = 8;

p_W = p0; p_E = p0; p_S = p0; p_N = p0;
```

GRID AND INITIALIZATION:

```
x = linspace(0, Lx, Nx+1);
y = linspace(0, Ly, Ny+1);
t = linspace(0, T_final, Nt+1);
[X, Y] = meshgrid(x, y);

bc_type = [0, 0, 0, 0];

Q_strength = -1e5;
Q_loc_x = Lx/2; Q_loc_y = Ly/2;
Q_radius = 5;

solutions = cell(length(alphas), 1);
p_evolution = zeros(length(alphas), Nt+1);
```

LOOP OVER FRACTIONAL ORDERS:

```
for idx = 1:length(alphas)
    alpha = alphas(idx);
    fprintf('Simulating with alpha = \%.2f\n', alpha);

    p = p0 * ones(Ny+1, Nx+1);
```

```
p_old = p;

center_x = round(Nx/2) + 1;
center_y = round(Ny/2) + 1;
p_evolution(idx, 1) = p(center_y, center_x);

sigma_d = 20e6;
```

TIME-STEPPING LOOP:

```
for n = 1:Nt
    dt_alpha = t(n+1)^alpha - t(n)^alpha;

    phi = phi_r + (phi0 - phi_r) * exp(-a * sigma_d);
    K = K0 * exp(c * (phi/phi0 - 1));

    A = sparse((Nx+1)*(Ny+1), (Nx+1)*(Ny+1));
    b = zeros((Nx+1)*(Ny+1), 1);
```

BUILD THE LINEAR SYSTEM:

LOOP OVER GRID AND APPLY BOUNDARY CONDITIONS

```
    for j = 1:Ny+1
        for i = 1:Nx+1
            idx_c = global_index(i, j, Nx+1);

            if (i == 1 && bc_type(1) == 0) || ...
               (i == Nx+1 && bc_type(2) == 0) || ...
               (j == 1 && bc_type(3) == 0) || ...
               (j == Ny+1 && bc_type(4) == 0)
                A(idx_c, idx_c) = 1;
                b(idx_c) = p0;
                continue;
            end
```

INTERIOR NODE CALCULATIONS:

```
            rho_g = (Mw/(R*T)) * p_old(j, i);
            b_param = sqrt(8*pi*R*T/Mw) * (1/rw) * (2/alpha_m
                 - 0.995) * mu_g;

            D_center = (rho_g/mu_g) * K * p_old(j, i) * (1 +
                b_param/p_old(j, i));

            A(idx_c, idx_c) = A(idx_c, idx_c) + alpha/
                dt_alpha;
            b(idx_c) = b(idx_c) + (alpha/dt_alpha) * p_old(j,
                 i);
```

SOURCE TERM:

```
            dist = sqrt((x(i) - Q_loc_x)^2 + (y(j) - Q_loc_y)
                ^2);
            if dist <= Q_radius
                Q = Q_strength;
                b(idx_c) = b(idx_c) + (dt_alpha/alpha) * Q;
            end
```

SOLVE AND UPDATE:

```
        p_vec = A \ b;

        for j = 1:Ny+1
            for i = 1:Nx+1
                idx_c = global_index(i, j, Nx+1);
                p(j, i) = p_vec(idx_c);
            end
        end

        p_evolution(idx, n+1) = p(center_y, center_x);
        p_old = p;
```

FINAL RESULTS:

```
fprintf('\nMinimum␣pressure␣values␣at␣final␣time:\n');
for idx = 1:length(alphas)
    fprintf('Alpha␣=␣\%.1f:␣\%.2e␣Pa\n', alphas(idx), min(min(
        solutions{idx})));
end
```

FUNCTION:

```
function idx = global_index(i, j, nx)
    idx = (j-1)*nx + i;
end
```

Figure 7.6 shows how pressure at the central point of the domain evolves over time for three different values of the fractional order α. The initial pressure is set at 10^7 Pa, and we observe a gradual decline over time due to gas extraction (a sink term). For $\alpha = 1.0$ (the classical case), the pressure drops smoothly and slowly, reflecting standard diffusive behavior. As α decreases to 0.8 and 0.6, the pressure drop becomes more pronounced, especially in the early stages of the simulation. This trend indicates that lower values of α correspond to a faster initial response of the system, which is characteristic of sub-diffusion or memory-retaining processes often found in complex porous media.

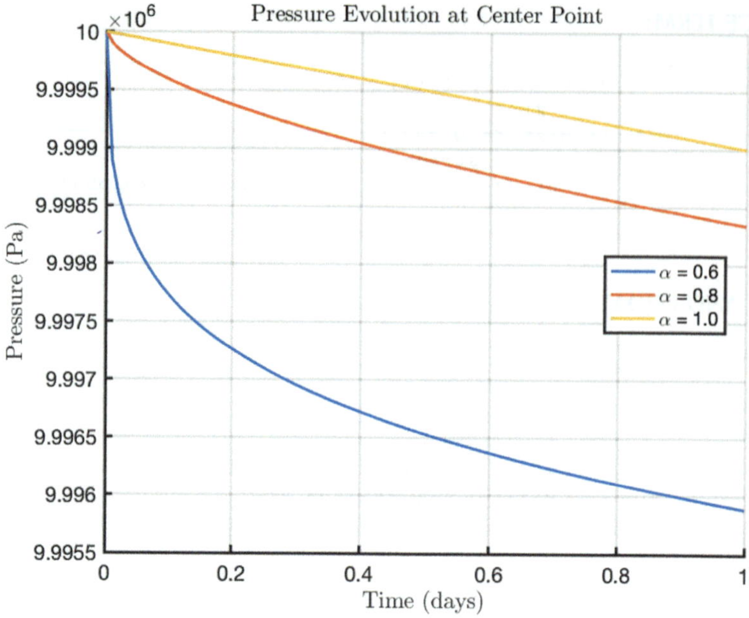

Figure 7.6 Pressure evolution at the center of the domain over time for different conformable fractional orders $\alpha = 0.6$, 0.8, and 1.0.

7.9 FINITE ELEMENT METHODS

This section outlines how the Finite Element Method (FEM) can be extended to solve fractional differential equations (FDEs). FEM is a powerful numerical approach used to approximate solutions to differential equations by discretizing the problem domain and using variational principles. When applied to FDEs, FEM must account for the non-local characteristics of fractional derivatives, which represent memory effects in physical systems.

Recent developments have produced fast and robust FEM-based solvers for various types of FDEs. These include methods for equations involving the Riesz fractional derivative [202], the Galerkin approach for multiterm linear FDEs [161], and the discontinuous Galerkin method, which handles irregular geometries and complex boundary conditions effectively [183].

In FEM, the solution domain is divided into finite elements, and within each element, polynomial shape functions approximate the unknown solution. For FDEs, the weak form of the equations is constructed by integrating the fractional derivatives against test functions, often requiring fractional integration by parts. The assembled system incorporates non-local interactions and is solved using specialized numerical techniques.

7.9.1 GALERKIN FINITE ELEMENT METHOD

To illustrate the application of FEM to FDEs, consider the space-fractional diffusion equation (SFDE) [176]:

$$\frac{\partial u(x,t)}{\partial t} = d_f \frac{\partial^{1+\alpha} u(x,t)}{\partial x^{1+\alpha}} + f(x,t), \tag{7.95}$$

where $0 < \alpha \leq 1$, and d_f is the diffusion coefficient. The classical second-order derivative is recovered when $\alpha = 1$. The problem is subject to:

$$\begin{aligned} u(x,0) = \psi(x), \quad a \leq x \leq b, \\ u(a,t) = 0, \quad u(b,t) = \phi_b(t), \quad 0 \leq t \leq T. \end{aligned} \tag{7.96}$$

Let $\Omega = [a,b]$ and define a uniform partition:

$$a = x_0 < x_1 < \cdots < x_m = b, \quad h = \frac{b-a}{m}. \tag{7.97}$$

Define piecewise-linear basis functions $\phi_j(x)$ over the intervals $\Omega_i = [x_{i-1}, x_i]$ and a solution space:

$$V_h = \{v \in C(\Omega) : v|_{\Omega_i} \in P_1(\Omega_i)\}. \tag{7.98}$$

The approximate solution at time t_n is written as:

$$u^n(x) = \sum_{j=0}^{m} u_j^n \phi_j(x). \tag{7.99}$$

The Galerkin weak form of the SFDE is:

$$\left(\frac{u^n - u^{n-1}}{\tau}, v\right) + d_f \left(\frac{\partial^\alpha u^n}{\partial x^\alpha}, \frac{\partial v}{\partial x}\right) = (f^n, v). \tag{7.100}$$

Choosing $v - \phi_i(x)$ and substituting the expansion for u^n yields the discrete form:

$$\begin{aligned} &\frac{h}{6}(u_{i-1}^n + 4u_i^n + u_{i+1}^n) + \frac{d_f \tau}{h}\left[\sum_j u_j^n \int_{x_{i-1}}^{x_i} \frac{\partial^\alpha \phi_j}{\partial x^\alpha} dx - \sum_j u_j^n \int_{x_i}^{x_{i+1}} \frac{\partial^\alpha \phi_j}{\partial x^\alpha} dx\right] \\ &= \frac{h}{6}(u_{i-1}^{n-1} + 4u_i^{n-1} + u_{i+1}^{n-1}) + (\psi, \phi_i) + \tau(f^n, \phi_i). \end{aligned} \tag{7.101}$$

Integrals of the fractional derivatives are approximated by:

$$\int_{x_{i-1}}^{x_i} \frac{\partial^\alpha \phi_j}{\partial x^\alpha} dx = \frac{h^{1-\alpha}}{\Gamma(3-\alpha)} r_{i,j}, \quad \int_{x_i}^{x_{i+1}} \frac{\partial^\alpha \phi_j}{\partial x^\alpha} dx = \frac{h^{1-\alpha}}{\Gamma(3-\alpha)} s_{i,j}, \tag{7.102}$$

with specific coefficients $r_{i,j}$ and $s_{i,j}$. The final update formula becomes:

$$\frac{h}{6}(u_{i-1}^n + 4u_i^n + u_{i+1}^n) + \lambda \sum_j (r_{i,j} - s_{i,j}) u_j^n = \frac{h}{6}(u_{i-1}^{n-1} + 4u_i^{n-1} + u_{i+1}^{n-1})$$
$$+ (\psi, \phi_i) + \tau (f^n, \phi_i), \quad \lambda = \frac{d_f \tau}{h^\alpha \Gamma(3-\alpha)}. \tag{7.103}$$

The initial and boundary conditions are:

$$u_i^0 = \psi(ih), \quad u_0^n = 0, \quad u_m^n = \phi_b(n\tau). \tag{7.104}$$

The scheme achieves a convergence rate of $O(\tau + h^2)$.

7.9.2 MIXED FINITE ELEMENT METHOD

Standard FEM can sometimes exhibit instability. Mixed finite element methods (MFEMs) address these challenges by introducing auxiliary variables, enabling stable and higher-order approximations. In MFEM, both the primary variable (e.g., pressure or temperature) and its flux are approximated in compatible finite element spaces.

Let $\Omega \subset \mathbb{R}^d$ be a polygonal domain. Define the Sobolev and Lebesgue spaces $L^p(\Omega)$ and $W^{k,p}(\Omega)$, with $H^k(\Omega) = W^{k,2}(\Omega)$. For vector and matrix-valued functions, define norms:

$$\|v\|_{(L^p(\Omega))^d} = \||v|\|_{L^p(\Omega)}, \quad \|A\|_{(L^p(\Omega))^{d\times d}} = \||A|\|_{L^p(\Omega)}. \tag{7.105}$$

Define spaces:

$$V = \{v \in (L^2(\Omega))^d : \nabla \cdot v \in L^2(\Omega)\}, \tag{7.106}$$

$$H(\text{div}; \Omega) = \{v \in (L^2(\Omega))^d : \nabla \cdot v \in L^2(\Omega)\}. \tag{7.107}$$

7.9.2.1 Fractional Diffusion-Reaction Equations

Liu et al. [186] applied MFEM to fractional diffusion-reaction equations:

$$D_{0,t}^\alpha u = \frac{\partial}{\partial x}\left(d_f \frac{\partial u}{\partial x}\right) + ku + f(x,t), \tag{7.108}$$

with auxiliary variable $p = d_f \frac{\partial u}{\partial x}$, the system becomes:

$$D_{0,t}^\alpha u = \frac{\partial p}{\partial x} + ku + f(x,t), \tag{7.109}$$

$$d_f \frac{\partial u}{\partial x} = p. \tag{7.110}$$

Testing with appropriate functions and applying the L1 scheme, the discrete MFEM formulation seeks u_h^n, p_h^n in finite-dimensional spaces X_h^r, X_{0h}^r such that:

$$({}_C\tilde{D}_{0,t}^{\alpha} p_h^n, w) = -\left(\frac{\partial p_h^n}{\partial x}, \frac{\partial w}{\partial x}\right) + k(p_h^n, w) - (f^n, \frac{\partial w}{\partial x}), \tag{7.111}$$

$$d_f\left(\frac{\partial u_h^n}{\partial x}, \frac{\partial v}{\partial x}\right) = (p_h^n, \frac{\partial v}{\partial x}), \tag{7.112}$$

for all $w \in X_h^r$, $v \in X_{0h}^r$. Liu et al. demonstrated convergence rates of $O(\Delta t^{2-\alpha} + h^{r+1}\Delta t^{-\alpha} + h^{r+1})$ in the L^2 norm.

7.10 CONCLUSION

In this chapter, we have comprehensively explored the application of fractional calculus in modeling fluid flow through porous and fractured media. By introducing and analyzing key fractional derivatives, including Riemann–Liouville, Caputo, and conformable definitions, we have established a robust theoretical foundation supporting their use in practical simulations. The discussion encompassed detailed developments of fractional models for single-phase and multiphase flows. Numerical methodologies such as finite difference and finite element were thoroughly examined, along with their practical implementations in MATLAB®. The analytical power-series solution and associated numerical examples provided valuable insights into how fractional parameters directly influence system behavior and solution dynamics.

REVIEW QUESTIONS

1. What are the main physical phenomena that fractional calculus can model more effectively than classical calculus? Provide at least three examples.
2. Differentiate between Riemann–Liouville and Caputo fractional derivatives in terms of their definitions and practical applications. Why is Caputo often preferred in engineering problems?
3. How does the concept of anomalous diffusion differ from classical diffusion, and why is it important in the context of porous media?
4. What is the significance of the fractional-order parameters α and β in the fractional Darcy's law and continuity equations? How do they affect model behavior?
5. Describe the role of the Mittag-Leffler function in solving fractional differential equations. How does it generalize the exponential function?
6. Provide the difference between dual-porosity and dual-permeability models in fractured porous media. What physical assumptions distinguish the two?
7. How does the conformable fractional derivative differ from other fractional derivatives in terms of definition and computational advantages?

8. What are the challenges in numerically solving fractional differential equations, and how do finite difference and finite element methods address these challenges?
9. In the analytical solution for gas flow in a radial reservoir, how does the introduction of a fractional order α affect the pressure profile, and what transformation helps in simplifying the governing equation?

PROBLEMS

1. Implement the Caputo fractional derivative using the Grünwald–Letnikov approximation in MATLAB® or Python. Test the implementation on a simple function such as $f(t) = t^2$ and compare the numerical result with the analytical expression of the fractional derivative.
2. Simulate the time-fractional diffusion equation

$$\frac{\partial^\alpha p}{\partial t^\alpha} = D\frac{\partial^2 p}{\partial x^2}$$

using the fractional finite difference method. Plot the pressure profiles for different values of $\alpha \in (0, 1]$.
3. Develop a numerical model for 1D fractional Darcy's law using

$$u = -\eta \nabla^\beta p.$$

Simulate and compare the velocity field for $\beta = 1$ (classical) and $\beta = 0.8$ (fractional) with a given pressure gradient.
4. Write a code to solve a fractional initial value problem using the conformable fractional Euler method for

$$D_t^\alpha u(t) = f(t, u(t)).$$

Use $f(t, u) = -u + t$ with a suitable initial condition, and analyze the effect of varying α.
5. Implement the analytical power-series solution for radial gas flow with fractional derivatives. Plot pressure $p(r, t)$ profiles and investigate how they change with different values of α.
6. Simulate the fractional continuity equation for wetting phase:

$$\phi \frac{\partial^\alpha S_w}{\partial t^\alpha} + \frac{t^{1-\alpha}}{\Gamma(2-\alpha)} \nabla \cdot \mathbf{u}_w = 0.$$

Analyze the evolution of S_w over time for various values of α.
7. Numerically compute the fractional derivative of the Mittag-Leffler function $E_{\alpha,\beta}(z)$. Plot its behavior for different combinations of α and β, and discuss its importance in solving fractional differential equations.
8. Develop a simplified dual-porosity model in MATLAB® or Python, incorporating fractional continuity equations to simulate matrix-fracture fluid exchange. Visualize pressure evolution in both domains.

9. Extend a classical 1D finite difference model for pressure diffusion to the fractional case. Compare simulation results, memory usage, and computational time for $\alpha = 1$ and $\alpha < 1$.
10. Use or modify the provided MATLAB® code to simulate and visualize 2D pressure distribution using contour plots. Analyze how different values of α influence diffusion behavior spatially and temporally.

DECLARATION

During the preparation of this work, the author used ChatGPT in order to improve the language and readability. After using this tool/service, the author reviewed and edited the content as needed and takes full responsibility for the content of the publication.

8 Non-isothermal Flow in Porous Media

8.1 HEAT TRANSFER IN POROUS MEDIA

Heat transfer in porous media governs a wide range of physical and engineering processes where fluids flow through solid structures with interconnected pores. These include geothermal energy extraction, oil recovery, thermal insulation, and environmental remediation. The complexity of these systems arises from the interaction between two thermally distinct phases, solid and fluid, each with unique transport properties and behavior under heat and mass transfer.

A complete mathematical model must account for multiple heat transfer mechanisms: conduction through the solid matrix, conduction and convection within the fluid, and interfacial heat exchange between the phases. These processes are influenced by porosity, thermal conductivity, specific heat capacities, and flow velocities, among other factors. Accurately capturing this interplay is crucial for predicting temperature distributions and designing effective systems.

In both natural and engineered porous systems, the dynamic coupling of thermal and fluid flow behaviors demands rigorous theoretical formulations and efficient numerical methods. The following sections present detailed models for single-phase and two-phase flow heat transfer, followed by advanced numerical schemes designed to handle the computational challenges associated with these coupled, multi-scale processes.

8.1.1 NON-ISOTHERMAL SINGLE-PHASE FLOW

Energy conservation in porous media with distinct solid matrix (m) and fluid phase (f) temperatures can be expressed as:

Solid phase:

$$\frac{\partial}{\partial t}\left[\rho_m C_{pm}(1-\phi)T_m\right] = \frac{\partial}{\partial x}\left[(1-\phi)\kappa_m \frac{\partial T_m}{\partial x}\right] + \dot{q}_m \tag{8.1}$$

where ρ_m is the solid matrix density [kg/m^3], C_{pm} is the specific heat capacity of the solid matrix [J/kg·K], ϕ is the porosity of the medium [-], T_m is the temperature of the solid matrix [K], κ_m is the thermal conductivity tensor of the solid matrix [W/m·K], and $\dot{q}_m$ is the heat source/sink term for the solid phase [W/m^3].

DOI: 10.1201/9781032724973-8

Fluid phase:

$$\frac{\partial}{\partial t}\left[\rho_f C_{pf} \phi T_f\right] + \frac{\partial}{\partial x}\left[\rho_f C_{pf} \mathbf{u} T_f\right] = \frac{\partial}{\partial x}\left[\kappa_f \phi \frac{\partial T_f}{\partial x}\right] + \dot{q} + q_T \tag{8.2}$$

where ρ_f is the fluid density [kg/m^3], C_{pf} is the specific heat capacity of the fluid [J/kg·K], T_f is the temperature of the fluid [K], $\mathbf{u}$ is the fluid velocity vector [m/s], κ_f is the thermal conductivity tensor of the fluid [W/m·K], $\dot{q}$ is the heat exchange between phases [W/m^3], and q_T is the external heat source/sink term [W/m^3]. Heat exchange between phases is given by:

$$\dot{q}_m = -\dot{q} = A_{sf} h (T_f - T_m) \tag{8.3}$$

where A_{sf} is the specific interfacial area between solid and fluid [m^2/m^3], and h is the heat transfer coefficient [W/m^2·K]. Heat conduction follows Fourier's law:

$$\mathbf{J}_c = -\kappa \nabla T \tag{8.4}$$

where $\mathbf{J}_c$ is the conductive heat flux vector [W/m^2], κ is the thermal conductivity tensor [W/m·K], and ∇T is the temperature gradient [K/m]. The thermal conductivities are defined as:

$$\kappa_m = k_m \mathbf{I} \tag{8.5}$$

$$\kappa_f = k_f \mathbf{I} + \rho_f C_{pf} D_m \tag{8.6}$$

where k_m is the solid thermal conductivity [W/m·K], k_f is the fluid thermal conductivity [W/m·K], $\mathbf{I}$ is the identity tensor, and D_m is the mechanical dispersion coefficient [m^2/s]

Local Thermal Equilibrium (LTE)

When temperature differences between phases are negligible ($T_f = T_m = T$), the combined energy equation under local thermal equilibrium simplifies to:

$$\frac{\partial}{\partial t}\left[\rho C_p T\right] + \frac{\partial}{\partial x}\left[\rho_f C_{pf} \mathbf{u} T\right] = \frac{\partial}{\partial x}\left[\kappa_e \frac{\partial T}{\partial x}\right] + q_T \tag{8.7}$$

where ρC_p is the effective volumetric heat capacity of the medium [J/m^3·K], T is the temperature of both phases [K], and κ_e is the effective thermal conductivity tensor [W/m·K]. Along with effective properties:

$$\rho C_p = \rho_m C_{pm}(1 - \phi) + \rho_f C_{pf} \phi \tag{8.8}$$

$$\kappa_e = \kappa_f \phi + (1 - \phi)\kappa_m \tag{8.9}$$

This simplified equation combines conduction, convection, and dispersion effects, providing a practical approach for many engineering applications.

Numerical Implementation

This section presents the numerical implementation of our thermal single-phase flow simulation. As a simplified case of the above model, the governing equations for non-isothermal single-phase flow in porous media consist of coupled pressure and temperature equations:

$$\nabla \cdot \left(-\frac{\mathbf{K}}{\mu}\nabla p\right) = 0 \tag{8.10}$$

$$\phi_T \frac{\partial T}{\partial t} + \nabla \cdot (\mathbf{u}T) = \nabla \cdot (D_T \nabla T) \tag{8.11}$$

where p is pressure, $\mathbf{K}$ is the permeability tensor, μ is viscosity, T is temperature, ϕ_T is the effective heat capacity, $\mathbf{u}$ is the Darcy velocity, and D_T is the effective thermal diffusivity.

The Darcy velocity includes a buoyancy term due to thermal expansion:

$$\mathbf{v} = -\frac{\mathbf{K}}{\mu}(\nabla p - \beta_w \nabla T) \tag{8.12}$$

where β_w is the thermal expansion coefficient.

The effective heat capacity and thermal diffusivity are calculated as volume-weighted averages:

$$\phi_T = (1-\phi)\rho_s c_{ps} + \phi(\rho_w c_{pw}) \tag{8.13}$$

$$D_T = (1-\phi)\kappa_s + \phi\kappa_w \tag{8.14}$$

where ϕ is porosity, ρ_s and ρ_w are the densities of solid and water phases, c_{ps} and c_{pw} are their specific heat capacities, and κ_s and κ_w are their thermal conductivities.

The main simulation code implements the non-isothermal single-phase flow in porous media. Using operator splitting techniques, this function coordinates the numerical solution of pressure and temperature fields over time. The code uses cell-centered finite difference methods to solve a coupled system of partial differential equations describing flow and heat transfer in porous media. The implementation is structured into a main function that handles the overall simulation logic and calls two specialized discretization functions for the advection and diffusion terms.

8.1.1.1 Simulation Parameters

The simulation begins by defining the necessary physical and numerical parameters:

```
% Domain discretization
nx = 20;   ny = 50;          % Number of cells in x and y
Lx = 2;    Ly = 4;           % Domain dimensions [m]

% Time stepping
nTSteps = 100;               % Number of time steps
injection = 0.01;            % Injection rate
tFinal = 0.5/injection;      % Final simulation time

% Thermal properties
T0 = 300;                    % Initial temperature [K]
Tw = 360;                    % Injection temperature [K]
betw = 0.005;                % Thermal expansion coefficient [1/K]

% Material properties
poroval = 0.3;               % Porosity

% Heat capacity
cps = 800;                   % Solid
cpw = 4000;                  % Water
cpn = 2300;                  % Non-wetting phase

% Density
densS = 2500;                % Solid
densW = 1000;                % Water
densN = 660;                 % Non-wetting phase

% Thermal conductivity
hs = 0.718;                  % Solid
hw = 0.6;                    % Water
hn = 0.2;                    % Non-wetting phase

% Viscosity
mu = 0.001;                  % Dynamic viscosity

% Iteration control
NIT = 1;                     % Maximum number of nonlinear
    iterations
subNT1 = 1;                  % Substeps for outer loop
subNT2 = 1;                  % Substeps for inner loop

% Unit conversions
y2s = 365*24*60*60;          % Year to seconds
bar = 1.0e+5;                % Bar to Pa
md = 9.86923e-16;
cp = 1.0e-3;                 % Centipoise to Pa.s
```

The simulation domain is discretized into 20×50 grid cells over a physical domain of 2×4 meters. We set the initial temperature T_0 to 300 K, with injection fluid at a higher temperature of 360 K to observe thermal effects. The thermal expansion coefficient $\beta_w = 0.005\ \mathrm{K}^{-1}$ couples the temperature field to the flow equations through buoyancy effects.

8.1.1.2 Grid and Boundary Conditions

The computational grid and boundary conditions are defined as follows:

```
% Grid generation
xs = Lx*(0:nx)'/nx;           % Cell interfaces in x-direction
ys = Ly*(0:ny)'/ny;           % Cell interfaces in y-direction
hxs = xs(2:end) - xs(1:end-1);  % Cell sizes in x-direction
hys = ys(2:end) - ys(1:end-1);  % Cell sizes in y-direction

% Time stepping
ht = tFinal / nTSteps;        % Time step size
hts = ht;                     % Step size for outer loop
htc = ht;                     % Step size for concentration
htt = ht;                     % Step size for temperature

% Pressure boundary conditions
vBW = zeros(ny, 1);
vBW(1:5) = (injection*Lx*porovai);  % Injection at west boundary
vBE = zeros(ny, 1);
vBS = zeros(nx, 1);
vBN = zeros(nx, 1);

% Pressure boundary type
pwBW = zeros(ny, 1);
pwBE = zeros(ny, 1);
pwBS = zeros(nx, 1);
pwBN = zeros(nx, 1);

pwDirW = false(ny, 1);        % Neumann at west
pwDirE = true(ny, 1);         % Dirichlet at east
pwDirS = false(nx, 1);        % Neumann at south
pwDirN = false(nx, 1);        % Neumann at north

% Temperature boundary conditions
TBW = T0 * ones(ny, 1);
TBW(1:5) = Tw;                % Injection temperature at west
TBE = T0 * ones(ny, 1);
TBS = T0 * ones(nx, 1);
TBN = T0 * ones(nx, 1);

% Temperature boundary type
TDirW = false(ny, 1);         % Neumann at west
TDirE = true(ny, 1);          % Dirichlet at east
TDirS = false(nx, 1);         % Neumann at south
TDirN = false(nx, 1);         % Neumann at north

% Initialize fields
cInit = zeros(nx, ny);
TInit = T0 * ones(nx, ny);
```

The boundary conditions implement an injection scenario where fluid enters through the lower portion of the west boundary (first 5 cells), maintained at a higher temperature T_w. The east boundary is kept at the initial temperature T_0 with a Dirichlet condition, while no-flow (Neumann) conditions are applied at the remaining boundaries.

8.1.2 MAIN SIMULATION LOOP

The core of the simulation is a time-stepping loop that solves the coupled pressure and temperature equations:

```
% Transport parameters
poro0 = poroval * ones(nx, ny);
K1x0 = PermeabilityField(nx, ny, xs, ys, md);
K1y0 = K1x0;

% Initialize solution
T = TInit;

% Simulation time loop
for iT = 1:nTSteps
    fprintf('Time step %d of %d\n', iT, nTSteps);

    % Get temperature field at previous time step
    Tp = reshape(T, [nx, ny]);

    % Setup properties
    poro = poro0;
    poroT = (1 - poro0)*cps + poro0*(cpw + cpn);
    difXXT = (1 - poro0)*hs + poro0*(hw + hn);
    difYYT = difXXT;
    K1x = K1x0;
    K1y = K1y0;

    % Initialize velocity boundary conditions
    vBW = zeros(ny, 1);
    vBE = zeros(ny, 1);
    vBS = zeros(nx, 1);
    vBN = zeros(nx, 1);

    poroBar = poro .* (hxs * hys');
    matB = sparse(diag(poroBar(:)));
    matBp = sparse(diag(poroBar(:)));
```

At each time step, we update the material properties based on the current temperature field. The effective heat capacity and thermal diffusivity are calculated as volume-weighted averages of solid and fluid properties.

8.1.3 PRESSURE EQUATION SOLUTION

The first step in each iteration is to solve the pressure equation, including thermal effects:

```
% Pressure permeabilities
Kx = K1x/mu;
Ky = K1y/mu;

% Compute matrix A and b for pressure
[pwmatM, pwmatA, pwrhs, ~, ~, ~, para] = ccfdDiffusion1(poro, Kx,
      Ky, ...
    pwBW, pwBE, pwBS, pwBN, vBW, vBE, vBS, vBN, ...
```

```
    pwDirW, pwDirE, pwDirS, pwDirN, cInit, xs, ys, tFinal, 1, [])
        ;

% Calculate temperature gradient
gradTX = ((para.W - para.E)*T(:) + para.BW*TBW - para.BE*TBE) ...
    + para.BW*(vBW.*para.hys) + para.BE*(vBE.*para.hys);
gradTX = reshape(gradTX, [nx+1, ny]);

gradTY = ((para.S - para.N)*T(:) + para.BS*TBS - para.BN*TBN) ...
    + para.BS*(vBS.*para.hxs) + para.BN*(vBN.*para.hxs);
gradTY = reshape(gradTY, [nx, ny+1]);

% Compute buoyancy term
gradT = betw * K1x .* (gradTX(1:end-1,:) + gradTY(:,1:end-1));

% Solve for pressure
totalA = pwmatA;
gradc = gradT(:);
Pw = totalA \ (pwrhs - gradT(:) - gradc);
```

The pressure equation is discretized using a cell-centered finite difference method, which accounts for the temperature-dependent buoyancy term. The temperature gradient ∇T is calculated at cell interfaces and then used to compute the buoyancy force $\beta_w \mathbf{K} \nabla T$. This term is added to the right-hand side of the pressure equation to account for thermally-induced flow.

After solving for pressure, we compute the Darcy velocity field:

```
% Compute flux
fx = para.dfxdash(:) .* ((para.W - para.E)*Pw + para.BW*para.cBW
    - para.BE*para.cBE) ...
    + para.BW*(para.vBW.*para.hys) + para.BE*(para.vBE.*para.hys)
        ;
fx = reshape(fx, [nx+1, ny]);
fx = fx ./ (ones(nx+1,1) * para.hys'); % Convert to flux per
    unit face area

fy = para.dfydash(:) .* ((para.S - para.N)*Pw + para.BS*para.cBS
    - para.BN*para.cBN) ...
    + para.BS*(para.vBS.*para.hxs) + para.BN*(para.vBN.*para.hxs)
        ;
fy = reshape(fy, [nx, ny+1]);
fy = fy ./ (para.hxs * ones(1, ny+1)); % Convert to flux per
    unit face area
```

The velocities are computed at the cell interfaces based on the pressure gradient, following Darcy's law. These fluxes represent the volumetric flow rate per unit area across each cell face.

8.1.4 HEAT EQUATION SOLUTION

Once the velocity field is known, we solve the temperature equation using operator splitting:

```
% Nonlinear iterations
for iK = 1:NIT
    Ti = T(:);  % Temperature at current iteration

    % Outer loop for substeps
    for seT1 = 1:subNT1
        % Inner loop for substeps
        for seT2 = 1:subNT2
            % Advection matrices for temperature
            [MT, MBw, rhsBwT, ~] = ccfdAdvection1(poroT, fx, fy,
                ...
                TBW, TBE, TBS, TBN, TInit, xs, ys, tFinal, 0, [])
                    ;

            % Diffusion matrices for temperature
            [MT, MBDT, rhsBDT, ~, ~, ~, ~] = ccfdDiffusion1(poro,
                 difXXT, difYYT, ...
                TBW, TBE, TBS, TBN, vBW, vBE, vBS, vBN, ...
                TDirW, TDirE, TDirS, TDirN, TInit, xs, ys, tFinal
                    , 0, []);

            % Total system matrix for temperature
            matBtT = matB + htt * (MBw + MBDT);

            % Right-hand side for temperature
            matQctT = htt * (rhsBwT + rhsBDT) + matBp * Tp(:);

            % Solve for temperature
            T = matBtT \ matQctT;
        end
    end

    % Check convergence
    if (norm(T - Ti) < 1e-7)
        fprintf('Converged in %d iterations\n', iK);
        break;
    end
end
```

The temperature equation is solved using a semi-implicit scheme with operator splitting. The advection term is discretized using an upwind scheme (in `ccfdAdvection`) to ensure stability, while the diffusion term uses a cell-centered finite difference approach with harmonic averaging (in `ccfdDiffusion`).

The nonlinear coupling between temperature and flow is handled through an iterative process. We first solve for the temperature using the previously computed velocity field, then check for convergence based on the change in temperature values between iterations. This approach allows us to capture the complex interplay between temperature-dependent fluid properties and flow dynamics.

The discrete form of the temperature equation can be written as:

$$(\phi_T + \Delta t(\mathbf{A}_{\mathbf{adv}} + \mathbf{A}_{\mathbf{diff}}))\,\mathbf{T}^{n+1} = \phi_T \mathbf{T}^n + \Delta t\mathbf{b} \tag{8.15}$$

where $\mathbf{A}_{\mathbf{adv}}$ and $\mathbf{A}_{\mathbf{diff}}$ are the discrete advection and diffusion operators, $\mathbf{T}^n$ is the temperature vector at time step n, and $\mathbf{b}$ contains boundary condition contributions.

8.1.5 POST-PROCESSING AND VISUALIZATION

After completing the time stepping, we prepare the results for analysis and visualization:

Listing 8.1: Post-processing and visualization

```
% Post-processing
Pw = reshape(Pw, [nx, ny]);
T = reshape(T, [nx, ny]);

% Output error
err2 = norm(T - Ti);
fprintf('Final temperature error: %e\n', err2);

% Plot results
x = Lx * (0:nx-1)' / nx;
y = Ly * (0:ny-1)' / ny;

figure(1);
[c, ~] = contourf(x, y, T', 20);
shading interp
xlabel('Length (meters)');
ylabel('Width (meters)');
title('Temperature Distribution');
set(c, 'LineStyle', 'none');
colorbar
```

As shown in Figure 8.1, the temperature field exhibits a smooth decay outward from the heat source, indicating effective thermal diffusion through the porous medium.

The implicit time discretization ensures unconditional stability, allowing for larger time steps compared to explicit schemes. Harmonic averaging of coefficients at interfaces properly handles discontinuities in material properties across cell boundaries. The upwind scheme for advection prevents numerical oscillations in advection-dominated regions. The cell-centered finite difference approach ensures local and global mass conservation. Iterative coupling between temperature and flow equations captures the nonlinear interplay between these physical processes.

8.1.6 NONISOTHERMAL TWO-PHASE FLOW

Consider the problem of immiscible incompressible two-phase flow with convective heat transfer in heterogeneous porous media. The velocity of the phase

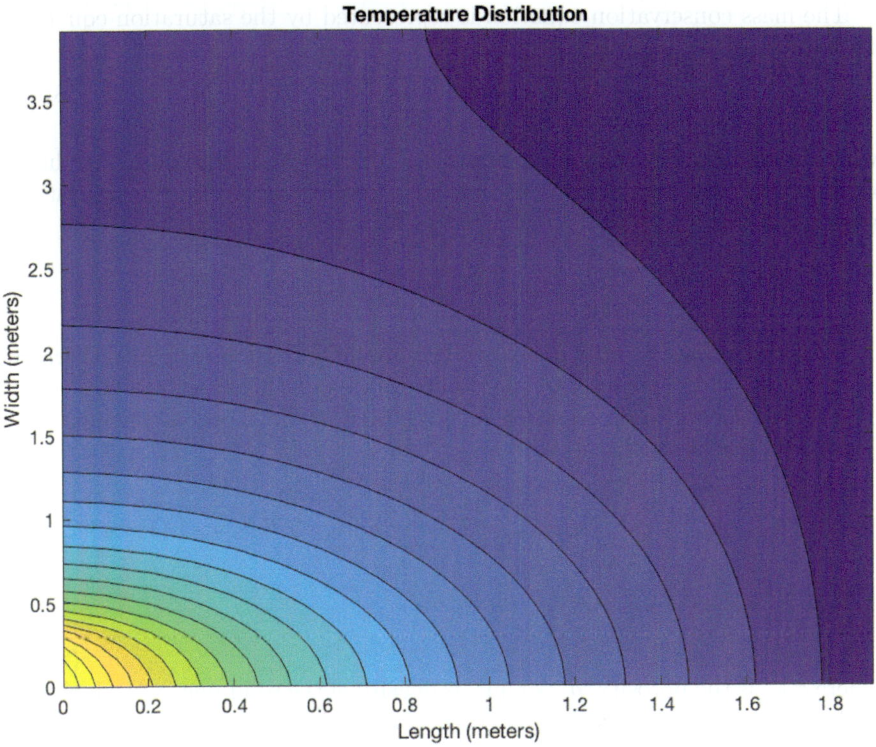

Figure 8.1 Temperature distribution in a porous domain showing the spatial thermal gradient resulting from a localized heat source.

α in porous media is given by Darcy's law,

$$\mathbf{u}_\alpha = -\frac{k_{r\alpha}\mathbf{K}}{\mu_\alpha}\nabla(p_\alpha + g\rho_{\alpha,0}(1-\beta_\alpha(T-T_r))\nabla z), \qquad \alpha = w, n, \tag{8.16}$$

where $\mathbf{K}$ is the permeability tensor, g is the gravitational acceleration, and z is the depth. $p_\alpha, \mu_\alpha, k_{r\alpha}$ and q_α, are, respectively, the pressure, the viscosity, the relative permeability and the external mass flow rate. w stands for the wetting phase, and n stands for the nonwetting phase. The temperature difference may cause a convective flow, which is represented by the Boussinesq approximation in the body force term of the previous equation (8.16). β is the thermal expansion coefficient and T_r is a temperature reference. In the Boussinesq approximation, the density in the body force term is calculated by,

$$\rho_\alpha = \rho_{\alpha,0}(1-\beta_\alpha(T-T_r)), \quad \beta_\alpha = -\frac{1}{\rho_{\alpha,0}}\frac{\partial \rho_\alpha}{\partial t}, \quad \alpha = w, n.$$

The mass conservation equation is represented by the saturation equation as follows,

$$\phi\frac{\partial s_\alpha}{\partial t}+\nabla\cdot\mathbf{u}_\alpha=q_\alpha,\quad \sum_\alpha s_\alpha=1,\quad \alpha=w,n, \tag{8.17}$$

where q_α is the external mass flow rate. In the flow equations, we have assumed that the contribution of time variation of porosity in the saturation equation is negligibly small ($\partial\phi/\partial t\approx 0$). Thus,

$$\frac{\partial(\phi s_\alpha)}{\partial t}=\phi\frac{\partial s_\alpha}{\partial t}+\frac{\partial\phi}{\partial t}s_\alpha=\phi\frac{\partial s_\alpha}{\partial t},$$

We assume that the solid phase and the two fluid phases are in local thermal equilibrium, which means that the temperatures of all phases are the same. The energy conservation equation in two-phase flow in a porous medium can be represented as [204],

$$\frac{\partial}{\partial t}\left[(1-\phi)\rho_s c_{p,s}+\phi\sum_\alpha s_\alpha\rho_\alpha c_{p,\alpha}\right]T+\sum_\alpha\rho_\alpha\mathbf{u}_\alpha\cdot\nabla T=\nabla\cdot\Big((1-\phi)\,h_s\nabla T$$
$$+\phi\sum_\alpha s_\alpha h_\alpha\nabla T\Big)+\rho Q_T,\quad \alpha=w,n, \tag{8.18}$$

where ϕ [–] is the porosity, ρ_s [kg/m^3] is density of the solid phase, c_p [J/Kg.K] is the heat capacity, T [K] is the temperature and h [J/K.m.s] is the thermal conductivity. $\mathbf{u}_\alpha$ [m/s], ρ_α [kg/m^3], s_α [–] are, respectively, the velocity, the density, and the saturation of the phase α. s stands for the solid phase. Q_T [m^3/s] is the heat source term.

Adding the saturation equation of the wetting phase to the saturation equation of the nonwetting phase (8.17), the pressure equation can be rewritten as,

$$\nabla\cdot\mathbf{u}_t=-\nabla\cdot\lambda_t\mathbf{K}\nabla(p_w+g\rho_{\alpha,0}(1-\beta_\alpha(T-T_r))\nabla z)-\nabla\cdot\lambda_n\mathbf{K}$$
$$\nabla\left(p_c+g\Delta\rho_{\alpha,0}(1-\beta_\alpha(T-T_r))z\right)=q_t. \tag{8.19}$$

where $\Delta\rho=\rho_n-\rho_w$. $p_c(s_w)=p_n-p_w$ is the capillary pressure, $\mathbf{u}_t=\mathbf{u}_w+\mathbf{u}_n$ is the total velocity, $f_w=\lambda_w/\lambda_t$ is the flow fraction, $\lambda_\alpha=k_{r\alpha}/\mu_\alpha$ is the phase mobility, $\lambda_t=\lambda_w+\lambda_n$ is the total mobility, and $q_t=q_w+q_n$ is the total source mass transfer. The above equation (8.19) will be coupled with the following saturation equation to compute the water pressure,

$$\phi\frac{\partial s_w}{\partial t}-q_w=-\nabla\cdot\lambda_w\mathbf{K}\nabla(p_w+g\rho_{\alpha,0}(1-\beta_\alpha(T-T_r))\nabla z). \tag{8.20}$$

Therefore, we can update the saturation by the following equation,

$$\phi\frac{\partial s_w}{\partial t}-q_w=-\nabla\cdot(f_w\mathbf{u}_a). \tag{8.21}$$

where $\mathbf{u}_w=f_w\mathbf{u}_a$ and $\mathbf{u}_a=-\lambda_t\mathbf{K}\nabla p_w+g\rho_{\alpha,0}(1-\beta_\alpha(T-T_r))\nabla z$.

If we select the formula $p_c = -p_e(T)\log S$, to represent the capillary pressure, $S = \frac{s_w - s_{wr}}{1 - s_{nr} - s_{wr}}$, is the normalized wetting phase saturation, such that, $0 \le S \le 1$. $p_e(T)$ is the entry pressure which may be a function of temperature [205],

$$p_e(T) = p_{e0}\frac{\sigma(T)}{\sigma(T_0)}$$

where $\sigma(T)$ is surface tension which is represented by the following correlation,

$$\sigma(T) = 0.3258\left(1 - \frac{T}{647.3}\right)^{1.256} - 0.148\left(1 - \frac{T}{647.3}\right)^{2.256}.$$

The relative permeabilities may be written as,

$$k_{rw} = k^0_{rw}S^2, \quad k_{rn} = k^0_{rn}\left(1 - S\right)^2,$$

where $k^0_{rw} = k_{rw}\left(S = 1\right), k^0_{rw} = k_{rw}\left(S = 1\right)$. p_e is the capillary pressure parameter, s_{wr} is the irreducible wetting phase saturation and s_{nr} is the residual nonwetting phase saturation.

8.2 MULTI-SCALE TIME SPLITTING METHOD

In the technique of multiscale time splitting, different time step sizes are used for each time derivative in its corresponding differential equation. In the current model, the pressure is coupled with the saturation via the capillary pressure and with temperature via the Boussinesq approximation. For example, the pressure time-step size can be taken larger than those of saturation or temperature. For pressure the total time interval, $[0, T]$, is divided into N_p, time-steps, namely, $0 = t_0 < t_1 < \cdots < t_{N_p = T}$, with a time-step of length, $\Delta t^k = t^{k+1} - t^k$. On the other hand, for saturation, each interval, $(t^k, t^{k+1}]$, is divided into $N_{p,s}$ subintervals, i.e. $(t^k, t^{k+1}] = \cup_{l=0}^{N_{p,s}-1}(t^{k,l}, t^{k,l+1}]$. Moreover, temperature varies more rapidly than the pressure (and may be saturation), so one may use smaller time-step sizes for the energy equation. Therefore, each subinterval $(t^{k,l}, t^{k,l+1}]$ can be partitioned into $N_{p,s,T}$ subsubintervals, $(t^{k,l}, t^{k,l+1}] = \cup_{m=0}^{N_{p,s,T}-1}(t^{k,l,m}, t^{k,l,m+1}]$. In this study, we use the backward Euler time discretization for the time derivatives of the equations of temperature and saturation.

We linearize the capillary pressure, p_c, in terms of saturation,

$$p_c\left(s^*_w\right) \cong p_c\left(s^k_w\right) + p'_c\left(s^k_w\right)\left[s^{k+1}_w - s^k_w\right], \tag{8.22}$$

where p'_c is derivative of p_c. The quantity, $[s^{k+1}_w - s^k_w]$, can be calculated from the saturation equation,

$$s^{k+1}_w - s^k_w = \frac{\Delta t^k}{\phi}\left[q^{k+1}_w - \nabla\cdot\lambda_t\left(s^k_w\right)\mathbf{K}\nabla(p^{k+1}_w + \rho_w g(T^k - T_r)\nabla z)\right]. \tag{8.23}$$

The pressure equation becomes,

$$-\nabla\cdot\lambda_t\left(s_w^k\right)\mathbf{K}\nabla(p_w^{k+1}+\rho_w g(T^k-T_r)\nabla z)-\nabla\cdot\lambda_n\left(s_w^k\right)\mathbf{K}\nabla(p_c(s_w^*)$$
$$+\rho_w g(T^k-T_r)\nabla z)=q_t^{k+1}. \tag{8.24}$$

The water pressure, p_w is obtained by solving the coupled system (8.37)–(8.39) implicitly. Then, the saturation is obtained explicitly using,

$$\phi\frac{s_w^{k,l+1}-s_w^{k,l}}{\Delta t^l}+\nabla\cdot\left(f_w^k\mathbf{u}_a^{k+1}\right)=q_w^{k,l+1}. \tag{8.25}$$

The energy equation is computed implicitly as follows,

$$\frac{1}{\Delta t^m}\left[\left((1-\phi)\rho_s c_{p,s}+\phi\sum_\alpha s_\alpha^{k+1}\rho_\alpha(T^k)c_{p,\alpha}\right)T^{k,l,m+1}\right.$$
$$\left.-\left((1-\phi)\rho_s c_{p,s}+\phi\sum_\alpha s_\alpha^{k}\rho_\alpha(T^k)c_{p,\alpha}\right)T^{k,l,m}\right]+$$
$$\sum_\alpha\rho_\alpha\mathbf{u}_\alpha^{k+1}\cdot\nabla T^{k+1}=\nabla\cdot\left((1-\phi)h_s+\phi\sum_\alpha s_\alpha^{k+1}h_\alpha\right)\nabla T^{k+1}+\rho Q_T^{k+1},$$
$$\alpha=w,n. \tag{8.26}$$

Finally, we update other parameters such as λ_w, λ_n, λ_t, and f_w.

Now, let us introduce the Courant–Friedrichs–Lewy (CFL) condition. The CFLs for the saturation equation are defined as,

$$\text{CFL}_{s,x}=\frac{\mathbf{u}_x\Delta t^{k,l}}{\Delta x},\quad \text{CFL}_{s,y}=\frac{\mathbf{u}_y\Delta t^{k,l}}{\Delta y}, \tag{8.27}$$

and for the temperature equations are,

$$\text{CFL}_{T,x}=\frac{\mathbf{u}_x\Delta t^{k,l,m}}{\Delta x},\quad \text{CFL}_{T,y}=\frac{\mathbf{u}_y\Delta t^{k,l,m}}{\Delta y}. \tag{8.28}$$

The initial time-step of the saturation equation is taken as the pressure time-step, i.e., $\Delta t^{k,0}=\Delta t^k$, and the initial time-step of the temperature and concentration equations is taken as the saturation time-step, i.e., $\Delta t^{k,l,0}=\Delta t^{k,l}$. Then, we check if $\text{CFL}_{s,x}>1$ or $\text{CFL}_{s,y}>1$, the saturation time-step will be divided into smaller steps, then, $\text{CFL}_{s,x}$ and $\text{CFL}_{s,y}$ will be recalculated and so on until we get $\text{CFL}_{s,x}<1$ and $\text{CFL}_{s,y}<1$. A similar procedure is employed for conditions, $\text{CFL}_{T,x}$ and $\text{CFL}_{T,y}$. The CCFD method, which is locally conservative, is employed for the spatial discretization.

8.3 ITERATIVE IMPTES ALGORITHM

The non-isothermal multiphase flow in porous media may be applied in many fields such as steam injection in hydrocarbon/geothermal reservoirs and other engineering/environmental applications [208, 209, 207]. There are two standard methods for solving flow and transport equations in porous media: the implicit-explicit and fully implicit methods. The IMPES (Implicit Pressure Explicit Saturation) method is a typical example of the implicit-explicit type. The solutions approximate unknown variables at each time step and each cell. The IMPES algorithm is less stable, but it is faster than the fully implicit one. The capillary pressure mobility terms are treated in the IMPES technique at the previous time stage n, so it needs a smaller time-step size. The fully implicit method, on the other hand, is unconditionally stable but takes more CPU memory and becomes much slower because it takes a bigger time-step size. Therefore, the IMPES ongoing research is focused on improving its stability and convergence to keep it faster and more stable. The iterative IMPES implicit temperature (I-IMPES-IMT) method was implemented by El-Amin [207] to solve the nonlinear governing equation of the problem of heat transfer and two-phase flow in porous media.

In the following, the iterative version of the IMPTES algorithm has been presented for the above governing equations, which are highly nonlinear and strongly coupled. Firstly, let us discretize the total time interval $[0, T]$ with the step-size $\Delta t^n = t^{n+1} - t^n$, such that N_T is the number of time-step, $n+1$ is the current time-step, n is the previous time-step. Secondly, we utilize a number of N_I iterations over each time step, and use the symbols:
$l+1$ is the current iteration, l is the previous iteration. The backward Euler scheme has been used to calculate the temporal discretization of time derivatives. Now, let us start with the pressure equation discretization, namely,

$$-\nabla \cdot \lambda_t \left(s_w^{n+1,l}\right) \mathbf{K} \nabla (p_w^{n+1,l+1} + g\rho_{w,0}(1 - \beta_w(T^{n+1,l} - T_r))\nabla z) - \nabla \cdot \lambda_n \left(s_w^{n+1,l}\right) \mathbf{K} \nabla \left(p_c^{n+1,l+1} + g\Delta\rho_0(1 - \beta_w(T^{n+1,l} - T_r))\nabla z\right) = q_t. \quad (8.29)$$

It is a key point to linearize the capillary pressure $p_c^{n+1,l+1}$ as follows,

$$p_c \left(s_w^{n+1,l+1}, T^{n+1,l}\right) \cong p_c \left(s_w^{n+1,l}, T^{n+1,l}\right) + \frac{\partial p_c}{\partial s_w}\Big|_{s_w^{n+1,l}, T^{n+1,l}} \left(s_w^{n+1,l+1} - s_w^n\right). \quad (8.30)$$

The saturation difference appears in (8.30) can be calculated as,

$$s_w^{n+1,l+1} - s_w^n = -\frac{\Delta t^n}{\phi} \left[\nabla \cdot \lambda_w^{n+1,l} \mathbf{K} \nabla (p_w^{n+1,l+1} + g\rho_{w,0}(1 - \beta_w(T^{n+1,l} - T_r))\nabla z)\right]. \quad (8.31)$$

The above equations (8.29)–(8.31) are solved to find the pressure. The following explicit equation is solved to find the saturation,

$$\phi \frac{s_w^{n+1,l+1} - s_w^n}{\Delta t^n} + \alpha_r \nabla \cdot \left(f_w^{n+1,l} \mathbf{u}_a^{n+1,l+1}\right) = q_w^{n+1}. \quad (8.32)$$

where

$$\mathbf{u}_a^{n+1,l+1} = -\lambda_t^{n+1,l}\mathbf{K}\nabla p_w^{n+1,l+1} + g\rho_{w,0}(1-\beta_w(T^{n+1,l}-T_r))\nabla z.$$

Similarly, the iterative implicit scheme of the energy equation can be written as,

$$\frac{1}{\Delta t^n}\left[\left((1-\phi)\rho_s c_{p,s} + \phi\sum_\alpha s_\alpha^{n+1,l+1}\rho_\alpha(T^{n+1,l})c_{p,\alpha}\right)T^{n+1,l+1} - \left((1-\phi)\rho_s c_{p,s} + \phi\sum_\alpha s_\alpha^{n}\rho_\alpha(T^{n})c_{p,\alpha}\right)T^{n}\right] + \alpha_r\sum_\alpha \rho_\alpha \mathbf{u}_\alpha^{n+1,l+1}\cdot\nabla T^{n+1,l+1} = \nabla\cdot \left((1-\phi)h_s + \phi\sum_\alpha \tau s_\alpha^{n+1,l+1}h_\alpha^{n+1,l+1}\right)\nabla T^{n+1,l+1} + Q_T^{n+1}, \quad \alpha = w, n. \tag{8.33}$$

where

$$h_\alpha^{n+1,l+1} = (\phi s_w^{n+1,l+1}\tau h_{\alpha,d} + d_{t,w}|\mathbf{u}_w|^{n+1,l+1})\mathbf{I} + (d_{l,w}-d_{t,w})\frac{\mathbf{u}_w^{n+1,l+1}\mathbf{u}_w^{n+1,l+1^T}}{|\mathbf{u}_w|^{n+1,l+1}}$$

Remark: In order to stabilize the numerical scheme, the advection term in equations (8.32) and (8.33) is multiplied with the relaxation factor, α_r. The CFL condition (CFL<1) has been employed to evaluate the relaxation factor. The relaxation factor is defined as,

$$\alpha_r = \min(\frac{c_1}{\max(\mathrm{CFL})}, c_2), \quad 0 \le c_1, c_2 \le 1, \quad \alpha_r \mathrm{CFL} < 1.$$

Finally, the functions, f_w, λ_t, λ_w, λ_n and ρ_α, are updated accordingly. This process is repeated until approaching the desired convergence.

8.4 SPATIAL DISCRETIZATION

The above scheme may be discretized using several methods such as the mixed finite element method, the finite volume method, and the cell-centered finite difference method, which are locally conservative and suit the governing partial differential equations. Now, let us apply the spatial discretization to the system of equations (8.29)–(8.33), namely,

$$\mathbf{M}_a\left(\mathbf{s}_w^{n+1,l},\mathbf{T}^{n+1,l}\right)\mathbf{p}_w^{n+1,l+1} + \mathbf{M}_c\left(\mathbf{s}_w^{n+1,l},\mathbf{T}^{n+1,l}\right)\mathbf{P}_d\left(\mathbf{s}_w^{n+1,l+1},\mathbf{T}^{n+1,l}\right) = \mathbf{M}_T\left(\mathbf{T}^{n+1,l}\right) + \mathbf{Q}_{ac}^{n+1}. \tag{8.34}$$

The two matrices $\mathbf{M}_a$ and $\mathbf{M}_c$ are functions of the two vectors $\mathbf{s}_w$ and $\mathbf{T}$ calculated at the iteration step l. Also, the matrix $\mathbf{M}_a$ is a function of the temperature vector $\mathbf{T}$ computed at the iteration step l. The algebraic version of the capillary pressure equation (8.30) can be written as,

$$\mathbf{P}_d\left(\mathbf{s}_w^{n+1,l+1},\mathbf{T}^{n+1,l}\right)=\mathbf{P}_c\left(\mathbf{s}_w^{n+1,l},\mathbf{T}^{n+1,l}\right)$$
$$+\,\mathbf{dP}_s\left(\mathbf{s}_w^{n+1,l},\mathbf{T}^{n+1,l}\right)\left[\mathbf{s}_w^{n+1,l+1}-\mathbf{s}_w^{n}\right], \tag{8.35}$$

where the matrix $\mathbf{dP}_s$ is diagonal for the total number of cells N_c, i.e.,

$$\mathbf{dP}_s\left(\mathbf{s}_w^{n+1,l},\mathbf{T}^{n+1,l}\right)=\text{diag}\left(\mathbf{dP}_s\left(\mathbf{s}_{w,h}^{n+1,l},\mathbf{T}_{w,h}^{n+1,l}\right)\right),\quad h=1,2,...,N_c, \tag{8.36}$$

The saturation equation, coupled to the pressure equation, is spatially discretization as,

$$\mathbf{\Phi}\frac{\mathbf{s}_w^{n+1,l+1}-\mathbf{s}_w^{n}}{\Delta t^n}+\mathbf{A}_w\left(\mathbf{s}_w^{n+1,l},\mathbf{T}^{n+1,l}\right)\mathbf{P}_w^{n+1,l+1}=\mathbf{Q}_w^{n+1}, \tag{8.37}$$

where $\mathbf{\Phi}$ is a diagonal matrix.
Substituting (8.36) and (8.37) into (8.35), the pressure equation becomes,

$$\mathbf{M}_t\left(\mathbf{s}_w^{n+1,l},\mathbf{T}^{n+1,l}\right)\mathbf{P}_w^{n+1,l+1}=\mathbf{Q}_t\left(\mathbf{s}_w^{n+1,l},\mathbf{T}^{n+1,l}\right), \tag{8.38}$$

where:

$$\mathbf{M}_t=\mathbf{M}_a-\Delta t^n\mathbf{M}_c\mathbf{dP}_s\mathbf{A}^{-1}\mathbf{M}_w,$$

and

$$\mathbf{Q}_t=\mathbf{M}_T+\mathbf{Q}_{ac}^{n+1}-\mathbf{M}_c\left[\mathbf{p}_c+\mathbf{dP}_s\left(\mathbf{s}_w^{n}-\mathbf{s}_w^{n+1,l}\right)\right]-\Delta t^n\mathbf{M}_c\mathbf{dP}_s\mathbf{A}^{-1}\mathbf{Q}_w^{n+1}.$$

It is worth mentioning that the upwind scheme has been used in the convection term, Eq. (8.32) to get,

$$\mathbf{A}\frac{\mathbf{s}_w^{n+1,l+1}-\mathbf{s}_w^{n}}{\Delta t^n}+\mathbf{M}_s\left(\mathbf{s}_w^{n+1,l},\mathbf{P}_w^{n+1,l+1}\right)\mathbf{f}_w\left(\mathbf{s}_w^{n+1,l}\right)=\mathbf{Q}_s^{n+1}. \tag{8.39}$$

Then, $\mathbf{f}_w^{n+1,l+1}$ and therefore, $\mathbf{u}_w^{n+1,l+1}$ are updated to be utilized in the energy equation (8.33),

$$\frac{1}{\Delta t^n}\mathbf{A}_T\left[\mathbf{F}_m\left(\mathbf{s}_w^{n+1,l+1},\mathbf{T}^{n+1,l}\right)\mathbf{T}^{n+1,l+1}-\mathbf{F}_m\left(\mathbf{s}_w^{n},\mathbf{T}^{n}\right)\mathbf{T}^{n}\right]+$$
$$\mathbf{F}_w\left(\mathbf{s}_w^{n+1,l+1},\mathbf{P}_w^{n+1,l+1},\mathbf{T}^{n+1,l}\right)\mathbf{T}^{n+1,l+1}- \tag{8.40}$$
$$\mathbf{F}_D\left(\mathbf{s}_w^{n+1,l+1},\mathbf{P}_w^{n+1,l+1}\right)\mathbf{T}^{n+1,l+1}=\mathbf{Q}_T^{n+1}.$$

Re-arrange Eq. (8.40) to take the form,

$$\mathbf{F}_t\left(\mathbf{s}_w^{n+1,l+1},\mathbf{P}_w^{n+1,l+1},\mathbf{T}^{n+1,l}\right)\mathbf{T}^{n+1,l+1}=\mathbf{Q}_{Tt}\left(\mathbf{s}_w^{n},\mathbf{T}^{n}\right). \tag{8.41}$$

where

$$\mathbf{F}_t=\mathbf{F}_m\left(\mathbf{s}_w^{n+1,l+1},\mathbf{s}_w^{n+1,l}\right)+\Delta t^n\mathbf{A}_T^{-1}\left(\mathbf{F}_w-\mathbf{F}_D\right),$$

and,

$$\mathbf{Q}_{Tt}=\Delta t^n\mathbf{A}_T^{-1}\mathbf{Q}_c^{n+1}+\mathbf{F}_m(\mathbf{s}_w^{n},\mathbf{s}_w^{n})\mathbf{T}^{n}.$$

MATLAB® CODE

The primary simulation code `thermal_multiphase_simulation()` provides the entire computational workflow. The code begins by defining simulation parameters, including spatial discretization (grid size), temporal discretization (time steps), physical domain dimensions, and injection rate parameter.

```
function [Sw, T, presW] = thermal_multiphase_simulation()
    %% Simulation Parameters and Initialization

    % Domain Discretization
    nx = 120;              % Number of grid cells in x-direction
    ny = 50;               % Number of grid cells in y-direction
    nTSteps = 3000;        % Total number of time steps

    % Physical Domain Characteristics
    Lx = 10;               % Domain length (m)
    Ly = 4;                % Domain width (m)
    injection = 0.1;       % Injection rate parameter
    tFinal = 7.5/injection; % Total simulation time
```

The initialization captures multiple physical properties, such as temperature boundaries, thermal expansion coefficient, specific heat capacities for different phases, and density variations.

```
    %% Physical Properties
    % Thermal Parameters
    T0 = 300;              % Initial temperature (K)
    Tw = 360;              % Boundary temperature (K)
    betw = 0.005;          % Thermal expansion coefficient

    % Specific Heat Capacities
    cps = 800;     % Solid specific heat
    cpw = 4000;    % Water specific heat
    cpn = 2300;    % Non-wetting phase specific heat

    % Densities
    densS = 2500;      % Solid density
    densW = 1000;      % Water density
    densN = 660;       % Non-wetting phase density
```

The following function computes the relative permeabilities:

```
function [krw, krn, lagt] = computeRelativePermeability(Sw, Srw,
    Srn)
    % Compute relative permeability
    permType = 2;  % Power-law exponent

    % Effective saturation
    Se = (Sw - Srw) ./ (1 - Srw - Srn);

    % Relative permeabilities
    krw = (Se.^permType) ./ visw;
    krn = ((1 - Se).^permType) ./ visn;
```

```
    % Total mobility
    lagt = krw + krn;
end
```

Thermal properties are computed using weighted averaging:

```
function [poroT, difXXT] = computeThermalProperties(Sw)
    % Compute effective thermal properties
    poro0 = 0.3;  % Porosity

    % Effective heat capacity
    poroT = (1 - poro0) * cps + poro0 .* (Sw * cpw + (1 - Sw) *
        cpn);

    % Effective thermal conductivity
    difXXT = (1 - poro0) * hs + poro0 .* (Sw * hw + (1 - Sw) * hn
        );
end
```

Main Simulation Loop to compute saturation is listed here:

```
for iT = 1:nTSteps
   Swp = reshape(Sw,[nx,ny]);
   Tp = reshape(T,[nx,ny]);

   % Compute effective properties
   poroT = (ones(size(poro0)) - poro0)*cps + ...
           poro0.*(Swp*cpw + (ones(size(Swp))-Swp)*cpn);

   difXXT = (ones(size(poro0)) - poro0)*hs + ...
            poro0.*(Swp*hw + (ones(size(Swp))-Swp)*hn);

   % Compute relative permeabilities
   Se = ( Sw - Srw )./( ones(size(Sw))-Srw - Srn );
   krw = ( Se.^permType ) ./ visw;
   krn = ( (ones(size(Sw)) - Se).^permType ) ./ visn;
   lagt = krw + krn;
   krt = krw./lagt;

   % Compute various permeability tensors
   pwpermXX = permXX .* lagt;
   pcpermXX = permXX .* krn;
   pspermXX = permXX .* krw;

   % Compute capillary pressure
   cap = - (log(Se)) .* BCFun(permXX./md);
   errcap = - sf*BCFun(permXX./md)./Se;
end

    %% Assemble and solve flow equations (ps, pc, pw)
    [psM, psA, psRHS, para] = ccfdDiffusion( poro, permXX.*krw,
        permYY.*krw, ...
        krwBdryW, zeros(ny,1), zeros(nx,1), zeros(nx,1), ...
        zeros(ny,1), zeros(ny,1), zeros(nx,1), zeros(nx,1), ...
        false(ny,1), false(ny,1), false(nx,1), false(nx,1), ...
```

```
    Sw_old, xs, ys, tFinal, 1, struct());

[pcM, pcA, pcRHS] = ccfdDiffusion(poro, permXX.*krn, permYY.*
    krn, ...
    zeros(ny,1), zeros(ny,1), zeros(nx,1), zeros(nx,1), ...
    zeros(ny,1), zeros(ny,1), zeros(nx,1), zeros(nx,1), ...
    false(ny,1), true(ny,1), false(nx,1), false(nx,1), ...
    Sw_old, xs, ys, tFinal, 1, struct());

[pwM, pwA, pwRHS] = ccfdDiffusion(poro, permXX.*(krw+krn),
    permYY.*(krw+krn), ...
    zeros(ny,1), zeros(ny,1), zeros(nx,1), zeros(nx,1), ...
    vpwBdryW, zeros(ny,1), zeros(nx,1), zeros(nx,1), ...
    pwisDirichlet, false(ny,1), false(nx,1), false(nx,1), ...
    Sw_old, xs, ys, tFinal, 1, struct());

%% Capillary pressure
cap     = -log(Se).*BCFun(permXX./9.86923e-16);
errcap  = -BCFun(permXX./9.86923e-16)./Se;
Derrcap = diag(sparse(ht*errcap(:)));

%% Solve for water pressure
totalA  = pwA - pcA*psA;
presW   = totalA \ (pwRHS - ht*errcap(:) + pcRHS - pcA*cap(:)
    );

%% Compute fluxes and advect saturation
fluxX = para.difxxBar(:).*((para.adjW-para.adjE)*presW + ...
        para.adjBW*para.cBdryW - para.adjBE*para.cBdryE);
fluxY = para.difyyBar(:).*((para.adjS-para.adjN)*presW + ...
        para.adjBS*para.cBdryS - para.adjBN*para.cBdryN);
fluxX = reshape(fluxX, [nx+1, ny])./ (hys');
fluxY = reshape(fluxY, [nx, ny+1])./ (hxs);

[M_adv, A_adv, RHS_adv, concp] = ccfdAdvection( ...
    poro, fluxX, fluxY, krwBdryW, zeros(ny,1), zeros(nx,1),
        zeros(nx,1), ...
    Sw_old, xs, ys, tFinal, 0, struct());

%% Update saturation (explicit in split step)
Sw = Sw_old + (M_adv \ (ht*(RHS_adv - A_adv*krt(:))));

%% Heat transport via diffusion + advection
[M_T, B_W, RHS_W] = ccfdAdvection(poroT, fluxX, fluxY, ...
    TBdryW, zeros(ny,1), zeros(nx,1), zeros(nx,1), T_old, xs,
        ys, tFinal, 0, struct());
[M_D, B_D, RHS_D] = ccfdDiffusion(poro, difXXT, difYYT, ...
    TBdryW, zeros(ny,1), zeros(nx,1), zeros(nx,1), ...
    zeros(ny,1), zeros(ny,1), zeros(nx,1), zeros(nx,1), ...
    TisDirichletW, true(ny,1), false(nx,1), false(nx,1), ...
    T_old, xs, ys, tFinal, 0, struct());

matBT = sparse(diag(poro(:))) + ht*(B_W + B_D);
vecQT = ht*(RHS_W + RHS_D) + sparse(diag(poro(:)))*T_old(:);
T     = reshape(matBT \ vecQT, [nx, ny]);
```

Table 8.1
Error estimations of saturation and temperature for different TSN and ITN [207]

TSN	ITN	$\|s_w^{n+1,l+1} - s_w^{n+1}\|$	$\|T^{n+1,l+1} - T^{n+1}\|$
2000	1	1.6205E-3	6.9198E-4
1000	1	3.2422E-3	1.3857E-3
500	1	6.4895E-3	2.7781E-3
100	1	3.2657E-2	1.4169E-2
100	10	1.3387E-2	5.8768E-6
100	30	1.2699E-2	8.7596E-8
100	50	1.2699E-2	8.7596E-8

```
    %% Convergence check (nonlinear split)
    if norm(Sw(:)-Sw_old(:)) < 1e-7 && norm(T(:)-T_old(:)) < 1e-7
        break
    end
end

%% Post-processing: reshape outputs
Sw    = reshape(Sw,    [nx, ny]);
presW = reshape(presW, [nx, ny]);


end

%% Sub-functions

function bc = BCFun(perm)
    % Capillary pressure scaling
    bc = 50 ./ sqrt(perm);
end
```

To demonstrate the usefulness of both the numerical scheme and the theory, as shown in Table 8.1, we present some error estimates for saturation and temperature for different TSN and ITN. We can find all of the parameters used in the calculation in our previous work, [207]. From Table 8.1, we can see that as TSN and ITN increase the error decreases and the temperature converges faster than saturation.

CONCLUSION

Non-isothermal flow in porous media involves heat-fluid interactions and multiphysics coupling. This chapter has detailed the governing equations for energy, pressure, and saturation in both single-phase and two-phase flows, emphasizing local thermal equilibrium and phase-dependent heat transfer. Advanced numerical schemes, such as operator splitting, multiscale time integration, and the iterative IMPTES method, enable stable and accurate simulation of strongly coupled thermal-hydraulic systems. Numerical experiments confirm the importance of iteration count and time step size in controlling error. This modeling framework can be extended to real-world problems such as geothermal reservoirs, thermal enhanced oil recovery, and contaminant transport under variable temperature fields.

REVIEW QUESTIONS

1. What is the significance of local thermal equilibrium (LTE) in modeling heat transfer in porous media?
2. How does the Boussinesq approximation influence the velocity field in non-isothermal flow?
3. Why is the multiscale time-splitting method advantageous for solving non-isothermal two-phase flow?
4. What are the stability trade-offs between fully implicit and IMPES methods in porous media simulations?
5. How does capillary pressure vary with temperature, and what is its impact on flow dynamics?

PROBLEMS

1. Derive the effective thermal conductivity κ_e and effective heat capacity ρC_p expressions under the local thermal equilibrium assumption, starting from the phase-specific heat equations.
2. Modify the single-phase simulation code to implement a constant heat source term and analyze its effect on temperature distribution across the domain.
3. The MATLAB® simulation reports non-convergence in temperature after several iterations. Suggest and implement techniques (e.g., under-relaxation, improved convergence criteria) to ensure numerical stability.
4. Write the fully coupled system of equations for two-phase flow including pressure, saturation, and temperature using implicit time stepping. Comment on the coupling terms and challenges in numerical solution.
5. Using the Boussinesq approximation, calculate the additional buoyancy force term when $\Delta T = 60$ K, $\beta_w = 0.005$ K^{-1}, and $\nabla z = -9.81$ m/s^2.
6. For the temperature-dependent capillary pressure function $p_c = -p_e(T) \log S$, where $S \in [0.2, 1]$, plot the variation of p_c for different temperatures (e.g., 300 K, 350 K, 400 K) and interpret the physical implications.

DECLARATION

During the preparation of this work, the author used ChatGPT in order to improve the language and readability. After using this tool/service, the author reviewed and edited the content as needed and takes full responsibility for the content of the publication.

9 Nanoparticles Transport in Porous Media

9.1 INTRODUCTION

Richard Feynman introduced the idea of nanotechnology in 1959 [227], describing it as the manipulation of materials at the nanoscale, where their behavior differs from bulk matter. Nanoparticles (NPs), typically 1–100 nanometers (nm) (1 nm = 10^{-9} m), exhibit unique properties due to their scale. According to the European Commission (2011/696/EU), a substance qualifies as a nanoparticle if at least 50% of its particles are under 100 nm. At this scale, surface forces like van der Waals interactions dominate, and gravity becomes negligible. These characteristics make nanotechnology valuable across multiple industries.

Nanofluids are formed by dispersing nanoparticles in base fluids such as water or oil. Their high surface area enhances molecular interactions, influencing chemical processes. Nanoparticles are classified by size, shape, and composition into five main types: carbon-based, ceramic, polymeric, metal, and semiconductor [236]. Carbon-based types include carbon nanotubes and fullerenes, which come in single- or multi-walled forms [240]. Ceramic nanoparticles, made of oxides or carbides, are stable and heat-resistant, ideal for drug delivery and photocatalysis. Polymeric nanoparticles, such as nanospheres and nanocapsules, are widely used in medicine. Metal nanoparticles, derived from metal salts, are useful in imaging and environmental applications. Semiconductor nanoparticles like gallium nitride (GaN) and zinc oxide (ZnO) are used in electronics and catalysis. Surface characteristics can also categorize polysilicon nanoparticles as hydrophilic-lipophobic (LHPN) or hydrophobic-lipophilic (HLPN) [232].

In secondary oil recovery, fluid injection maintains reservoir pressure to push oil toward production wells. Tertiary recovery, or EOR, involves injecting fluids that interact with reservoir rocks and fluids to enhance recovery. However, conventional EOR faces issues like low sweep efficiency and formation damage. Nanoparticles address many of these problems [247]. They are used in nanofluids, nanoemulsions, and nanocatalysts to modify wettability, lower viscosity, prevent asphaltene deposition, and reduce interfacial tension. They also support thermal, chemical, and gas flooding techniques. By altering rock and fluid properties, nanoparticles can shift wettability from oil-wet to water-wet, improving waterflooding performance. Water-wet conditions allow better oil displacement. Nanoparticles also improve viscosity control, interfacial tension, and emulsion stability, outperforming conventional surfactant-polymer systems [236]. Their stability under high-temperature conditions makes them

DOI: 10.1201/9781032724973-9

especially suitable for deep reservoir EOR methods [240]. Detailed numerical modeling of nanoparticle transport in porous media is provided by El-Amin [25], offering insights into their behavior under reservoir conditions.

In many natural formations, especially sedimentary rocks, nanoparticles are small relative to the average pore throat size, making physicochemical interactions the primary retention mechanism. To describe nanoparticle transport behavior, two common modeling approaches are used: one based on colloid filtration theory, which focuses on particle capture efficiency, and the other based on chemical adsorption models, which consider energy-driven attachment due to differences in chemical potential between the fluid and solid phases. Filtration theory, originally developed to describe the transport and retention of colloidal particles in porous media, has been widely adapted to model the behavior of nanoparticles in subsurface environments. Filtration theory has its roots in studies of colloid and particle transport in porous media [242, 381]. As nanoparticles are injected into porous formations, such as in enhanced oil recovery or contaminant remediation, their movement is governed by physical, chemical, and mechanical interactions with the solid matrix. Understanding these interactions is critical to predicting nanoparticle mobility and their long-term impact on permeability and transport efficiency.

9.2 MODEL DEVELOPMENT

Nanoparticle transport in porous media can be described using the colloid filtration theory, which accounts for the processes of attachment and detachment. Cushing and Lawler's filtration theory is applied to describe nanoparticle transport in porous media [381]. The transport is typically represented by the advection-diffusion/dispersion equation with an additional colloid filtration term, given by:

$$\frac{\partial C}{\partial t} + \frac{\rho_b}{\phi}\frac{\partial C_s}{\partial t} + u\frac{\partial C}{\partial x} = D\frac{\partial^2 C}{\partial x^2} \tag{9.1}$$

where C is the concentration of nanoparticles in the fluid phase [M/L^3], C_s is the concentration of nanoparticles deposited onto the porous matrix, ρ_b is the bulk density of the porous medium [M/L^3], ϕ is the medium porosity, u is the interstitial velocity [L/t], and D is the diffusion/dispersion coefficient [L^2/t]. The three-dimensional advection-diffusion-dispersion equation with a filtration term is expressed as:

$$\frac{\partial C}{\partial t} + \frac{\rho_b}{\phi}\frac{\partial C_s}{\partial t} + \mathbf{u}\cdot\nabla C = \nabla\cdot(D\nabla C), \tag{9.2}$$

where $\mathbf{u}$ is the interstitial velocity vector [L/T].

The second term in the equation represents the retention of nanoparticles through deposition, which can be written as:

$$\frac{\rho_b}{\phi}\frac{\partial C_s}{\partial t} = k_{\text{dep}}C \tag{9.3}$$

where k_{dep} is the particle deposition (attachment) rate coefficient [1/t]. According to the colloid filtration model, if the nanoparticle concentration C remains greater than zero, the retention concentration C_s will continue to increase indefinitely, meaning the effluent concentration will not reach the injection concentration after a full breakthrough.

To model detachment, Bradford et al. [375] introduced a particle release term into the equation, resulting in:

$$\frac{\rho_b}{\phi}\frac{\partial C_s}{\partial t} = k_{\text{dep}}C - \frac{\rho_b}{\phi}k_{\text{det}}C_s \tag{9.4}$$

Here, k_{det} is the detachment rate coefficient [1/t]. The modified model now accounts for reversible adsorption, with detachment from the solid matrix.

Cullen et al. [380] later refined this model by adding a term for maximum retention capacity, $C_{s,\max}$, to reflect the limitation on attachment sites. The modified equation becomes:

$$\frac{\rho_b}{\phi}\frac{\partial C_s}{\partial t} = k_{\text{dep}}C\left(1 - \frac{C_s}{C_{s,\max}}\right) \tag{9.5}$$

This formulation assumes that as C_s approaches $C_{s,\max}$, the deposition rate decreases, representing site blocking. Further modifications by Wang et al. [414] added both site blocking and detachment, resulting in:

$$\frac{\rho_b}{\phi}\frac{\partial C_s}{\partial t} = k_{\text{dep}}C\left(1 - \frac{C_s}{C_{s,\max}}\right) - \frac{\rho_b}{\phi}k_{\text{det}}C_s \tag{9.6}$$

This combined model accounts for both reversible adsorption and site-blocking effects, providing a more realistic description of nanoparticle transport in porous media.

Experiments have shown that for nanoparticles smaller than 10 nm, chemical potential gradients drive adsorption. Park et al. [406] demonstrated that such nanoparticles adhere to surfaces according to the Langmuir isotherm, implying a balance of adsorption and desorption at equilibrium. Bradford et al. [375] proposed the following equation to model the attachment-detachment behavior of nanoparticles:

$$\frac{\rho_b}{\phi}\frac{\partial C_s}{\partial t} = k_{\text{att}}C - \frac{\rho_b}{\phi}k_{\text{det}}C_s \tag{9.7}$$

where k_{att} is the attachment rate coefficient, and k_{det} is the detachment rate coefficient. In this context, the advection-diffusion equation is updated to include terms for both attachment and detachment:

$$\frac{\partial C}{\partial t} + v\frac{\partial C}{\partial x} = D\frac{\partial^2 C}{\partial x^2} - k_{\text{att}}C + \frac{\rho_b}{\phi}k_{\text{det}}C_s \tag{9.8}$$

Validation of these models is often achieved through laboratory experiments. Nanoparticles, such as those made from silica or iron oxide, are injected into water-saturated porous columns, which are typically packed with sand grains or glass beads. In one of Murphy's experiments [404], the simulation results were compared against experimental data. The model showed reasonable agreement with the experimental outcomes, validating its ability to predict nanoparticle transport in such systems.

9.3 MODEL VALIDATION

The numerical algorithm begins with the initialization phase, where the spatial and temporal domains are defined. The spatial domain, denoted by x, is divided into discrete intervals with grid size Δx and total length L. Similarly, the time domain is discretized into time steps Δt with a total simulation time T. Matrices are initialized to store the concentration of nanoparticles in the fluid phase, $c(i, n)$, where i refers to the spatial position and n to the time step. Additionally, matrices for the adsorbed nanoparticle concentrations on primary adsorption sites ($s1(i, n)$) and secondary adsorption sites ($s2(i, n)$) are initialized. The boundary conditions are specified at the inlet ($x = 0$), where the concentration is set to a constant value during the injection period. Initial conditions are applied such that the nanoparticle concentration c and adsorbed concentrations $s1$ and $s2$ are zero across the spatial domain at $t = 0$. The algorithm proceeds by iterating over time steps from $n = 2$ to the total number of time steps N_t. For each time step, the algorithm loops over spatial grid points from $i = 2$ to $N_x - 1$, where N_x is the number of spatial grid points. The main equation governing nanoparticle transport in the fluid phase is the advection-diffusion equation with terms for deposition and detachment. In its continuous form, this equation is:

$$\frac{\partial c}{\partial t} + u\frac{\partial c}{\partial x} = D\frac{\partial^2 c}{\partial x^2} - k_{\text{irr}}\left(1 - \frac{s1}{s1_{\text{max}}}\right)c + k_{\text{ra}}\left(1 - \frac{s2}{s2_{\text{max}}}\right)c - \frac{\rho_b k_{\text{rd}}}{\phi}s2 \tag{9.9}$$

To solve this equation numerically, it is discretized using finite differences. The time derivative $\frac{\partial c}{\partial t}$ is approximated by a forward difference:

$$\frac{\partial c}{\partial t} \approx \frac{c(i, n+1) - c(i, n)}{\Delta t} \tag{9.10}$$

The advection term $u\frac{\partial c}{\partial x}$ is discretized using a backward difference:

$$u\frac{\partial c}{\partial x} \approx u\frac{c(i,n) - c(i-1,n)}{\Delta x} \tag{9.11}$$

The diffusion term $D\frac{\partial^2 c}{\partial x^2}$ is discretized using a central difference:

$$D\frac{\partial^2 c}{\partial x^2} \approx D\frac{c(i+1,n) - 2c(i,n) + c(i-1,n)}{\Delta x^2} \tag{9.12}$$

Substituting these discretized terms into the continuous equation, we obtain the following discretized equation for the nanoparticle concentration at each spatial point i and time step n:

$$\begin{aligned} c(i,n+1) = c(i,n) + \frac{D\Delta t}{\Delta x^2}\left(c(i+1,n) - 2c(i,n) + c(i-1,n)\right) \\ -\frac{u\Delta t}{\Delta x}\left(c(i,n) - c(i-1,n)\right) - k_{\text{irr}}\Delta t\left(1 - \frac{s1(i,n)}{s1_{\text{max}}}\right)c(i,n) \\ +k_{\text{ra}}\Delta t\left(1 - \frac{s2(i,n)}{s2_{\text{max}}}\right)c(i,n) - \frac{\rho_b k_{\text{rd}}}{\phi}\Delta t\, s2(i,n) \end{aligned} \tag{9.13}$$

This equation is used to update the concentration of nanoparticles at each spatial point and time step.

Next, the adsorbed concentrations on the solid matrix are updated. The adsorbed concentration on primary adsorption sites, $s1(i,n)$, is governed by an adsorption equation. In its continuous form, this equation is:

$$\frac{\partial s1}{\partial t} = \frac{\phi}{\rho_b}k_{\text{irr}}\left(1 - \frac{s1}{s1_{\text{max}}}\right)c \tag{9.14}$$

Discretizing the time derivative using a forward difference gives:

$$\frac{\partial s1}{\partial t} \approx \frac{s1(i,n+1) - s1(i,n)}{\Delta t} \tag{9.15}$$

Substituting this into the continuous equation, we obtain the discretized form for $s1(i,n)$:

$$s1(i,n+1) = s1(i,n) + \frac{\phi\Delta t}{\rho_b}k_{\text{irr}}\left(1 - \frac{s1(i,n)}{s1_{\text{max}}}\right)c(i,n) \tag{9.16}$$

Similarly, the adsorbed concentration on secondary adsorption sites, $s2(i,n)$, is governed by both adsorption and desorption processes. The continuous form of the equation is:

$$\frac{\partial s2}{\partial t} = \frac{\phi}{\rho_b}k_{\text{ra}}\left(1 - \frac{s2}{s2_{\text{max}}}\right)c - k_{\text{rd}}s2 \tag{9.17}$$

Discretizing the time derivative and substituting into the equation gives:

$$s2(i, n+1) = s2(i, n) + \frac{\phi \Delta t}{\rho_b} k_{\text{ra}} \left(1 - \frac{s2(i, n)}{s2_{\text{max}}}\right) c(i, n) - \Delta t k_{\text{rd}} s2(i, n) \quad (9.18)$$

This equation is used to update the secondary adsorbed concentration $s2$ at each spatial and time point.

After solving the system for all time steps and spatial positions, the concentration $c(i, n)$ is normalized. This normalization is performed by dividing the concentration by the maximum concentration value obtained during the simulation:

$$c(i, n) = \frac{c(i, n)}{c_{\text{max}}} \quad (9.19)$$

The numerical solution of the above finite difference scheme, which is implemented in MATLAB®, is listed below:

```
% PARAMETERS
L_core = 0.3;                    % Core length (m)
Q_flow = 8.33 * 0.01667 * 10^-6; % Volumetric flow rate (m^3/s)
A_area = 4.49 * 10^-5;           % Cross-sectional area (m^2)
v_velocity = Q_flow / A_area; % Fluid velocity (m/s)
D_diff = 2.1 * 10^-7;            % Diffusion coefficient (m^2/s)
porosity = 0.347;                % Porosity of the medium
rho_bulk = 1.28 * 10^3;          % Bulk density (kg/m^3)
PVI_total = 1.2456;              % Total pore volume injected slug
pore_volume = 14.8 * 10^-6; % Pore volume (m^3)
PVI_time = PVI_total * pore_volume / Q_flow; % Injection time (s)
K_adsorp = 0.135;                % Adsorption coefficient
Kap_eff = 1 + (rho_bulk / porosity) * K_adsorp * eta;
k_irrev = 7.00 * 10^-3;          % Irreversible adsorption rate
    constant
k_ra = 1.20 * 10^-4;             % Reversible adsorption rate constant
k_rd = 5.00 * 10^-4;             % Desorption rate constant
s1_max = 5.75 * 10^-4;           % Maximum intrinsic adsorption
    capacity
s2_max = 2.87 * 10^-1;           % Maximum secondary adsorption
    capacity

% NUMERICAL PARAMETERS
time_steps = 600;                % Total time steps
dx_space = 0.01;                 % Spatial step (m)
dt_time = 0.001;                 % Time step (s)

% DISCRETIZATION OF SPACE AND TIME
x_space = 0:dx_space:L_core;
t_time = 0:dt_time:time_steps;
nx = length(x_space);
nt = length(t_time);

% INITIALIZATION OF CONCENTRATION AND ADSORPTION VARIABLES
conc = zeros(nx, nt);           % Concentration (C)
```

```
s1_adsorp = zeros(nx, nt);  % Adsorbed concentration (S1)
s2_adsorp = zeros(nx, nt);  % Secondary adsorption (S2)
pv_injected = zeros(1, nt); % Pore volumes injected (PV)

% INITIAL CONDITION FOR CONCENTRATION AT THE BOUNDARY
for i = 2:nt-1
    if t_time(i) <= PVI_time
        conc(1, i) = 0.01;  % Set boundary concentration
    end
end

% CALCULATE PORE VOLUMES INJECTED (PV)
for n = 2:nt
    dpv = v_velocity / L_core * dt_time;
    pv_injected(n) = pv_injected(n-1) + dpv;
end

% MAXIMUM CONCENTRATION FOR NORMALIZATION
c_max = max(max(conc));

% FINITE DIFFERENCE SCHEME FOR TRANSPORT AND ADSORPTION
for n = 2:nt
    for i = 2:nx-1
        % Concentration equation (C)
        conc(i, n) = conc(i, n-1) + (D_diff * dt_time / dx_space
            ^2) * ...
                    (conc(i+1, n-1) - 2 * conc(i, n-1) + conc(i
                       -1, n-1)) ...
                    - (v_velocity * dt_time / dx_space) * (conc(
                       i, n-1) - conc(i-1, n-1)) ...
                    - k_irrev * (1 - s1_adsorp(i, n-1) / s1_max)
                       * conc(i, n-1) ...
                    + k_ra * (1 - s2_adsorp(i, n-1) / s2_max) *
                       conc(i, n-1) ...
                    - (rho_bulk * k_rd / porosity) * s2_adsorp(i
                       , n-1);

        % Adsorption equation (S1)
        s1_adsorp(i, n) = s1_adsorp(i, n-1) + (porosity /
            rho_bulk) * dt_time * ...
                          k_irrev * (1 - s1_adsorp(i, n-1) /
                             s1_max) * conc(i, n-1);

        % Secondary adsorption equation (S2)
        s2_adsorp(i, n) = s2_adsorp(i, n-1) + (porosity /
            rho_bulk) * dt_time * ...
                          k_ra * (1 - s2_adsorp(i, n-1) / s2_max)
                             * conc(i, n-1) ...
                          - dt_time * k_rd * s2_adsorp(i, n-1);
    end
end

% NORMALIZATION OF CONCENTRATION
conc = conc / c_max;

% PLOT RESULTS - Numerical vs Experimental Data
```

```
figure(1)
plot(pv_injected, conc(nx-1, :), 'k', 'LineWidth', 2);
set(gca, 'fontsize', 12)
legend('Numerical')
hold on;

% Load experimental data from Excel file
Exp_data = xlsread('Exprm.xls');
pv_exp = Exp_data(:, 1);
c_exp = Exp_data(:, 2);

% Plot experimental data
plot(pv_exp, c_exp, 'ko', 'DisplayName', 'Experimental', '
    LineWidth', 1);
hold off;
grid on;
xlabel('PV');
ylabel('C/C0');
```

Figure 9.1 shows a comparison between the experimental data [404] and the numerical results for nanoparticle transport in porous media. The black solid line represents the numerical solution obtained using the advection-diffusion equation with attachment and detachment processes, while the open circles indicate the experimental data.

The experimental data demonstrate the breakthrough curve of nanoparticles, where the concentration of nanoparticles C/C_0 in the effluent increases, peaks, and eventually decreases as pore volumes (PV) are injected. The experimental data indicate a breakthrough at around $PV = 2$, with a peak near $PV = 3.5$, followed by a decline in concentration. The model successfully captures the key dynamics of nanoparticle transport in porous media, including the rise and fall of nanoparticle concentrations in the effluent as pore volumes are injected. The use of a Gaussian curve to represent the concentration distribution provides a reasonable approximation of the experimental data, suggesting that the dominant mechanisms of advection, diffusion, and reversible adsorption are well-represented in the numerical model.

9.4 ANALYTICAL SOLUTIONS

Beyond the standard advection-dispersion framework, nanoparticle transport is influenced by a range of retention processes. These include reversible and irreversible deposition, time-dependent detachment, and blocking effects that alter available deposition sites. Models that incorporate these processes provide a more realistic depiction of transport behavior in porous systems.

The classical one-dimensional advection-dispersion equation (ADE) in porous media is given by:

$$\phi\,\frac{\partial c}{\partial t} + v\,\frac{\partial c}{\partial x} = D\,\frac{\partial^2 c}{\partial x^2} \tag{9.20}$$

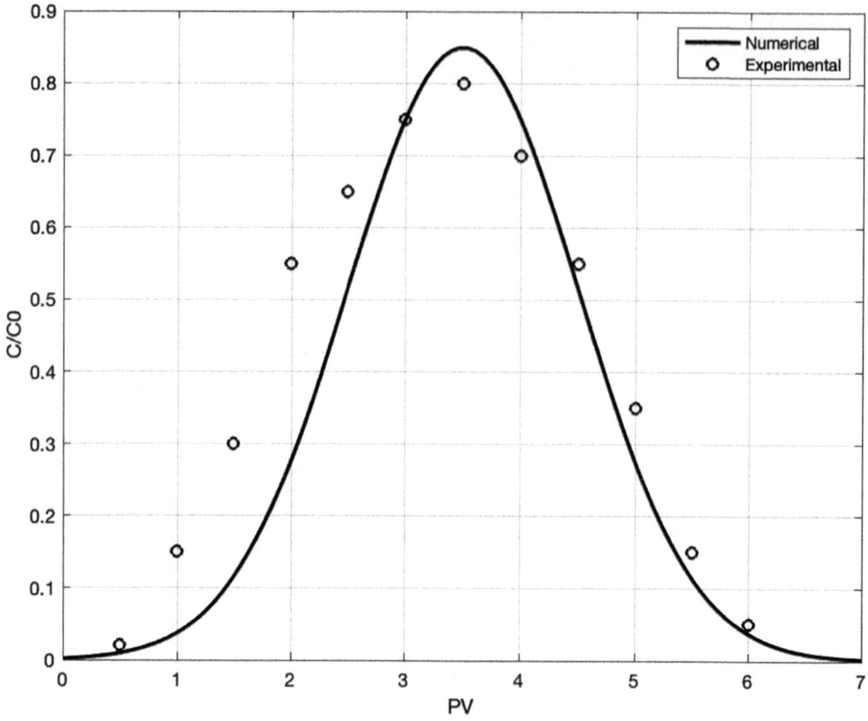

Figure 9.1 Comparison of numerical and experimental data for nanoparticle transport.

where:

c is the nanoparticle concentration in the fluid phase [M/L^3],
v is the average pore velocity [L/T],
D is the hydrodynamic dispersion coefficient [L^2/T], and
ϕ is the porosity of the porous medium.

ATTACHMENT MODEL

One widely used approach to model nanoparticle behavior in porous media relies on classical filtration theory, which inherently couples transport with deposition. In this framework, particle accumulation in the solid phase is captured by adding a deposition term to the mass balance equation [215, 216, 217]. The resulting governing equation is:

$$\frac{\partial c}{\partial t} + \frac{\rho_b}{\phi}\frac{\partial c_s}{\partial t} + v\frac{\partial c}{\partial x} = D\frac{\partial^2 c}{\partial x^2} \tag{9.21}$$

where c_s is the solid-phase concentration of retained nanoparticles (mass per unit mass of porous matrix), ρ_b is the bulk density of the medium [M/L^3].

The second term captures deposition onto the porous matrix and is typically modeled as:

$$\frac{\rho_b}{\phi}\frac{\partial c_s}{\partial t} = k_{\text{dep}}\, c \tag{9.22}$$

where k_{dep} is the first-order deposition rate constant $[\mathrm{T}^{-1}]$. Substituting this into the previous equation yields:

$$\phi\,\frac{\partial c}{\partial t} + v\,\frac{\partial c}{\partial x} = D\,\frac{\partial^2 c}{\partial x^2} - R\,c \tag{9.23}$$

Here, $R = k_{\text{dep}}\,\rho_b/\phi$ represents the effective sink term due to particle deposition.

The problem is typically solved under the following initial and boundary conditions:

$$\begin{cases} c(x,0) = 0, \\ c(0,t) = c_0, \\ \dfrac{\partial c}{\partial x}(\infty,t) = 0, \end{cases} \tag{9.24}$$

The analytical solution for this system, derived through Laplace transform techniques, is given by [69]:

$$c(x,t) = \frac{c_0}{2}\,\exp\left[\frac{(v-u)x}{2D}\right]\,\mathrm{erfc}\left(\frac{\phi\,x - u\,t}{2\sqrt{D\,\phi\,t}}\right) + \frac{c_0}{2}\,\exp\left[\frac{(v+u)x}{2D}\right]\,\mathrm{erfc}\left(\frac{\phi\,x + u\,t}{2\sqrt{D\,\phi\,t}}\right) \tag{9.25}$$

with the auxiliary velocity term:

$$u = v\sqrt{1 + \frac{4\,k_{\text{dep}}\,D}{v^2}} \tag{9.26}$$

This solution highlights how deposition alters effective transport speed and concentration profiles.

Notably, this analytical framework assumes a constant flow velocity. In real-world porous systems, flow is governed by Darcy's Law:

$$u = -\frac{k}{\mu}\frac{dp}{dx} \tag{9.27}$$

where u is the Darcy velocity [L/T], p is the pressure [Pa], k is the intrinsic permeability $[\mathrm{L}^2]$, μ is the dynamic viscosity $[\mathrm{M}/(\mathrm{L}\cdot\mathrm{T})]$.

Introducing a spatially varying source term $q(x)$, the continuity equation becomes:

$$\frac{du}{dx} = q(x) \tag{9.28}$$

Combining with Darcy's law leads to:

$$\frac{d}{dx}\left[-\frac{k}{\mu}\frac{dp}{dx}\right] = q(x) \tag{9.29}$$

Integrating both sides gives:

$$Q(x) = \int_0^x q(\xi)\, d\xi \quad \text{with} \quad Q(0) = u_0 \tag{9.30}$$

Solving for pressure yields a linear profile [69]:

$$p(x) = p_h + \frac{u_0\, \mu}{k}\,(h - x) \tag{9.31}$$

This result confirms that, under steady-state conditions and uniform flow, pressure decreases linearly with distance, and velocity remains constant, justifying the assumption used in the analytical solution.

ONE-SITE NANOPARTICLE MODEL

For a more generalized scenario, consider a non-zero initial concentration c_i and a pulse injection at the left boundary, where the nanoparticle concentration is maintained at c_0 until time t_0, after which it drops to zero. The initial and boundary conditions are then given by:

$$\begin{cases} c(x,0) = c_i, \\ c(0,t) = \begin{cases} c_0, & 0 < t < t_0, \\ 0, & t > t_0, \end{cases} \\ \dfrac{\partial c}{\partial x}(\infty, t) = 0. \end{cases} \tag{9.32}$$

The corresponding analytical solution, based on the formulations of Lapidus and Amundson [239], and van Genuchten [251], is expressed as [225]:

$$c(x,t) = \begin{cases} c_i\, A(x,t) + c_0\, B(x,t), & 0 < t < t_0, \\ c_i\, A(x,t) + c_0\, \left[B(x,t) - B(x, t - t_0)\right], & t > t_0, \end{cases} \tag{9.33}$$

where:

$$A(x,t) = \exp\left(\frac{k_{\text{dep}}\, t}{\phi}\right) \left\{1 - \frac{1}{2}\,\text{erfc}\left[\frac{\phi x - vt}{2\sqrt{D\phi t}}\right] - \frac{1}{2}\exp\left(\frac{vx}{D}\right)\text{erfc}\left[\frac{\phi x + vt}{2\sqrt{D\phi t}}\right]\right\} \tag{9.34}$$

$$B(x,t) = \frac{1}{2}\exp\left(\frac{(v-u)x}{2D}\right)\text{erfc}\left[\frac{\phi x - ut}{2\sqrt{D\phi t}}\right] + \frac{1}{2}\exp\left(\frac{(v+u)x}{2D}\right)\text{erfc}\left[\frac{\phi x + ut}{2\sqrt{D\phi t}}\right] \tag{9.35}$$

and the auxiliary velocity term u is defined as:

$$u = v\sqrt{1 + \frac{4\, k_{\text{dep}}\, D}{v^2}}. \tag{9.36}$$

This solution accounts for both the initial background concentration and the finite-duration injection, capturing transient nanoparticle behavior under one-site deposition conditions.

9.5 GENERAL MODEL FOR SINGLE-PHASE FLOW

Nanoparticles are transported through the pore spaces, where their behavior is influenced by factors such as adsorption onto pore surfaces, entrapment within pore throats, and Brownian motion [225]. The interaction between nanoparticles and the porous medium results in complex dynamics that are governed by a combination of diffusion, advection, and retention processes. In this context, a mathematical model is essential to capture the underlying physics and predict the transport behavior. In the case where the nanoparticles are of a uniform size, the governing transport equation is formulated as follows:

$$\phi\frac{\partial c}{\partial t} + \frac{\partial}{\partial z}\left(uc - \phi D\frac{\partial c}{\partial z}\right) = R, \tag{9.37}$$

where the reaction term R is given by:

$$R = \frac{\partial c_{s1}}{\partial t} + \frac{\partial c_{s2}}{\partial t}. \tag{9.38}$$

Here, c represents the concentration of nanoparticles within the water phase, c_{s1} denotes the concentration of nanoparticles adhered to pore surfaces, and c_{s2} accounts for the concentration of nanoparticles trapped in pore throats due to plugging or bridging effects. The rate of surface deposition is governed by:

$$\frac{\partial c_{s1}}{\partial t} = \begin{cases} \gamma_d|u|c, & \text{if } u_w \leq u_c \\ \gamma_d|u|c - \gamma_e|u - u_c|c_{s1}, & \text{if } u_w > u_c \end{cases} \tag{9.39}$$

where γ_d $[\mathrm{m}^{-1}]$ is the surface retention rate of nanoparticles, γ_e $[\mathrm{m}^{-1}]$ is the entrainment rate, and u_c $[\mathrm{m/s}]$ is the critical velocity for the water phase. The rate at which nanoparticles become trapped in pore throats is expressed as:

$$\frac{\partial c_{s2}}{\partial t} = \gamma_{pt}|u|c, \tag{9.40}$$

with γ_{pt} $[\mathrm{m}^{-1}]$ representing the constant for pore throat blockage.

Due to their submicron size, nanoparticles experience pronounced Brownian motion. The diffusion-dispersion tensor D $[\mathrm{m}^2/\mathrm{s}]$ is the sum of molecular diffusion D_{diff} and hydrodynamic dispersion D_{disp}, given by:

$$D = D_{\text{diff}} + D_{\text{disp}}. \tag{9.41}$$

The molecular diffusion coefficient D_{diff} is derived from the Stokes-Einstein equation:

$$D_{\text{diff}} = \tau\frac{k_B T}{3\pi\mu d_p}, \tag{9.42}$$

where τ is the tortuosity factor, k_B is the Boltzmann constant, T [K] is the absolute temperature, and d_p is the nanoparticle diameter.

The suspension velocity u follows Darcy's law, incorporating gravitational effects:

$$u = -\frac{K}{\mu}\left(\frac{\partial p}{\partial z} - \rho g\right), \tag{9.43}$$

where ρ [kg/m^3] is the fluid density, K [m^2] is the medium's permeability, p [Pa] is the fluid pressure, g [m/s^2] is gravitational acceleration, and μ [Pa s] is the fluid's viscosity.

Initially, the nanoparticle concentrations are zero:

$$c = c_{s1} = c_{s2} = 0 \text{ at } t = 0,\ 0 \leq z \leq H, \tag{9.44}$$

where H is the depth of the rock. The boundary condition at the inlet is:

$$c = c_0, \quad c_{s1} = c_{s2} = 0 \text{ for } t > 0,\ z = 0, \tag{9.45}$$

where c_0 is the injection concentration of the nanoparticle suspension. At the far boundary, a no-flow condition is applied:

$$\frac{\partial c}{\partial z} = \frac{\partial c_{s1}}{\partial z} = \frac{\partial c_{s2}}{\partial z} = 0 \text{ for } t > 0,\ z = H. \tag{9.46}$$

MATLAB® "PDEPE" IMPLEMENTATION

The following MATLAB® code solves the one-dimensional partial differential equations (PDEs) governing the transport of nanoparticles in a porous medium under single-phase flow conditions. The problem is simplified to focus on nanoparticle concentration C, surface deposition C_{s1}, and pore throat blocking C_{s2}. The code employs MATLAB's built-in pdepe solver, which uses a combination of finite difference methods for spatial discretization and an adaptive time-stepping method for time integration.

MATLAB® Code

```
function OneD_SinglePhase_Nano
    clc;
    clear all;
    close all;

    % Geometry type (m = 0 for Cartesian)
    m = 0;

    % Discretization
    x = linspace(0, 0.2, 200);                % Space discretization
    t = linspace(0, 200*24*3600, 200);        % Time discretization
        (200 time steps)

    % Main function for solving PDE
    sol = pdepe(m, @pdex4pde, @pdex4ic, @pdex4bc, x, t);
```

```
    % Extract results for nanoparticle concentration and deposit
    varphi = sol(:, :, 1);  % Nanoparticle concentration (previously C)
    phi_s1 = sol(:, :, 2);  % Surface deposition (previously Cs1)
    phi_s2 = sol(:, :, 3);  % Pore throat blocking (previously Cs2)

    % Find the indices for specific times (20 hrs, 100 hrs, 200 hrs)
    t_hours = t / 3600;  % Convert time from seconds to hours
    idx_20hrs = find(t_hours >= 20, 1);    % Find index closest to 20 hours
    idx_100hrs = find(t_hours >= 100, 1); % Find index closest to 100 hours
    idx_200hrs = find(t_hours >= 200, 1); % Find index closest to 200 hours

    % Visualization of results

    figure(1);
    plot(x, varphi(idx_20hrs, :), 'b', 'LineWidth', 2); hold on;
    plot(x, varphi(idx_100hrs, :), 'k', 'LineWidth', 2); hold on;
    plot(x, varphi(idx_200hrs, :), 'r', 'LineWidth', 2); hold off;
    legend('20 hrs', '100 hrs', '200 hrs');
    title('Nanoparticle Concentration Profile (\varphi)', 'FontSize', 14);
    xlabel('x'); ylabel('\varphi');

    figure(2);
    plot(x, phi_s1(idx_20hrs, :), 'b', 'LineWidth', 2); hold on;
    plot(x, phi_s1(idx_100hrs, :), 'k', 'LineWidth', 2); hold on;
    plot(x, phi_s1(idx_200hrs, :), 'r', 'LineWidth', 2); hold off;
    legend('20 hrs', '100 hrs', '200 hrs');
    title('Surface Deposition Profile (\phi_{s1})', 'FontSize', 14);
    xlabel('x'); ylabel('\phi_{s1}');

    figure(3);
    plot(x, phi_s2(idx_20hrs, :), 'b', 'LineWidth', 2); hold on;
    plot(x, phi_s2(idx_100hrs, :), 'k', 'LineWidth', 2); hold on;
    plot(x, phi_s2(idx_200hrs, :), 'r', 'LineWidth', 2); hold off;
    legend('20 hrs', '100 hrs', '200 hrs');
    title('Blocking Concentration Profile (\phi_{s2})', 'FontSize', 14);
    xlabel('x'); ylabel('\phi_{s2}');
end

% Define the PDE system
function [c, f, s] = pdex4pde(x, t, u, DuDx)
    % Parameters
    phi = 0.3;                     % Porosity
    D = 5.6E-8;             % Diffusion coefficient
    alphad = 16;            % Deposition constant
```

```
    alphae = 30;            % Entrainment constant
    alphapt = 1.28;         % Blocking constant
    vc = 0.0000046;         % Critical velocity

    % PDE coefficients for nanoparticle concentration
    c1 = phi;
    f1 = phi * D * DuDx(1);

    % Source/sink terms
    vw = -DuDx(1); % Simplified velocity for single-phase flow

    % Surface deposition and pore throat blocking rates
    if abs(vw) < vc
        s3 = alphad * vw * u(1);  % Surface deposition
    else
        s3 = alphad * vw * u(1) - alphae * u(2) * (vw - vc); %
            Adjusted for velocity
    end

    s4 = alphapt * vw * u(1);  % Pore throat blocking

    % Total source term
    R = s3 + s4;

    % Time derivatives and fluxes
    s1 = -R - vw * DuDx(1);  % For concentration
    s2 = s3;                    % For surface deposition
    s3 = s4;                    % For pore throat blocking

    c = [c1; 1; 1];
    f = [f1; 0; 0];
    s = [s1; s2; s3];
end

% Initial conditions
function u0 = pdex4ic(x)
    u0 = [0.0; 0.0; 0.0];  % Initial values for \varphi, \phi_{s1
        }, and \phi_{s2}
end

% Boundary conditions
function [pl, ql, pr, qr] = pdex4bc(xl, ul, xr, ur, t)
    % Left boundary (inlet)
    pl = [ul(1) - 0.99; ul(2); ul(3)];
    ql = [0; 0; 0];

    % Right boundary (no-flow condition)
    pr = [0; 0; 0];
    qr = [1; 1; 1];
end
```

The figure illustrates the concentration profiles of nanoparticles in porous media at different times (2 hours, 1 day, 10 days, 20 days, and 40 days). Initially, the concentration decreases rapidly due to advection-dominated transport, followed by a slower diffusion-dominated phase as the system approaches

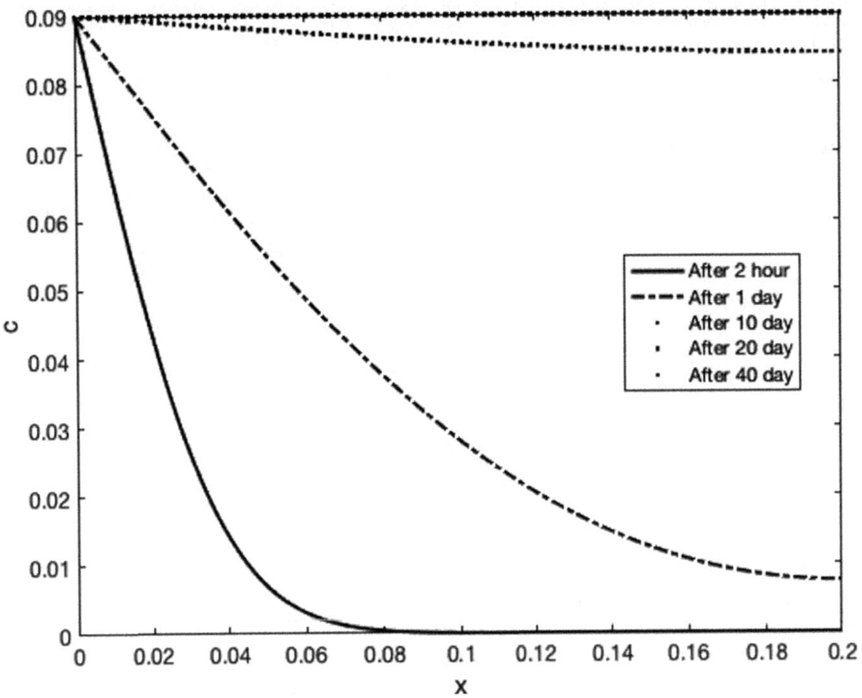

Figure 9.2 Concentration profiles of nanoparticles over time in porous media.

equilibrium. Over time, the concentration profile flattens, indicating that the nanoparticles are being retained or deposited onto the pore surfaces and within the pore throats. This behavior reflects the combined effects of diffusion, advection, and retention mechanisms in nanoparticle transport.

9.6 NUMERICAL ITERATIVE SCHEMES

9.6.1 IMPLICIT PRESSURE–CONCENTRATION SCHEME

Before discussing the numerical method in depth, let us briefly review the mathematical model, which encompasses the governing partial differential equations along with the associated algebraic relationships and initial/boundary conditions. The model for nanoparticle transport in single-phase flow in porous media comprises:

1. A pressure equation arising from the combination of Darcy's law and the continuity equation,
2. A nanoparticle concentration equation,
3. An equation for deposited nanoparticle concentration on pore walls,
4. An equation for entrapped nanoparticle concentration within pore throats.

The pressure equation, obtained by integrating Darcy's law with the continuity equation, is

$$-\nabla \cdot \left(\frac{K}{\mu}\nabla p\right) = q, \tag{9.47}$$

while the nanoparticle concentration equation is

$$\phi\,\frac{\partial C}{\partial t} + \nabla \cdot \left(uC - D\nabla C\right) = R + Q_c, \tag{9.48}$$

The equation governing the deposited nanoparticle concentration on the pore walls is given by

$$\frac{\partial C_{s1}}{\partial t} = \begin{cases} \gamma_d\,|u|\,C, & u \le u_r, \\ \gamma_d\,|u|\,C - \alpha_r\,\gamma_e\,|u-u_r|\,C, & u > u_r, \end{cases} \tag{9.49}$$

and the entrapped nanoparticle concentration within the pore throats is governed by

$$\frac{\partial C_{s2}}{\partial t} = \gamma_{pt}\,|u|\,C. \tag{9.50}$$

9.6.2 ITERATIVE IMPC SCHEME

To discretize the governing PDEs numerically, the total simulation time interval $[0,T]$ is divided into N_T time steps of size $\Delta t^n = t^{n+1} - t^n$. Furthermore, within each time step, N_I iterative cycles are employed. By linearizing capillary pressure with respect to saturation, the coupled system is solved implicitly to obtain the pressure potential. Subsequently, the saturation is updated explicitly within the current iteration, and the nanoparticle concentration, the deposited nanoparticle concentration on pore walls, and the entrapped nanoparticle concentration within pore throats are computed implicitly. Parameters such as permeability, porosity, and relative permeabilities are refreshed after each iteration. This process is repeated until a specified error tolerance is met.

A backward Euler time discretization is applied to the pressure and concentration equations, leading to

$$-\nabla \cdot \left(K(C_{s1}^{n+1,k}, C_{s2}^{n+1,k})\nabla P^{n+1,k+1}\right) = q^{n+1}, \tag{9.51}$$

where $n+1$ refers to the current time step, n to the previous time step, $k+1$ to the current iteration, and k to the previous iteration.

The three concentration equations are discretized as follows:

$$\begin{aligned} &\phi(C_{s1}^{n+1,k}, C_{s2}^{n+1,k})\,\frac{C^{n+1,k+1} - C^n}{\Delta t^n} + \\ &\quad \nabla \cdot \left\{\alpha_r\, u^{n+1,k+1}\, C^{n+1,k+1} - D(p^{n+1,k+1}, C_{s1}^{n+1,k}, C_{s2}^{n+1,k})\nabla C^{n+1,k+1}\right\} \\ &\qquad = R(p^{n+1,k+1}, C^{n+1,k}, C_{s1}^{n+1,k}) + Q_c^{n+1}, \end{aligned} \tag{9.52}$$

$$\frac{C_{s1}^{n+1,k+1} - C_{s1}^{n}}{\Delta t^n} = \begin{cases} \alpha_r\, \gamma_d \left|u^{n+1,k+1}\right| C^{n+1,k+1}, & u^{n+1,k+1} \le u_r, \\ \alpha_r\, \gamma_d \left|u^{n+1,k+1}\right| C^{n+1,k+1} & u^{n+1,k+1} > u_r, \\ \quad -\alpha_r\, \gamma_e \left|u^{n+1,k+1} - u_r\right| C_{s1}^{n+1,k+1}, & \end{cases} \tag{9.53}$$

$$\frac{C_{s2}^{n+1,k+1} - C_{s2}^{n}}{\Delta t^n} = \alpha_r\, \gamma_{pt} \left|u^{n+1,k+1}\right| C^{n+1,k+1}. \tag{9.54}$$

Note that the advection term in Eqs. (9.52)–(9.54) is multiplied by the relaxation factor α_r. Since the velocity u depends on the pressure p, and both the permeability K and the porosity ϕ depend on p and C, while D and R are functions of p, C_{s1}, and C_{s2}, the values of K and ϕ from the previous iteration (using C_{s1} and C_{s2}) are used. In contrast, D and R are evaluated at the current iteration for p, C_{s1}, and C_{s2}.

In algebraic form, the pressure equation is expressed as

$$A_a\big(C_{s1}^{n+1,k}, C_{s2}^{n+1,k}\big)p^{n+1,k+1} = Q_a^{n+1}, \tag{9.55}$$

indicating that A_a depends on C_{s1} and C_{s2}.

The velocity $u^{n+1,k+1}$ is then updated and substituted into the algebraic equation for the concentration, yielding

$$B_t\big(p^{n+1,k+1}, C_{s1}^{n+1,k}, C_{s2}^{n+1,k}\big)C^{n+1,k+1} = Q_{ct}\big(p^{n+1,k+1}, C^{n+1,k}, C_{s1}^{n+1,k}, C_{s2}^{n+1,k}\big), \tag{9.56}$$

where

$$B_t\big(p^{n+1,k+1}, C_{s1}^{n+1,k}, C_{s2}^{n+1,k}\big) = \Delta t^n\, M\big(C_{s1}^{n+1,k}, C_{s2}^{n+1,k}\big)^{-1} \Big[B_w\big(p^{n+1,k+1}\big) - B_D\big(p^{n+1,k+1}, C_{s1}^{n+1,k}, C_{s2}^{n+1,k}\big)\Big], \tag{9.57}$$

$$Q_{ct}\big(p^{n+1,k+1}, C^{n+1,k}, C_{s1}^{n+1,k}, C_{s2}^{n+1,k}\big) = \Delta t^n\, M\big(C_{s1}^{n+1,k}, C_{s2}^{n+1,k}\big)^{-1} \Big[Q_c^{n+1} - R\big(p^{n+1,k+1}, C^{n+1,k}, C_{s1}^{n+1,k}\big)\Big] C^n. \tag{9.58}$$

Here, $C_{s1}^{n+1,k}$ from the previous iteration is used in the term R because R depends on both C and C_{s1}; in the case $u > u_r$, R is linearized to preserve the consistency of the scheme.

The algebraic form of the C_{s1} equation is

$$C_{s1}^{n+1,k+1} = \begin{cases} R_{1t}\big(p^{n+1,k+1}, C^{n+1,k+1}, C_{s1}^{n}\big), & u^{n+1,k+1} \le u_r, \\ \Big[I + \Delta t^n\, R_2\big(p^{n+1,k+1}\big)\Big]^{-1} & u^{n+1,k+1} > u_r, \\ \quad R_{1t}\big(p^{n+1,k+1}, C^{n+1,k+1}, C_{s1}^{n}\big), & \end{cases} \tag{9.59}$$

where

$$R_{1t}\left(p^{n+1,k+1},C^{n+1,k+1},C_{s1}^{n}\right)=\Delta t^{n}\,R_{1}\left(p^{n+1,k+1},C^{n+1,k+1}\right)+C_{s1}^{n}. \quad (9.60)$$

Since R_2 is positive, the matrix $\left[I+\Delta t^{n}\,R_{2}\right]$ is invertible for sufficiently small Δt^{n}.

Finally, the algebraic equation for the C_{s2} equation is

$$C_{s2}^{n+1,k+1}=R_{3t}\left(p^{n+1,k+1},C^{n+1,k+1},C_{s2}^{n}\right), \quad (9.61)$$

where

$$R_{3t}\left(p^{n+1,k+1},C^{n+1,k+1},C_{s2}^{n}\right)=\Delta t^{n}\,R_{3}\left(p^{n+1,k+1},C^{n+1,k+1}\right)+C_{s2}^{n}, \quad (9.62)$$

and

$$R_{3}\left(p^{n+1,k+1},C^{n+1,k+1}\right)=\gamma_{pt}\left|u^{n+1,k+1}\right|C^{n+1,k+1}. \quad (9.63)$$

9.7 NANOPARTICLES TRANSPORT IN TWO-PHASE FLOW

The injection of nanoparticles into an oil reservoir can alter fluid properties such as mobility and wettability. Incorporating several simplifying assumptions includes:

The fluids and the rock matrix are incompressible.
There is no mass transfer between the two phases.
The flow occurs under isothermal conditions.
The fluids are Newtonian with constant viscosities.
Nanoparticles exist only in the water phase.

The two-phase immiscible, incompressible flow in porous media is governed by Darcy's law and the mass conservation equation. These can be written as:

$$\phi\frac{\partial S_{\alpha}}{\partial t}+\nabla\cdot\mathbf{u}_{\alpha}=0,\quad \alpha=w,o, \quad (9.64)$$

$$\mathbf{u}_{\alpha}=-\frac{k_{r\alpha}}{\mu_{\alpha}}\mathbf{K}\left(\nabla P_{\alpha}-\rho_{\alpha}\mathbf{g}\right),\quad \alpha=w,o, \quad (9.65)$$

Nanoparticles are assumed to be transported only in the water phase. The transport equation, assuming a single-size interval of nanoparticles, is given by:

$$\phi\frac{\partial(S_{w}C)}{\partial t}+\nabla\cdot\left(\mathbf{u}_{w}C-\phi S_{w}\mathbf{D}\nabla C\right)=R, \quad (9.66)$$

where C is the nanoparticle concentration in the water phase, $\mathbf{D}$ is the diffusion-dispersion tensor, and R is the net rate of nanoparticle retention, which can be expressed as:

$$R = \frac{\partial C_{s1}}{\partial t} + \frac{\partial C_{s2}}{\partial t}, \tag{9.67}$$

where C_{s1} is the nanoparticle concentration retained on pore surfaces, and C_{s2} is the nanoparticle concentration trapped in pore throats due to bridging. The diffusion-dispersion tensor, $\mathbf{D}$, accounts for molecular diffusion ($\mathbf{D}_{\text{diff}}$) and hydrodynamic dispersion ($\mathbf{D}_{\text{disp}}$):

$$\mathbf{D} = \mathbf{D}_{\text{diff}} + \mathbf{D}_{\text{disp}}. \tag{9.68}$$

The surface retention of nanoparticles, considering a critical velocity threshold, is modeled as:

$$\frac{\partial C_{s1}}{\partial t} = \begin{cases} \gamma_d |\mathbf{u}_w| C, & |\mathbf{u}_w| \le u_c, \\ \gamma_d |\mathbf{u}_w| C - \gamma_e \big(|\mathbf{u}_w| - u_c\big) C_{s1}, & |\mathbf{u}_w| > u_c, \end{cases} \tag{9.69}$$

where γ_d is the surface retention coefficient, γ_e is the entrainment coefficient, and u_c is the critical velocity. Entrapment in pore throats is expressed as:

$$\frac{\partial C_{s2}}{\partial t} = \gamma_{\text{pt}} |\mathbf{u}_w| C, \tag{9.70}$$

where γ_{pt} is the pore throat blocking constant.

Nanoparticle deposition and bridging can alter the porosity and permeability of the porous medium. The porosity variation is given by

$$\phi = \phi_0 - \delta\phi, \quad \delta\phi = C_{s1} + C_{s2}, \tag{9.71}$$

where ϕ_0 is the initial porosity. The permeability variation can be expressed as:

$$K = K_0 \Big[(1 - f) k_f + f \frac{\phi}{\phi_0}\Big]^l, \tag{9.72}$$

where K_0 is the initial permeability, k_f is a constant for fluid seepage through plugged pores, $f = 1 - \gamma_f C_{s2}$ is the flow efficiency factor, γ_f is the flow efficiency coefficient, and l is an exponent typically ranging from 2.5 to 3.5.

Nanoparticles can influence fluid flow in porous media by modifying relative permeability. When nanoparticles are retained on pore walls, they alter the available flow pathways, leading to changes in permeability characteristics. This section outlines how such changes are quantified and modeled.

An empirical equation used to calculate the specific surface area of a rock is:

$$a_{\text{sp}} = A\phi \sqrt{\frac{\phi}{K}} \tag{9.73}$$

where A is a constant, ϕ is porosity, K is permeability.

The total surface area of retained nanoparticles per unit bulk volume is:

$$a_{\text{tot}} = \frac{6\beta\delta\phi}{d} \tag{9.74}$$

where d is nanoparticle diameter, $\delta\phi$ is the change in porosity, β is a constant.

If $a_{\text{tot}} \geq a_{\text{sp}}$, the pore surfaces are fully covered by nanoparticles; otherwise, only partial coverage occurs. The relative permeability for phase α with particle coverage is modeled as:

$$k_{r\alpha,p} = k_{r\alpha} + \frac{a_{\text{tot}}}{a_{\text{sp}}}(k_{r\alpha,c} - k_{r\alpha}) \tag{9.75}$$

where $k_{r\alpha,c}$ is the relative permeability under full surface coverage.

Ju and Fan [232] reported that effective permeability improved by 1.627 to 2.136 times with nanoparticles, despite a 10% drop in absolute permeability. The wetting-phase relative permeability can also be expressed as:

$$k_{r\alpha,p} = \left[1 + r_a(\theta_\alpha - 1)\right] k_{r\alpha}, \quad \alpha = w, n \tag{9.76}$$

with:

$$\theta_\alpha = \frac{k_{r\alpha,c}}{k_{r\alpha}}, \quad r_a = \frac{a_{\text{tot}}}{a_{\text{sp}}}$$

9.7.1 NUMERICAL EXAMPLE

At the start of the simulation (time $t = 0$), the porous medium is assumed to contain no nanoparticles. The water saturation is set to its initial value throughout the domain, while the nanoparticle concentration in all phases is zero. Mathematically, this is expressed as:

$$S_w = S_{iw}, \quad c = c_{s1} = c_{s2} = 0, \quad \text{at } t = 0,\ 0 \leq z \leq H \tag{9.77}$$

At the injection boundary $z = 0$, the saturation and nanoparticle concentration are fixed at prescribed values for all times $t > 0$, while the deposited and blocking concentrations remain zero:

$$S_w = S_{Ow}, \quad c = c_0, \quad c_{s1} = c_{s2} = 0, \quad \text{at } t > 0,\ z = 0 \tag{9.78}$$

At the far end of the domain $z = H$, a no-flow (Neumann) boundary condition is imposed, indicating that there is no gradient (flux) of any quantity across the boundary. This can be described as:

$$\frac{dS_w}{dz} = \frac{dc}{dz} = \frac{dc_{s1}}{dz} = \frac{dc_{s2}}{dz} = 0, \quad \text{at } t > 0,\ z = H \tag{9.79}$$

MATLAB® "PDEPE" IMPLEMENTATION

This MATLAB® code simulates nanoparticle transport in a porous medium, considering two fluid phases (oil and water). The model is governed by a system of partial differential equations (PDEs) solved using MATLAB's built-in PDE solver `pdepe`.

MATLAB® CODE

The MATLAB® script consists of:

1. **Main function**: Sets up parameters, solves PDEs, and visualizes results.
2. **Sub-functions**:
   - `pdeProblem`: Defines PDE equations.
   - `initialConditions`: Sets initial conditions.
   - `boundaryConditions`: Establishes boundary conditions.
3. **Helper functions**:
   - `getModelParameters`: Contains physical parameters.
   - `plotSolution`: Visualizes simulation results.

```
function OneD_TwoPhase_Nano()
    % Clear workspace and figures
    clc
    clear
    close all

    % Spatial and temporal discretization
    domainLength = 0.2;                    % Length of domain [m]
    simulationTime = 200*24*3600;          % Total simulation time [
        s]
    nx = 200;                              % Number of spatial nodes
    nt = 200*24*60;                        % Number of time steps

    % Coordinates and discretization
    m = 0;                                 % Coordinate system (0 =
        Cartesian)
    x = linspace(0, domainLength, nx);     % Spatial grid
    t = linspace(0, simulationTime, nt); % Time grid

    %% Solve PDE System
    sol = pdepe(m, @pdeProblem, @initialConditions,
        @boundaryConditions, x, t);

    % Extract solution components
    Sw  = sol(:,:,1);  % Water saturation
    C   = sol(:,:,2);  % Nanoparticle concentration in water
    Cs1 = sol(:,:,3);  % Deposited nanoparticle concentration
    Cs2 = sol(:,:,4);  % Blocking nanoparticle concentration

    %% Visualization
    % Define time points for plotting (hours)
```

```
    plotTimes = [20, 100, 200];                 % Hours
    timeIndices = plotTimes * 60;               % Convert to array
        indices

    % Plot settings
    lineColors = {'b', 'k', 'r'};
    lineWidth = 2;

    % Figure 1: Water Saturation
    plotSolution(x, Sw, timeIndices, lineColors, lineWidth, ...
                 'Saturation␣Profile', 'x␣[m]', 'S_w␣[-]',
                     plotTimes);

    % Figure 2: Nanoparticle Concentration
    plotSolution(x, C, timeIndices, lineColors, lineWidth, ...
                 'Concentration␣Profile', 'x␣[m]', 'C␣[kg/m$^3$]'
                     , plotTimes);

    % Figure 3: Deposited Concentration
    plotSolution(x, Cs1, timeIndices, lineColors, lineWidth, ...
                 'Deposited␣Concentration␣Profile', 'x␣[m]', 'C_{
                     s1}␣[kg/m$^3$]', plotTimes);

    % Figure 4: Blocking Concentration
    plotSolution(x, Cs2, timeIndices, lineColors, lineWidth, ...
                 'Blocking␣Concentration␣Profile', 'x␣[m]', 'C_{
                     s2}␣[kg/m$^3$]', plotTimes);
end

%% PDE System Definition
function [c, f, s] = pdeProblem(x, t, u, DuDx)

    % Extract solution components
    Sw = u(1);      % Water saturation
    C = u(2);       % Nanoparticle concentration
    Cs1 = u(3);     % Deposited concentration
    Cs2 = u(4);     % Blocking concentration

    % Physical parameters
    params = getModelParameters();

    % Porosity and permeability calculations
    f0 = 1 - params.gamaf * Cs2;
    phi = params.phi0 - (Cs1 + Cs2);
    k = params.k0 * ((1-f0)*params.kf + f0*phi/params.phi0).^
        params.nn;

    % Normalized saturation
    Sw_norm = (Sw - params.Swr) / (1 - params.Swr - params.Sor);

    % Surface area calculations
    a_sp = 0.7 * phi .* sqrt(phi./k);
    a_tot = params.bet0 * 6/params.d_np * (Cs1 + Cs2);
    r_a = a_tot ./ a_sp;

    % Relative permeability with nanoparticle effects
```

```
krw = params.krw0 * Sw_norm.^params.a;
kro = params.kro0 * (1-Sw_norm).^params.b;

krw_mod = krw .* (1 + r_a * (params.theta_w - 1));
kro_mod = kro .* (1 + r_a * (params.theta_o - 1));

% Mobility calculations
lamw = krw_mod / params.muw;
lamo = kro_mod / params.muo;
lamt = lamw + lamo;
m = lamw * lamo / lamt;

% Capillary pressure and its derivative
pc = params.cw / ((Sw - params.Swr)/(1 - params.Swr)).^params
    .aw + ...
     params.co / ((1 - Sw - params.Sor)/(1 - params.Sor)).^
         params.ao;

AA1 = -params.aw * params.cw / (1 - params.Swr);
AA2 = ((Sw - params.Swr) / (1 - params.Swr)).^(1 + params.aw)
    ;
AA3 = params.ao * params.co / (1 - params.Sor);
AA4 = ((1 - Sw - params.Sor) / (1 - params.Sor)).^(1 + params
    .ao);
dpcz = DuDx(1) * (AA1 / AA2 + AA3 / AA4);

% Water velocity
drho = params.rhow - params.rhoo;
vw = k * m * (drho * params.g + dpcz);

% PDE coefficients
% Saturation equation
c1 = phi;
f1 = k * m * (drho * params.g - dpcz);
s1 = 0;

% Concentration equation
c2 = phi * Sw;
f2 = phi * params.D * Sw * DuDx(2);

% Retention terms
if abs(vw) < params.vc
    s3 = params.alphad * vw * C;
else
    s3 = params.alphad * vw * C - params.alphae * Cs1 * (vw -
        params.vc);
end

s4 = params.alphapt * vw * C;

% Total retention
R = s3 + s4;

% Concentration equation source term
s2 = -R - vw * DuDx(2);
```

```
    % Deposited and blocking concentration equations
    c3 = 1;
    c4 = 1;
    f3 = 0;
    f4 = 0;

    % Return coefficients
    c = [c1; c2; c3; c4];
    f = [f1; f2; f3; f4];
    s = [s1; s2; s3; s4];
end

%% Initial Conditions
function u0 = initialConditions(x)

    u0 = [0.01;     % Initial water saturation
          0.0;      % Initial nanoparticle concentration
          0.0;      % Initial deposited concentration
          0.0];     % Initial blocking concentration
end

%% Boundary Conditions
function [pl, ql, pr, qr] = boundaryConditions(xl, ul, xr, ur, t)
    % Left boundary (x = 0): Dirichlet conditions
    pl = [ul(1) - 0.99;    % Water saturation = 0.99
          ul(2) - 0.0009;  % Concentration = 0.0009
          ul(3);           % No condition on Cs1
          ul(4)];          % No condition on Cs2
    ql = [0; 0; 0; 0];     % pl = 0 means ul = specified value

    % Right boundary (x = L): Neumann conditions
    pr = [0; 0; 0; 0];
    qr = [1; 1; 1; 1];     % dC/dx = 0 (zero flux)
end

%% Helper Functions
function params = getModelParameters()

        % Basic medium properties
    params.phi0 = 0.3;          % Initial porosity [-]
    params.gamaf = 0.01;        % Formation damage coefficient [-]
    params.k0 = 0.02e-12;       % Initial permeability
    params.nn = 3.0;            % Porosity-permeability exponent
        [-]
    params.kf = 0.3;            % Formation coefficient [-]

    % Saturation parameters
    params.Sor = 0.001;         % Residual oil saturation [-]
    params.Swr = 0.001;         % Irreducible water saturation [-]

    % Capillary pressure parameters
    params.aw = 0.5;            % Water exponent in capillary
        pressure [-]
    params.ao = 0.5;            % Oil exponent in capillary
        pressure [-]
    params.cw = 500;            % Water capillary coefficient [Pa]
```

```
    params.co = -500;           % Oil capillary coefficient [Pa]

    % Relative permeability parameters
    params.krw0 = 1;            % Water endpoint relative
        permeability [-]
    params.kro0 = 1;            % Oil endpoint relative
        permeability [-]
    params.a = 4;               % Water relative permeability
        exponent [-]
    params.b = 4;               % Oil relative permeability
        exponent [-]
    params.theta_w = 0.6;       % Water wettability parameter [-]
    params.theta_o = 3;         % Oil wettability parameter [-]

    % Fluid properties
    params.muw = 0.001;         % Water viscosity [Pa.s]
    params.muo = 0.001;         % Oil viscosity [Pa.s]
    params.rhow = 1000;         % Water density
    params.rhoo = 660;          % Oil density
    params.g = 9.80;            % Gravitational acceleration

    % Nanoparticle properties
    params.d_np = 40e-9;        % Nanoparticle diameter [m]
    params.bet0 = 0.8;          % Retention coefficient [-]
    params.vc = 0.0000046;      % Critical velocity [m/s]
    params.D = 5.6e-8;          % Diffusion coefficient

    % Retention kinetics parameters
    params.alphad = 16;         % Deposition rate coefficient [1/m]
    params.alphae = 30;         % Entrainment rate coefficient [1/m
        ]
    params.alphapt = 1.28;      % Pore throat blocking coefficient
        [1/m]
end

function plotSolution(x, solution, timeIndices, colors, lineWidth
    , titleText, xLabel, yLabel, timePoints)


    figure;
    hold on;

    % Plot solution at different times
    for i = 1:length(timeIndices)
        plot(x, solution(timeIndices(i), :), colors{i}, '
            LineWidth', lineWidth);
    end

    % Add labels and legend
    xlabel(xLabel);
    ylabel(yLabel);
    title(titleText, 'FontSize', 14);

    % Create legend entries
    legendEntries = cell(size(timePoints));
    for i = 1:length(timePoints)
```

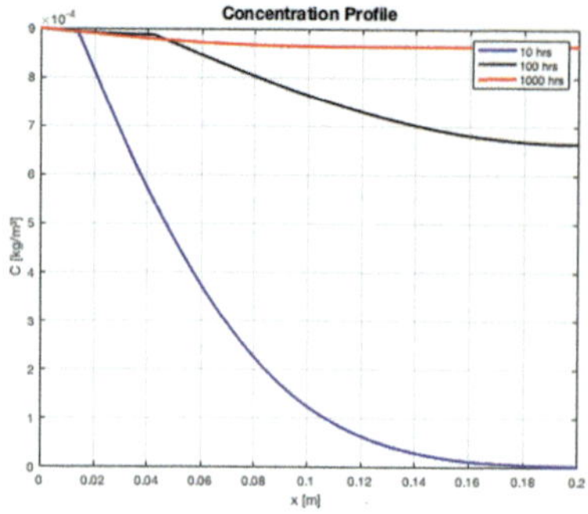

(a) Nanoparticle concentration C

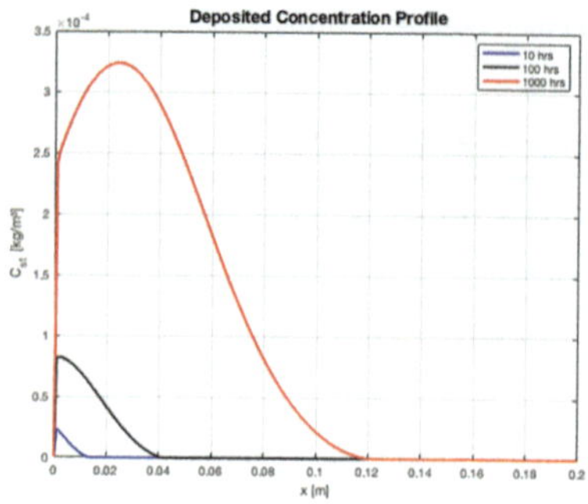

(b) Deposited concentration C_{s1}

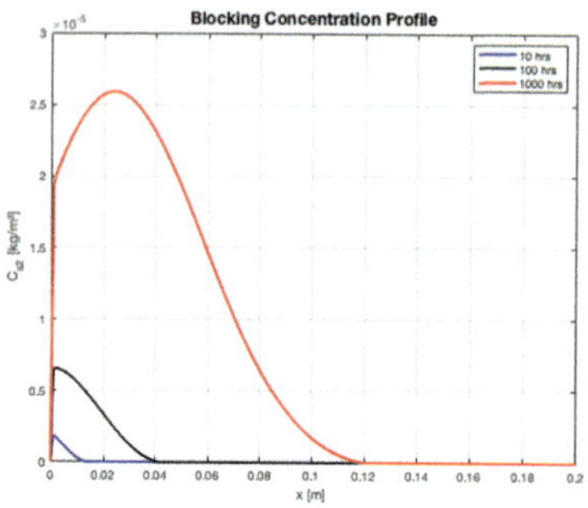

(c) Blocking concentration C_{s2}

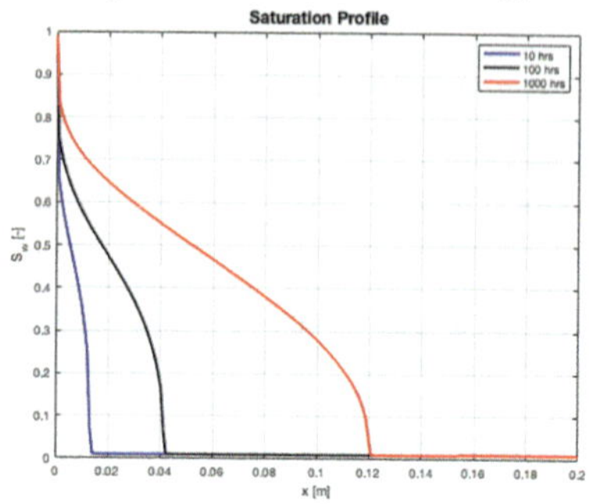

(d) Water saturation S_w

Figure 9.3 Simulation results at 10, 100, and 1000 hours. The trends show how nanoparticles penetrate, deposit, and block pores in a two-phase porous medium.

```
        legendEntries{i} = sprintf('%d hrs', timePoints(i));
    end
    legend(legendEntries);

    hold off;
    box on;
    grid on;
end
```

Simulation results are visualized at specified intervals (10, 100, and 1000 hours), showing spatial profiles of:

In Figure (9a), the nanoparticle concentration C profile shows higher concentrations retained near the inlet as time increases, reflecting limited nanoparticle transport due to retention mechanisms. Figure (9b) shows the evolution of deposited nanoparticles C_{s1}, with a peak forming and growing closer to the inlet, indicating surface adsorption behavior. In Figure (9c), the blocking concentration C_{s2} accumulates more slowly but follows a similar spatial trend, suggesting a cumulative pore-blocking effect over long periods. The saturation profile in Figure (9d) demonstrates how the water front progresses gradually into the domain. Over time, the front shifts deeper as injection continues.

ITERATIVE IMPES–IMC TECHNIQUE

The iterative Implicit-Pressure Explicit-Saturation Implicit-Concentration (IMPES–IMC) technique was developed for nanoparticle transport in two-phase flow. IMPES is a well-known approach for solving two-phase flow problems, and the IMPES–IMC strategy extends this method by incorporating nanoparticle concentration equations.

Assume the total time interval $[0, T]$ is divided into N_T steps of length

$$\Delta t^n = t^{n+1} - t^n,$$

where $n+1$ denotes the current time step and n the previous time step. Over each time step, N_I iterations are performed. We denote the current iteration by $k+1$ and the previous iteration by k.

A backward Euler method is used to discretize the time derivatives in the saturation and concentration equations. In order to compute the pressure, the following two equations are solved together:

$$\begin{aligned} &-\nabla\cdot\left[\lambda_t\left(s_w^{(n+1,k)}\right)\mathbf{K}\left(c_{s1}^{(n+1,k)},c_{s2}^{(n+1,k)}\right)\nabla P_w^{(n+1,k+1)}\right] \\ &-\nabla\cdot\left[\lambda_n\left(s_w^{(n+1,k)}\right)\mathbf{K}\left(c_{s1}^{(n+1,k)},c_{s2}^{(n+1,k)}\right)\nabla P_c\left(s_w^{(n+1,k+1)}\right)\right] = q_t^{(n+1)}, \end{aligned} \tag{9.80}$$

$$\begin{aligned} &\phi\left(c_{s1}^{(n+1,k)},c_{s2}^{(n+1,k)}\right)\frac{s_w^{(n+1,k+1)}-s_w^n}{\Delta t^n} - q_w^{(n+1)} \\ &= \nabla\cdot\left[\lambda_t\left(s_w^{(n+1,k)}\right)\mathbf{K}\left(c_{s1}^{(n+1,k)},c_{s2}^{(n+1,k)}\right)\nabla P_w^{(n+1,k+1)}\right]. \end{aligned} \tag{9.81}$$

The capillary pressure P_c is linearized as follows [90]:

$$P_c\left(s_w^{(n+1,k+1)}\right) \approx P_c\left(s_w^{(n+1,k)}\right) + P_c'\left(s_w^{(n+1,k)}\right)\left[s_w^{(n+1,k+1)} - s_w^{(n+1,k)}\right], \tag{9.82}$$

where P_c' denotes the first derivative of P_c.

To update the saturation, an explicit equation is used:

$$\phi\left(c_{s1}^{(n+1,k)},c_{s2}^{(n+1,k)}\right)\frac{s_w^{(n+1,k+1)}-s_w^n}{\Delta t^n} + \alpha_{r1}\nabla\cdot\left(f_w^{(n+1,k)}\,\mathbf{u}_a^{(n+1,k+1)}\right) = q_w^{(n+1)}, \tag{9.83}$$

where α_{r1} is a relaxation factor.

Next, the equations for the nanoparticle concentration and the deposited/released nanoparticle concentrations are discretized. For the nanoparticles transported in the water phase, we have

$$
\begin{aligned}
&\phi\big(c_{s1}^{(n+1,k)}, c_{s2}^{(n+1,k)}\big) \frac{s_w^{(n+1,k+1)}\, c^{(n+1,k+1)} - s_w^n\, c^n}{\Delta t^n} \\
&\quad + \nabla\cdot\Big\{\alpha_r\, \mathbf{u}_w^{(n+1,k+1)}\, c^{(n+1,k+1)} - D\big(s_w^{(n+1,k+1)}, P_w^{(n+1,k+1)}, c_{s1}^{(n+1,k)}, \\
&\quad c_{s2}^{(n+1,k)}\big)\, \nabla c^{(n+1,k+1)}\Big\} = R\Big(s_w^{(n+1,k+1)}, P_w^{(n+1,k+1)}, c^{(n+1,k)}, c_{s1}^{(n+1,k)}\Big) \\
&\quad + Q_c^{(n+1)},
\end{aligned}
\tag{9.84}
$$

$$
\frac{c_{s1}^{(n+1,k+1)} - c_{s1}^n}{\Delta t^n} = \begin{cases} \alpha_{r2}\, \gamma_d\, |\mathbf{u}_w^{(n+1,k+1)}|\, c^{(n+1,k+1)}, & \mathbf{u}_w^{(n+1,k+1)} \le \mathbf{u}_r, \\ \alpha_{r2}\, \gamma_d\, |\mathbf{u}_w^{(n+1,k+1)}|\, c^{(n+1,k+1)} & \mathbf{u}_w^{(n+1,k+1)} > \mathbf{u}_r, \\ \quad - \alpha_{r2}\, \gamma_e\, |\mathbf{u}_w^{(n+1,k+1)} - \mathbf{u}_r|\, c_{s1}^{(n+1,k+1)}, & \end{cases}
\tag{9.85}
$$

$$
\frac{c_{s2}^{(n+1,k+1)} - c_{s2}^n}{\Delta t^n} = \alpha_{r3}\, \gamma_{pt}\, |\mathbf{u}_w^{(n+1,k+1)}|\, c^{(n+1,k+1)}.
\tag{9.86}
$$

In the above equations, both the permeability $\mathbf{K}$ and the porosity ϕ are taken from the previous iteration (i.e., using $c_{s1}^{(n+1,k)}$ and $c_{s2}^{(n+1,k)}$). In contrast, D and R are evaluated at the current iteration $(k+1)$ for s_w and P_w while still using the previous iteration values for c_{s1} and c_{s2}. Moreover, the variables λ_w, λ_n, λ_t, and f_w are computed as functions of $s_w^{(n+1,k)}$.

Applying a cell-centered finite difference (CCFD) spatial discretization scheme, the discretized pressure equation takes the form

$$
\begin{aligned}
&A_a\big(s_w^{(n+1,k)}, c_{s1}^{(n+1,k)}, c_{s2}^{(n+1,k)}\big)\, P_w^{(n+1,k+1)} \\
&\quad + A_c\big(s_w^{(n+1,k)}, c_{s1}^{(n+1,k)}, c_{s2}^{(n+1,k)}\big)\, P_d\big(s_w^{(n+1,k+1)}\big) = Q_{ac}^{(n+1)},
\end{aligned}
\tag{9.87}
$$

with

$$
P_d\big(s_w^{(n+1,k+1)}\big) = P_c\big(s_w^{(n+1,k)}\big) + P'\big(s_w^{(n+1,k)}\big)\left[s_w^{(n+1,k+1)} - s_w^{(n+1,k)}\right],
\tag{9.88}
$$

and

$$
P'\big(s_w^{(n+1,k)}\big) = \mathrm{diag}\Big(P_c'\big(s_{w,h}^{(n+1,k)}\big)\Big), \quad h = 1, 2, \ldots, N_c,
\tag{9.89}
$$

where N_c is the total number of cells.

Similarly, the spatial discretization of the saturation equation is expressed as

$$
\begin{aligned}
&M\big(c_{s1}^{(n+1,k)}, c_{s2}^{(n+1,k)}\big) \frac{s_w^{(n+1,k+1)} - s_w^n}{\Delta t^n} \\
&\quad + A_w\big(s_w^{(n+1,k)}, c_{s1}^{(n+1,k)}, c_{s2}^{(n+1,k)}\big)\, P_w^{(n+1,k+1)} = Q_w^{(n+1)},
\end{aligned}
\tag{9.90}
$$

where M is a diagonal matrix representing the porosity as a function of $c_{s1}^{(n+1,k)}$ and $c_{s2}^{(n+1,k)}$.

The coupled pressure–saturation system is written in matrix form as

$$A_t\big(s_w^{(n+1,k)}, c_{s1}^{(n+1,k)}, c_{s2}^{(n+1,k)}\big)\, P_w^{(n+1,k+1)} = Q_t\big(s_w^{(n+1,k)}, c_{s1}^{(n+1,k)}, c_{s2}^{(n+1,k)}\big), \tag{9.91}$$

where

$$\begin{aligned} A_t\big(s_w^{(n+1,k)}, c_{s1}^{(n+1,k)}, c_{s2}^{(n+1,k)}\big) &= A_a\big(s_w^{(n+1,k)}, c_{s1}^{(n+1,k)}, c_{s2}^{(n+1,k)}\big) \\ &- \Delta t^n\, A_c\big(s_w^{(n+1,k)}, c_{s1}^{(n+1,k)}, c_{s2}^{(n+1,k)}\big)\, P'\big(s_w^{(n+1,k)}\big) M^{-1} \\ &A_w\big(s_w^{(n+1,k)}, c_{s1}^{(n+1,k)}, c_{s2}^{(n+1,k)}\big), \end{aligned} \tag{9.92}$$

and

$$\begin{aligned} Q_t\big(s_w^{(n+1,k)}, c_{s1}^{(n+1,k)}, c_{s2}^{(n+1,k)}\big) &= Q_{ac}^{(n+1)} - A_c\big(s_w^{(n+1,k)}, c_{s1}^{(n+1,k)}, c_{s2}^{(n+1,k)}\big) \\ &\times \Big[P_c\big(s_w^{(n+1,k)}\big) + P'\big(s_w^{(n+1,k)}\big)\big(s_w^n - s_w^{(n+1,k)}\big)\Big] \\ &- \Delta t^n\, A_c\big(s_w^{(n+1,k)}, c_{s1}^{(n+1,k)}, c_{s2}^{(n+1,k)}\big) \\ &P'\big(s_w^{(n+1,k)}\big) M^{-1}\, Q_w^{(n+1)}. \end{aligned} \tag{9.93}$$

The spatial discretization is

$$\begin{aligned} &M\big(c_{s1}^{(n+1,k)}, c_{s2}^{(n+1,k)}\big)\, \frac{s_w^{(n+1,k+1)} - s_w^n}{\Delta t^n} \\ &+ A_s\big(s_w^{(n+1,k)}, P_w^{(n+1,k+1)}\big) f_w\big(s_w^{(n+1,k)}\big) = Q_s^{(n+1)}. \end{aligned} \tag{9.94}$$

The matrix-form of the discretized concentration equations is

$$\begin{aligned} &M\big(c_{s1}^{(n+1,k)}, c_{s2}^{(n+1,k)}\big)\left[B_m\big(s_w^{(n+1,k+1)}\big)\frac{c^{(n+1,k+1)}}{\Delta t^n} - B_m\big(s_w^n\big)\frac{c^n}{\Delta t^n}\right] \\ &+ B_w\big(s_w^{(n+1,k+1)}, P_w^{(n+1,k+1)}\big)\, c^{(n+1,k+1)} - B_D\big(s_w^{(n+1,k+1)}, P_w^{(n+1,k+1)}, \\ &c_{s1}^{(n+1,k)}, c_{s2}^{(n+1,k)}\big)\, c^{(n+1,k+1)} = R\big(s_w^{(n+1,k+1)}, P_w^{(n+1,k+1)}, \\ &c^{(n+1,k)}, c_{s1}^{(n+1,k)}\big) + Q_c^{(n+1)}. \end{aligned} \tag{9.95}$$

Here, B_m is a diagonal matrix that may have zero entries (if some cells have zero saturation); to guarantee invertibility, such zeros can be replaced by a small positive value (e.g., $\min(s_{w,1}^{(n+1,k+1)}, s_w^n) > 0$). Note that in the term R, $c_{s1}^{(n+1,k)}$ is taken from the previous iteration to maintain linearity when $\mathbf{u}_w > \mathbf{u}_r$ (it is always linear for $\mathbf{u}_w \leq \mathbf{u}_r$).

In compact form, Eq. (9.95) can be rewritten as

$$B_t\Big(s_w^{(n+1,k+1)}, P_w^{(n+1,k+1)}, c_{s1}^{(n+1,k)}, c_{s2}^{(n+1,k)}\Big)\, c^{(n+1,k+1)} = \tag{9.96}$$

$$Q_{ct}\Big(s_w^{(n+1,k+1)}, P_w^{(n+1,k+1)}, c^{(n+1,k)}, c_{s1}^{(n+1,k)}, c_{s2}^{(n+1,k)}\Big), \tag{9.97}$$

where

$$B_t = B_m\big(s_w^{(n+1,k+1)}\big) + \Delta t^n\, M\big(c_{s1}^{(n+1,k)}, c_{s2}^{(n+1,k)}\big)^{-1} \tag{9.98}$$

$$\Big[B_w\big(s_w^{(n+1,k+1)}, P_w^{(n+1,k+1)}\big) - B_D\big(s_w^{(n+1,k+1)}, P_w^{(n+1,k+1)}, c_{s1}^{(n+1,k)}, c_{s2}^{(n+1,k)}\big)\Big], \tag{9.99}$$

and

$$\begin{aligned} Q_{ct} &= \Delta t^n\, M\big(c_{s1}^{(n+1,k)}, c_{s2}^{(n+1,k)}\big)^{-1} \\ &\quad \Big[Q_c^{(n+1)} - R\big(s_w^{(n+1,k+1)}, P_w^{(n+1,k+1)}, c^{(n+1,k)}, c_{s1}^{(n+1,k)}\big)\Big] \\ &\quad + B_m\big(s_w^n\big)\, c^n. \end{aligned} \tag{9.100}$$

The CCFD discretization of the c_{s1} equation is given by

$$\frac{c_{s1}^{(n+1,k+1)} - c_{s1}^n}{\Delta t^n} = \tag{9.101}$$

$$\begin{cases} R_1\big(s_w^{(n+1,k+1)}, P_w^{(n+1,k+1)}, c^{(n+1,k+1)}\big), & \mathbf{u}_w^{(n+1,k+1)} \le \mathbf{u}_r, \\ R_1\big(s_w^{(n+1,k+1)}, P_w^{(n+1,k+1)}, c^{(n+1,k+1)}\big) & \mathbf{u}_w^{(n+1,k+1)} > \mathbf{u}_r, \\ \quad + R_2\big(s_w^{(n+1,k+1)}, P_w^{(n+1,k+1)}\big)\, c_{s1}^{(n+1,k+1)}, & \end{cases} \tag{9.102}$$

where

$$R_1\big(s_w^{(n+1,k+1)}, P_w^{(n+1,k+1)}, c^{(n+1,k+1)}\big) = \gamma_d \left|\mathbf{u}_w\big(s_w^{(n+1,k+1)}, P_w^{(n+1,k+1)}\big)\right| c^{(n+1,k+1)},$$

and

$$R_2\big(s_w^{(n+1,k+1)}, P_w^{(n+1,k+1)}\big) = \gamma_e \left|\mathbf{u}_w\big(s_w^{(n+1,k+1)}, P_w^{(n+1,k+1)}\big) - \mathbf{u}_r\right|.$$

Equation (9.102) can be rewritten as

$$c_{s1}^{(n+1,k+1)} = \begin{cases} R_{1t}\Big(s_w^{(n+1,k+1)}, P_w^{(n+1,k+1)}, c^{(n+1,k+1)}, c_{s1}^n\Big), & \mathbf{u}_w^{(n+1,k+1)} \le \mathbf{u}_r, \\ \Big[I + \Delta t^n\, R_2\big(s_w^{(n+1,k+1)}, P_w^{(n+1,k+1)}\big)\Big]^{-1} & \mathbf{u}_w^{(n+1,k+1)} > \mathbf{u}_r, \\ \quad R_{1t}\Big(s_w^{(n+1,k+1)}, P_w^{(n+1,k+1)}, c^{(n+1,k+1)}, c_{s1}^n\Big), & \end{cases} \tag{9.103}$$

with

$$\begin{aligned} & R_{1t}\Big(s_w^{(n+1,k+1)}, P_w^{(n+1,k+1)}, c^{(n+1,k+1)}, c_{s1}^n\Big) \\ &\quad = \Delta t^n\, R_1\Big(s_w^{(n+1,k+1)}, P_w^{(n+1,k+1)}, c^{(n+1,k+1)}\Big) + c_{s1}^n. \end{aligned} \tag{9.104}$$

Since R_2 is positive, the matrix $I + \Delta t^n\, R_2$ is invertible for sufficiently small Δt^n.

Finally, the matrix form of the discretized c_{s2} equation is

$$c_{s2}^{(n+1,k+1)} = R_{3t}\left(s_w^{(n+1,k+1)}, P_w^{(n+1,k+1)}, c^{(n+1,k+1)}, c_{s2}^n\right), \tag{9.105}$$

where

$$R_{3t}\left(s_w^{(n+1,k+1)}, P_w^{(n+1,k+1)}, c^{(n+1,k+1)}, c_{s2}^n\right) = \Delta t^n\, R_3\left(s_w^{(n+1,k+1)}, P_w^{(n+1,k+1)}, c^{(n+1,k+1)}\right) + c_{s2}^n, \tag{9.106}$$

$$R_3\left(s_w^{(n+1,k+1)}, P_w^{(n+1,k+1)}, c^{(n+1,k+1)}\right) = \gamma_{pt}\left|\mathbf{u}_w\left(s_w^{(n+1,k+1)}, P_w^{(n+1,k+1)}\right)\right| c^{(n+1,k+1)}. \tag{9.107}$$

This completes the description of the iterative IMPES–IMC method for nanoparticles in two-phase flow.

9.8 MULTIPLE SIZES OF NANOPARTICLES

For multiple size intervals of nanoparticles, the transport equation for each size i in phase α is

$$\phi\frac{\partial(S_\alpha C_{i\alpha})}{\partial t} + \nabla\cdot\left(\mathbf{u}_\alpha C_{i\alpha} - \phi S_\alpha \mathbf{D}_{i\alpha}\nabla C_{i\alpha}\right) = R_{i\alpha}, \tag{9.108}$$

where $R_{i\alpha}$ represents the net rate of loss for size interval i in phase α, including retention and entrainment dynamics.

Nanoparticle movement is governed by diffusion and dispersion. The total diffusion-dispersion tensor for a nanoparticle size interval i in phase α is:

$$D_{i\alpha} = D_{i\alpha}^{\text{diff}} + D_{i\alpha}^{\text{disp}} \tag{9.109}$$

Molecular diffusion is given by the Stokes-Einstein equation:

$$D_{i\alpha}^{\text{diff}} = \tau_\alpha d_{i\alpha}^{\text{mol}}\boldsymbol{I} \tag{9.110}$$

$$d_{i\alpha}^{\text{mol}} = \frac{k_B T}{3\pi\mu d_\rho} \tag{9.111}$$

where τ_α is tortuosity, k_B is the Boltzmann constant, T is temperature, μ is fluid viscosity, d_ρ is particle diameter.

Dispersion due to flow is:

$$\phi s_\alpha D_{i\alpha}^{\text{disp}} = d_{t,\alpha}|u_\alpha|\boldsymbol{I} + (d_{l,\alpha} - d_{t,\alpha})\frac{u_\alpha u_\alpha^T}{|u_\alpha|} \tag{9.112}$$

So the full tensor becomes:

$$\phi s_\alpha D_{i\alpha} = (\phi s_\alpha \tau_\alpha d_{i\alpha}^{\text{mol}} + d_{t,\alpha}|u_\alpha|)\boldsymbol{I} + (d_{l,\alpha} - d_{t,\alpha})\frac{u_\alpha u_\alpha^T}{|u_\alpha|} \tag{9.113}$$

The net loss rate of nanoparticles in interval i, phase α is:

$$R_{i\alpha} = \frac{\partial}{\partial t}(\delta\phi)_{i\alpha} \tag{9.114}$$

$$(\delta\phi)_{i\alpha} = C_{s1,i\alpha} + C_{s2,i\alpha}$$

where $C_{s1,i\alpha}$: particles attached to pore surfaces, $C_{s2,i\alpha}$: particles trapped in pore throats.

The surface deposition is modeled by:

$$\frac{\partial C_{s1,i\alpha}}{\partial t} = \begin{cases} \gamma_{d,i\alpha} u_\alpha C_{i\alpha}, & u_\alpha \le u_{\alpha c} \\ \gamma_{d,i\alpha} u_\alpha C_{i\alpha} - \gamma_{e,i\alpha}(u_\alpha - u_{\alpha c})C_{s1,i\alpha}, & u_\alpha > u_{\alpha c} \end{cases} \tag{9.115}$$

This can be simplified as:

$$\frac{\partial C_{s1,i\alpha}}{\partial t} = \gamma_{d,i\alpha} u_\alpha C_{i\alpha} - \gamma_{e,i\alpha}(u_\alpha - u_{\alpha c})C_{s1,i\alpha}, \quad \gamma_{e,i\alpha} = 0 \text{ if } u_\alpha \le u_{\alpha c} \tag{9.116}$$

The entrapment rate is:

$$\frac{\partial C_{s2,i\alpha}}{\partial t} = \gamma_{pt,i\alpha} u_\alpha C_{i\alpha} \tag{9.117}$$

So the overall loss rate becomes:

$$R_{i\alpha} = (\gamma_{d,i\alpha} + \gamma_{pt,i\alpha})u_\alpha C_{i\alpha} - \gamma_{e,i\alpha}(u_\alpha - u_{\alpha c})C_{s1,i\alpha} \tag{9.118}$$

Porosity is updated by:

$$\phi = \phi_0 - \sum_{i\alpha}(\delta\phi)_{i\alpha} \tag{9.119}$$

Permeability variation is given by:

$$K = K_0 \left[(1 - f)k_f + f\frac{\phi}{\phi_0}\right]^l \tag{9.120}$$

with flow efficiency:

$$f = 1 - \sum_i \gamma_{f,i} \left(\sum_\alpha C_{s2,i\alpha}\right) \tag{9.121}$$

where the exponent l typically ranges from 2.5 to 3.5. Retained nanoparticles may be remobilized due to hydrodynamic forces, potentially leading to further redistribution or pore blockage elsewhere in the medium.

9.8.1 MATLAB® CODE

```
    function [Sw, presW, concp, cs1, cs2, poro, permXX, permYY] =
        SplitNanoScheme()
%% 1. Discretization
nx     = 120;      ny    = 50;            % grid cells
Lx     = 1.2;      Ly    = 0.5;           % domain size [m]
nT     = 500;                            % time steps
tFinal= 0.5/0.1;                         % final time [s]
ht     = tFinal / nT;                    % time step

xs = linspace(0, Lx, nx+1)';             % x-nodes
ys = linspace(0, Ly, ny+1)';             % y-nodes
dx = xs(2)-xs(1);  dy = ys(2)-ys(1);
hxs= diff(xs);     hys= diff(ys);

%% 2. Physical & numerical parameters
poroval = 0.3;     serr = 1e-4;          % porosity, residual
    saturation
permBase = perm()*9.86923e-16;           % base permeability
visW    = 1e-3;    visN = 0.45e-3;       % viscosities
rhoW    = 1000;    rhoN = 660;           % densities
permType=2;                              % Corey exponent
sf        = 1;                           % split factor
NIT       = 1;                           % max nonlinear iterations

%% 3. Initial conditions
Sw     = serr   * ones(nx,ny);
concp  = zeros(nx,ny);
cs1    = zeros(nx,ny);
cs2    = zeros(nx,ny);

poro      = poroval * ones(nx,ny);
permXX    = permBase;
permYY    = permBase;

%% 4. Boundary conditions
% 4.1 Relative-perm for Sw
Seb       = (1-serr)/(1-2*serr);
krwEdge  = (Seb^permType)/visW;
krnEdge  = ((1-Seb)^permType)/visN;
krtEdge  = krwEdge/(krwEdge+krnEdge);
krwW      = krtEdge * ones(ny,1);  % west face

% 4.2 Water-phase flow
injRate = 0.1;
vpwW      = injRate * Lx * poroval * ones(ny,1);
pwisW     = false(ny,1); pwisW(1:5)=true;

% 5. Time-stepping loop
for tStep = 1:nT
    Sw_old = Sw; cs1_old = cs1; cs2_old = cs2;

    % 5.1 Update porosity & permeability
    gamafi = 1e-3; kf = 0.5;
    poro     = poroval - (cs1_old + cs2_old);
```

```
coeff    = 1 - gamafi*cs2_old;
coef     = (kf*(1-coeff) + coeff.*poro/poroval).^3;
permXX   = permBase .* coef;
permYY   = permXX;

% 5.2 Mobilities
Se   = (Sw_old-serr)/(1-2*serr);
krw  = Se.^permType/visW;
krn  = (1-Se).^permType/visN;
krt  = krw./(krw+krn);

% 5.3 Solve ps, pc, pw via CCFD diffusion (stubbed)
[~,  psA,  psRHS, ~, ~, ~, para] = ccfdDiffusion(poro, permXX
    .*krw, permYY.*krw, ...
    krwW, zeros(ny,1), zeros(nx,1), zeros(nx,1), ...
    zeros(ny,1), zeros(ny,1), zeros(nx,1), zeros(nx,1), ...
    false(ny,1), false(ny,1), false(nx,1), false(nx,1), ...
    Sw_old, xs, ys, tFinal, 1, struct());
[~,  pcA,  pcRHS                ] = ccfdDiffusion(poro, permXX
    .*krn, permYY.*krn, ...
    zeros(ny,1), zeros(ny,1), zeros(nx,1), zeros(nx,1), ...
    zeros(ny,1), zeros(ny,1), zeros(nx,1), zeros(nx,1), ...
    false(ny,1), true(ny,1), false(nx,1), false(nx,1), ...
    Sw_old, xs, ys, tFinal, 1, struct());
[~,  pwA,  pwRHS                ] = ccfdDiffusion(poro, permXX
    .*(krw+krn), permYY.*(krw+krn), ...
    zeros(ny,1), zeros(ny,1), zeros(nx,1), zeros(nx,1), ...
    vpwW,    zeros(ny,1), zeros(nx,1), zeros(nx,1), ...
    pwisW,   false(ny,1), false(nx,1), false(nx,1), ...
    Sw_old, xs, ys, tFinal, 1, struct());

% 5.4 Capillary pressure linearization
cap     = -log(Se).*BCFun(permXX);
dpc     = -BCFun(permXX)./Se;
Derr    = diag(sparse(ht*dpc(:)));

% 5.5 Water pressure solve
totalA= pwA - pcA*psA;
presW = totalA \ (pwRHS + pcRHS - pcA*cap(:));

% 5.6 Compute Darcy fluxes
[fluxX, fluxY] = computeFluxes(presW, para);

% 5.7 Saturation update via CCFD advection (stubbed)
[Madv, Aadv, Radv, ~] = ccfdAdvection(poro, fluxX, fluxY,
    krwW, zeros(ny,1), zeros(nx,1), zeros(nx,1), Sw_old, xs,
    ys, tFinal, 0, struct());
Sw  = Sw_old + reshape(Madv \ (ht*(Radv - Aadv*krt(:))), [nx,
    ny]);

% 5.8 Nanoparticle transport
[~, Bwa, Rwa] = ccfdAdvection(poro, fluxX, fluxY, zeros(ny,1)
    , zeros(ny,1), zeros(nx,1), zeros(nx,1), concp, xs, ys,
    tFinal, 0, struct());
[~, Bda, Rda] = ccfdDiffusion(poro, diffXX(poro,Sw_old),
    diffYY(poro,Sw_old), ...
```

```
        zeros(ny,1), zeros(ny,1), zeros(nx,1), zeros(nx,1), ...
        zeros(ny,1), zeros(ny,1), zeros(nx,1), zeros(nx,1), ...
        false(ny,1), true(ny,1), false(nx,1), false(nx,1), ...
        concp, xs, ys, tFinal, 0, struct());

    Unorm = computeNorm(fluxX, fluxY, hxs, hys);
    matR  = computeReaction(Unorm, cs1_old, concp);

    Btot = diag(poro(:).*Sw_old(:)) + ht*(Bwa + Bda - matR);
    Qtot = ht*(Rwa + Rda - matR*concp(:)) + diag(poro(:).*Sw_old
        (:))*concp(:);
    concp= reshape(Btot\Qtot, [nx,ny]);

    % 5.9 Species update (stubbed ODE)
    [cs1, cs2] = updateSpecies(concp, Unorm, cs1_old, cs2_old, ht
        );

    % 5.10 Convergence check
    if max(abs(Sw(:)-Sw_old(:)))<1e-6, break; end
end

%% 6. Visualization example
figure; contourf(xs(1:end-1), ys(1:end-1), Sw', 20, 'LineStyle','
    none');
axis image; colorbar; title('Water␣Saturation');

end

%% --- Subfunctions ---

function bc = BCFun(perm)
    bc = 50 ./ sqrt(perm);
end

function [matM, matA, rhs, conc, fluxX, fluxY, para] =
    ccfdDiffusion(varargin)
    % Stub: implement CCFD diffusion as in main code, sectioned
        and commented
    matM = []; matA = []; rhs = []; conc = []; fluxX = []; fluxY
        = []; para = struct();
end

function [matM, matA, rhs, conc] = ccfdAdvection(varargin)
    % Stub: implement CCFD advection analogously
    matM = []; matA = []; rhs = []; conc = [];
end

function [fluxX, fluxY] = computeFluxes(presW, para)
    fx = para.difxxBar(:).*((para.adjW-para.adjE)*presW + para.
        adjBW*para.cBdryW - para.adjBE*para.cBdryE);
    fy = para.difyyBar(:).*((para.adjS-para.adjN)*presW + para.
        adjBS*para.cBdryS - para.adjBN*para.cBdryN);
    fluxX = reshape(fx, [para.nx+1, para.ny])./para.hys';
    fluxY = reshape(fy, [para.nx, para.ny+1])./para.hxs;
end
```

```
function Unorm = computeNorm(fx, fy, hxs, hys)
    Ux = 0.5*(fx(1:end-1,:)+fx(2:end,:));
    Uy = 0.5*(fy(:,1:end-1)+fy(:,2:end));
    Unorm = sqrt(Ux.^2 + Uy.^2);
end

function matR = computeReaction(Unorm, cs1, concp)
    Ucrit = 4.6e-6; gamad=16; gamapt=1.28; gamae=30;
    Rval = (Unorm<=Ucrit) .* (gamad+gamapt) + (Unorm>Ucrit) .* ((
        gamad+gamapt).*Unorm - gamae.*cs1.*Unorm);
    matR = spdiags(Rval(:), 0, numel(Rval), numel(Rval));
end

function [cs1, cs2] = updateSpecies(concp, Unorm, cs1_old,
    cs2_old, ht)
    % Placeholder for reaction-ODE update
    R1 = 16; R3 = 1.28;
    cs1 = cs1_old + ht*R1.*concp;
    cs2 = cs2_old + ht*R3.*concp;
end
```

9.9 APPLICATIONS OF NPS IN POROUS MEDIA

Nanoparticle applications in porous media are increasingly significant due to their unique properties and potential to revolutionize various industrial processes, especially in enhanced oil recovery (EOR), heat transfer enhancement, carbon dioxide (CO_2) capture, and geological CO_2 sequestration.

In enhanced oil recovery, nanoparticles address critical limitations of conventional chemical agents, such as degradation under harsh reservoir conditions. Nanofluids, suspensions containing nanoparticles, offer improved stability and functional durability. Silica nanoparticles, for instance, exhibit environmental safety and structural similarity to sandstone reservoirs, significantly enhancing recovery efficiency through improved wettability and reduced interfacial tension. Nanoflooding, a cost-effective EOR method, involves direct injection of nanoparticle suspensions, facilitating substantial improvements in oil recovery. Investigations reveal the influence of nanoparticle characteristics like size, shape, and surface properties on their transport and retention within reservoir rocks, demonstrating that smaller nanoparticles with optimized surface charges enhance mobility, reduce pore throat blockage, and effectively alter rock wettability to improve oil displacement efficiency.

Nanofluids also play a vital role in heat transfer applications within porous media due to their superior thermal properties. Their enhanced thermal conductivity makes them ideal for various thermal management applications, such as solar collectors, heat exchangers, industrial cooling, microelectromechanical systems (MEMS), and biomedical engineering. Nanoparticles significantly boost thermal efficiency and contribute to more compact, energy-efficient designs.

Furthermore, nanoporous materials have shown promising capabilities in capturing carbon dioxide, a leading contributor to global warming. These materials possess a high surface area and affinity for CO_2 adsorption, vastly exceeding conventional absorbents. Metal oxide nanoparticles, particularly copper oxide, have demonstrated remarkable efficacy due to their strong electron donor-acceptor interactions with CO_2 molecules. Advanced materials such as metal-organic frameworks (MOFs) further enhance CO_2 capture efficiency due to their porous structure and large surface area, paving the way for sustainable carbon management solutions.

Lastly, the incorporation of nanoparticles in geological CO_2 sequestration introduces innovative mechanisms to enhance storage security and efficiency. Injecting nanoparticles with CO_2 into deep saline aquifers accelerates the onset of density-driven instabilities, enhancing convective mixing and improving sequestration safety. Mathematical modeling and simulation tools, such as the Implicit Pressure Explicit Saturation-Concentration (IMPESC) numerical schemes, are crucial for predicting nanoparticle-augmented CO_2 transport dynamics, thereby optimizing sequestration processes.

CONCLUSION

Nanoparticle transport in porous media exhibits rich, multi-scale physics driven by advection, diffusion, and complex retention mechanisms. Classical colloid-filtration theory, enhanced with reversible adsorption and site-blocking modifications, provides a sound foundation for modeling one-dimensional breakthrough curves. Analytical solutions offer insight into how deposition and detachment kinetics shape concentration profiles, while numerical schemes enable simulation of more realistic geometries, heterogeneous properties, and multi-phase flows. The integration of dynamic porosity and permeability feedback captures how nanoparticle accumulation alters pore structure and flow behavior, critical for forecasting long-term performance in reservoir and remediation contexts. Validation against experimental data confirms the predictive capability of these models, and extensions to include multiple particle sizes and iterative IMPES-IMC coupling further expand their applicability. As nanoparticles continue to transform enhanced oil recovery, thermal management, and carbon-capture technologies, robust transport models and efficient numerical implementations will be indispensable for design optimization and scale-up. The methodologies presented here lay a comprehensive groundwork for both fundamental research and engineering practice in nanoparticle-enhanced porous-media processes.

REVIEW QUESTIONS

1. Define the key differences between colloid-filtration theory and reversible adsorption models for nanoparticle transport in porous media.
2. Derive the three-dimensional advection–dispersion–filtration equation, explaining the physical meaning of each term.
3. Explain how site-blocking and detachment modify the classical filtration model, and write the corresponding retention rate expressions.

4. Outline the steps of a Laplace-transform solution for the one-dimensional ADE with first-order deposition.
5. Describe how dynamic porosity and permeability feedback is incorporated into two-phase flow–particle transport models.
6. Discuss the role of the relaxation factor in IMPES–IMC iterative schemes for coupled pressure–saturation–concentration problems.
7. Identify at least three emerging applications of nanofluids in porous media and the unique benefits nanoparticles provide in each case.

PROBLEMS

1. Starting from

$$\phi \frac{\partial c}{\partial t} + v \frac{\partial c}{\partial x} = D \frac{\partial^2 c}{\partial x^2} - k_{\text{dep}}\, c,$$

derive the steady-state breakthrough concentration $c(x)$ for a semi-infinite column with constant inlet concentration c_0, assuming $\partial c/\partial t = 0$ and no detachment.
2. Implement a MATLAB® script to solve the pulse-injection analytical solution

$$c(x,t) = c_i\, A(x,t) + c_0\big[B(x,t) - B(x, t - t_0)\big]$$

at three spatial locations $x = 0.1, 0.2, 0.3$ m over $0 < t < 2t_0$. Plot c/c_0 versus time and verify mass conservation.
3. Using experimental column data (e.g. silica nanoparticles in sand), fit both the irreversible-only filtration model and the site-blocking model to the observed breakthrough curve. Report the best-fit k_{dep} and $C_{s,\text{max}}$ parameters and discuss which model better captures tailing behavior.
4. Modify the provided pdepe code for single-phase transport to include a detachment term $-\frac{\rho_b}{\phi}\, k_{\text{det}}\, c_s$ in the adsorbed-concentration PDE. Run simulations with $k_{\text{det}} = 10^{-4}$ /s and compare concentration profiles at 50 h and 200 h with the no-detachment case.
5. For a simple layered medium with two horizontal layers of equal thickness but different permeabilities (K_1, K_2), analyze how nanoparticle retention in the water phase alters the mobility ratio and slug-bank stability in a water-flood EOR scenario. Assume linear capillary pressure, Corey permeabilities, and first-order deposition. Compute the change in shock-front speed due to nanoparticle injection.

DECLARATION

During the preparation of this work, the author used ChatGPT in order to improve the language and readability. After using this tool/service, the author reviewed and edited the content as needed and takes full responsibility for the content of the publication.

10 Underground Hydrogen Storage in Depleted Reservoirs

10.1 INTRODUCTION

Hydrogen is the lightest element, and while abundant on Earth, it is usually found in compounds like water or methane, necessitating the extraction of pure hydrogen. Known as a cleaner alternative to methane, hydrogen's versatility stems from its various production sources, marking it as an essential player in the future energy landscape. Hydrogen boasts the highest energy per unit mass among fuels, outstripping hydrocarbons in efficiency, particularly when utilized in fuel cells that can exceed 60% efficiency, as opposed to the 25%-30% range for internal combustion engines. However, hydrocarbons hold the upper hand in volumetric energy density and current infrastructure suitability, with their liquid state at room temperature allowing for more straightforward storage and transportation than hydrogen, which requires compression or liquefaction. While hydrogen combustion yields only water, making it a cleaner option with no carbon emissions, the challenge lies in its production, which is energy-intensive and often dependent on hydrocarbon energy sources. The convenience of hydrocarbons as a readily available energy source contrasts with hydrogen's potential for a greener footprint and higher weight-specific energy yield, positioning the latter as an appealing candidate for sustainable energy systems, provided advancements in technology and infrastructure continue to evolve. Hydrogen color coded classification—gray, blue, and green—helps delineate its source, distinguishing fossil-based hydrogen from renewable hydrogen (Figure 10.1). Grey hydrogen is derived from fossil fuels via thermal processes like steam-methane reformation, emitting greenhouse gases. Blue hydrogen, produced similarly, employs carbon capture and sequestration (CCS) to reduce emissions. Green hydrogen is produced from water electrolysis powered by renewable energy sources like solar and wind. Green hydrogen presents a sustainable and cost-effective solution for storing excess clean energy from renewable sources. Additionally, pink hydrogen, produced similarly to green, utilizes nuclear energy for electrolysis, leveraging nuclear reactors' high temperatures for efficient electrolysis or alternative hydrogen synthesis methods. Yellow hydrogen, produced through solar-powered electrolysis, and biomass-derived hydrogen are other notable types, each showcasing diverse production methodologies.

DOI: 10.1201/9781032724973-10

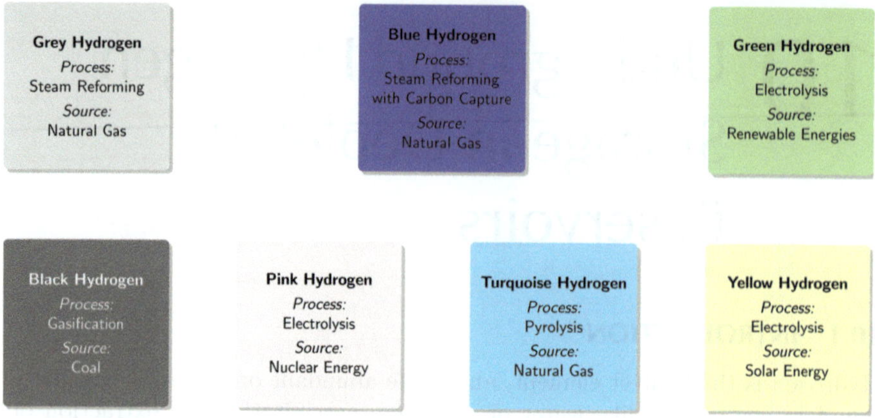

Figure 10.1 Hydrogen production classification.

Hydrogen has been gaining prominence as a clean energy source, with the potential to play a pivotal role in global decarbonization efforts. It can serve as an intermediary for energy transition, being producible from renewable sources and re-electrifiable to provide electricity and heat, besides being storable for future use. Various methods for hydrogen production from renewable sources have been explored, including water splitting and biomass mechanisms, each having its own set of limitations [253].

Hydrogen storage methods encompass gas compression under high pressures, cryogenic liquid storage, and physical or chemical absorption in solid or liquid materials. Each technique has specific requirements, with compressed gas storage necessitating tanks capable of withstanding pressures between 350 and 700 bar and liquid hydrogen requiring cryogenic temperatures of -252.8°C along with insulated storage facilities to prevent evaporation. Material-based storage, such as metal hydrides, offers a promising avenue for hydrogen storage, leveraging chemical reactions to achieve high hydrogen absorption capacities. One of the most promising methods for hydrogen storage is subsurface (underground) hydrogen storage (Figure 10.2). During the process of underground hydrogen storage in depleted gas reservoirs, rock-fluid interactions occur in multiple dimensions, primarily through geochemical interactions, termed abiotic interactions, and microbial activities, referred to as biotic reactions. These interactions are significantly influenced by various physicochemical processes unfolding within the reservoirs amidst hydrogen storage, encompassing alterations in wettability, interfacial tension, diffusion, adsorption, and solubility. In the abiotic reactions, hydrogen engages with rock minerals, instigating mineral dissolution and precipitation primarily within the reservoir's reactive zone. These interactions predominantly manifest under elevated pressures and temperatures during hydrogen storage. Conversely, biotic reactions are recognized as a principal contributor to hydrogen loss

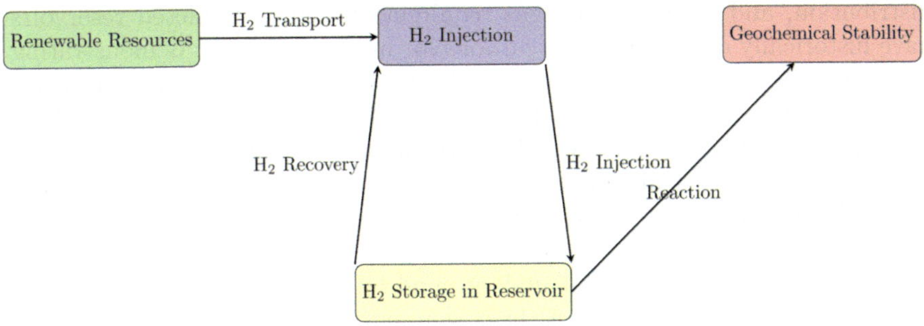

Figure 10.2 Depleted reservoir and its transformation into a hydrogen storage facility.

throughout the storage phase, as they transpire under typical reservoir conditions (below 130°C temperature and 35 MPa pressure). An instance of such a reaction is the sulfate-reducing reaction, which engenders an acidic milieu and sulfide precipitation near the wellbore, thereby inducing material corrosion and permeability diminution.

Factors such as wettability, interfacial tension, adsorption, diffusivity, and solubility are instrumental in determining rock-fluid interactions in hydrogen storage. For instance, hydrogen exhibits noticeably lower wettability in reservoir rocks than other gases do. In reservoirs with expansive surface areas, like coal beds, the adsorption and desorption properties of hydrogen are paramount and vary significantly based on pressure changes and the type of cementing material present. Although hydrogen has limited solubility in brine, its buoyancy and diffusivity surpass those of many other gases. The dynamics of these rock-fluid interactions play an essential role in safety considerations for hydrogen storage, especially concerning the integrity of caprock and wellbores. The prospect of using depleted hydrocarbon reservoirs for hydrogen storage remains largely untapped, presenting an opportunity to address existing constraints. However, transitioning to large scale storage in these reservoirs introduces fresh challenges that demand innovative solutions. This emphasizes the importance of advanced modeling techniques for underground hydrogen storage, not only to harness its distinct attributes but also to navigate its specific challenges, setting the stage for a greener energy landscape. In parallel, integrating artificial intelligence (AI) into the modeling and simulation of hydrogen behavior within these geological formations offers a promising approach. This innovation seeks to mitigate some of these hurdles by enhancing the precision and speed of predictive models, potentially propelling the development of hydrogen storage aligned with global sustainability objectives.

The advantages of UHS in depleted reservoirs can be summarized as the established geological integrity, pre-existing infrastructure, wide geographical distribution, and lower cushion gas requirements [362]. Depleted reservoirs have historically contained hydrocarbons for millions of years, demonstrating their capability for secure gas storage. Existing pipelines and wells can be repurposed for hydrogen injection, reducing development costs and time. These reservoirs are abundant and widely distributed, providing access to diverse regions for storage. Unlike other storage media, depleted reservoirs require minimal cushion-gas due to the residual hydrocarbons, enhancing operational efficiency.

One significant challenge for underground hydrogen storage (UHS) in depleted reservoirs is maintaining hydrogen purity. Residual hydrocarbons present in these reservoirs can mix with stored hydrogen, reducing its purity and requiring the development of advanced gas processing infrastructure to ensure usability. Additionally, hydrogen reactivity poses another major concern. It can react with reservoir minerals, such as pyrite, or interact with microbial communities, leading to hydrogen loss and potentially destabilizing the reservoir over time. Moreover, infrastructure upgrades are essential for safely and effectively storing hydrogen. Existing pipelines and well casings may need to be replaced or reinforced with materials specifically designed to resist hydrogen embrittlement, ensuring long-term integrity and safety.

Several pilot projects and studies are addressing these challenges and advancing the field of UHS [362]. The HyStorPor initiative in the UK is focused on understanding hydrogen purity issues and the interactions between hydrogen and reservoir materials. Austria's Underground Sun Storage project has explored storing mixtures of hydrogen and natural gas, aiming to achieve pure hydrogen storage by 2030. Similarly, the Hychico project in Argentina investigates integrating hydrogen storage with renewable energy sources, demonstrating the feasibility of UHS in depleted fields. These initiatives pave the way for the broader adoption of hydrogen storage in geological formations.

10.2 HAZARD OF HYDROGEN LEAKAGE

The safety concerns surrounding hydrogen, especially its storage and transmission, stem from its physical properties, which make it susceptible to ignition and explosion under certain conditions. These concerns are accentuated by historical events, notably the 1937 Hindenburg disaster, which has irretrievably linked hydrogen with extreme danger in the public's mind. The Hindenburg Syndrome, named after the disaster, casts a long shadow over hydrogen's image as a safe fuel. The accident showcased hydrogen's flammability when it ignited and caused a fatal explosion on board the airship Hindenburg. This incident significantly contributed to the perception of hydrogen as highly dangerous, a perception that persists to some extent today. Leaks of hydrogen are hazardous because they can catch fire if they come in contact with air [275, 276, 277, 278, 279, 280, 281, 282, 283, 284, 285, 286].

In terms of storage, hydrogen presents a distinct set of challenges compared to other gases. Its low volumetric energy density necessitates innovative storage solutions to make its use viable for various applications. Additionally, the knowledge regarding the consequences of storing hydrogen in different geological media is still in its infancy, signifying a need for further research and understanding. The hazards associated with hydrogen storage, such as accidental releases leading to fires or explosions, underscore the importance of robust safety measures and risk analysis to ensure its safe handling and usage in energy systems [300, 310, 327, 330, 334, 333].

10.3 PROPERTIES OF HYDROGEN

The unique physical properties (atomic number 1, atomic mass 1.008) and combustion characteristics of hydrogen distinguish it from other fuels. Hydrogen leaks are particularly challenging for human senses to detect as they are colorless, odorless, and tasteless. When burned in a clean atmosphere, hydrogen produces an invisible flame. It is also prone to leakage, and hydrogen fires can potentially escalate into explosions. However, one of hydrogen's key safety advantages is its high buoyancy compared to other gases, enabling it to rapidly disperse from an incident or accident scene and dilute to safe concentration levels in the air. Hydrogen's hazards stem from its unique properties, such as its atomic and molecular structure, influencing storage (e.g., ortho-para hydrogen conversion) and liquefaction risks. Its high buoyancy enables rapid dispersion in open environments. Key parameters include vapor density, diffusivity, viscosity, thermal conductivity, specific heat, and specific heat ratio. Flammability and ignition risks include detonability limits of hydrogen–air and hydrogen–oxygen mixtures, flame visibility, auto-ignition temperature, effects of diluents and inhibitors on flammability range, adiabatic flame temperature, thermal radiation, and quenching/blow-off limits. These characteristics guide safe handling and fire suppression practices.

Gaseous hydrogen (GH_2) is 14 times lighter than air and can cause embrittlement of materials. Liquid hydrogen (LH_2), a cryogenic fluid, causes frostbite or hypothermia upon contact and expands by a ratio of 1:848 when vaporized. LH_2 is stored in vacuum-insulated containers with pressure relief devices (PRDs), eliminating the risk of boiling liquid expanding vapor explosion (BLEVE).

Figure 10.3 shows the hydrogen phase diagram illustrating its transitions among solid, liquid, and gaseous states under varying pressures and temperatures. It includes the triple point, critical point, and behaviors such as boiling, melting, and sublimation. These properties are crucial for designing safe storage and handling systems.

Hydrogen exists in three primary forms, including gaseous, liquid, and slush. Liquid hydrogen (LH_2) is a clear liquid with a faint blue tint, while slush hydrogen is a mixture of solid and liquid phases at the triple point. Phase transitions between hydrogen's gaseous, liquid, and solid states are heavily influenced by its low temperature properties. LH_2 has a density of 70.78 kg/m^3, making it approximately 14 times less dense than water, with

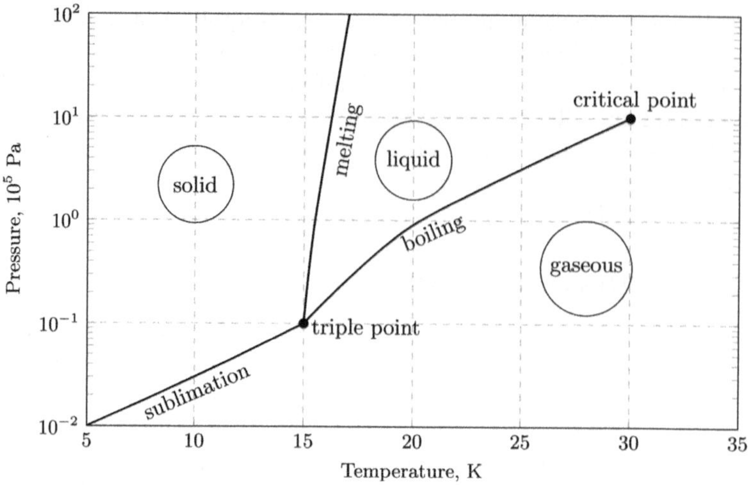

Figure 10.3 Phase diagram of hydrogen.

a specific gravity of 0.071 compared to 1 for water. In the event of an LH_2 spill or when gas below 193 K is vented, the higher density of saturated hydrogen vapor at low temperatures may cause the hydrogen cloud to spread horizontally or even flow downward. This behavior is a critical consideration for first responders managing accident scenes [328].

10.3.1 PHYSICAL AND CHEMICAL CHARACTERISTICS

Gaseous hydrogen has a density of 0.0838 kg/m^3 at normal temperature and pressure (NTP), making it over 14 times less dense than air (1.205 kg/m^3) under the same conditions. The specific gravities of hydrogen and air are 0.07 and 1.0, respectively, as shown in Figure 10.4. This makes hydrogen significantly lighter than air, causing it to rise and disperse rapidly in open environments [328].

Compared to hydrogen, other fuels like propane and petrol vapor are denser than air. At the same time, methane (natural gas) is approximately twice as light as air but still about 8 times heavier than hydrogen. The low vapor density of hydrogen gives it the highest buoyancy among gases on Earth, which is a key safety feature. In the event of hydrogen release, the gas rises and disperses quickly, reducing the risk of large combustible clouds, especially in open or partially confined spaces without significant accumulation.

In contrast, heavier hydrocarbon-based fuels like propane and petrol vapor can form extensive combustible clouds, contributing to catastrophic explosions, such as those at Flixborough [291] and Buncefield [264]. Hydrogen, on the other hand, is positively buoyant above 22 K, covering nearly the entire temperature range of its gaseous state [305].

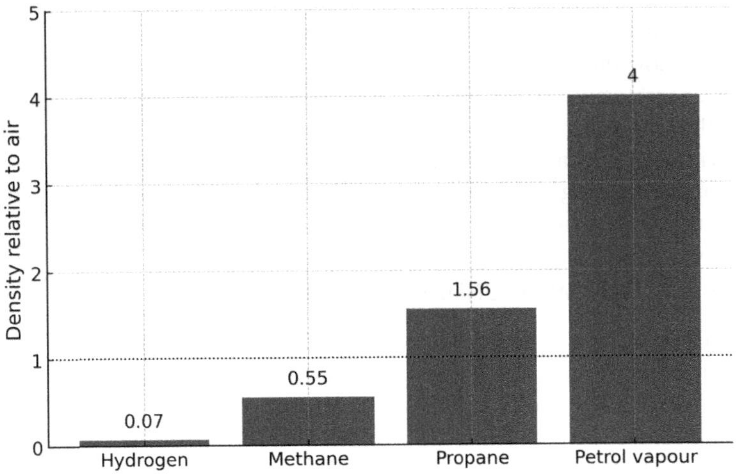

Figure 10.4 Densities relative to air for hydrogen and other common fuels. At normal temperature and pressure (NTP), hydrogen has a density of 0.0838 kg/m^3, which is approximately 14 times lower than the density of air (1.205 kg/m^3).

10.3.1.1 Flammability and Ignition Properties

As noted in [328], hydrogen has unique safety considerations that distinguish it from other fuels. Reports such as [306, 329] provide comprehensive guidelines for hydrogen safety systems. Hydrogen's flammability and ignition characteristics are critical for safety analysis, as detailed below. Hydrogen has an auto-ignition temperature of approximately 510°C in air, higher than hydrocarbons like propane and gasoline. Objects at 500–580°C can ignite hydrogen-air mixtures at atmospheric pressure, while hot air jet ignition occurs at 670°C. Hydrogen requires extremely low energy to ignite, with sparks from static electricity (e.g., human body discharge) being sufficient. Hydrogen's flammability and detonability ranges are wider than other fuels, enabling easier ignition and broader explosion conditions. Hydrogen has a wide flammability range of 4.1–75 vol.%, allowing it to ignite easily under varying conditions. Its detonability range is 18.3–59 vol.%, indicating its potential for explosive reactions within this concentration range. The stoichiometric mixture for hydrogen in air is 29.59 vol.%, representing the optimal ratio for complete combustion [328]. Hydrogen burns with a pale-blue flame, emitting no visible light in daylight nor producing smoke (pure water is the byproduct). Its maximum flame temperature is 2,400 K, capable of causing severe burns. Hydrogen flames emit less radiant heat compared to hydrocarbons, necessitating advanced sensors for detection and heat flux assessment. Hydrogen is a colorless, non-toxic gas that is 14 times lighter than air. It has an energy content 2.8 times higher than petrol by weight but 4 times lower by volume. These characteristics highlight its high energy efficiency per unit mass while emphasizing the challenges of its low energy density per unit volume. Hydrogen is easily ignited by weak energy sources, requiring strict control of ignition sources. Wide flammability and detonability ranges make hydrogen both versatile and hazardous. Hydrogen's pale blue flame is less detectable, necessitating advanced sensors for monitoring. The high buoyancy and rapid dispersion of hydrogen provide safety advantages by reducing combustible cloud formation.

10.3.2 REACTIVITY WITH UNDERGROUND SUBSTANCES

Hydrogen's reactivity with geological and chemical substances plays a crucial role in underground storage systems. Its interactions with hydrocarbons, brine, and rock formations significantly impact storage integrity, hydrogen purity, and overall system safety. In geological formations, hydrogen interacts with residual hydrocarbons through hydrogenation and methanation reactions. These chemical processes can alter the stored hydrogen's composition and reduce its purity. These include hydrogenation reactions with unsaturated hydrocarbons, such as ethene (C_2H_4), and reactions with hydrogen to form saturated hydrocarbons, like ethane (C_2H_6). Also, methanation reactions, such that hydrogen reacts with carbon dioxide (CO_2) or carbon monoxide (CO) in the presence of catalytic surfaces, such as iron or nickel, produce methane (CH_4). These reactions are exothermic and can generate localized heat, potentially destabilizing storage conditions [273]. Saline aquifers commonly contain brine, which interacts with hydrogen in several ways, such as the dissolution of hydrogen, where it dissolves into the brine, reducing the amount of recoverable hydrogen. Moreover, electrochemical reactions in which the dissolved hydrogen reacts with salts in the brine could potentially cause corrosion of well casings or storage equipment. Additionally, bubble formation and migration in brine can affect pressure dynamics and storage efficiency [306, 329].

Hydrogen interacts chemically and physically with the minerals in rock formations, leading to potential changes in reservoir properties. Hydrogen diffuses into metallic or mineral structures, causing weakening through embrittlement. Rock formations containing metal oxides or sulfides can catalyze hydrogen-related reactions. Hydrogen molecules adsorb onto rock surfaces and diffuse into micropores, altering permeability and storage capacity. These interactions can compromise the geological integrity of storage sites and increase the risk of hydrogen leakage [328, 309, 351].

10.4 NUMERICAL MODELING DEVELOPMENT

The study by Feldmann et al. [290] explores the hydrodynamics of hydrogen and its interaction with residual fluids in the reservoir, which is mathematically described using a compositional two-phase flow model. Through a numerical example in a depleted gas reservoir, the authors examined a scenario where hydrogen storage is charged for 5 years, followed by 5 years of seasonal cyclic operation to estimate injection and production rates, pressure response, and the composition of the produced gas stream. Hagemann et al. [300] presented a mathematical model to elucidate the hydrodynamic effects in underground hydrogen storage, focusing on the convective and diffusive fluxes of four chemical components across two mobile phases, utilizing the open-source code DuMux for numerical implementation. Their simulation reveals notable differences in gas injections into a reservoir based on the injection rate. A GPSFLOW (General Purpose Subsurface Flow Simulator) simulation was developed to address the grid-scale underground hydrogen and gas mixture (e.g., H2-CO2-CH4-N2) storage in various geological formations such as caverns, deep saline aquifers, and depleted gas fields. This development aimed to fill the gap in numerical simulation tools dedicated to designing and optimizing underground hydrogen storage technology at the grid scale. Ganzer et al. [295] embarked on a comprehensive exploration of hydrogen's impact on fundamental reservoir properties and mineralogy, both experimentally and through numerical simulations. Utilizing materials and facilities supplied by industrial partners, the project was meticulously structured to resonate with real-world reservoir conditions. They found that simulating the geochemical reactions ensuing from gas mixture interactions with rock minerals, organic substances, and formation water. In addition to investigating the interplay between bio-dynamic processes and reactive transport in porous media. Lubon and Tarkowski [324] leveraged numerical modeling, particularly employing the PetraSim program with a TOUGH2 deposit simulator, to assess the feasibility of seasonal (cyclic) hydrogen storage in a deep aquifer located in the porous rocks of a well-identified geological structure, Suliszewo. The findings from the modeling exercise were encouraging, suggesting that underground hydrogen storage in a deep aquifer could be pursued with reasonable gas recovery parameters, presenting a viable scenario for hydrogen storage. The potential of UHS in saline aquifers as a viable solution for large-scale hydrogen energy storage, distinct due to hydrogen's unique thermodynamic and physical properties, has been studied numerically by Zhao et al. [361]. By developing a two-phase, three-component reservoir simulator under the Delft Advanced Reservoir Simulation (DARSim) framework and employing the GERG-2008 equation of state (EoS) for fluid mixture properties, the researchers delve into the impact of different cushion gases (CO2, CH4, and N2) on UHS. A study by Wang et al. [356] focused on the flow behavior of hydrogen storage in subsurface porous media, particularly when carbon dioxide (CO2) is used as a cushion gas. The research examines the interactions among various factors, such as permeability heterogeneity, viscous instability, and the balance of viscous and gravity forces, with a special focus on the impact of CO2 solubility in water on H2 purity. Through detailed 2D vertical cross-sectional simulations at the decameter scale, the study unveils that CO2 solubility could have varying impacts on H2 recovery performance depending on the flow regimes in the system.

Boon and Hajibeygi [262] conducted core-scale experiments using a medical X-ray CT scanner to investigate hydrogen transport properties in porous Berea sandstone for underground hydrogen storage (UHS). Their study measured key properties like capillary pressure and relative permeability, revealing complex displacement patterns, including gravity segregation, capillary barrier effects, and significant hydrogen trapping during imbibition. These findings enhance understanding of hydrogen/water flow behavior in heterogeneous geological reservoirs, critical for optimizing UHS.

Ghaed et al. [298] focused on mixing dynamics and factors affecting hydrogen recovery factors during cyclic hydrogen storage in such reservoirs.

Albadan et al. [255] focused on numerical simulations of underground hydrogen storage in depleted gas reservoirs and saline aquifers, analyzing operational parameters like well count and bottom-hole pressure limits to optimize hydrogen storage capacity, deliverability, and purity. Doan et al. [271] presented an analytical framework for estimating hydrogen storage capacity in depleted gas reservoirs using a material balance approach, integrating reservoir engineering and thermodynamics, but does not provide a review of mathematical modeling and numerical simulation.

10.5 HYDROGEN INJECTION IN POROUS MEDIA

Hydrogen, as a light and highly diffusive gas, plays a critical role in various engineering applications, including energy systems, combustion processes, and environmental studies. Modeling its behavior in a porous medium under compressible flow conditions is essential for understanding scenarios where pressure, temperature, and density variations significantly influence the transport and dispersion mechanisms. This document presents the governing equations for compressible hydrogen flow, derived from conservation laws for mass, momentum, and energy. The equations incorporate critical physical properties of hydrogen, such as its low density and high buoyancy compared to air, and are tailored for scenarios where Darcy's law governs the flow in porous media. The goal is to provide a comprehensive framework for simulating hydrogen dispersion in practical applications while considering compressibility effects.

To model the dispersion of hydrogen in a gas mixture (air and hydrogen) through a porous medium under compressible conditions, the governing equations must account for variations in density ρ due to pressure and temperature changes. The compressible continuity equation is:

$$\frac{\partial \rho}{\partial t} + \nabla \cdot (\rho \mathbf{u}) = 0. \tag{10.1}$$

where ρ is the fluid density, and $\mathbf{u}$ is the velocity vector.

In a porous medium, the momentum equation is based on Darcy's law, which can be written as:

$$\mathbf{u} = -\frac{K}{\mu}\nabla p + \frac{K\,\beta(C)}{\mu}\rho\mathbf{g}, \tag{10.2}$$

where $\beta(C)$ remains the buoyancy coefficient, dependent on hydrogen concentration C.

The energy equation accounts for heat transfer and compressibility effects:

$$\frac{\partial(\rho e)}{\partial t} + \nabla \cdot (\rho e\mathbf{u}) = -p\nabla \cdot \mathbf{u} + \nabla \cdot (k\nabla T), \tag{10.3}$$

where e is the specific internal energy, k is the thermal conductivity, and T is the temperature.

Hydrogen concentration C evolves according to the compressible advection-diffusion equation:

$$\frac{\partial(\rho C)}{\partial t} + \nabla \cdot (\rho C\mathbf{u}) = \nabla \cdot (D\nabla C) + S, \tag{10.4}$$

where D is the diffusion coefficient, and S is the source term for hydrogen (e.g., leakage or injection).

The density ρ is determined using the ideal gas law:

$$\rho = \frac{p}{RT}, \tag{10.5}$$

where R is the specific gas constant for the mixture, computed as:

$$R = \frac{R_{\text{hydrogen}}C + R_{\text{air}}(1 - C)}{M}, \tag{10.6}$$

with M as the molar mass of the gas mixture, p is the pressure, and T is the temperature.

For compressible flow, properties like viscosity and buoyancy coefficient remain temperature-dependent. Key parameters include dynamic viscosity

$$\mu(T) = \mu_0 \left(\frac{T}{T_0}\right)^{0.7}, \tag{10.7}$$

where $\mu_0 = 8.76 \times 10^{-6}$ Pa·s at $T_0 = 300\,\text{K}$.

Thermal conductivity takes the form

$$k(T) = k_0 \left(\frac{T}{T_0}\right)^{0.7}, \tag{10.8}$$

where $k_0 \approx 0.186\,\text{W/m·K}$ at $T_0 = 300\,\text{K}$.

Table 10.1
Physical properties of hydrogen at standard temperature and pressure

Property	Value	Notes
Dynamic viscosity (μ)	8.76×10^{-6} Pa·s	At 300 K
Thermal conductivity (k)	0.186 W/m·K	At 300 K
Buoyancy coefficient ($\beta(C)$)	$0.926 \cdot C$	Hydrogen-air mixtures

The buoyancy coefficient $\beta(C)$ accounts for density differences between hydrogen and air:

$$\beta(C) = \frac{\rho_{\text{air}} - \rho_{\text{hydrogen}}}{\rho_{\text{air}}} \cdot C \approx 0.926 \cdot C, \tag{10.9}$$

where $\rho_{\text{air}} \approx 1.225\,\text{kg/m}^3$ (air density), $\rho_{\text{hydrogen}} \approx 0.08988\,\text{kg/m}^3$ (hydrogen density). For a pure hydrogen domain, $\beta(C) = 1$.

Boundary conditions for compressible flow typically involve inflow pressure p_{in} and temperature T_{in}. Hydrogen concentration at inflow ($C = C_{\text{in}}$). No-Slip condition $\mathbf{u} = 0$ on impermeable walls. The outflow condition, which is zero normal gradients for p, T, and C at the outflow.

Initial conditions are set for pressure, temperature, and hydrogen concentration:

$$p(\mathbf{x}, 0) = p_0(\mathbf{x}), \quad T(\mathbf{x}, 0) = T_0(\mathbf{x}), \quad C(\mathbf{x}, 0) = C_0(\mathbf{x}), \tag{10.10}$$

for $\mathbf{x} \in \Omega$.

10.5.1 MIXED FINITE ELEMENT FRAMEWORK

Mixed finite element approach allows simultaneous approximation of interdependent variables such as velocity ($\mathbf{u}$), pressure (p), temperature (T), and concentration (C) while maintaining compatibility across discretized equations. The mixed formulation is particularly well-suited for compressible flows in porous media, as it ensures accurate enforcement of conservation laws (mass, momentum, energy, and species transport). In addition to the strong coupling between variables such as density (ρ), pressure, and velocity. It provides compatibility with boundary conditions across different fields.

The mixed finite element method couples velocity, pressure, and scalar fields, ensuring consistent numerical behavior. The choice of compatible spaces (e.g., $\mathbf{H}(\text{div}; \Omega)$ for velocity and $L^2(\Omega)$ for pressure) satisfies the LBB condition, enhancing numerical stability. The approach accommodates nonlinearities from compressibility and buoyancy effects, making it suitable for complex multi-physics problems.

The velocity field is defined in the space $\mathbf{V}_h \subset \mathbf{H}(\text{div}; \Omega)$, which ensures divergence constraints are respected and normal continuity across element boundaries. The pressure is approximated in $P_h \subset L^2(\Omega)$, allowing discontinuous or continuous finite element approximations depending on the problem requirements. Hydrogen concentration (C) and temperature (T) are defined in $Q_h \subset H^1(\Omega)$, which provides continuity and smoothness across element boundaries.

The velocity and pressure spaces are chosen to satisfy the Ladyzhenskaya–Babuka-Brezzi (LBB) condition (see Chapter 3 for more details). This condition guarantees stability by avoiding spurious pressure modes and ensures proper coupling between velocity and pressure fields. The scalar fields (C and T) are approximated in compatible spaces that ensure continuity and accuracy across the domain.

The weak form of the continuity equation, which enforces mass conservation, is:

$$\int_\Omega \frac{\partial \rho}{\partial t} w \, d\Omega - \int_\Omega (\rho \mathbf{u}) \cdot \nabla w \, d\Omega = 0, \tag{10.11}$$

where $w \in P_h$ is a test function for pressure, ρ is the density, and $\mathbf{u}$ is the velocity field.

This formulation ensures compatibility between the density and velocity fields in the mixed finite element framework.

The momentum equation (Darcy's law) relates the velocity field $\mathbf{u}$ to the pressure gradient and buoyancy forces. Its weak form is:

$$\int_\Omega \mathbf{u} \cdot \mathbf{v} \, d\Omega + \int_\Omega \frac{K}{\mu} \nabla p \cdot \mathbf{v} \, d\Omega = \int_\Omega \frac{K\beta(C)}{\mu} \rho \mathbf{g} \cdot \mathbf{v} \, d\Omega, \tag{10.12}$$

where $\mathbf{v} \in \mathbf{V}_h$ is a test function for velocity.

This formulation directly couples pressure and velocity in the porous medium.

The energy equation, accounting for compressibility and heat transfer, is expressed in weak form as:

$$\int_\Omega \frac{\partial(\rho e)}{\partial t} v \, d\Omega - \int_\Omega (\rho e \mathbf{u}) \cdot \nabla v \, d\Omega = -\int_\Omega p(\nabla \cdot \mathbf{u}) v \, d\Omega - \int_\Omega k \nabla T \cdot \nabla v \, d\Omega, \tag{10.13}$$

where $v \in Q_h$ is a test function for temperature.

This weak form provides a natural way to solve for temperature T in the energy equation, incorporating its coupling with velocity and pressure.

The weak form of the species transport equation, describing hydrogen dispersion, is:

$$\int_\Omega \frac{\partial(\rho C)}{\partial t} \phi \, d\Omega - \int_\Omega (\rho C \mathbf{u}) \cdot \nabla \phi \, d\Omega + \int_\Omega D \nabla C \cdot \nabla \phi \, d\Omega = \int_\Omega S \phi \, d\Omega, \tag{10.14}$$

where $\phi \in Q_h$ is a test function for concentration.

This equation couples the concentration field with velocity and density, ensuring species conservation in the porous medium. The equation of state provides a pointwise relationship, which is solved algebraically and used to update ρ in the other equations.

10.6 GOVERNING EQUATIONS

The mathematical framework for modeling hydrogen (H_2) storage in depleted hydrocarbon reservoirs is based on the principles of multiphase flow and energy conservation. This section presents the primary equations governing mass, momentum, and energy within the porous medium, which are essential for understanding the behavior of hydrogen injection, withdrawal, and storage.

10.6.1 MULTIPHASE MASS BALANCE

Mass conservation for a component k (e.g., H_2, water, or residual hydrocarbons) in the liquid (l) and gas (g) phases is expressed as:

$$\frac{\partial}{\partial t}\left(\phi s_l \rho_l x_l^k + \phi s_g \rho_g x_g^k\right) - \nabla \cdot \left(\rho_l x_l^k f_l + \rho_g x_g^k f_g\right) = q^k, \tag{10.15}$$

where ϕ is the porosity, s_l and s_g are the phase saturations, and ρ_l and ρ_g are the phase densities. The terms x_l^k and x_g^k represent the mass fractions of component k in the liquid and gas phases, while f_l and f_g are the phase fluxes governed by Darcy's law. The term q^k accounts for sources or sinks, such as injection or withdrawal of component k.

10.6.2 MULTIPHASE DARCY'S LAW

The fluxes f_l and f_g appearing in the mass balance equation are described by the generalized multiphase form of Darcy's law:

$$f_\beta = -\frac{k k_{r,\beta}}{\mu_\beta}\nabla\varphi_\beta, \quad \nabla\varphi_\beta = \nabla p_\beta - \rho_\beta \mathbf{g}, \tag{10.16}$$

where k is the intrinsic permeability of the porous medium, and $k_{r,\beta}$ is the relative permeability for phase $\beta \in \{l, g\}$. The viscosity of phase β is denoted by μ_β, p_β is the pressure, and ρ_β represents the density. Gravitational effects are captured by the term $\rho_\beta \mathbf{g}$. For hydrogen storage, the low viscosity and density of H_2 can significantly impact the mobility ratio, influencing the efficiency of the injection and withdrawal processes.

10.6.3 ENERGY (HEAT) BALANCE

Thermal effects, such as changes in gas density or viscosity with temperature, play a critical role in the performance of H_2 storage. The energy balance equation is given by:

$$\frac{\partial}{\partial t}\left[(1-\phi)\rho_r u_r(\theta) + \phi(s_l \rho_l u_l + s_g \rho_g u_g)\right] - \nabla \cdot \left(-\lambda_t \nabla\theta + h_l f_l + h_g f_g\right) = q^\theta, \tag{10.17}$$

where ρ_r is the rock density, and $u_r(\theta)$ represents the rock's internal energy, which is temperature-dependent. The specific internal energies of the liquid and gas phases are denoted by u_l and u_g, respectively. The term λ_t is the effective thermal conductivity, while h_l and h_g are the specific enthalpies of the liquid and gas phases. The term q^θ represents volumetric heat sources or sinks, which may arise due to temperature differences in the injected fluid or exothermic/endothermic reactions.

The internal energies (u_l, u_g) and enthalpies (h_l, h_g) that appear in the energy equation are intrinsically related to temperature through the thermodynamic properties of the fluid phases. For gases, these properties are derived from the equation of state (EOS), which provides a functional relationship between pressure (P), temperature (T), and density (ρ). The specific internal energy (u) can be expressed as a temperature-dependent quantity derived from the EOS using the thermodynamic relation:

$$u = \int c_v(T)\, dT + u_{\text{ref}}, \tag{10.18}$$

where $c_v(T)$ is the specific heat capacity at constant volume, and u_{ref} is the reference internal energy at a specific temperature. Similarly, the specific enthalpy (h) is linked to internal energy and pressure by:

$$h = u + \frac{p}{\rho}, \tag{10.19}$$

providing a direct connection to the EOS. This linkage ensures that the energy equation accounts for temperature-dependent variations in fluid properties, such as phase densities and heat capacities, as governed by the EOS.

10.6.4 HYDROGEN DIFFUSIVITY

The diffusivity of hydrogen (D_{H_2}) varies significantly, where it increases with rising temperature and decreases with increasing pressure. This behavior is attributed to the influence of temperature on molecular momentum, Brownian motion, and intermolecular forces, which collectively affect diffusivity [267]. Furthermore, the diffusivity of H_2 in hydrocarbons decreases with increasing carbon number. While it would be insightful to relate hydrogen diffusivity to viscosity using the Einstein relationship:

$$D = \frac{k_B T}{6\pi\mu r}, \tag{10.20}$$

where D is the diffusion coefficient, μ is the viscosity, k_B is Boltzmann's constant, T is the temperature, and r is the particle radius, such an analysis is limited by the lack of data reporting H_2 viscosity and diffusivity under identical thermophysical conditions.

These observations indicate that H_2 diffusion could lead to significant losses in deep aquifers and depleted hydrocarbon reservoirs. Numerical reservoir simulations have estimated that approximately 1% of H_2 is lost after 15 years in a 7 m high reservoir with 20% porosity due to diffusion into the surrounding aquifer [267]. However, experimental data on H_2 diffusion coefficients in both aqueous and non-aqueous liquids under reservoir conditions remain scarce and warrant further investigation [334].

10.7 EQUATION OF STATE

Equations of state (EOS) are essential for modeling the thermodynamic behavior of gas mixtures, including hydrogen. They relate pressure (p), temperature (T), and density (ρ) of the gas, while also providing a foundation to derive thermodynamic properties such as internal energy (u) and enthalpy (h).

10.7.1 PENG-ROBINSON EQUATION

The Peng-Robinson (PR) EOS is expressed as:

$$p = \frac{RT}{V_m - b} - \frac{a\alpha}{V_m^2 + 2bV_m - b^2}, \tag{10.21}$$

where:

$$a = \frac{0.45724R^2T_C^2}{p_C}, \quad b = \frac{0.07780RT_C}{p_C}, \tag{10.22}$$

$$\alpha = \left[1 + \left(0.37464 + 1.54226\omega - 0.26992\omega^2\right)\left(1 - T_r^{0.5}\right)\right]^2, \quad T_r = \frac{T}{T_C}. \tag{10.23}$$

Here, T is the temperature (K), p is the pressure (Pa), R is the gas constant (8.314413 J/mol·K), V_m is the gas molar volume (m^3/mol), ω is the acentric factor, and T_C and p_C are the critical temperature and pressure, respectively.

10.7.2 SOAVE-REDLICH-KWONG EQUATION

The Soave-Redlich-Kwong (SRK) EOS is given by:

$$p = \frac{RT}{V_m - b} - \frac{a\alpha}{V_m(V_m + b)}, \tag{10.24}$$

where:

$$a = \frac{0.42748R^2T_C^2}{p_C}, \quad b = \frac{0.08664RT_C}{p_C}, \tag{10.25}$$

$$\alpha = \left[1 + \left(0.48508 + 1.55171\omega - 0.15613\omega^2\right)\left(1 - T_r^{0.5}\right)\right]^2, \quad T_r = \frac{T}{T_C}. \tag{10.26}$$

This equation modifies the original Redlich-Kwong formulation to better predict the phase behavior and density of real gases.

10.7.3 Z-CUBIC EQUATIONS

Central to the compressibility formulation is the compressibility factor, Z, which is obtained by solving a cubic equation of the form

$$Z^3 - (1 - B)Z^2 + (A - 2B - 3B^2)Z - (AB - B^2 - B^3) = 0 \tag{10.27}$$

Here, Z is the compressibility factor, a dimensionless quantity that relates the real gas behavior to that of an ideal gas. The dimensionless parameters A and B are defined as:

$$A = \frac{aP}{(RT)^2}, \qquad B = \frac{bP}{RT} \tag{10.28}$$

These depend on the temperature T, pressure P, and the critical properties of the substance. The coefficients a and b account for attractive and repulsive interactions, respectively:

$$a = 0.45724\frac{R^2T_c^2}{P_c} \cdot \alpha(T_r), \qquad b = 0.07780\frac{RT_c}{P_c} \tag{10.29}$$

The function $\alpha(T_r)$ introduces the temperature dependency of intermolecular attractions:

$$\alpha(T_r) = \left[1 + k(1 - \sqrt{T_r})\right]^2, \qquad T_r = \frac{T}{T_c} \tag{10.30}$$

The constant k depends on the acentric factor ω of the substance:

$$k = 0.37464 + 1.54226\omega - 0.26992\omega^2 \tag{10.31}$$

The PR-EOS enables more accurate prediction of gas-phase and vapor-liquid equilibrium (VLE) properties by using this cubic equation in Z, which typically yields three roots. Among these, the physically meaningful root corresponds to the gas or liquid phase, depending on conditions. The inclusion of $\alpha(T_r)$ ensures that the equation accounts for the weakening of attractive forces at higher temperatures, enhancing its applicability to a wide range of fluids and conditions.

For comparison, the Soave–Redlich–Kwong (SRK) EOS also yields a cubic equation in Z:

$$Z^3 - Z^2 + (C - D - D^2)Z - CD = 0 \tag{10.32}$$

where the parameters are defined as:

$$C = \frac{mP}{(RT)^2}, \qquad D = \frac{nP}{RT} \tag{10.33}$$

with

$$m = 0.42748\frac{R^2T_c^2}{P_c}\cdot\beta(T_r), \qquad n = 0.08664\frac{RT_c}{P_c} \tag{10.34}$$

$$\beta(T_r) = \left[1 + q(1-\sqrt{T_r})\right]^2, \qquad q = 0.48508+1.55171\phi-0.15613\phi^2 \tag{10.35}$$

While the SRK EOS provides better performance for light gases such as hydrogen, the PR EOS remains more robust for hydrocarbons and general-purpose applications due to its improved formulation of the attractive term.

10.7.4 GERG-2008 MODEL

The GERG-2008 EOS is a more advanced model for gas mixtures and is expressed using the reduced Helmholtz free energy:

$$F(\delta,\tau,\bar{x}) = F^0(\rho,T,\bar{x}) + F^r(\delta,\tau,\bar{x}), \tag{10.36}$$

where:

$$\delta = \frac{\rho}{\rho_r(\bar{x})}, \qquad \tau = \frac{T_r(\bar{x})}{T}. \tag{10.37}$$

Here F^0 represents the ideal gas contribution while F^r accounts for the residual free energy. The reducing functions $\rho_r(\bar{x})$ and $T_r(\bar{x})$ are composition-dependent. From this model, all thermodynamic properties can be derived. For example, the pressure is related to density and temperature as:

$$\frac{p(\delta,\tau,\bar{x})}{\rho RT} = 1 + \delta F^r_\delta, \tag{10.38}$$

where $F^r_\delta = \frac{\partial F^r}{\partial \delta}\Big|_{\tau,\bar{x}}$.

The EOS models not only describe the p-T-ρ relationship but also provide the basis for calculating internal energy and enthalpy. Using the relationships:

$$u = \int c_v(T)\,dT + u_{\text{ref}}, \qquad h = u + \frac{p}{\rho}, \tag{10.39}$$

the internal energies (u_l, u_g) and enthalpies (h_l, h_g) used in the energy equation can be directly linked to temperature. This ensures consistency between the energy conservation equation and the EOS.

10.8 RELATIVE PERMEABILITIES

Relative permeability models are fundamental for understanding multiphase flow dynamics in porous media. Over the years, several key frameworks have emerged, offering distinct capabilities for capturing flow behavior under various conditions. The Brooks-Corey (BC) model [76] is widely used for modeling drainage processes across various porous media. Its power-law relationship has been validated for gas-water systems in tight formations by Dong et al. [272], making it a standard framework for modeling relative permeability behavior. The Van Genuchten (VG) model [297], originally developed for soil water retention curves, has been adapted for relative permeability applications. Lenhard and Parker extended this model to account for hysteresis effects in three-phase flow systems [316]. The Brooks-Corey variable Corey (BC-vC) model introduced additional parameters to better represent the complexities of multiphase flow, including hysteresis during imbibition processes [322]. Recent studies emphasize the importance of selecting an appropriate relative permeability model. For example, Juanes et al. [97] demonstrated that neglecting hysteresis effects can lead to significant errors in gas recovery predictions. Similarly, Bo et al. [261] highlighted that the choice of relative permeability model significantly impacts simulation outcomes for underground hydrogen storage (UHS).

Yekta et al. [357] explored hydrogen-water interactions in sedimentary formations, emphasizing the role of relative permeability and capillary pressure in improving the design and performance of UHS systems.

10.8.1 BROOKS-COREY (BC) MODEL

The Brooks-Corey model is widely adopted to describe the relative permeability of two-phase flow in porous media, based on the relationship with saturation [76]. For the wetting phase (Water During Drainage), the relative permeability:

$$k_{r,w} = k_{r0}(1 - S)^{3+\frac{2}{\lambda}}, \tag{10.40}$$

where $k_{r,w}$ is the relative permeability of the wetting phase, k_{r0} is the endpoint relative permeability, S is the normalized saturation, and λ is the pore size distribution index. While for the non-wetting phase (Hydrogen During Drainage), we have:

$$k_{r,\mathrm{H}_2} = k_{r0}(1 - S)^{\frac{1}{2}} \left[1 \quad S_w^{\frac{1}{2\lambda}}\right], \tag{10.41}$$

where k_{r,H_2} is the relative permeability of the non-wetting phase, k_{r0} is the endpoint relative permeability (typically assumed to be 1), S is the normalized water saturation (dimensionless, $0 \leq S \leq 1$), S_w is the water saturation, and λ is the pore size distribution index.

10.8.2 VAN GENUCHTEN (VG) MODEL

The Van Genuchten model offers a comprehensive approach to describing the relationship between relative permeability and saturation [297]. The relative permeability of the non-wetting Phase (Hydrogen) can be written as:

$$k_{r,\mathrm{H}_2} = k_{r0} S_g^{\frac{1}{2}} \left[1 - \left(S_w^{\frac{1}{m}} \right)^m \right]^2, \tag{10.42}$$

On the other hand, the relative permeability of the wetting phase (Water) has the following form:

$$k_{r,w} = k_{r0} S_w^{\frac{1}{2}} \left[1 - \left(1 - S_w^{\frac{1}{2}} \right)^m \right]^2, \tag{10.43}$$

where k_{r0} is the endpoint relative permeability, S_g is the gas saturation, S_w is the water saturation, and m is a fitting parameter related to the pore-size distribution of the medium.

10.8.3 BROOKS-COREY-VAN GENUCHTEN (BC-VC) MODEL

The BC-vC model combines the strengths of the Brooks-Corey and Van Genuchten models, providing a robust framework for capturing relative permeability hysteresis effects during imbibition [322]. The hydrogen relative permeability takes the form:

$$k_{r,\mathrm{H}_2} = k_{r0} S_g^2 \left[1 - S_w^{N_g} \right], \tag{10.44}$$

while water relative permeability becomes:

$$k_{r,w} = k_{r0} S_w^{N_g}, \tag{10.45}$$

where k_{r0} is the endpoint relative permeability, S_w is the water saturation, and N_g is the power-law exponent. This hybrid model enhances the predictive capability of relative permeability models by incorporating additional parameters to capture the complex interactions between phases.

10.9 EXAMPLE

In this example, hydrogen is injected into a depleted underground reservoir that is initially saturated with water or brine. The goal is to store excess renewable energy while understanding the complex two-phase flow dynamics that govern hydrogen migration in porous media. Key physical processes include viscous flow, capillary effects, and buoyancy-driven segregation. By integrating experimentally calibrated models into a robust numerical framework, this study aims to predict the evolution of hydrogen plumes and optimize storage performance.

The capillary pressure uses a logarithmic form:

$$p_c(S_w) = -p_{\mathrm{entry}} \ln\Big(\max\big(\min(S_w,\, 1 - S_{gr}),\, S_{wi} \big) \Big), \tag{10.46}$$

with parameters $p_{\text{entry}} = 1000$ Pa, $S_{wi} = 0.2$ (irreducible water saturation), and $S_{gr} = 0$.

A VG-based model is used for water relative permeability:

$$k_{r,w}(S_w) = S_e^{\epsilon}\left[1 - \left(1 - S_e^{1/m}\right)^m\right]^2, \tag{10.47}$$

with the effective saturation computed as

$$S_e = \frac{S_w - 0.14}{1 - 0.14}, \tag{10.48}$$

and parameters ϵ and m obtained through calibration.

The gas phase relative permeability is modeled as:

$$k_{r,g}(S_g) = (1 - S_e)^{\alpha}\left[1 - S_e^{1+2m}\right], \tag{10.49}$$

where $S_w = 1 - S_g$.

Hydrogen density is approximated by a polynomial:

$$\rho_g(p) - \max\Big(\min\big(\text{polyval}(p_{rho}, p), 100\big), 0.1\Big), \tag{10.50}$$

with fitted coefficients $p_{rho} = [6.5 \times 10^{-7}, 0.013643]$ and a reference pressure $p_{\text{ref}} = 10^5$ Pa.

10.9.1 BOUNDARY AND INITIAL CONDITIONS

Initial Conditions

Pressure: The entire domain is initialized at a reference pressure of 1×10^5 Pa.
Saturation: The reservoir is initially water-saturated, for example, $S_w = 0.9$ and $S_g = 0.1$, reflecting a depleted reservoir state.

Boundary Conditions

An injection well is placed on a segment of the left boundary In the code, nearly pure hydrogen is injected with a composition $[S_w, S_g] = [0.01, 0.99]$ at a rate (e.g., 17×10^{-5} m^3/s). The injection using the Peaceman model:

$$WI = \frac{2\pi K h}{\ln(r_e/r_w)}, \tag{10.51}$$

with a more conservative injection rate of 10^{-5} m^3/s and 100% hydrogen ($S_w = 0$, $S_g = 1$).

Production Boundary: The right boundary is maintained at a fixed pressure of 1×10^5 Pa with a prescribed water saturation.

No-Flow Boundaries: The top and bottom boundaries are treated as impermeable, with no flux conditions:

$$\frac{\partial p}{\partial n} = 0 \quad \text{and} \quad \frac{\partial S_g}{\partial n} = 0.$$

Gravity is activated as:

$$\mathbf{g} = \begin{bmatrix} 0 \\ -9.80665 \\ 0 \end{bmatrix} \text{ m/s}^2, \tag{10.52}$$

which drives the buoyancy-induced vertical migration of hydrogen.

10.10 NUMERICAL SIMULATION

The numerical simulation of gas flow in porous media was conducted using the MATLAB® Reservoir Simulation Toolbox (MRST), a powerful open-source software developed by SINTEF. MRST is widely used in academia and research for simulating multiphase fluid flow in complex reservoir systems. One of its key strengths lies in its modularity and extensibility, which allows users to implement custom physics, discretization schemes, and solution strategies. This makes MRST particularly suitable for developing and testing new numerical methods and models in reservoir engineering and porous media flow.

The computational domain considered in this study is a two-dimensional rectangular region with a length of $L_x = 20\,\text{m}$ in the horizontal direction and a height of $L_y = 30\,\text{m}$ in the vertical direction. This domain was discretized using a structured Cartesian grid consisting of $n_x \times n_y = 160 \times 220$ cells, resulting in a total of 35,200 control volumes. Each grid cell is assigned uniform dimensions of $\Delta x = 0.2\,\text{m}$ and $\Delta y = 0.25\,\text{m}$, ensuring regular spacing across the domain.

The mesh and associated geometric properties, such as cell centroids, face normals, face areas, and volumes, were computed using the `computeGeometry` function in MRST. This step is essential for defining the spatial discretization of the governing equations, particularly when using finite volume methods.

To ensure the reliability and accuracy of the simulation results, a mesh independence study was performed. This analysis evaluates whether the numerical solution is significantly affected by the grid resolution. Seven different grid configurations were tested, ranging from a coarse mesh of 40×100 cells to the finest resolution of 160×220 cells.

For each grid, simulations were run, and the gas saturation S_g profiles were extracted at three vertical cross-sections corresponding to $y = 0\,\text{m}$, $15\,\text{m}$, and $30\,\text{m}$. These profiles were analyzed at two specific simulation times: $t = 15{,}440\,\text{s}$ and $t = 30{,}900\,\text{s}$. Representative results at the top boundary $y = 30\,\text{m}$.

10.11 MRST CODE:

PREPROCESSING AND MODULE INITIALIZATION

The code begins by cleaning the workspace and loading the necessary MRST modules:

```
clear all
close all
clc

% Clear and load MRST modules
mrstModule clear
mrstModule add ad-core ad-props ad-blackoil
```

Workspace Cleanup: Commands such as `clear all`, `close all`, and `clc` remove previous variables and close figures.
Module Loading: The `mrstModule` calls clear existing modules and loads essential modules (e.g., `ad-core`, `ad-props`, `ad-blackoil`) required for multiphase flow simulations.

GRID GENERATION AND GEOMETRY COMPUTATION

The computational grid for the reservoir is defined:

```
% Grid dimensions
nx = 100;    % Number of cells in the x-direction
ny = 120;    % Number of cells in the y-direction
Lx = 20;     % Domain length in x [m]
Ly = 30;     % Domain length in y [m]

% Create a 2D Cartesian grid
G = cartGrid([nx, ny], [Lx, Ly]);

% Compute grid geometry (centroids, face areas, cell volumes)
G = computeGeometry(G);
```

Grid Creation: `cartGrid` creates a 2D Cartesian grid with the specified dimensions.
Geometry Computation: `computeGeometry` calculates cell centroids, face areas, and volumes for spatial discretization.

ROCK AND FLUID PROPERTY INITIALIZATION

Rock properties and fluid properties are defined next.

```
% Define rock properties (permeability in m^2 and porosity)
rock = makeRock(G, 44 * 9.869233e-13, 0.19);
```

Rock Properties: makeRock initializes the rock with a given permeability and porosity.

```
% Define fluid properties
rho_w = 1000;      % Water density
mu_w = 1e-3;       % Water viscosity
mu_g = 2e-5;       % Gas viscosity

% Gas density parameters and reference pressure
p_rho = [6.5e-7, 0.013643];
p_ref = 1e5;     % Reference pressure [Pa]
calculate_rho_n = @(P) max(min(polyval(p_rho, P), 100), 0.1);
rho_g_ref = calculate_rho_n(p_ref);

% Activate gravity
gravity on;
gravity([0; -9.80665; 0]);

% Initialize the fluid model (water-gas system)
fluid = initSimpleADIFluid('phases', 'WG', ...
                          'mu', [mu_w, mu_g], ...
                          'rho', [rho_w, rho_g_ref], ...
                          'n', [2, 2]);
```

Fluid Properties: Densities and viscosities for water and hydrogen are defined. A density function is provided for hydrogen, though the SRK EOS is recommended for improved accuracy.
Gravity: Gravity is enabled and set to act downward.
Fluid Initialization: initSimpleADIFluid sets up the two-phase fluid model, using Automatic Differentiation (ADI) to handle nonlinearities.

The code also integrates Van Genuchten-based functions for capillary pressure and relative permeability to describe the water-gas interaction.

BOUNDARY CONDITIONS AND WELL SPECIFICATION

Boundary conditions are specified next.

```
% Identify boundary faces
boundaryFaces = find(any(G.faces.neighbors==0, 2));
xf = G.faces.centroids(boundaryFaces,1);
left_faces = boundaryFaces(abs(xf) < eps);

% Sort left boundary faces and select injection cells
yf_left = G.faces.centroids(left_faces, 2);
[~, sort_idx] = sort(yf_left);
left_faces_sorted = left_faces(sort_idx);
dy = Ly / ny;
n_cells_injection = ceil(0.01 / dy);
inj_faces = left_faces_sorted(1:n_cells_injection);
```

```
% Identify production faces (right boundary)
right_faces = boundaryFaces(abs(xf - Lx) < eps);

% Set production boundary condition (fixed pressure, pure water
    saturation)
p_prod = 1e5;    % Production pressure [Pa]
bc_sat = repmat([1, 0], numel(right_faces), 1);
bc = [];
bc = addBC(bc, right_faces, 'pressure', p_prod, 'sat', bc_sat);
```

Boundary Identification: The code finds faces at the domain boundaries.
Injection Well: Injection cells are selected on the left boundary.
Production Boundary: The right boundary is set to a fixed pressure with pure water saturation.

```
% Determine injection cells and add the injection well
W = [];
inj_cells = unique(G.faces.neighbors(inj_faces, :));
inj_cells = inj_cells(inj_cells > 0);
injection_rate = 17e-5; % Injection rate
W = addWell(W, G, rock, inj_cells, 'Type', 'rate', 'Val',
    injection_rate, ...
            'comp_i', [0.01, 0.99], 'name', 'Injector', 'Sign',
                1, 'Radius', 0.001);
```

Injection Well: A well is added on the selected injection cells, injecting nearly pure hydrogen at the prescribed rate.

SOLVER CONFIGURATION AND TIME STEPPING

The simulation uses an implicit time-stepping scheme with adaptive steps:

```
% Simulation time configuration
totTime = 30900;  % Total simulation time [s]
dt = [repmat(1, 900, 1); repmat(10, 1000, 1); repmat(20, 1000, 1)
    ; repmat(50, 0, 1); repmat(100, 0, 1)];

% Verify total time consistency
totTime_actual = sum(dt);
if abs(totTime_actual - totTime) > 1e-6
    error('Time␣steps␣do␣not␣add␣up␣to␣the␣total␣simulation␣time.
        ');
end

% Nonlinear solver parameters
solver = NonLinearSolver('maxIterations', 50, 'tolerance', 1e-6,
    'verbose', true);
```

Time Stepping: The total simulation time and time step vector are defined.
Solver Setup: A Newton-type nonlinear solver is configured for the implicit system.

SIMULATION EXECUTION AND POSTPROCESSING

The simulation is run and the results are stored:

```
[wellSols, states] = simulateScheduleAD(state, model, schedule, '
    NonLinearSolver', solver);
```

Simulation Execution: `simulateScheduleAD` solves the coupled pressure–saturation system over the defined time schedule.
Postprocessing: Simulation states are then postprocessed to verify mass conservation, plot pressure and saturation fields, and analyze dimensionless parameters (e.g., capillary, Bond, and buoyancy numbers).

10.12 HYDROGEN STORE IN SALT CAVERNS

Salt caverns have emerged as a promising solution for large-scale hydrogen storage, leveraging their favorable geomechanical properties and existing expertise from natural gas storage operations (Figure 10.5). Recent advancements include successful drilling operations and mechanical integrity tests in the U.S., which signify strides toward decarbonizing the energy grid [348].

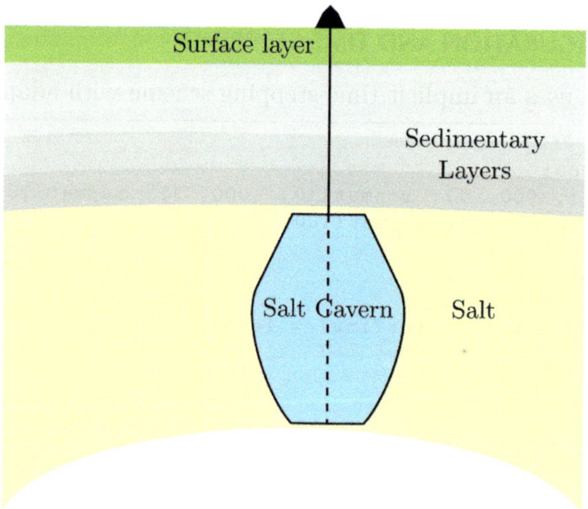

Figure 10.5 A salt cavern used for hydrogen storage.

Studies like those by Kumar [311] and Minougou [327] emphasize the importance of optimizing injection and production cycles for better performance, while Ozarslan [333] highlights the design and safety considerations in salt cavern operations. These caverns provide stability and scalability for hydrogen storage due to their impermeability and resistance to deformation under high pressure, as also explored by Kumar and Hajibeygi [312].

Mathematical modeling has played a pivotal role in optimizing the leaching process and understanding the geomechanical behavior of salt formations. Li et al. [318] developed models for cavern shape evolution in high-insoluble salt formations, laying the groundwork for simulating the effects of brine-gas interactions and the accumulation of insoluble sediments on cavern performance. These models account for dynamic processes such as dissolution rates, sediment redistribution, and brine flow, enabling precise predictions of cavern stability under operational conditions. Advances in modeling approaches also provide insights into phase behavior and the thermodynamic properties of stored hydrogen, further enhancing the safety and efficiency of hydrogen storage systems.

CONCLUSION

Underground hydrogen storage in depleted reservoirs combines economic and environmental benefits by using proven geological formations and existing infrastructure. The unique properties of hydrogen—low density, high buoyancy, and wide flammability range—necessitate careful modeling of multiphase flow, rock–fluid interactions, and safety hazards such as leakage and embrittlement. Advanced numerical methods, including mixed finite-element and compositional simulators coupled with robust equations of state, enable accurate prediction of storage performance and purity. Relative-permeability and capillary-pressure models capture two-phase displacement dynamics, while AI and machine-learning tools show promise in optimizing wettability predictions and operational parameters. Future research should focus on long-term field validation, improved microbial and geochemical reaction models, and integration of real-time monitoring data to support the safe, efficient, and scalable deployment of UHS as a key enabler of a low-carbon energy system.

REVIEW QUESTIONS

1. Explain the main differences between gray, blue, green, pink, and yellow hydrogen in terms of production pathways and environmental impact.
2. What are the primary abiotic and biotic reactions affecting hydrogen during subsurface storage, and how do they influence hydrogen loss?
3. Compare the Brooks–Corey and Van Genuchten relative-permeability models. Under what conditions might you prefer one over the other for UHS simulations?

4. Describe how the Peng–Robinson and GERG-2008 equations of state differ in predicting gas-phase behavior for pure hydrogen and hydrogen–gas mixtures.
5. Outline the key safety hazards associated with hydrogen leakage in underground storage, and discuss mitigation strategies presented in this chapter.

PROBLEMS

1. A depleted reservoir has a pore volume of 1×10^6 m^3, porosity $\phi = 0.15$, and initial water saturation $S_w = 0.2$. If reservoir pressure is $P = 50$ bar and temperature $T = 350$ K, estimate the maximum mass of hydrogen that can be stored, assuming ideal-gas behavior and hydrogen molar mass $M = 2 \times 10^{-3}$ kg/mol.
2. Given a hydrogen diffusivity in brine $D = 1 \times 10^{-9}$ m^2/s, reservoir thickness $L = 7$ m, and porosity $\phi = 0.20$, estimate the fraction of hydrogen lost by diffusion into the surrounding aquifer over 10 years. Assume one-dimensional diffusion and provide a brief discussion of the limitations of this approach.
3. Using the Peng–Robinson EOS (Eqs. (10.21)–(10.21) parameters) with hydrogen critical properties $T_c = 33.2$ K, $P_c = 1.296$ MPa, and acentric factor $\omega = 0.011$, solve for the compressibility factor Z at $P = 100$ bar and $T = 350$ K. Then compute the hydrogen density and discuss the impact of non-ideality on UHS capacity calculations.

DECLARATION

During the preparation of this work, the author used ChatGPT in order to improve the language and readability. After using this tool/service, the author reviewed and edited the content as needed and takes full responsibility for the content of the publication.

11 AI Applications in Fluid Flow in Porous Media

11.1 INTRODUCTION

Traditional mathematical models used to simulate hydrocarbon reservoirs are often complex and time-consuming. Parallel computing can reduce some of these limitations, but an alternative solution is machine learning, which has shown strong potential for prediction tasks across various fields. By training on historical oil field data and outputs from conventional simulators, machine learning models can offer faster and more efficient reservoir simulations.

Artificial Intelligence (AI), particularly its subfield Machine Learning (ML), enables computers to recognize patterns and make predictions by learning from data. ML algorithms, such as Artificial Neural Networks (ANNs), Random Forests (RF), Gradient Boosting Regressors (GBR), and Decision Trees (DT) can handle regression, classification, clustering, and forecasting tasks. Over the past decade, these techniques have improved significantly in modeling complex systems, including subsurface flow.

Unlike traditional models, ML can capture nonlinear relationships without predefined assumptions. It works well with large, noisy, or incomplete datasets and can detect meaningful patterns with high accuracy. These models also excel at automating repetitive tasks with fewer errors than humans.

This chapter focuses on applying machine and deep learning methods to simulate nanoparticle injection in oil reservoirs. The implementation is done in Python using Jupyter Notebook, an open-source tool for writing live code and building data-driven models.

Machine learning is a subset of artificial intelligence (AI) that enables systems to learn from data and make predictions or decisions. Unlike traditional programming, where explicit instructions are provided, machine learning algorithms improve their performance through experience, allowing them to handle complex and dynamic tasks. In the following, we list some standard ML algorithms:

Supervised Learning: Learning from labeled data, where the model learns to predict an output based on input data.
Unsupervised Learning: Learning from unlabeled data, where the model tries to find patterns and relationships within the data.
Reinforcement Learning: Learning by interacting with an environment, where the model learns to make decisions based on rewards and penalties.

DOI: 10.1201/9781032724973-11

11.2 MACHINE LEARNING MODELS

Once we have grasped the basics of machine learning, we can start exploring more advanced models that can handle more complex data and provide better predictions.

11.2.1 LINEAR REGRESSION

Linear regression (LR) is a core method in statistical modeling, commonly used to quantify relationships between a dependent variable and one or more independent variables [390]. In the context of fluid flow in porous media, LR provides a straightforward way to estimate how parameters such as pressure, saturation, or flow rate vary in response to changes in physical properties like porosity and permeability.

The simplest form, known as simple linear regression, models the relationship between a single predictor and a response variable using the least-squares method. Given a dataset of N points $(x_1, y_1), (x_2, y_2), \ldots, (x_N, y_N)$, the model fits a straight line:

$$y = mx + c \tag{11.1}$$

Here, y is the response, x is the predictor, m is the slope, and c is the intercept. The slope and intercept are estimated as:

$$m = \frac{N\sum x_i y_i - (\sum x_i)(\sum y_i)}{N\sum x_i^2 - (\sum x_i)^2} \tag{11.2}$$

$$c = \frac{\sum y_i - m\sum x_i}{N} \tag{11.3}$$

These values minimize the sum of squared residuals, ensuring the best linear approximation of the observed data.

For more complex systems involving multiple predictors, multiple linear regression (MLR) is applied. This is particularly relevant for porous media where several physical variables simultaneously affect the response. Consider a dataset $\{y_i, x_{i1}, \ldots, x_{ip}\}_{i=1}^{N}$, where each y_i depends on p predictors. The model is:

$$y_i = \beta_0 + \beta_1 x_{i1} + \cdots + \beta_p x_{ip} + \epsilon_i = \mathbf{x}_i^\top \boldsymbol{\beta} + \epsilon_i, \quad i = 1, \ldots, N \tag{11.4}$$

In matrix notation, this becomes:

$$\mathbf{y} = X\boldsymbol{\beta} + \boldsymbol{\epsilon} \tag{11.5}$$

with:

$$\boldsymbol{\beta} = \begin{pmatrix} \beta_0 \\ \beta_1 \\ \vdots \\ \beta_p \end{pmatrix}, \quad \boldsymbol{\epsilon} = \begin{pmatrix} \epsilon_1 \\ \epsilon_2 \\ \vdots \\ \epsilon_N \end{pmatrix}, \quad \mathbf{y} = \begin{pmatrix} y_1 \\ y_2 \\ \vdots \\ y_N \end{pmatrix}, \tag{11.6}$$

$$X = \begin{pmatrix} 1 & x_{11} & \cdots & x_{1p} \\ 1 & x_{21} & \cdots & x_{2p} \\ \vdots & \vdots & \ddots & \vdots \\ 1 & x_{N1} & \cdots & x_{Np} \end{pmatrix} \tag{11.7}$$

This matrix-based formulation allows for efficient computation and interpretation. In fluid flow modeling, it can be used to capture how flow-related outputs respond to simultaneous variations in multiple geological and physical variables.

The performance of a linear regression model is closely tied to the quality of the input data. Reliable datasets—derived from experiments, simulations, or field observations—are essential for producing accurate and meaningful predictions. Although LR cannot capture complex nonlinear behaviors inherent in many fluid flow processes, it remains a valuable tool for initial modeling, sensitivity analysis, and as a benchmark for evaluating more advanced machine learning models.

11.2.2 K-NEAREST NEIGHBORS

The k-Nearest Neighbors (k-NN) algorithm is a simple yet powerful non-parametric method used for both classification and regression tasks [365]. In fluid flow modeling, k-NN can be employed to estimate quantities such as permeability, saturation, or velocity at unknown locations by leveraging the similarity with neighboring samples in a high-dimensional feature space.

In k-NN regression, the output value for a new sample is computed as the weighted average of the values of its k closest neighbors in the training set. The choice of k and the distance metric greatly affect the model's performance. A small k results in a model sensitive to noise (high variance), while a large k smooths predictions but may underfit local variations (high bias).

The most common distance metric is the Euclidean distance:

$$d(\mathbf{x}, \mathbf{x}') = \sqrt{\sum_{j=1}^{p} (x_j - x'_j)^2} \tag{11.8}$$

Alternative metrics include the Manhattan distance:

$$d(\mathbf{x}, \mathbf{x}') = \sum_{j=1}^{p} |x_j - x'_j| \tag{11.9}$$

and the Hamming distance for categorical data:

$$d(\mathbf{x}, \mathbf{x}') = \sum_{j=1}^{p} \mathbb{I}(x_j \neq x'_j) \tag{11.10}$$

Here, $\mathbb{I}$ is the indicator function, equal to 1 when the condition is true and 0 otherwise.

To improve computational efficiency, especially with large datasets, spatial indexing structures such as k-d trees and ball trees are used. These structures partition the feature space hierarchically, reducing the number of distance computations during neighbor searches. Ball trees, in particular, perform well in high-dimensional spaces and are commonly used in fluid flow studies where multiple features influence flow behavior.

The k-NN method does not require any training phase, which makes it suitable for problems where model interpretability and quick setup are important. However, its prediction time can be slow on large datasets since it must compute distances to all training points at query time. Despite this, k-NN remains a valuable method for baseline modeling, validation, and data-driven interpolation in porous media applications.

11.2.3 DECISION TREES

Decision Trees (DTs) are supervised learning algorithms that can be applied to both classification and regression problems [397]. A decision tree recursively splits the input space into regions by selecting features and thresholds that maximize the homogeneity of the output within each region. Each internal node represents a decision based on a feature, each branch corresponds to an outcome of the decision, and each leaf node assigns a predicted value.

In regression trees, the prediction at a leaf node is typically the average of the response values in that region. The tree construction process aims to minimize a loss function, often the mean squared error (MSE), defined over the resulting partitions.

One of the key benefits of decision trees is their interpretability. The hierarchical structure of decisions can be visualized and understood without complex mathematics. This makes them useful in practical scenarios such as model validation, sensitivity analysis, and explaining predictions to stakeholders.

However, decision trees are prone to overfitting, especially when grown to full depth. Techniques such as pruning, setting a minimum number of samples per leaf, or limiting tree depth are commonly used to mitigate this issue.

11.2.4 RANDOM FOREST

Random Forest (RF) is an ensemble learning method that builds multiple decision trees and aggregates their predictions to improve accuracy and reduce overfitting [376]. Each decision tree in the forest is trained on a random subset of the data, both in terms of samples and features (a process called bootstrap aggregating or "bagging"). This randomness promotes diversity among trees, and the aggregation mean for regression or majority vote for classification reduces variance and improves generalization.

In regression applications, the RF prediction for an input $\mathbf{x}$ is the average of predictions from all T trees:

$$\hat{y} = \frac{1}{T}\sum_{t=1}^{T} h_t(\mathbf{x}) \tag{11.11}$$

where $h_t(\mathbf{x})$ is the prediction from the t-th tree.

RF models naturally handle high-dimensional datasets and multicollinearity among predictors, making them suitable for complex subsurface flow problems. They also provide measures of feature importance, which can help identify key variables influencing the flow.

Despite their strengths, RF models can be computationally expensive and less interpretable than single decision trees. However, their robustness and predictive power make them a valuable tool for analyzing fluid behavior in porous media, especially when interpretability is balanced with performance.

11.2.4.1 Implementation of DT and RF

Decision Trees are intuitive tree-based models used for both classification and regression tasks. Random Forests are an ensemble method that combines multiple decision trees to improve accuracy.

```
from sklearn.ensemble import RandomForestRegressor

# Creating and training a Random Forest model
forest_model = RandomForestRegressor(n_estimators=100,
    random_state=42)
forest_model.fit(X_train, y_train)

# Making predictions
forest_pred = forest_model.predict(X_test)
```

`Random Forest Regressor()` creates a random forest model with 100 decision trees. `fit()` trains the model on the training data. `predict()` makes predictions on the testing data.

11.2.5 SUPPORT VECTOR MACHINES

Support Vector Machines (SVMs) are powerful supervised learning models used primarily for classification tasks but also effective for regression, particularly when data exhibits complex, nonlinear relationships [379]. SVMs work by mapping input features into a higher-dimensional space using a kernel function and then identifying the optimal hyperplane that maximizes the margin between classes or minimizes error in regression. The use of kernel functions, such as linear, polynomial, radial basis function (RBF), or sigmoid, allows SVMs to operate in transformed feature spaces without explicitly computing the transformation, a technique known as the "kernel trick."

For regression tasks, the SVM is adapted into Support Vector Regression (SVR), which fits a function within a specified margin of tolerance ε from the true output. The SVR model aims to minimize the norm of the weights $\|\mathbf{w}\|^2$ while allowing some slack ξ_i for deviations larger than ε, regulated by a penalty parameter C:

$$\min_{\mathbf{w},b} \frac{1}{2}\|\mathbf{w}\|^2 + C\sum_{i=1}^{N}(\xi_i + \xi_i^*) \tag{11.12}$$

subject to:

$$\begin{cases} y_i - (\mathbf{w}^\top \phi(\mathbf{x}_i) + b) \leq \varepsilon + \xi_i \\ (\mathbf{w}^\top \phi(\mathbf{x}_i) + b) - y_i \leq \varepsilon + \xi_i^* \\ \xi_i, \xi_i^* \geq 0 \end{cases} \tag{11.13}$$

where $\phi(\cdot)$ is the kernel mapping function.

In porous media applications, SVR can model complex relationships between physical inputs (e.g., porosity, permeability, fluid properties) and outputs like pressure or saturation. Its robustness to overfitting and ability to model nonlinear patterns make it valuable when working with limited but high-quality data.

However, training SVMs on large datasets can be computationally demanding due to the quadratic programming involved. Proper selection of kernel type and hyperparameters (C, ε, γ) is crucial for optimal performance.

Despite these challenges, SVMs and SVRs offer strong generalization ability and are an important tool in the fluid flow modeling toolbox, particularly when data complexity demands models that go beyond linear or ensemble approaches.

11.2.6 GRADIENT BOOSTING REGRESSION

Gradient Boosting Regression (GBR) is a powerful ensemble learning technique that builds predictive models in a stage-wise fashion by sequentially adding decision trees to correct the errors made by previous models [391, 392]. Unlike Random Forests, which train trees in parallel, GBR constructs each tree based on the residual errors of the combined ensemble of prior trees. This iterative process allows the model to progressively minimize a chosen loss function, commonly the MSE for regression tasks.

Given a target function $F(x)$, the model is built as:

$$F_T(x) = \sum_{t=1}^{T} \lambda h_t(x) \tag{11.14}$$

where $h_t(x)$ is the prediction of the t-th tree, T is the total number of trees, and λ is the learning rate that controls the contribution of each tree.

At each iteration, a new tree is fitted to the negative gradient of the loss function (i.e., the residuals) with respect to the current prediction:

$$r_i^{(t)} = -\left[\frac{\partial \mathcal{L}(y_i, F_{t-1}(x_i))}{\partial F_{t-1}(x_i)}\right] \tag{11.15}$$

This mechanism allows GBR to converge toward a model that minimizes prediction error. GBR models often outperform simpler regression models, but they require careful tuning of hyperparameters such as learning rate, number of estimators, tree depth, and subsampling ratio. Overfitting can be mitigated through regularization techniques and early stopping.

Despite the computational cost of training, GBR remains a top choice in fluid modeling tasks when predictive accuracy is prioritized and sufficient computational resources are available.

11.2.7 ARTIFICIAL NEURAL NETWORKS

Artificial Neural Networks (ANNs) are a class of machine learning models inspired by the structure and function of the human brain [400, 393]. They are particularly well-suited for modeling complex, nonlinear relationships, making them highly applicable in fluid flow prediction within porous media.

An ANN consists of layers of interconnected nodes or "neurons." These include an input layer, one or more hidden layers, and an output layer. Each connection between neurons carries a weight that is adjusted during training to minimize prediction error. The neuron computes a weighted sum of its inputs, applies an activation function, and passes the result to the next layer.

Mathematically, the output of a single neuron is expressed as:

$$a = g\left(\sum_{j=1}^{n} w_j x_j + b\right) \tag{11.16}$$

where x_j are input features, w_j are the weights, b is the bias term, and $g(\cdot)$ is the activation function such as sigmoid, tanh, or ReLU.

The training process involves adjusting weights using algorithms like stochastic gradient descent (SGD) to minimize a loss function, typically MSE for regression tasks. Backpropagation is used to compute the gradients required for weight updates.

ANNs are particularly valuable when dealing with large datasets and when the relationship between inputs and outputs is highly nonlinear and difficult to model using traditional equations. Their ability to approximate any continuous function (as guaranteed by the universal approximation theorem) gives them a unique advantage in modeling subsurface flow behavior.

However, ANNs also come with challenges. They require careful tuning of hyperparameters (e.g., learning rate, number of layers, number of neurons per layer), are prone to overfitting if not regularized, and can be computationally intensive to train.

Despite these challenges, ANNs are widely used in scientific and engineering applications for their flexibility and predictive performance, making them a valuable tool in data-driven fluid flow analysis in porous media.

11.2.8 PHYSICS-INFORMED NEURAL NETWORKS

Physics-Informed Neural Networks (PINNs) are an extension of traditional ANNs designed to incorporate governing physical laws directly into the learning process [407]. This is especially valuable in fluid flow modeling, where physical laws such as conservation of mass, momentum, and energy, typically represented as partial differential equations (PDEs) must be respected.

PINNs combine data-driven learning with physics-based constraints by embedding the PDE residuals into the loss function during training. The network not only learns from data points (e.g., observed pressures or velocities) but is also penalized when its predictions violate the underlying physics.

For a PDE of the form $\mathcal{N}[u(x,t)] = 0$, the loss function in a PINN typically includes:

$$\mathcal{L}_{\text{total}} = \mathcal{L}_{\text{data}} + \lambda_{\text{pde}}\mathcal{L}_{\text{physics}} + \mathcal{L}_{\text{bc}} + \mathcal{L}_{\text{ic}} \tag{11.17}$$

where $\mathcal{L}_{\text{data}}$ is the supervised data loss, $\mathcal{L}_{\text{physics}}$ measures how well the model satisfies the PDE, and $\mathcal{L}_{\text{bc}}$, $\mathcal{L}_{\text{ic}}$ enforce boundary and initial conditions, respectively. λ_{pde} is a weight balancing the importance of the physics loss.

The key advantage of PINNs is that they can infer solutions at unmeasured points by leveraging physics constraints, even when data is sparse or noisy. This makes them especially attractive for modeling subsurface flow where direct measurements are often limited.

Training PINNs requires automatic differentiation to compute derivatives of the neural network outputs with respect to the input variables, which are then used to evaluate the residuals of the governing equations.

Applications in porous media include forward modeling (e.g., predicting pressure or saturation evolution) and inverse problems (e.g., estimating permeability fields from observed data). While PINNs can be computationally expensive and sensitive to hyperparameter tuning, they represent a promising direction for integrating machine learning with established physical modeling frameworks.

11.2.9 CONVOLUTIONAL NEURAL NETWORKS

Convolutional Neural Networks (CNNs) are a class of deep learning models particularly effective for capturing spatial hierarchies in data [398]. Though originally developed for image processing, CNNs have found increasing application in fluid flow modeling, especially when working with structured spatial data from simulations or experiments.

In CNNs, the key building block is the convolutional layer, which applies learnable filters to small patches of the input data. This operation detects localized features such as edges or gradients. Pooling layers then reduce spatial dimensionality, enhancing computational efficiency and helping prevent overfitting.

A typical CNN architecture includes:

Convolutional layers for feature extraction
Nonlinear activation functions (e.g., ReLU)
Pooling layers (e.g., max pooling) for downsampling
Fully connected layers for final predictions

Mathematically, a 2D convolution for an output feature map Y is given by:

$$Y(i,j) = \sum_{m=1}^{M}\sum_{n=1}^{N} X(i+m-1, j+n-1) \cdot K(m,n) \tag{11.18}$$

where X is the input, K is the kernel, and (i,j) indexes the output location.

CNNs are computationally efficient compared to fully connected networks when handling high-dimensional spatial data. However, they require substantial training data and careful design of the network architecture (e.g., kernel size, depth, stride) to capture flow-specific features effectively.

Despite these requirements, CNNs offer a robust solution for pattern recognition and regression tasks in spatial domains, making them a powerful option in the machine learning toolkit for fluid flow modeling in porous systems.

11.2.10 RECURRENT NEURAL NETWORKS

Recurrent Neural Networks (RNNs) [394] are a class of neural networks designed for processing sequential data. Unlike feedforward networks, RNNs maintain an internal memory that captures information about previous time steps. Each RNN unit takes input not only from the current data point but also from the output of the previous unit. The hidden state h_t at time step t is computed as:

$$h_t = \sigma(W_x x_t + W_h h_{t-1} + b) \tag{11.19}$$

where x_t is the input at time t, W_x and W_h are weight matrices, b is the bias term, and σ is an activation function such as tanh or ReLU.

Variants like Long Short-Term Memory (LSTM) and Gated Recurrent Units (GRU) are commonly used to overcome limitations of standard RNNs, particularly issues with vanishing or exploding gradients. These architectures include gating mechanisms that regulate information flow, enabling the network to retain relevant signals over longer sequences. While RNNs require significant computational resources and careful tuning, their ability to model dynamic systems makes them a strong candidate for time-dependent flow prediction in porous media.

11.2.11 COMPARATIVE INSIGHTS

This document has explored a range of machine learning algorithms applied to fluid flow modeling in porous media, each with distinct characteristics, advantages, and limitations. Linear Regression (LR) provides a transparent and interpretable model, best suited for systems with strong linear relationships. However, it lacks flexibility for capturing complex interactions. k-Nearest Neighbors (k-NN) is a simple and intuitive method effective for localized prediction and interpolation. Yet, its performance declines with high-dimensional data or large datasets due to computational inefficiency. Decision Trees (DT) offer rule-based interpretability and can model nonlinear interactions, but are prone to overfitting unless carefully pruned or regularized. Random Forest (RF) enhances DTs through ensemble learning, reducing variance and improving accuracy at the cost of interpretability and computational demand. Support Vector Machines (SVMs) are well-suited for capturing complex decision boundaries and nonlinear relationships, especially in small to medium-sized datasets. However, they can be computationally expensive to train. Gradient Boosting Regression (GBR) is a powerful ensemble approach that excels in predictive accuracy but requires careful hyperparameter tuning to avoid overfitting. ANNs are flexible and highly expressive, ideal for capturing intricate patterns in data. Their main challenges lie in training complexity and the need for large, labeled datasets. PINNs integrate domain knowledge through governing equations, making them ideal for data-scarce environments where physics must be honored. However, they require advanced implementation and computational resources. CNNs are effective for learning spatial patterns and are especially useful with gridded data from simulations. They are limited in modeling temporal dynamics on their own. RNNs, including LSTM and GRU variants, are designed for sequential data and temporal modeling. They are essential for capturing dynamic fluid behaviors over time.

In practice, the choice of algorithm depends on the nature of the data, modeling objectives, computational resources, and the degree to which interpretability and physics fidelity are required. Hybrid and ensemble models, along with domain-informed architectures like PINNs, are increasingly popular as they combine strengths from multiple approaches for robust and physically consistent modeling of fluid flow in porous media.

11.3 EVALUATION METRICS

In modeling fluid flow through porous media, selecting appropriate evaluation metrics is critical to determine how well a model replicates physical observations such as pressure, saturation, or flow rate. Since these models are often used to guide decisions in hydrogeology, petroleum engineering, or environmental remediation, even small prediction errors can lead to costly or hazardous consequences.

Evaluation metrics provide quantitative measures of the agreement between simulated outputs and real-world observations or high-fidelity reference models. This process typically involves a test dataset of n data points, where each point has an observed (true) value and a predicted value generated by the model. The goal of the evaluation is to capture the magnitude and nature of the prediction errors. Below, we discuss four widely used metrics for continuous output evaluation, adapted for the context of fluid flow in porous media.

MEAN ABSOLUTE ERROR (MAE)

Mean Absolute Error (MAE) provides a direct measure of the average deviation between predicted and observed values, regardless of direction:

$$\text{MAE} = \frac{1}{n}\sum_{i=1}^{n} |\text{actual}_i - \text{predicted}_i| \tag{11.20}$$

In the context of porous media, MAE might be used to quantify the average absolute error in pressure head predictions, water saturation levels, or solute concentrations. Its straightforward interpretation how much the model is wrong, on average makes it a useful diagnostic for early-stage model evaluation or communication with non-technical stakeholders.

However, MAE treats all errors equally, which can be a limitation in cases where larger deviations (e.g., underestimating breakthrough time in a contaminant plume) carry more serious implications than smaller ones.

MEAN SQUARED ERROR (MSE)

Mean Squared Error (MSE) penalizes larger deviations more heavily by squaring each error term before averaging:

$$\text{MSE} = \frac{1}{n}\sum_{i=1}^{n} (\text{actual}_i - \text{predicted}_i)^2 \tag{11.21}$$

MSE is particularly useful in flow simulations where large errors in pressure or velocity fields can significantly affect downstream predictions, such as breakthrough curves or reservoir drainage times. For example, if a model underestimates pressure in a high-gradient zone, the squared error helps highlight the severity of that miss.

Additionally, MSE is commonly used during the training of surrogate models (e.g., neural networks or reduced-order models) because it is differentiable and sensitive to outliers. However, its sensitivity can also exaggerate the importance of a few large errors at the expense of the overall fit.

ROOT MEAN SQUARED ERROR (RMSE)

Root Mean Squared Error (RMSE) is the square root of MSE and provides an error estimate in the same units as the target variable:

$$\text{RMSE} = \sqrt{\frac{1}{n}\sum_{i=1}^{n}(\text{actual}_i - \text{predicted}_i)^2} \tag{11.22}$$

In porous media applications, RMSE is especially valuable when comparing models predicting variables like pore pressure (in Pascals), fluid saturation (fractional), or hydraulic conductivity (m/s). Because it returns errors in physical units, RMSE helps contextualize whether a model's error is significant relative to natural variability or measurement noise.

For instance, an RMSE of 0.03 in water saturation may be negligible in some formations but critical in others near phase transition thresholds. RMSE also provides a good overall measure of model performance but, like MSE, can be influenced heavily by a few large errors.

R-SQUARED (R^2)

The coefficient of determination, R^2, measures how well the model captures the variance in the observed data:

$$R^2 = 1 - \frac{\text{MSE}_{\text{model}}}{\text{MSE}_{\text{base}}} \tag{11.23}$$

Here, MSE_{base} refers to the error from a baseline model that simply predicts the mean of the observed values. An R^2 value close to 1 indicates that the model explains most of the variability in the data, while a value near 0 suggests poor performance.

In porous media modeling, R^2 is useful when comparing models across different field conditions or test scenarios. For example, when evaluating multiple models predicting pressure fields in a heterogeneous aquifer, R^2 helps identify which model captures the spatial trends most effectively. However, R^2 can be misleading in the presence of strong local deviations, e.g., near wells or faults, so it should be used in combination with localized error metrics.

11.4 IMPLEMENTATION SETUP

Before building a machine learning model, it is crucial to preprocess the data and perform feature engineering. This ensures that the data is in a suitable format for modeling and can significantly impact the performance of the model.

11.4.1 HANDLING MISSING DATA

One common issue in real-world data is missing values. These can be handled by either removing the rows with missing data or imputing the missing values using techniques such as mean, median, or mode imputation.

```
import pandas as pd
from sklearn.impute import SimpleImputer

# Load the dataset
df = pd.read_csv("fluid_flow_data.csv")

# Handling missing values by imputation
imputer = SimpleImputer(strategy="mean")
df["Porosity"] = imputer.fit_transform(df[["Porosity"]])

print(df.head())
```

`SimpleImputer(strategy="mean")` creates an imputer that replaces missing values with the mean of the column. `fit_transform()` applies the imputation to the porosity column.

11.4.2 FEATURE SCALING AND NORMALIZATION

Feature scaling is important for algorithms that rely on distance metrics. Common techniques include Min-Max scaling and Standardization.

```
from sklearn.preprocessing import StandardScaler

# Standardizing features
scaler = StandardScaler()
df[["Porosity", "Permeability", "Depth"]] = scaler.fit_transform(
    df[["Porosity", "Permeability", "Depth"]])

print(df.head())
```

`StandardScaler()` standardizes features by removing the mean and scaling to unit variance. `fit_transform()` scales the specified columns.

BUILDING AND EVALUATING ML MODELS

Once the data is preprocessed, we can move on to building and evaluating a machine learning model. We will start with a linear regression model, which is one of the most fundamental algorithms in supervised learning.

11.4.3 SPLITTING DATA INTO TRAINING AND TESTING SETS

It is important to split the data into training and testing sets to evaluate the model's performance on unseen data.

```
from sklearn.model_selection import train_test_split

# Splitting the data into training and testing sets
X = df[["Porosity", "Permeability", "Depth"]]
y = df["Pressure"]

X_train, X_test, y_train, y_test = train_test_split(X, y,
    test_size=0.2, random_state=42)
```

`train_test_split()` splits the data into training and testing sets. `test_size=0.2` means 20% of the data is used for testing, and 80% for training.

11.5 IMPLEMENTATION STEPS

Let us explore how to implement the linear regression (LR) algorithm using the Python "Scikit-Learn" library, one of Python's most popular machine learning libraries. We will walk through the steps of importing necessary libraries, preparing the data, and performing the linear regression analysis.

Step 1: Import Libraries

First, we need to import the essential libraries. We will use `numpy` for numerical operations, `pandas` for data manipulation, and `scikit-learn` for implementing the linear regression model.

```
import numpy as np
import pandas as pd
from sklearn.model_selection import train_test_split
from sklearn.linear_model import LinearRegression
from sklearn.metrics import mean_squared_error, r2_score
```

Step 2: Load the Data

Next, we will load the data using *pandas*. Suppose our dataset is stored in a CSV file named *dataset.csv*.

```
# Load the dataset
data = pd.read_csv('dataset.csv')

# Display the first few rows of the dataset
data.head()
```

Step 3: Data Preparation

In the dataset, we have n columns; we want to predict one concentration based on $n-1$ features. The dependent variable (label) is col_n, and the independent variables (attributes) are:

col_1
col_2
...
...
col_{n-1}

We will separate the features and the label accordingly.

```
# Define the feature set (independent variables)
X = data[['col_1', 'col_1', ..., 'col_(n-1)']]

# Define the label (dependent variable)
y = data['col_n']
```

Step 4: Split the Data

We must split the data into training and testing sets to evaluate our model.

```
# Split the data into training and testing sets
X_train, X_test, y_train, y_test = train_test_split(X, y,
    test_size=0.2, random_state=42)
```

Step 5: Train the Linear Regression Model

We will now create an instance of the `Linear Regression` class and train it using the training data.

```
# Create a Linear Regression model
model = LinearRegression()

# Train the model
model.fit(X_train, y_train)
```

Step 6: Make Predictions

After training the model, we can make predictions on the test data.

```
# Make predictions on the test data
y_pred = model.predict(X_test)
```

Step 7: Evaluate the Model

Finally, we will evaluate the model's performance using metrics such as Mean Squared Error (MSE) and the Coefficient of Determination (R^2 score).

```
# Calculate Mean Squared Error (MSE)
mse = mean_squared_error(y_test, y_pred)

# Calculate R-squared score
r2 = r2_score(y_test, y_pred)

print(f"Mean␣Squared␣Error:␣{mse}")
print(f"R-squared␣Score:␣{r2}")
```

Step 8: Visualize the Results

To better understand the model's performance, we can plot the predicted values against the actual values.

```
import matplotlib.pyplot as plt

# Plot the predicted vs actual values
plt.scatter(y_test, y_pred)
plt.xlabel('Actual␣Values')
plt.ylabel('Predicted␣Values')
plt.title('Actual␣vs␣Predicted␣Nanoparticles␣Concentration')
plt.show()
```

11.6 APPLICATIONS

In the following sections, we present several case studies that demonstrate the application of machine learning (ML) techniques to various fluid flow problems in porous media. These include permeability estimation and formation classification, prediction of nanoparticle transport, and forecasting of scaled oil recovery.

11.6.1 PERMEABILITY AND FORMATION CLASSIFICATION

Understanding the relationship between porosity and permeability is central to effective reservoir characterization and the prediction of fluid flow in subsurface formations. This case study integrates insights from core analysis standards and a machine learning workflow applied to a real dataset to explore these rock properties and evaluate classification performance.

The API Recommended Practice 40 (RP 40) outlines best practices for core analysis in petroleum reservoirs [368]. It addresses core retrieval, preservation, and laboratory procedures aimed at minimizing alterations and maintaining representative properties (Fig 11.2):

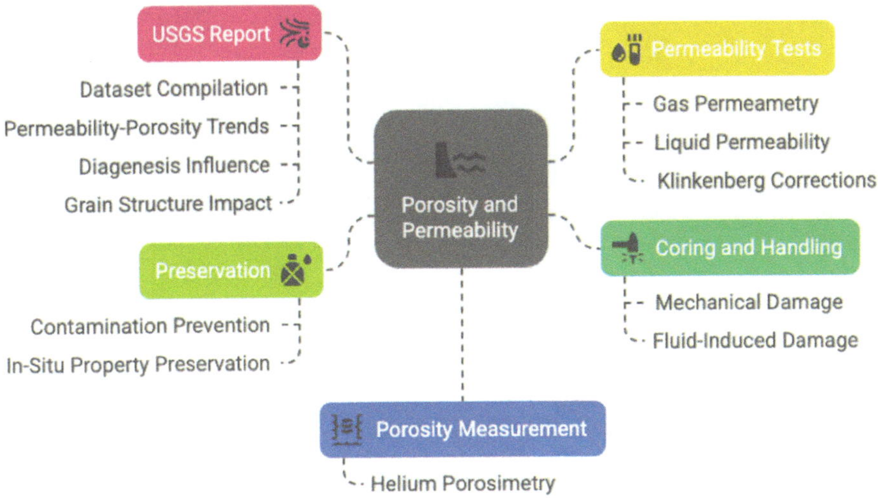

Figure 11.1 Permeability and porosity measurements and preparation.

Coring and handling: Guidelines ensure minimal mechanical or fluid-induced damage during retrieval.
Preservation: Techniques are described to prevent contamination and preserve in-situ properties before analysis.
Permeability tests: Gas permeametry, liquid permeability, and Klinkenberg corrections are detailed.
Porosity measurement: Methods like helium porosimetry are recommended for accurate pore volume estimation.

Complementing RP 40, the U.S. Geological Survey (USGS) report provides an extensive dataset on porosity and permeability in siliciclastic rocks [405]. The report includes:

A literature-based compilation of 70 datasets from various formations.
Analysis of permeability-porosity trends across depositional environments and lithologies.
Emphasis on how diagenesis and grain structure influence permeability reduction under reservoir pressure.
Observations on how quartz cementation or clay coatings alter flow pathways.

Both sources highlight the dynamic interplay between depositional and diagenetic processes, such as compaction, cementation, dissolution, and clay migration, that modify pore structures over geological time. These processes determine whether porosity is preserved or reduced and how effectively fluid can flow through a rock matrix. A robust interpretation of lab measurements must account for these evolving rock properties and their behavior

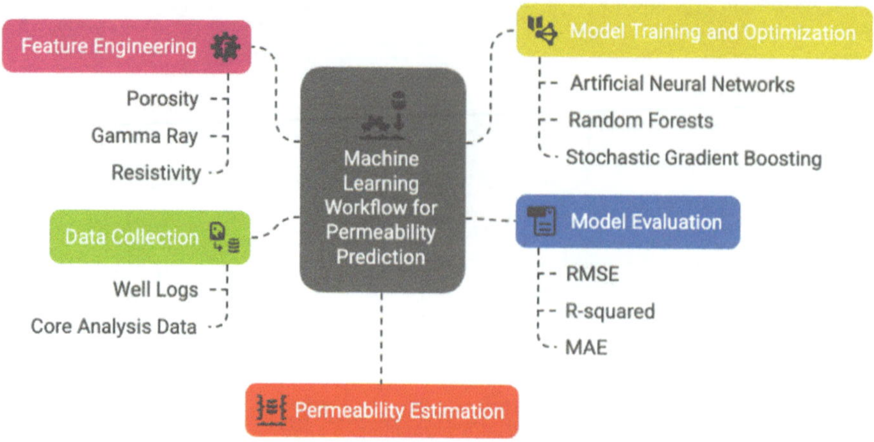

Figure 11.2 Porosity and permeability measurements in geological formation.

under in-situ conditions, where stress sensitivity can significantly alter flow performance.

MACHINE LEARNING WORKFLOW

A machine learning pipeline was designed to predict permeability using core and well-log data. As outlined in Figure 11.2, the pipeline includes data collection, preprocessing, feature selection, model training, and performance evaluation using regression metrics such as RMSE and R-squared.

DATASET DESCRIPTION AND CLEANING

The dataset includes 339 entries with features like depth, porosity, permeability, and formation labels. Initial preprocessing involved removing header rows and columns with excessive missing values. Attributes such as `Depth (m)`, `Porosity (%)`, and `Permeability (md)` were imputed using median values. Missing `Formation` values were replaced with the mode: "sandstone."

The heatmap in Figure 11.3 shows a moderate correlation between porosity and permeability (approximately 0.6), which is consistent with physical expectations—higher porosity generally enables higher fluid transmissivity. However, the moderate strength of this correlation also underscores that porosity alone does not fully determine permeability. Microstructural features such as pore throat size, connectivity, and tortuosity also play critical roles. Weak correlation with depth may suggest localized depositional or diagenetic influences rather than a consistent depth trend across the formation.

Table 11.1
Descriptive statistics of the cleaned dataset

Statistic	Depth (m)	Porosity (%)	Permeability (md)
Count	339	339	339
Mean	4060.87	13.45	6.98
Std. dev.	57.03	6.14	10.67
Min	4000.00	0.10	0.01
25th percentile	4025.00	8.05	1.30
Median	4067.00	13.30	4.00
75th percentile	4100.00	19.50	8.95
Max	4130.00	31.00	60.00

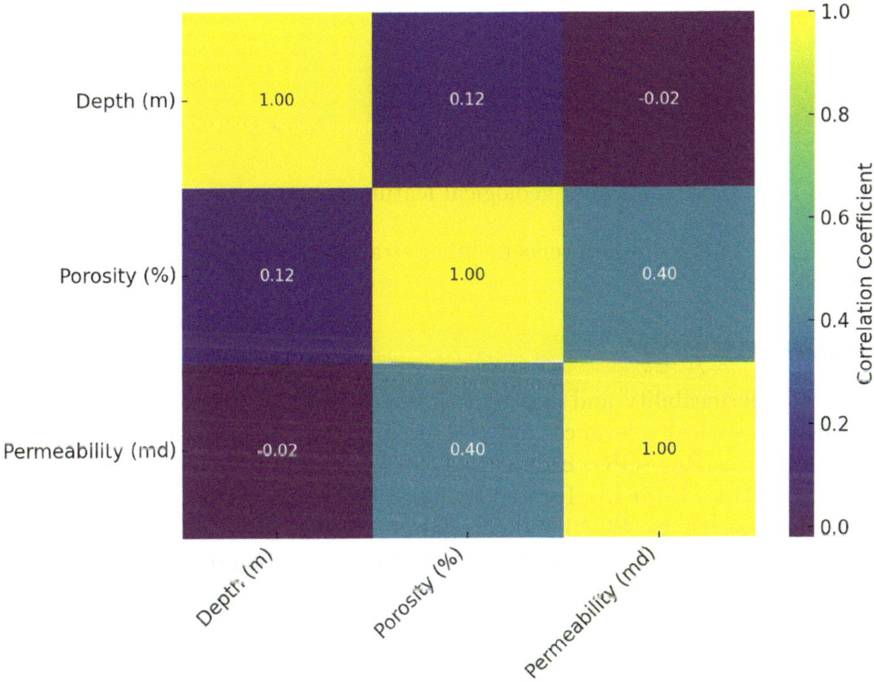

Figure 11.3 Correlation of features.

Figure 11.4 reveals strong right-skewness in permeability, indicating a dataset dominated by low-permeability samples. This suggests tight sandstone environments or partially cemented rocks, where only a small subset exhibits significantly higher flow capacity. The near-normal distribution of porosity suggests a more uniform depositional or compactional control. The

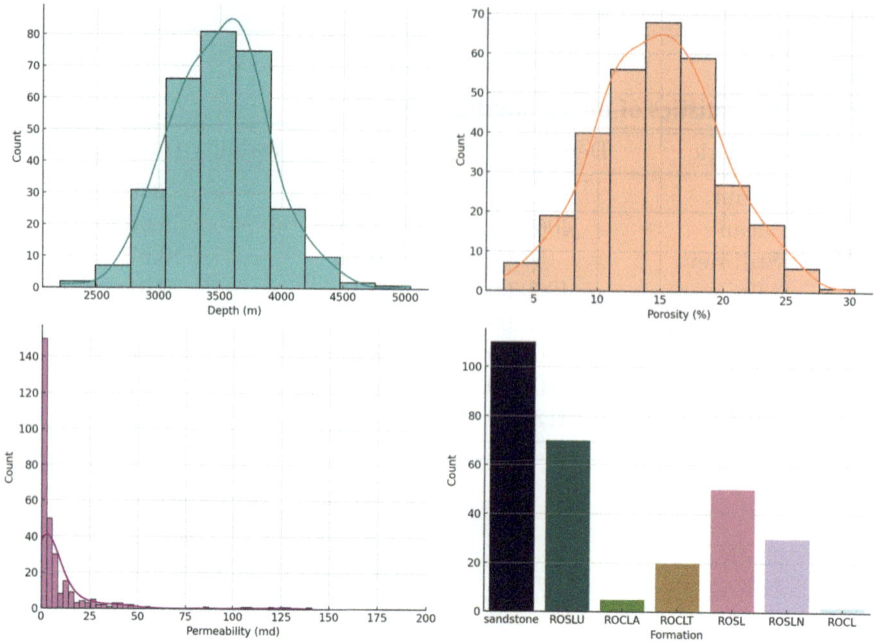

Figure 11.4 Distribution of geological features.

bimodal depth distribution reflects either stratigraphic targeting or core sampling strategy.

FEATURE SELECTION

To model permeability and classify formations, features were selected based on their geological relevance and statistical significance. The final set included Gamma Ray (GR), Bulk Density (RHOB), Neutron Porosity (NPHI), Resistivity (Rt, RMSFL), Total Porosity (PHIT), and Depth (D). Principal Component Analysis (PCA) and correlation filtering were used to remove redundant variables. These features reflect lithological composition, pore space characteristics, and fluid presence—key drivers of permeability behavior.

FORMATION CLASSIFICATION

Three models were compared, selected for their interpretability, efficiency, and ability to capture nonlinear relationships:

1. Logistic Regression: Accuracy = 47.06%. This model, which assumes linear separability, underperformed particularly for minority classes—due to the inherently nonlinear and imbalanced nature of the dataset.

2. Decision Tree: Accuracy = 69.12%. By segmenting feature space based on decision thresholds, this model better captured nonlinearity and variable interactions, especially for dominant classes like sandstone.

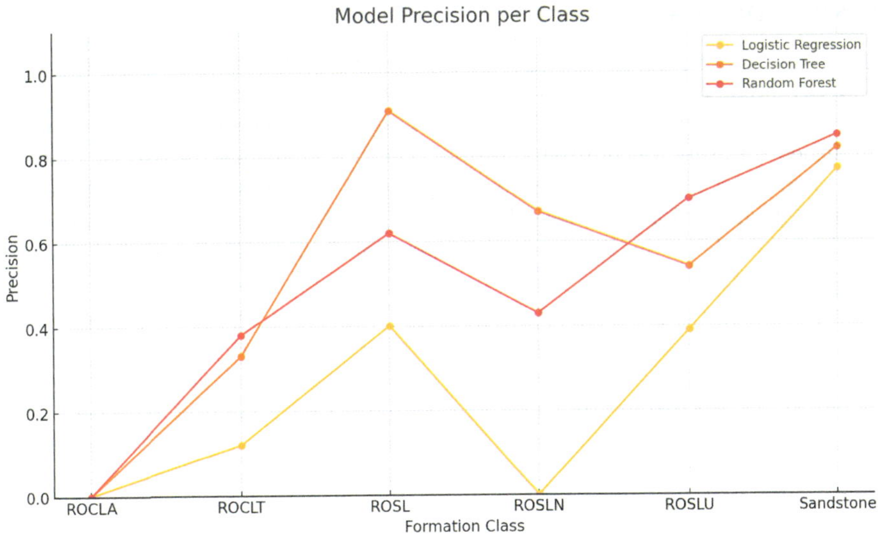

Figure 11.5 Precision comparison across models and classes.

3. Random Forest: Accuracy = 66.18%. This ensemble method offered good generalization by aggregating multiple decision trees, though further tuning or feature engineering may improve class-specific performance.

Figure 11.5 visually compares the precision scores of each model across the formation classes. It clearly illustrates the performance gaps: Logistic Regression performs poorly on most classes except sandstone, where it achieves acceptable precision. Decision Tree and Random Forest exhibit more consistent precision, especially for majority classes. The Decision Tree model shows notable strength in distinguishing ROSL and sandstone formations, while Logistic Regression fails to differentiate smaller classes like ROCLA and ROSLN. This reinforces the advantage of tree-based models for handling class imbalance and nonlinearity in geological data.

The superior performance of the Decision Tree model suggests it is particularly well-suited for problems where input-output relationships are governed by thresholds and categorical splits common in geological classification. Future improvements may include hyperparameter optimization and advanced models like XGBoost to address class imbalance and overfitting.

11.6.2 NANOPARTICLES PREDICTION

This section outlines the process of building predictive ML models using synthetic datasets generated from validated numerical simulations, focusing on key variables such as nanoparticle concentration, fluid saturations, and relative permeabilities.

11.6.2.1 Datasets

In machine and deep learning applications for petroleum engineering, datasets typically stem from experimental setups and well log data. However, most real-world datasets are proprietary and inaccessible to researchers. As a result, published datasets are limited. To overcome this challenge, researchers [367, 387, 388] have increasingly relied on synthetic datasets generated through numerical simulations. These datasets are validated against experimental results available in the literature to ensure credibility. Given the scarcity of data on nanoparticle transport in porous media, this study creates artificial datasets using a rigorously validated numerical model. Validation is carried out by comparing the simulation outputs to empirical data.

We generate artificial datasets derived from a validated numerical model, including preprocessing steps and feature importance analysis. Focusing on the case of nanoparticle transport in two-phase flow, we use a MATLAB® simulation code (previously discussed) to create the dataset. The dataset captures multiple target variables such as nanoparticle concentration C, water saturation S_w, relative permeability of water k_{rw}, and relative permeability of oil k_{ro}. Corresponding features include those defined in the two-phase flow mathematical framework: time t, spatial coordinate x, initial porosity ϕ_0, current porosity ϕ, oil and water densities (ρ_o, ρ_w), capillary pressure p_c, initial permeability k_0, absolute permeability K, initial relative permeabilities k_{ro0} and k_{rw0}, modified relative permeabilities under nanoparticle surface coverage (k_{rwp}, k_{rop}), adhesion impact ratios (θ_w, θ_o), gravitational acceleration g, nanoparticle volume deposited on pore walls C_{s1} and entrapped in pore throats C_{s2}, diffusion-dispersion coefficient D, mobility ratios (λ_{wp}, λ_{op}), viscosity μ, oil flow fraction f_o, residual oil saturation S_{or}, irreducible water saturation S_{iw}, nanoparticle diameter d_{np}, flow efficiency coefficient γ_f, and constants a, b, k_f, a_w, and a_o.

Data Preprocessing and Feature Correlation

Preprocessing includes cleansing, standardizing, and analyzing the dataset. Because the simulated data is complete, no imputation is necessary. All features were retained for initial modeling efforts. The variables C, S_w, k_{rw}, and k_{ro} are treated as targets, one at a time.

Implementation was done using Jupyter Notebook with Python. The `StandardScaler` from the `Scikit-Learn` library normalized the input features. Feature selection was applied based on correlation analysis. Feature importance helps prioritize the most influential variables for predicting a target, reducing dimensionality and boosting efficiency. Using the `feature_importance_` attribute from `Scikit-Learn`, we evaluated input features across different models. Figure 11.6 displays importance scores for RF and GBR models. Results show that C_{s1}, C_{s2}, and t consistently rank as top predictors of nanoparticle concentration, though the exact scores vary between algorithms.

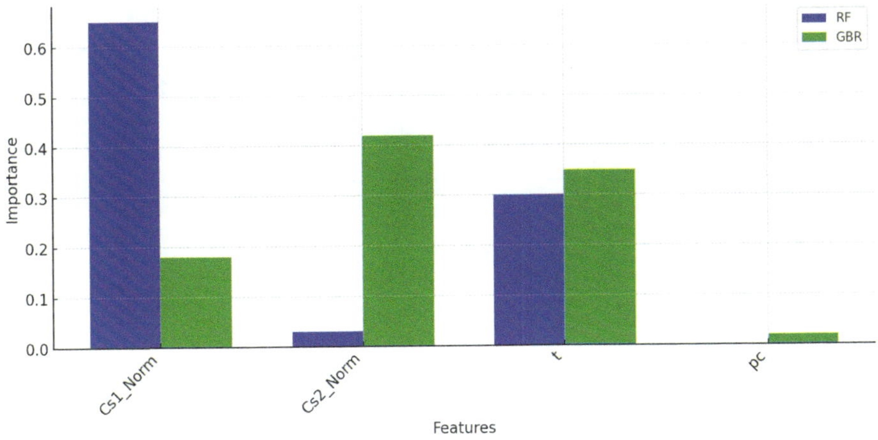

Figure 11.6 Relative feature importance using RF and GBR models for nanoparticle concentration prediction.

Machine Learning Implementation

The synthetic dataset was split into training (80%) and testing (20%) sets using `train_test_split` from `Scikit-Learn`. Four sets were generated: x_{train}, y_{train}, x_{test}, and y_{test}. All were scaled using the `StandardScaler` function.

We applied four machine learning models: Decision Tree (DT), Random Forest (RF), Gradient Boosting Regressor (GBR), and Artificial Neural Network (ANN). The dataset contained 288,000 samples. Scatter plots in Figure 11.7 compare actual vs. predicted values. GBR and ANN showed the highest accuracy, with ANN using ReLU activation performing best.

Table 11.2 summarizes performance using metrics: Root Mean Squared Error (RMSE), Mean Squared Error (MSE), Mean Absolute Error (MAE), and R^2 score. The ANN model outperformed the others.

Table 11.2
Performance evaluation of machine learning models

Model	RMSE	MSE	MAE	R^2
DT	0.018620	0.000347	0.002308	0.995142
RF	0.004788	0.000023	0.000251	0.999679
GBR	0.001669	0.000003	0.001128	0.999961
ANN	**0.000216**	**0.000000**	**0.000202**	**0.999999**

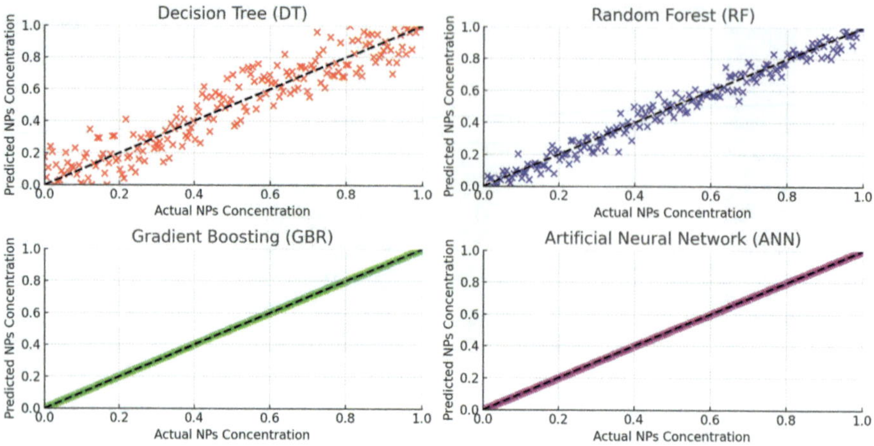

Figure 11.7 Predicted vs. actual nanoparticle concentrations using DT, RF, GBR, and ANN models.

11.6.3 PREDICTING SCALED OIL RECOVERY WITH SGB

Predicting oil recovery in fractured hydrocarbon reservoirs is critical for efficient resource management. Traditional scaling laws use dimensionless time to relate lab experiments to field behavior, but struggle with complex, real-world conditions. This study applies a Stochastic Gradient Boosting (SGB) regression model to improve the prediction of oil recovery and dimensionless time using basic physical properties of rocks and fluids [174].

Scaling Law Framework

The dimensionless time T is a vital variable in the scaling of oil recovery, defined by several models ([402, 370, 371, 412, 416, 411, 399, 403]) depending on physical parameters like permeability (K), porosity (ϕ), interfacial tension (γ), viscosity (μ), and contact angle (θ):

$$T = \left(\frac{K}{\phi}\right)^{1/2} \frac{\gamma}{\mu_w L^2} t \tag{11.24}$$

$$T = \left(\frac{K}{\phi}\right)^{1/2} \frac{\gamma \cos\theta}{\mu_w L^2} t \tag{11.25}$$

$$T = \left(\frac{K}{\phi}\right)^{1/2} \frac{\gamma}{\sqrt{\mu_w \mu_o} L^2} t \tag{11.26}$$

To generalize both spontaneous and forced imbibition, a power-law form was introduced [382]:

$$T = \frac{U_c^m U_i^{1-m}}{L\phi} t \tag{11.27}$$

Here, U_c is the characteristic velocity, U_i is the injection velocity, and m is the power-law index.

Machine Learning Methodology

Several machine learning models were evaluated: k-Nearest Neighbors (k-NN), Artificial Neural Networks (ANN), Support Vector Machines (SVM), Random Forests (RF), and Stochastic Gradient Boosting (SGB). These models were trained to predict oil recovery and dimensionless time based on rock and fluid properties.

Model performance was assessed using correlation coefficient (R), mean absolute error (MAE), root mean squared error (RMSE), relative absolute error (RAE), and root relative squared error (RRSE). SGB consistently outperformed other models [174]:

For the intermediate water-wet case in predicting oil recovery, the Stochastic Gradient Boosting (SGB) model achieved the best performance with a correlation coefficient of $R = 0.9869$, a mean absolute error (MAE) of 3.46, and a root mean square error (RMSE) of 5.69. The k-Nearest Neighbors (k-NN) model also performed well, yielding $R = 0.9799$, MAE $= 3.87$, and RMSE $= 6.64$. In the strong water-wet case, focused on dimensionless time prediction, the SGB model again outperformed other methods with $R = 0.9956$, MAE $= 1.25$, and RMSE $= 2.01$, followed closely by the Random Forest (RF) model, which attained $R = 0.9939$, MAE $= 1.69$, and RMSE $= 2.53$. For the weak water-wet condition, the SGB model remained the top performer with $R = 0.9989$, MAE $= 8.08$, and RMSE $= 35.13$, while the Support Vector Machine (SVM) model achieved a slightly lower performance with $R = 0.998$, MAE $= 10.48$, and RMSE $= 40.07$.

The results shown in Figure 11.8 confirm the effectiveness of the trained machine learning model in simulating critical reservoir responses. The strong correlation between predicted and actual values in both subplots, with R^2 values exceeding 0.92, indicates a high degree of accuracy. This suggests that the model can be reliably used to estimate dimensionless time and scaled oil recovery, which are fundamental parameters in porous media flow modeling and enhanced oil recovery (EOR) analysis.

The results indicate that SGB regression is effective at capturing nonlinear relationships between inputs and outcomes, especially under varying wettability conditions. It showed robust performance even when the data was incomplete or noisy, confirming its suitability for real-world reservoir forecasting.

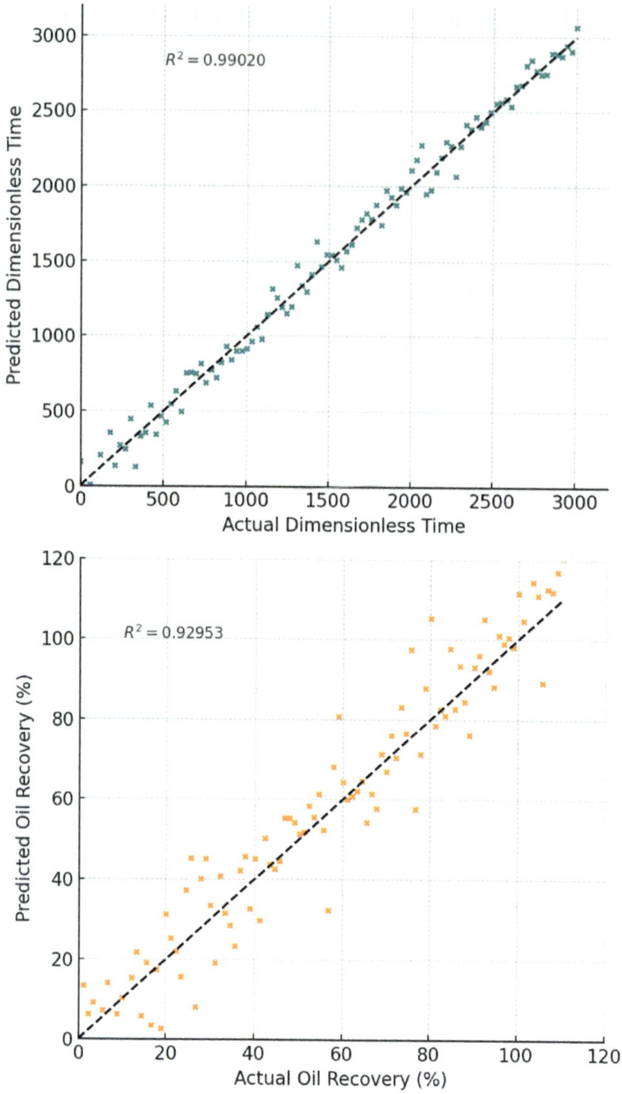

Figure 11.8 Performance of the machine learning model in predicting (top) dimensionless time and (bottom) oil recovery.

11.7 AI IN HYDROGEN STORAGE

Tariq et al. [349] studied hydrogen geo-storage, focusing on enhancing wettability prediction in the presence of organics within rock-H2-brine systems. They employed data-driven machine learning models to accurately predict wettability, a critical factor in understanding fluid behavior in subsurface reservoirs. This approach not only augments the understanding of hydrogen storage dynamics but also presents optimizing hydrogen geo-storage by ensuring better predictability and management of subsurface interactions, thereby contributing to the broader objective of advancing hydrogen as a reliable energy storage medium. Kanaani et al. [310] explored the integration of UHS and Carbon Capture and Sequestration (CCS) as a dual solution to balance energy production and consumption while reaping environmental benefits. The complex challenge of parameter optimization was tackled using artificial intelligence, specifically machine learning methods, to develop a proxy model for the UHS process. Employing the NSGA-II algorithm, the operational parameters were optimized considering three objective functions: Net Present Value (NPV), hydrogen recovery factor, and CCS. A set of innovative machine learning models for predicting the deliverability of underground natural gas storage (UNGS) in salt caverns has been introduced by Ali [364], addressing the inherent demand and supply imbalances in the natural gas supply chain. The paper marks a preliminary effort in employing machine learning algorithms to enhance the predictability of UNGS deliverability in salt caverns, by evaluating the performance of three machine learning algorithms: Artificial Neural Network (ANN), Support Vector Machine (SVM), and Random Forest (RF). Utilizing diverse monthly field storage data samples from 36 active UNGS salt caverns in the United States across different years, the study ascertains the predictive efficacy of these models.

CONCLUSION

AI and ML methods have emerged as transformative tools for modeling fluid flow in porous media. Data-driven surrogates can approximate complex multiphase flow behavior with high speed and acceptable accuracy when properly trained and validated. Ensemble methods (random forests, gradient boosting) offer robustness and feature-importance insights, while deep learning models (ANNs, CNNs, RNNs) excel at capturing nonlinear, high-dimensional relationships and temporal dynamics. Physics-informed neural networks bridge the gap between pure data-driven and equation-based models by enforcing governing laws and reducing data requirements. Effective deployment requires careful attention to data quality, feature engineering, hyperparameter tuning, and the choice of evaluation metrics (MAE, MSE, RMSE, R^2). Hybrid approaches that combine traditional simulators with ML corrections or embed physical constraints into learning frameworks promise the best of both worlds: speed, interpretability, and physical fidelity. As computational resources and data availability continue to grow, AI-driven workflows will play an increasingly central role in reservoir engineering, contaminant transport, and environmental remediation applications.

REVIEW QUESTIONS

1. Explain the key differences between supervised, unsupervised, and reinforcement learning in the context of porous-media flow modeling.
2. Describe how physics-informed neural networks (PINNs) incorporate governing partial differential equations into the training process. What are the main advantages and challenges?
3. Discuss the trade-offs between tree-based ensemble methods (e.g., random forest, gradient boosting) and deep neural networks (e.g., CNNs, RNNs) for predicting multiphase flow properties.
4. What data preprocessing steps are critical before training an ML model on reservoir simulation outputs? Why are scaling and handling missing data important?
5. Choose one evaluation metric (MAE, MSE, RMSE, or R^2) and discuss its suitability and limitations for assessing fluid flow predictions in heterogeneous reservoirs.

PROBLEMS

1. Given a dataset of core samples with features $\{\phi, K, \text{GR}, \text{RHOB}, \text{NPHI}\}$ and measured permeability, split the data into training and testing sets, train a random forest regressor, and report RMSE and R^2. Perform feature-importance analysis and discuss which features dominate.
2. Formulate a physics-informed neural network to solve the 1D diffusivity equation
$$\frac{\partial p}{\partial t} = D\frac{\partial^2 p}{\partial x^2},$$
with Dirichlet boundary conditions $p(0,t) = p_L$, $p(L,t) = p_R$ and initial condition $p(x,0) = p_0(x)$. Implement and train the PINN using automatic differentiation in any deep-learning framework, then compare the PINN solution to the analytical solution.
3. Using simulated two-phase flow output (pressure time series at a fixed sensor location), train an LSTM network to forecast the next 10 time steps. Evaluate the forecasting error (MAE) for different sequence lengths and hidden-unit sizes, and identify the optimal architecture.
4. Develop a hybrid surrogate that combines a physics-based two-phase flow simulator with an ANN correction term to predict nanoparticle concentration $C(x,t)$. Describe how you would generate training data, structure the correction network, and evaluate the hybrid model's performance against pure simulator and pure ANN baselines.

DECLARATION

During the preparation of this work, the author used ChatGPT in order to improve the language and readability. After using this tool/service, the author reviewed and edited the content as needed and takes full responsibility for the content of the publication.

References

1. Scheidegger, A. E. 1974. *The physics of flow through porous media.* Toronto: University of Toronto Press.
2. Dullien, F. A. L. 1992. *Porous media: Fluid transport and pore structure.* San Diego: Academic Press.
3. Darcy, H. 1856. *Les fontaines publiques de la ville de Dijon.* Paris: Dalmont.
4. Nield, D. A., and A. Bejan. 2017. Mechanics of fluid flow through a porous medium. In *Convection in porous media.* Cham: Springer. https://doi.org/10.1007/978-3-319-49562-0˙1.
5. Peaceman, D. W. 1977. *Fundamentals of numerical reservoir simulation.* Elsevier.
6. Bear, J. 1972. *Dynamics of fluids in porous media.* New York: Dover Publications.
7. Atkins, P. W., and J. De Paula. 2006. *Atkins' physical chemistry.* 8th ed. New York: W.H. Freeman and Company.
8. Redlich, O., and J. N. S. Kwong. 1949. On the thermodynamics of solutions. V. An equation of state. Fugacities of gaseous solutions. *Chemical Reviews* 44, no. 1:233–44.
9. Soave, G. 1972. Equilibrium constants from a modified Redlich-Kwong equation of state. *Chemical Engineering Science* 27, no. 6:1197–1203.
10. Peng, D.-Y., and D. B. Robinson. 1976. A new two-constant equation of state. *Industrial & Engineering Chemistry Fundamentals* 15, no. 1:59–64.
11. Stryjek, R., and J. H. Vera. 1986. PRSV: An improved Peng–Robinson equation of state for pure compounds and mixtures. *The Canadian Journal of Chemical Engineering* 64, no. 2:323–333.
12. Patel, N. C., and A. S. Teja. 1982. A new cubic equation of state for fluids and fluid mixtures. *Chemical Engineering Science* 37, no. 3:463–73.
13. Hassanpour, R. M., J. G. Manchuk, O. Leuangthong, and C. V. Deutsch. 2010. Calculation of permeability tensors for unstructured gridblocks. *Journal of Canadian Petroleum Technology* 49, no. 10:65–74.
14. Liakopoulos, A. C. 1965. Darcy's coefficient of permeability as symmetric tensor of a second rank. *International Association of Scientific Hydrology. Bulletin* 10:41–8.
15. Rice, P. A., D. J. Fontugne, R. G. Latini, and A. J. Barduhn. 1969. Anisotropic permeability in porous media. *Industrial and Engineering Chemistry* 62:23–31.
16. Salama, A., A. Negara, M. F. El-Amin, and S. Sun. 2015. Numerical investigation of nanoparticles transport in anisotropic porous media. *Journal of Contaminant Hydrology* 181:114–30.

17. Hsieh, P. A., and S. P. Neuman. 1985. Field determination of the three-dimensional hydraulic conductivity tensor of anisotropic media. *Water Resources Research* 21:1655–65.
18. Kasap, E., and L. W. Lake. 1990. Calculating the effective permeability tensor of a gridblock. *SPE Formation Evaluation* 5:192–200.
19. Higham, D. J., and N. J. Higham. 2016. *MATLAB guide*. 3rd ed. Philadelphia: SIAM.
20. Hahn, B. D., and D. T. Valentine. 2021. *Essential MATLAB for engineers and scientists*. 7th ed. Cambridge: Academic Press.
21. Hanselman, D., and B. Littlefield. 2021. *Mastering MATLAB*. Harlow: Pearson.
22. McKinney, W. 2022. *Python for data analysis*. 3rd ed. USA: O'Reilly Media.
23. Oliphant, T. E. 2006. *A guide to NumPy*. USA: Trelgol Publishing.
24. Skeel, R. D., and M. Berzins. 1990. A method for the spatial discretization of parabolic equations in one space variable. *SIAM Journal on Scientific and Statistical Computing* 11, no. 1:1–32.
25. El-Amin, M. F. 2023. *Numerical modeling of nanoparticle transport in porous media: MATLAB/Python approach*. Amsterdam: Elsevier.
26. El-Amin, M. F., A. Salama, and S. Sun. 2015. Numerical and dimensional analysis of nanoparticles transport with two-phase flow in porous media. *Journal of Petroleum Science and Engineering* 128:53–64.
27. El-Amin, M. F., A. Salama, and S. Sun. 2017. Numerical and dimensional investigation of two-phase countercurrent imbibition in porous media. *Journal of Computational and Applied Mathematics* 242:285–96.
28. Lie, K.-A. 2019. An introduction to the MATLAB Reservoir Simulation Toolbox (MRST). *Computational Geosciences* 23:1193–219.
29. SINTEF. 2024. MRST MATLAB Reservoir Simulation Toolbox. https://www.sintef.no/projectweb/mrst/ (accessed November 20, 2024).
30. Arbogast, T., C. N. Dawson, P. T. Keenan, et al. 1998. Enhanced cell-centered finite differences for elliptic equations on general geometry. *SIAM Journal on Scientific Computing*. 19(2):404–425.
31. Sun, S., A. Salama, and M. F. El-Amin. 2012. Matrix-oriented implementation for the numerical solution of the partial differential equations governing flows and transport in porous media. *Computers & Fluids* 68:38–46.
32. Sun, S., A. Salama, and M. F. El Amin. 2012. An equation-type approach for the numerical solution of the partial differential equations governing transport phenomena in porous media. *Procedia Computer Science* 9:661–69.
33. Salama, A., S. Sun, and M. F. El Amin. 2013. A multipoint flux approximation of the steady-state heat conduction equation in anisotropic media. *Journal of Heat Transfer* 135, no. 4:041302.

34. Salama, A., S. Sun, and M. F. El Amin. 2015. Investigation of thermal energy transport from an anisotropic central heating element to the adjacent channels: A multipoint flux approximation. *Annals of Nuclear Energy* 76:100–12.
35. Negara, A., A. Salama, and S. Sun. 2013. 3D numerical investigation of subsurface flow in anisotropic porous media using multipoint flux approximation method. Paper presented at the SPE Reservoir Characterization Simulation Conference and Exhibition, Abu Dhabi, Sept. 16–18, SPE 165960.
36. Zienkiewicz, O. C., and R. L. Taylor. 2000. *The finite element method.* Butterworth-Heinemann.
37. Donea, J. 1982. A Taylor-Galerkin method for convective transport problems. *International Journal for Numerical Methods in Engineering* 20:101–19. https://doi.org/10.1002/nme.1620200108.
38. Douglas, J., and T. Dupont. 1970. Galerkin methods for parabolic problems. *SIAM Journal on Numerical Analysis* 7, no. 4:575–626. https://doi.org/10.1137/0707052.
39. Douglas, J., Jr., and T. Dupont. 1970. Galerkin methods for parabolic problems. *SIAM Journal on Numerical Analysis* 7, no. 4:575–626. https://doi.org/10.1137/0707052.
40. Alber, C., C. Ma, and R. Scheichl. 2024. A mixed multiscale spectral generalized finite element method. arXiv preprint. https://doi.org/10.48550/arXiv.2403.16714.
41. Ly, H.-B., and T.-A. Nguyen. 2023. Accelerating fluid flow simulations through doubly porous media using a FEM-assisted machine learning approach. *Results in Physics* 54:107036. https://doi.org/10.1016/j.rinp.2023.107036.
42. Monaghan, J. J. 1992. Smoothed particle hydrodynamics. *Annual Review of Astronomy and Astrophysics* 30:543–74.
43. Fasshauer, G. E. 2007. *Meshfree approximation methods with MATLAB.* World Scientific.
44. Atluri, S. N., and T. Zhu. 1998. A new meshless local Petrov-Galerkin (MLPG) approach in computational mechanics. *Computational Mechanics* 22:117–27.
45. Belytschko, T., Y. Y. Lu, and L. Gu. 1994. Element-free Galerkin methods. *International Journal for Numerical Methods in Engineering* 37:229–56.
46. LeVeque, R. J. 1992. *Numerical methods for conservation laws.* Birkhauser.
47. Lucy, L. B. 1977. A numerical approach to the testing of the fission hypothesis. *The Astronomical Journal* 82:1013–24. https://doi.org/10.1086/112164.
48. Gingold, R. A., and J. J. Monaghan. 1977. Smoothed particle hydrodynamics: Theory and application to non-spherical stars.

Monthly Notices of the Royal Astronomical Society 181:375–89. https://doi.org/10.1093/mnras/181.3.375.

49. Liu, G. R., and Y. T. Gu. 2003. *An introduction to meshfree methods and their programming.* Springer.
50. Liu, G. R., and Y. T. Gu. 2007. *Meshfree methods: Moving beyond the finite element method.* CRC Press.
51. Bonet, J., and T.-S. L. Lok. 2002. A simplified approach to enhance the performance of smooth particle hydrodynamics methods. *Applied Mathematics and Computation* 126, no. 2-3:133–55.
52. Kansa, E. J. 1990. Multiquadricsa scattered data approximation scheme with applications to computational fluid-dynamics: Surface approximations and partial derivative estimates. *Computers & Mathematics with Applications* 19, no. 8-9:127–45.
53. Arnaldsson, A., J.-C. Berthet, S. Kjaran, and S. P. Sigurosson. 2014. Numerical scheme to simulate flow through anisotropic rocks in TOUGH2. *Computers & Geosciences* 65:37–45.
54. Negara, A., A. Salama, and S. Sun. 2015. Multiphase flow simulation with gravity effect in anisotropic porous media using multipoint flux approximation. *Computers & Fluids* 114:66–74.
55. Aavatsmark, I., T. Barkve, and T. Mannseth. 1998. Control-volume discretization methods for 3D quadrilateral grids in inhomogeneous, anisotropic reservoirs. *SPE Journal* 3:146–54.
56. Aavatsmark, I. 2002. An introduction to multipoint flux approximations for quadrilateral grids. *Computational Geosciences* 6:405–32.
57. Brezzi, F., J. Douglas, and L. D. Marini. 1985. Two families of mixed finite elements for second order elliptic problems. *Numerische Mathematik* 47:217–35.
58. Wheeler, M. F., and I. Yotov. 2006. A multipoint flux mixed finite element method. *SIAM Journal on Numerical Analysis* 44:2082–106.
59. Mekonnen, M., and A. Hoekstra. 2016. Four billion people facing severe water scarcity. *Science Advances* 2:e1500323–e1500323. https://doi.org/10.1126/sciadv.1500323.
60. Gleick, P. H. 1993. Water and conflict: Fresh water resources and international security. *International Security* 18, no. 1:79–112. https://doi.org/10.2307/2539033.
61. Wahlgren, R. V. 2001. Atmospheric water vapour processor designs for potable water production: A review. *Water Research* 35, no. 1:1–22. https://doi.org/10.1016/S0043-1354(00)00247-5.
62. Beysens, D., I. Milimouk, and A. Schweitzer. 2000. *The case for alternative fresh water sources.* Environmental Science.
63. Kim, H., et al. 2017. Water harvesting from air with metal-organic frameworks powered by natural sunlight. *Science* 356, no. 6336:430–34. https://doi.org/10.1126/science.aam8743.

64. Li, R., Y. Shi, L. Shi, M. Alsaedi, and P. Wang. 2018. Harvesting water from air: Using anhydrous salt with sunlight. *Environmental Science & Technology* 52, no. 9:5398–406. https://doi.org/10.1021/acs.est.7b06373.
65. Schemenauer, R. S., and P. I. Joe. 1989. The collection efficiency of a massive fog water collector. *Atmospheric Research* 24:53–69. https://doi.org/10.1016/0169-8095(89)90018-2.
66. ASHRAE. 2017. *ASHRAE handbook - fundamentals.* Atlanta, GA: American Society of Heating, Refrigerating and Air-Conditioning Engineers.
67. Atkins, P., and J. de Paula. 2006. *Atkins' physical chemistry.* 8th ed. Oxford University Press.
68. Gregg, S. J., and K. S. W. Sing. 1982. *Adsorption, surface area and porosity.* 2nd ed. Academic Press.
69. El-Amin, M. F. 2023. Analytical solutions for harvesting atmospheric water using desiccant materials. *Sustainable Water Resources Management* 9:167. https://doi.org/10.1007/s40899-023-00942-y.
70. Kabbaj, N., and M. F. El-Amin. 2024. Prediction of adsorption and desorption isotherms for atmospheric water harvesting. Paper presented at the 2024 21st Learning and Technology Conference. IEEE.
71. Almasarani, A., I. K. Ahmad, M. F. El-Amin, and T. Brahimi. 2022. Experimental investigations and modeling of atmospheric water generation using a desiccant material. *Energies* 15(6):6834. https://doi.org/10.3390/en15186834.
72. Sibie, S. K., M. F. El-Amin, and S. Sun. 2021. Modeling of water generation from air using anhydrous salts. *Energies* 14, no. 13:3822.
73. Kabbaj, N., and M. F. El-Amin. 2025. Numerical modeling with thermodynamic aspects of harvesting atmospheric water using porous desiccant materials. *Sustainable Water Resources Management* 11, no. 3:1–18.
74. Muskat, M. 1937. *The flow of homogeneous fluids through porous media.* McGraw-Hill.
75. Corey, A. T. 1954. The interrelation between gas and oil relative permeabilities. *Producers Monthly* 19, no. 1:38–41.
76. Brooks, R. H., and A. T. Corey. 1964. *Hydraulic properties of porous media.* Hydrology Paper No. 3. Colorado State University.
77. Leverett, M. C. 1941. Capillary behavior in porous solids. *Transactions of AIME* 142:152–69.
78. Lenhard, R. J., and J. C. Parker. 1989. Measurement and prediction of two-phase relative permeability during three-phase immiscible displacements. *Journal of Contaminant Hydrology* 4, no. 1:49–70.
79. Whitaker, S. 1986. Flow in porous media I: A theoretical derivation of Darcy's law. *Transport in Porous Media* 1, no. 1:3–25.
80. Hassanizadeh, S. M., and W. G. Gray. 1993. Fundamentals of multiphase flow in porous media including interfacial phenomena. *Advances in Water Resources* 16, no. 5:87–113.

81. van Genuchten, M. T. 1980. A closed-form equation for predicting the hydraulic conductivity of unsaturated soils. *Soil Science Society of America Journal* 44, no. 5:892–98.
82. Mualem, Y. 1976. A new model for predicting the hydraulic conductivity of unsaturated porous media. *Water Resources Research* 12, no. 3:513–22.
83. Birkholzer, J. T., and Q. Zhou. 2009. Modeling the impact of CO2 leakage on groundwater resources. *Ground Water* 47, no. 5:653–62.
84. Pruess, K., and C.-F. Tsang. 1999. Numerical modeling of multiphase fluid flow in fractured porous media. *Advances in Water Resources* 21, no. 2:73–86.
85. Bennion, B., and S. Bachu. 2008. Drainage and imbibition relative permeability relationships for supercritical CO_2/brine and H_2S/brine systems in intergranular sandstone, carbonate, shale, and anhydrite rocks. *SPE Reservoir Evaluation & Engineering* 11, no. 3:487–96.
86. Aziz, K., and A. Settari. 1979. *Petroleum reservoir simulation.* London: Applied Science Publishers.
87. Chen, Z., G. Huan, and Y. Ma. 2006. *Computational methods for multiphase flows in porous media.* Philadelphia: SIAM Computational Science and Engineering.
88. Sheldon, J. W., and W. T. Cardwell Jr. 1959. One-dimensional, incompressible, noncapillary, two-phase fluid flow in a porous medium. *Transactions of the AIME* 216, no. 1:290–6.
89. Stone, H. L., and A. O. Gardner Jr. 1961. Analysis of gas-cap or dissolved-gas reservoirs. *TAPE AIME* 222:92–104.
90. Kou, J., and S. Sun. 2010. A new treatment of capillarity to improve the stability of IMPES two-phase flow formulation. *Computers and Fluids* 39, no. 10:1923–31.
91. El-Amin, M. F., J. Kou, and S. Sun. 2017. Convergence analysis of the nonlinear iterative method for two-phase flow in porous media associated with nanoparticle injection. *International Journal of Numerical Methods for Heat & Fluid Flow* 27, no. 10:2289–317.
92. Eymard, R., and V. Schleper. 2014. Study of a numerical scheme for miscible two-phase flow in porous media. *Numerical Methods for Partial Differential Equations* 30, no. 3:723–48.
93. Lake, L. W. 1989. *Enhanced oil recovery.* Prentice Hall.
94. Dia, B. M., B. Saad, and M. Saad. 2021. Modeling and simulation of partially miscible two-phase flow with kinetics mass transfer. *Mathematics and Computers in Simulation* 186:29–40.
95. Benson, S. M., and D. R. Cole. 2005. Carbon dioxide capture and storage. *MRS Bulletin* 30, no. 4:303–7.
96. Metz, B., O. Davidson, H. de Coninck, M. Loos, and L. Meyer, eds. 2005. *IPCC special report on carbon dioxide capture and storage.* Cambridge University Press.

97. Juanes, R., E. J. Spiteri, F. M. Orr Jr., and M. J. Blunt. 2006. Impact of relative permeability hysteresis on geological CO2 storage. *Water Resources Research* 42, no. 12.
98. Krevor, S., R. Pini, L. Zuo, and S. M. Benson. 2012. Capillary trapping for geologic carbon dioxide storage–From pore scale physics to field scale implications. *International Journal of Greenhouse Gas Control* 10:374–92.
99. Bachu, S. 2008. CO2 storage in geological media: Role, means, status and barriers to deployment. *Progress in Energy and Combustion Science* 34, no. 2:254–73.
100. Ennis-King, J., and L. Paterson. 2005. Role of convective mixing in the long-term storage of carbon dioxide in deep saline formations. *SPE Journal* 10, no. 3:349–56.
101. Matter, J. M., and P. B. Kelemen. 2009. Permanent storage of carbon dioxide in geological reservoirs by mineral carbonation. *Nature Geoscience* 2, no. 12:837–41.
102. Xu, T., J. A. Apps, and K. Pruess. 2004. Numerical simulation of CO_2 disposal by mineral trapping in deep aquifers. *Applied Geochemistry* 19, no. 6:917–36.
103. Pruess, K. 2008. On CO2 fluid flow and heat transfer behavior in the subsurface, following leakage from a geologic storage reservoir. *Environmental Geology* 54:1677–86.
104. Class, H., A. Ebigbo, R. Helmig, H. K. Dahle, J. M. Nordbotten, M. A. Celia, P. Audigane, M. Darcis, J. Ennis-King, Y. Fan, and B. Flemisch. 2009. A benchmark study on problems related to CO_2 storage in geologic formations. *Computational Geosciences* 13, no. 4:409–34.
105. Hurter, S., D. Labregere, and J. Berge. 2007. Simulations for CO_2 injection projects with compositional simulator. Paper presented at the Offshore Europe Conference. Society of Petroleum Engineers.
106. Nordbotten, J. M., and M. A. Celia. 2011. *Geological storage of CO_2: Modeling approaches for large-scale simulation*. John Wiley & Sons.
107. Nordbotten, J. M., B. Flemisch, S. E. Gasda, H. M. Nilsen, Y. Fan, G. E. Pickup, B. Wiese, M. A. Celia, H. K. Dahle, G. T. Eigestad, and K. Pruess. 2012. Uncertainties in practical simulation of CO_2 storage. *International Journal of Greenhouse Gas Control* 9:234–42.
108. Bielinski, A. 2007. Numerical simulation of CO_2 sequestration in geological formations. PhD diss., University of Stuttgart.
109. Hassanzadeh, H., M. Pooladi-Darvish, and D. W. Keith. 2007. Scaling behavior of convective mixing, with application to geological storage of CO_2. *AIChE Journal* 53, no. 5:1121–31.
110. Riaz, A., M. Hesse, H. A. Tchelepi, and F. M. Orr. 2006. Onset of convection in a gravitationally unstable diffusive boundary layer in porous media. *Journal of Fluid Mechanics* 548:87–111.

111. Gershenzon, N. I., R. W. Ritzi Jr., D. F. Dominic, E. Mehnert, and R. T. Okwen. 2017. Comparison of CO_2 trapping in highly heterogeneous reservoirs with Brooks-Corey and van Genuchten type capillary pressure curves. *Advances in Water Resources* 108:277–90.
112. Spycher, N., K. Pruess, and J. Ennis-King. 2003. CO_2-H_2O mixtures in the geological sequestration of CO_2. I. Assessment and calculation of mutual solubilities from 12 to 100° C and up to 600 bar. *Geochimica et Cosmochimica Acta* 67, no. 16:3015–31.
113. Duan, Z., and R. Sun. 2006. A model to predict phase equilibrium of CH_4 and CO_2 clathrate hydrate in aqueous electrolyte solutions. *American Mineralogist* 91, no. 8-9:1346–54.
114. Duan, Z., R. Sun, C. Zhu, and I. M. Chou. 2003. An improved model for the calculation of CO_2 solubility in aqueous solutions containing Na^+, K^+, Ca^{2+}, Mg^{2+}, Cl^-, and SO_4^{2-}. *Marine Chemistry* 98, no. 2-4:131–9.
115. Spycher, N., and K. Pruess. 2005. CO_2-H_2O mixtures in the geological sequestration of CO_2. II. Partitioning in chloride brines at 12–100° C and up to 600 bar. *Geochimica et Cosmochimica Acta* 69, no. 13:3309–20.
116. Span, R., and W. Wagner. 1996. A new equation of state for carbon dioxide covering the fluid region from the triple-point temperature to 1100 K at pressures up to 800 MPa. *Journal of Physical and Chemical Reference Data* 25, no. 6:1509–96.
117. Fenghour, A., W. A. Wakeham, and V. Vesovic. 1998. The viscosity of carbon dioxide. *Journal of Physical and Chemical Reference Data* 27, no. 1:31–44.
118. Batzle, M., and Z. Wang. 1992. Seismic properties of pore fluids. *Geophysics* 57, no. 11:1396–408.
119. Garcia, J. E. 2001. *Density of aqueous solutions of CO_2*. Technical Report LBNL-49023. Lawrence Berkeley National Laboratory.
120. Berg, S., S. Oedai, and H. Ott. 2013. Displacement and mass transfer between saturated and unsaturated CO_2-brine systems in sandstone. *International Journal of Greenhouse Gas Control* 12:478–92.
121. Xu, X., S. Chen, and D. Zhang. 2006. Convective stability analysis of the long-term storage of carbon dioxide in deep saline aquifers. *Advances in Water Resources* 29, no. 3:397–407.
122. MacMinn, C. W., M. L. Szulczewski, and R. Juanes. 2011. CO_2 migration in saline aquifers. Part 2. Capillary and solubility trapping. *Journal of Fluid Mechanics* 688:321–51.
123. Emami-Meybodi, H., H. Hassanzadeh, C. P. Green, and J. Ennis-King. 2015. Convective dissolution of CO_2 in saline aquifers: Progress in modeling and experiments. *International Journal of Greenhouse Gas Control* 40:238–66.
124. Neufeld, J. A., M. A. Hesse, A. Riaz, M. A. Hallworth, H. A. Tchelepi, and H. E. Huppert. 2010. Convective dissolution of carbon dioxide in saline aquifers. *Geophysical Research Letters* 37, no. 22.

125. Gaus, I., P. Audigane, L. Andre, J. Lions, N. Jacquemet, P. Durst, I. Czernichowski-Lauriol, and M. Azaroual. 2008. Geochemical and solute transport modelling for CO_2 storage, what to expect from it– *International Journal of Greenhouse Gas Control* 2, no. 4:605–25.
126. Xu, T., N. Spycher, E. Sonnenthal, G. Zhang, L. Zheng, and K. Pruess. 2011. TOUGHREACT Version 2.0: A simulator for subsurface reactive transport under non-isothermal multiphase flow conditions. *Computers & Geosciences* 37, no. 6:763–74.
127. Kharaka, Y. K., D. R. Cole, S. D. Hovorka, W. D. Gunter, K. G. Knauss, and B. M. Freifeld. 2006. Gas-water-rock interactions in Frio Formation following CO_2 injection: Implications for the storage of greenhouse gases in sedimentary basins. *Geology* 34, no. 7:577–80.
128. Li, C. P., F. H. Zeng, and F. Liu. 2012. Spectral approximations to the fractional integral and derivative. *Fractional Calculus and Applied Analysis* 15:383–406.
129. Amaefule, J. O., L. L. Handy, and M. T. Ityokumbul. 1988. Capillary number and its influence on scaling of laboratory measured displacement efficiencies. *SPE Reservoir Engineering* 3, no. 1:110–6.
130. Tarhuni, M. N., W. R. Sulaiman, M. Z. Jaafar, M. Milad, and A. M. Alghol. 2021. A review of the dynamic modeling approaches for characterizing fluid flow in naturally fractured reservoirs. *Energy Engineering* 118, no. 4:71–95.
131. Barenblatt, G. I., I. P. Zheltov, and I. N. Kochina. 1960. Basic concepts in the theory of seepage of homogeneous liquids in fissured rocks. *Journal of Applied Mathematics and Mechanics* 24, no. 5:1286–1303.
132. Warren, J. E., and P. J. Root. 1963. The behavior of naturally fractured reservoirs. *Society of Petroleum Engineers Journal* 3, no. 03:245–55.
133. Kazemi, H. 1969. Pressure transient analysis of naturally fractured reservoirs with uniform fracture distribution. *Society of Petroleum Engineers Journal* 9, no. 04 (December):451–62.
134. Pirson, S. J. 1953. Performance of fractured oil reservoirs. *AAPG Bulletin* 37, no. 2 (February):232–44.
135. Noorishad, J., and M. Mehran. 1982. An upstream finite element method for solution of transient transport equation in fractured porous media. *Water Resources Research* 18, no. 3:588–96.
136. Hoteit, H., and A. Firoozabadi. 2008. An efficient numerical model for incompressible two-phase flow in fractured media. *Advances in Water Resources* 31, no. 6 (June):891–905.
137. Lee, S. H., C. L. Jensen, and M. F. Lough. 2000. Efficient finite-difference model for flow in a reservoir with multiple length-scale fractures. *SPE Journal* 3, no. 5:268–75.
138. Lee, S. H., M. F. Lough, and C. L. Jensen. 2001. Hierarchical modeling of flow in naturally fractured formations with multiple length scales. *Water Resources Research* 37:443–55.

139. Li, L., and S. H. Lee. 2008. Efficient field-scale simulation of black oil in naturally fractured reservoir through discrete fracture networks and homogenized media. Paper presented at the SPE Reservoir Simulation Symposium, 750–58.
140. Amir, S. Z., and S. Sun. 2017. An efficient two-scale hybrid embedded fracture model for shale gas simulation. *Journal of Petroleum Science and Engineering* 152:683–714.
141. Salama, A., M. F. El Amin, K. Kumar, and S. Sun. 2017. Flow and transport in tight and shale formations: A review. *Geofluids* 2017:4251209. https://doi.org/10.1155/2017/4251209.
142. Chen, H., and S. Sun. 2016. A two-scale reduced model for Darcy flow in fractured porous media. Paper presented at the Proceedings of the International Conference on Computational Science (ICCS) 80:1324–33.
143. Douglas, J., and T. Arbogast. 1989. *Dual-porosity models for flow in naturally fractured reservoirs.* Technical Report. Purdue University.
144. Peaceman, D. W. 1978. Interpretation of well-block pressure in numerical reservoir simulation. Paper presented at the SPE-AIME 52nd Annual Fall Technical Conference and Exhibition, 183–94, SPE 6893.
145. El-Amin, M. F. 2017. Analytical solution of the apparent-permeability gas-transport equation in porous media. *The European Physical Journal Plus* 132:129.
146. Bustin, A. M. M., X. Cui, D. J. K. Ross, and R. M. Bustin. 2010. Impact of volumetric gas storage capacity on production in shale gas reservoirs. *Journal of Petroleum Science and Engineering* 71, no. 1-2:118–33. https://doi.org/10.1016/j.petrol.2010.01.007.
147. Ozkan, E., R. Raghavan, and H. Kazemi. 1999. Modeling of naturally fractured reservoirs with dual-porosity and dual-permeability models. *SPE Reservoir Evaluation & Engineering* 2, no. 5:438–46. https://doi.org/10.2118/56574-PA.
148. Moridis, G. J., and E. D. Sloan. 2007. Gas production from unconventional reservoirs: Mechanisms and models. *Energy Conversion and Management* 48, no. 6:1834–58. https://doi.org/10.1016/j.enconman.2007.01.017.
149. Wu, Y. S., and P. Fakcharoenphol. 2005. A generalized dual-continuum approach for modeling flow and transport processes in fractured reservoirs. *Transport in Porous Media* 58, no. 1-2:121–49. https://doi.org/10.1007/s11242-004-2199-2.
150. Guo, T., T. Xu, and J. Zhang. 2015. Modeling gas flow in tight shale reservoirs with a mathematical model incorporating multiphysics mechanisms. *Journal of Natural Gas Science and Engineering* 22:695–706. https://doi.org/10.1016/j.jngse.2015.02.003.
151. El Amin, M. F., S. Amir, A. Salama, D. Urozayev, and S. Sun. 2017. Comparative study of shale-gas production using single- and

dual-continuum approaches. *Journal of Petroleum Science and Engineering* 157:894–905. https://doi.org/10.1016/j.petrol.2017.07.029.

152. Arbogast, T., J. Douglas, and U. Hornung. 1993. Derivation of the dual-porosity model for single-phase and multiphase flow. *SIAM Journal on Applied Mathematics* 53, no. 6:1598–618. https://doi.org/10.1137/0153079.
153. Showalter, R. E. 2000. Mathematical properties of dual-porosity models for flow in fractured media. *Mathematical Geosciences* 32, no. 1:137–50. https://doi.org/10.1023/A:1007545829273.
154. Biot, M. A., and D. G. Willis. 1957. The elastic coefficients of the theory of consolidation. *Journal of Applied Mechanics* 24:594–601.
155. El-Amin, M. F., J. Kou, and S. Sun. 2018. Numerical modeling and simulation of shale-gas transport with geomechanical effect. *Transport in Porous Media.* https://doi.org/10.1007/s11242-018-1206-z.
156. El-Amin, M. F., J. Kou, and S. Sun. 2018. Mixed finite element simulation with stability analysis for gas transport in low-permeability reservoirs. *Energies* 11, no. 1:208–26. https://doi.org/10.3390/en11010208.
157. El-Amin, M. F., J. Kou, and S. Sun. 2020. Theoretical stability analysis of mixed finite element model of shale-gas flow with geomechanical effect. *Oil & Gas Science and Technology - Rev. IFP Energies nouvelles* 75:33. https://doi.org/10.2516/ogst/2020026.
158. El-Amin, M. F., and M. A. El-Beltagy. 2021. Stochastic estimation of the slip factor in apparent permeability model of gas transport in porous media. *Transport in Porous Media* 137:433–49. https://doi.org/10.1007/s11242-021-01575-5.
159. Atanackovic, T. M., S. Pilipovic, B. Stankovic, and D. Zorica. 2014. *Fractional calculus with applications in mechanics: Wave propagation, impact and variational principles.* John Wiley & Sons.
160. Azar, A. T., A. G. Radwan, and S. Vaidyanathan. 2021. *Fractional order systems: Optimization, control, circuit realizations and applications.* Academic Press.
161. Badr, A. A. 2012. Finite element method for linear multiterm fractional differential equations. *Journal of Applied Mathematics* 2012:482890. https://doi.org/10.1155/2012/482890.
162. Benson, D. A., S. W. Wheatcraft, and M. M. Meerschaert. 2000. Fractional dispersion, Levy motion, and the MADE tracer tests. *Transport in Porous Media* 42:211–40.
163. Benson, D. A., R. Schumer, M. M. Meerschaert, and S. W. Wheatcraft. 2001. The fractional-order governing equation of Levy motion. *Water Resources Research* 37, no. 6:1413–23.
164. Benson, D. A., M. M. Meerschaert, and J. Revielle. 2004. Applications of fractional advection-dispersion equations. *Advances in Water Resources* 27, no. 5:497–510.

165. Caputo, M. 1967. Linear models of dissipation whose Q is almost frequency independent–II. *Geophysical Journal International* 13, no. 5:529–39.
166. Carella, A. R. 2012. Spectral finite element methods for solving fractional differential equations with applications in anomalous transport. PhD diss., Norwegian University of Science and Technology.
167. Chang, A., H. Sun, Y. Zhang, C. Zheng, and F. Min. 2019. Spatial fractional Darcy's law to quantify fluid flow in natural reservoirs. *Physica A* 519:119–26.
168. Chen, W., and S. Holm. 2009. *Fractional derivatives for physicists and engineers*. Springer.
169. Clemence-Mkhope, D., and B. Clemence-Mkhope. 2021. The limited validity of the conformable Euler finite difference method and an alternate definition of the conformable fractional derivative to justify modification of the method. *Mathematical and Computational Applications* 26, no. 4:66. https://doi.org/10.3390/mca26040066.
170. Cushman, J. H., and T. R. Ginn. 1993. Nonlocal dispersive transport theory. *Water Resources Research* 29, no. 2:391–401.
171. Dehghan, M., and M. Abbaszadeh. 2017. Spectral element technique for nonlinear fractional evolution equation, stability and convergence analysis. *Applied Numerical Mathematics* 119:51–66.
172. Dehghan, M., and M. Abbaszadeh. 2018. A Legendre spectral element method (SEM) based on the modified bases for solving neutral delay distributed-order fractional damped diffusion-wave equation. *Mathematical Methods in the Applied Sciences* 41, no. 9:3476–94.
173. El-Amin, M. F., A. G. Radwan, and S. Sun. 2017. Analytical solution for fractional derivative gas-flow equation in porous media. *Results in Physics* 7:2432–8.
174. El-Amin, M. F. 2021. Derivation of fractional-derivative models of multiphase fluid flows in porous media. *Journal of King Saud University - Science* 33, no. 6:101562.
175. El-Amin, M. F. 2025. *Fractional modeling of fluid flow and transport phenomena*. Elsevier.
176. Feng, L. B., P. Zhuang, F. Liu, I. Turner, and Y. T. Gu. 2016. Finite element method for space-time fractional diffusion equation. *Numerical Algorithms* 72:749–67. https://doi.org/10.1007/s11075-015-0065-8.
177. Gorenflo, R., A. A. Kilbas, F. Mainardi, and S. V. Rogosin. 2013. *Fractional calculus*. Springer.
178. Hilfer, R. 2000. *Applications of fractional calculus in physics*. World Scientific.
179. Jiang, B.-n. 1998. *The least-squares finite element method: Theory and applications in computational fluid dynamics and electromagnetics*. Berlin/Heidelberg: Springer-Verlag.

180. Khalil, R., M. Al Horani, A. Yousef, and M. Sababheh. 2014. A new definition of fractional derivative. *Journal of Computational and Applied Mathematics* 264:65–70.
181. Kharazmi, E., M. Zayernouri, and G. E. Karniadakis. 2017. A Petrov–Galerkin spectral element method for fractional elliptic problems. *Computer Methods in Applied Mechanics and Engineering* 324:512–36.
182. Kilbas, A. A., H. M. Srivastava, and J. J. Trujillo. 2006. *Theory and applications of fractional differential equations.* Elsevier.
183. Li, C., et al. 2020. The discontinuous Galerkin finite element method for fractional differential equations. *Advances in Computational Mathematics* 46, no. 1:26.
184. Li, Q., and F. Song. 2020. Splitting spectral element method for fractional reaction-diffusion equations. *Journal of Algorithms & Computational Technology* 14:1748302620966705.
185. Lin, Y. M., and C. J. Xu. 2007. Finite difference/spectral approximations for the time-fractional diffusion equation. *Journal of Computational Physics* 225:1533–52.
186. Liu, Y., Y. Du, H. Li, and J. Wang. 2015. An H^1-Galerkin mixed finite element method for time fractional reaction-diffusion equation. *Journal of Applied Mathematics and Computing* 47:103–17. https://doi.org/10.1007/s12190-014-0764-7.
187. Machado, J. T., V. Kiryakova, and F. Mainardi. 2011. A review of definitions for fractional derivatives and integral. *Mathematical Problems in Engineering* 2011:1–9.
188. Meerschaert, M. M., and C. Tadjeran. 2004. Finite difference approximations for fractional advection–dispersion flow equations. *Journal of Computational and Applied Mathematics* 172:65–77.
189. Metzler, R., and J. Klafter. 2000. The random walk's guide to anomalous diffusion: A fractional dynamics approach. *Physics Reports* 339:1–77.
190. Metzler, R., and J. Klafter. 2004. The restaurant at the end of the random walk: Recent developments in the description of anomalous transport by fractional dynamics. *Journal of Physics A: Mathematical and General* 37:R161–R208.
191. Obembe, A. D., M. E. Hossain, and S. A. Abu-Khamsin. 2017. Variable-order derivative time fractional diffusion model for heterogeneous porous media. *Journal of Petroleum Science and Engineering* 152:391–405.
192. Oldham, K. B., and J. Spanier. 1974. *The fractional calculus.* Academic Press.
193. Ortigueira, M. D. 2011. *Fractional calculus for scientists and engineers.* Springer.
194. Podlubny, I. 1998. *Fractional differential equations.* Vol. 198. Academic Press.
195. Podlubny, I. 1999. Fractional-order systems and PID controllers. *IEEE Transactions on Automatic Control* 44, no. 1:208–14.

196. Press, W. H., S. A. Teukolsky, W. T. Vetterling, and B. P. Flannery. 1993. *Numerical recipes in C: The art of scientific computing*. 2nd ed. Cambridge University Press.
197. Schumer, R., D. A. Benson, M. M. Meerschaert, and S. W. Wheatcraft. 2003. Fractal mobile/immobile solute transport. *Water Resources Research* 39, no. 10:1296.
198. Sheng, C., and J. Shen. 2017. A hybrid spectral element method for fractional two-point boundary value problems. *Numerical Mathematics: Theory, Methods and Applications* 10, no. 2:437–64.
199. Tian, W. Y., W. H. Deng, and Y. J. Wu. 2014. Polynomial spectral collocation method for space fractional advection-diffusion equation. *Numerical Methods for Partial Differential Equations* 30:514–35.
200. Zaslavsky, G. M. 2002. Chaos, fractional kinetics, and anomalous transport. *Physics Reports* 371:461–580.
201. Zayernouri, M., W. Cao, Z. Zhang, and G. E. Karniadakis. 2014. Spectral and discontinuous spectral element methods for fractional delay equations. *SIAM Journal on Scientific Computing* 36, no. 6:B904–B929.
202. Zhao, X. 2017. Adaptive finite element method for fractional differential equations. *Computational Methods in Applied Mechanics and Engineering* 304:1–77.
203. Zhong, W., C. Li, and J. Kou. 2013. Numerical fractional-calculus model for two-phase flow in fractured media. *Advanced Topics in Fractional Dynamics* 2013:429835. https://doi.org/10.1155/2013/429835.
204. Kolditz, O., and J. De Jonge. 2004. Non-isothermal two-phase flow in low-permeable porous medium. *Computational Mechanics* 33:345–64.
205. Singh, A. K., W. Wang, C. H. Park, U. J. Gorke, and O. Kolditz. 2010. Non-isothermal effects of multi-phase flow in porous medium. In *XVIII International Conference on Water Resources (CMWR 2010)*, ed. J. Carrera. Barcelona.
206. Cortellessa, G., F. Arpino, S. D. Fraia, and M. Scungio. 2018. Two-phase explicit CBS procedure for compressible viscous flow transport in porous materials. *International Journal of Numerical Methods for Heat & Fluid Flow* 28:336–60.
207. El-Amin, M. F. 2019. Iterative numerical scheme for non-isothermal two-phase flow in heterogeneous porous media. *Algorithms* 12, no. 6:117.
208. Falta, R. W., K. Pruess, I. Javandel, and P. A. Witherspoon. 1992. Numerical modelling of steam injection for the removal of nonaqueous phase liquids from the subsurface. 2. Code validation and application. *Advances in Water Resources* 28:451–65.
209. Forsyth, P. A. 1993. A positivity preserving method for simulation of steam injection for NAPL site remediation. *Advances in Water Resources* 16:351–70.
210. Afif, A., and B. Amaziane. 2001. Convergence of finite volume schemes for a degenerate convection-diffusion equation in porous media. *Journal of Porous Media* 5, no. 1:45–60.

211. Afif, A., and B. Amaziane. 2002. Further analysis of finite volume methods for convection-diffusion equations in porous media. *Computational Geosciences* 6, no. 2:123–37.
212. Baca, R. G., R. C. Arnett, and D. W. Langford. 1984. Modelling fluid flow in fractured-porous rock masses by finite-element techniques. *International Journal for Numerical Methods in Fluids* 4, no. 4:337–48.
213. Bai, M., D. Elsworth, and J.-C. Roegiers. 1993. Multiporosity/-multipermeability approach to the simulation of naturally fractured reservoirs. *Water Resources Research* 29, no. 6:1621–33. https://doi.org/10.1029/92WR02746.
214. Bear, J., and B. Berkowitz. 1987. Groundwater flow and pollution in fractured rock aquifers. In *Development of hydraulic engineering*, vol. 4, ed. P. Novak. Oxford, England: Elsevier Applied Science.
215. Bradford, S. A., J. Simunek, M. Bettahar, M. T. Van Genuchten, and S. R. Yates. 2003. Modeling colloid attachment, straining, and exclusion in saturated porous media. *Environmental Science & Technology* 37, no. 10:2242–50. https://doi.org/10.1021/es025899u.
216. Bradford, S. A., and M. Bettahar. 2004. Straining and attachment of colloids in physically heterogeneous porous media. *Vadose Zone Journal* 3, no. 2:384–94. https://doi.org/10.2113/3.2.384.
217. Bradford, S. A., S. Torkzaban, H. Kim, and J. Simunek. 2012. Modeling colloid and microorganism transport and release with transients in solution ionic strength. *Water Resources Research* 48, no. 9. https://doi.org/10.1029/2012WR012468.
218. Courdiere, C., et al. 2004. Convergence investigations of finite volume schemes for convection-diffusion equations. *Numerical Methods for Partial Differential Equations* 21, no. 4:455–72.
219. El-Amin, M. F., A. Salama, and S. Sun. 2015. Numerical and dimensional analysis of nanoparticles transport with two-phase flow in porous media. *Journal of Petroleum Science and Engineering* 128:53–64.
220. El-Amin, M. F., A. Salama, and S. Sun. 2017. Numerical and dimensional investigation of two-phase countercurrent imbibition in porous media. *Journal of Computational and Applied Mathematics* 242:285–96.
221. El-Amin, M. F., S. Amir, A. Salama, D. Urozayev, and S. Sun. 2017. Comparative study of shale-gas production using single- and dual-continuum approaches. *Journal of Petroleum Science and Engineering* 157:894–905.
222. El-Amin, M. F., J. Kou, and S. Sun. 2018. Discrete-fracture model of multi-scale time-splitting two-phase flow including nanoparticles transport in fractured porous media. *Journal of Computational and Applied Mathematics* 333:327–49. https://doi.org/10.1016/j.cam.2017.11.008.

223. El-Amin, M. F., J. Kou, S. Sun, and J. Li. 2019. Mixed finite element solution for the natural-gas dual-mechanism model. In *Computational science – ICCS 2019.* Lecture Notes in Computer Science, vol. 11540. Springer, Cham. https://doi.org/10.1007/978-3-030-22750-0_35.
224. El-Amin, M. F., J. Kou, and S. Sun. 2020. Numerical investigation of solute transport in fractured porous media using the discrete fracture model. In *Computational science – ICCS 2020.* Lecture Notes in Computer Science, vol. 12143. Springer, Cham. https://doi.org/10.1007/978-3-030-50436-6_8.
225. El-Amin, M. F. 2023. Analytical solutions for harvesting atmospheric water using desiccant materials. *Sustainable Water Resources Management* 9:167. https://doi.org/10.1007/s40899-023-00942-y.
226. Eymard, R., and T. Gallouet. 2007. Convergence of a numerical technique for immiscible two-phase flow in porous media. *SIAM Journal on Numerical Analysis* 45, no. 4:1320–41.
227. Feynman, R. P. 1992. There's plenty of room at the bottom [data storage]. *Journal of Microelectromechanical Systems* 1, no. 1:60–66. https://doi.org/10.1109/84.128057.
228. Foppen, J. W., M. van Herwerden, and J. Schijven. 2007. Transport of Escherichia coli in saturated porous media: Dual mode deposition and intra-population heterogeneity. *Water Research* 41, no. 8:1743–53. https://doi.org/10.1016/j.watres.2006.12.041.
229. Gargiulo, G., S. Bradford, J. Simunek, P. Ustohal, H. Vereecken, and E. Klumpp. 2007. Bacteria transport and deposition under unsaturated conditions: The role of the matrix grain size and the bacteria surface protein. *Journal of Contaminant Hydrology* 92, no. 3:255–73. https://doi.org/10.1016/j.jconhyd.2007.01.009.
230. Jenny, P., et al. 2003. An unconditionally convergent nonlinear solution for hyperbolic conservation laws with S-shaped flux functions. *SIAM Journal on Numerical Analysis* 40, no. 3:789–808.
231. Johnson, P. R., and M. Elimelech. 1995. Dynamics of colloid deposition in porous media: Blocking based on 456 sequential adsorption. *Langmuir* 11, no. 3:801–12. https://doi.org/10.1021/la00003a023.
232. Ju, B., and T. Fan. 2009. Experimental study and mathematical model of nanoparticle transport in porous media. *Powder Technology* 192, no. 2:195–202. https://doi.org/10.1016/j.powtec.2008.12.017.
233. Kasel, D., S. A. Bradford, J. Sim–nek, M. Heggen, H. Vereecken, and E. Klumpp. 2013. Transport and retention of multi-walled carbon nanotubes in saturated porous media: Effects of input concentration and grain size. *Water Research* 47, no. 2:933–44. https://doi.org/10.1016/j.watres.2012.11.019.
234. Kazemi, H., and L. S. Merrill. 1979. Numerical simulation of water imbibition in fractured cores. *Society of Petroleum Engineers AIME Journal* 19, no. 3:175–82. https://doi.org/10.2118/6895-pa.

235. Kazemi, H., J. R. Gilman, and A. M. Elsharkawy. 1992. Analytical and numerical solution of oil recovery from fractured reservoirs with empirical transfer functions (includes associated papers 25528 and 25818). *SPE Reservoir Engineering* 7, no. 2:219–27. https://doi.org/10.2118/19849-PA.

236. Kazemzadeh, Y., S. Shojaei, and M. Riazi. 2019. Review on application of nanoparticles for EOR purposes: A critical review of the opportunities and challenges. *Chinese Journal of Chemical Engineering* 27, no. 2:237–46.

237. Kou, X., and Y. Sun. 2010. An iterative IMPES scheme for two-phase flow with capillarity in heterogeneous porous media. *Journal of Petroleum Science and Engineering* 75, no. 1-2:56–68.

238. Lacroix, S., et al. 2008. An iterative fully implicit method for multiphase flow in porous media. *Journal of Computational Physics* 227, no. 1:290–312.

239. Lapidus, L., and N. R. Amundson. 1952. Mathematics of adsorption in beds. VI. The effect of longitudinal diffusion in ion exchange and chromatographic columns. *Journal of Physical Chemistry* 56, no. 8:984–88. https://doi.org/10.1021/j150500a014.

240. Lashari, N., and T. Ganat. 2020. Emerging applications of nanomaterials in chemical enhanced oil recovery: Progress and perspective. *Chinese Journal of Chemical Engineering* 28, no. 8:1995–2009.

241. Lu, Y., and M. F. Wheeler. 2009. An iterative numerical simulation for reservoir flow on high performance computers. *SIAM Journal on Scientific Computing* 31, no. 2:1234–50.

242. McDowell-Boyer, L. M., J. R. Hunt, and N. Sitar. 1986. Particle transport through porous media. *Water Resources Research* 22, no. 13:1901–21. https://doi.org/10.1029/WR022i013p01901.

243. Pruess, K. 1985. A practical method for modeling fluid and heat flow in fractured porous media. *Society of Petroleum Engineers Journal* 25, no. 1:14–26. https://doi.org/10.2118/10509-PA.

244. Saad, M., and N. Saad. 2006. Convergence of a combined finite volume and nonconforming finite element scheme on generic meshes for miscible two-phase flow in anisotropic porous media. *Advances in Water Resources* 29, no. 3:357–70.

245. Schijven, J. F., S. M. Hassanizadeh, and R. H. A. M. de Bruin. 2002. Two-site kinetic modeling of bacteriophages transport through columns of saturated dune sand. *Journal of Contaminant Hydrology* 57, no. 3:259–79. https://doi.org/10.1016/S0169-7722(01)00215-7.

246. Slough, K. J., E. A. Sudicky, and P. A. Forsyth. 1999. Grid refinement for modeling multiphase flow in discretely fractured porous media. *Advances in Water Resources* 23, no. 3:261–9. https://doi.org/10.1016/S0309-1708(99)00009-3.

247. Sun, X., Y. Zhang, G. Chen, and Z. Gai. 2017. Application of nanoparticles in enhanced oil recovery: A critical review of recent progress. *Energies* 10, no. 3:345. https://doi.org/10.3390/en10030345.
248. Tufenkji, N., and M. Elimelech. 2004. Deviation from the classical colloid filtration theory in the presence of repulsive DLVO interactions. *Langmuir* 20, no. 25:10818–28. https://doi.org/10.1021/la0486638.
249. Tufenkji, N., and M. Elimelech. 2004. Correlation equation for predicting single-collector efficiency in physicochemical filtration in saturated porous media. *Environmental Science & Technology* 38, no. 2:529–36. https://doi.org/10.1021/es034049r.
250. Tufenkji, N., and M. Elimelech. 2005. Spatial distributions of Cryptosporidium oocysts in porous media: Evidence for dual mode deposition. *Environmental Science & Technology* 39, no. 10:3620–9. https://doi.org/10.1021/es048289y.
251. Van Genuchten, M. Th. 1981. Analytical solutions for chemical transport with simultaneous adsorption, zero-order production and first-order decay. *Journal of Hydrology* 49, no. 3:213–33. https://doi.org/10.1016/0022-1694(81)90214-6.
252. Abohamzeh, E., F. Salehi, M. Sheikholeslami, R. Abbassi, and F. Khan. 2021. Review of hydrogen safety during storage, transmission, and applications processes. *Journal of Loss Prevention in the Process Industries* 72:104569. https://doi.org/10.1016/j.jlp.2021.104569.
253. Agyekum, E. B., C. Nutakor, A. M. Agwa, and S. Kamel. 2022. A critical review of renewable hydrogen production methods: Factors affecting their scale-up and its role in future energy generation. *Membranes* 12, no. 2:173. https://doi.org/10.3390/membranes12020173.
254. American Institute of Aeronautics and Astronautics. 2004. *Guide to safety of hydrogen and hydrogen systems.* AIAA standard G-095-2004. Reston, VA, USA.
255. Albadan, D. J., M. Delshad, and K. Sepehrnoori. 2024. Insights and strategies for underground hydrogen storage optimization. Paper presented at APOGCE 2024, Perth, Australia. https://doi.org/10.2118/221245-MS.
256. Alcock, J. L., L. C. Shirvill, and R. F. Cracknell. 2001. Compilation of existing safety data on hydrogen and comparative fuels. Deliverable report of European FP5 project EIHP2. http://www.eihp.org/public/documents/CompilationExistingSafetyData.pdf.
257. Ali, A. 2021. Data-driven based machine learning models for predicting the deliverability of underground natural gas storage in salt caverns. *Energy* 229:120648. https://doi.org/10.1016/j.energy.2021.120648.
258. APS. n.d. One man's crusade to exonerate hydrogen for Hindenburg disaster. https://www.aps.org.
259. Bai, M., et al. 2014. An overview of hydrogen underground storage technology and prospects in China. *Journal of Petroleum Science and Engineering* 122:38–495.

260. Blondel, V. D., J. L. Guillaume, R. Lambiotte, and E. Lefebvre. 2008. Fast unfolding of communities in large networks. *Journal of Statistical Mechanics: Theory and Experiment* 2008:P10008.
261. Bo, Z., M. Boon, H. Hajibeygi, and S. Hurter. 2024. Impact of experimentally measured relative permeability hysteresis on reservoir-scale performance of underground hydrogen storage (UHS). *Energy Storage.*
262. Boon, M., and H. Hajibeygi. 2022. Experimental characterization of H2/water multiphase flow in heterogeneous sandstone rock at the core scale relevant for underground hydrogen storage (UHS). *Scientific Reports* 12:14604. https://doi.org/10.1038/s41598-022-18759-8.
263. Brunel. n.d. 3 main types of hydrogen. https://www.brunel.net/en/blog/renewable-energy/3-main-types-of-hydrogen.
264. Buncefield Major Incident Investigation Board. 2005. Buncefield incident investigation report. UK Health and Safety Executive.
265. 2010. The Buncefield major incident investigation board. https://www.hse.gov.uk/comah/buncefield/policyproceduresreport.pdf.
266. Cai, Z., K. Zhang, and C. Guo. 2022. Development of a novel simulator for modelling underground hydrogen and gas mixture storage. *International Journal of Hydrogen Energy* 47, no. 14:8929–42.
267. Carden, P. O., and L. Paterson. 1979. Physical, chemical and energy aspects of underground hydrogen storage. *International Journal of Hydrogen Energy* 4:559–69. https://doi.org/10.1016/0360-3199(79)90083-1.
268. Chen, Q., T. T. Wu, and M. Fang. 2013. Detecting local community structure in complex networks based on local degree central nodes. *Physica A* 392:529–37.
269. Clauset, A., M. E. J. Newman, and C. Moore. 2004. Finding community structure in very large networks. *Physical Review E* 70:066111.
270. Danon, L., A. Diaz-Guilera, J. Duch, and A. Arenas. 2005. Comparing community structure identification. *Journal of Statistical Mechanics: Theory and Experiment* 2005:P09008.
271. Doan, Q. T., A. Keshavarz, P. Behrenbruch, and S. Iglauer. 2024. Using molecular dynamics simulation to estimate storage capacity in depleted gas reservoirs for underground hydrogen storage. Paper presented at APOGCE 2024, Perth, Australia, SPE-221244-MS. https://doi.org/10.2118/221244-MS.
272. Dong, M., F. Dullien, and I. Dai. 2015. Immiscible displacement in the interacting capillary bundle model part I. Development of interacting capillary bundle model. *Transport in Porous Media* 59, no. 1:1–18.
273. See Alcock, J. L., L. C. Shirvill, and R. F. Cracknell. 2001.
274. El-Amin, M. F. 2022. Modeling, analysis, and simulation of hydrogen leakage jet in the air. In *Industrial transformation*, 129–42. Taylor & Francis.

275. El-Amin, M. F., and H. Kanayama. 2008. Boundary layer theory approach to the concentration layer adjacent to a ceiling wall at impinging region of a hydrogen leakage. *International Journal of Hydrogen Energy* 33, no. 21:6393–400.
276. El-Amin, M. F., M. Inoue, and H. Kanayama. 2008. Boundary layer theory approach to the concentration layer adjacent to a ceiling wall of a hydrogen leakage: Far region. *International Journal of Hydrogen Energy* 33, no. 24:7642–7.
277. El-Amin, M. F., and H. Kanayama. 2009. Similarity consideration of the buoyant jet resulting from hydrogen leakage. *International Journal of Hydrogen Energy* 34:5803–9.
278. El-Amin, M. F. 2009. Non-Boussinesq turbulent buoyant jet resulting from hydrogen leakage in air. *International Journal of Hydrogen Energy* 34:7873–82.
279. El-Amin, M. F., and H. Kanayama. 2009. Integral solutions for selected turbulent quantities of small-scale hydrogen leakage: A non-buoyant jet or momentum-dominated buoyant jet regime. *International Journal of Hydrogen Energy* 34, no. 3:1607–12.
280. El-Amin, M. F., and Shuyu Sun. 2012. Horizontal H2-Air turbulent buoyant jet resulting from hydrogen leakage. *International Journal of Hydrogen Energy* 37, no. 4:3949–57.
281. El-Amin, M. F., Shuyu Sun, and Amgad Salama. 2012. Theoretical analysis and semi-analytical solutions for a turbulent buoyant hydrogen-air jet. *Journal of Applied Mathematics* 2012:1–18.
282. El-Amin, M. F., Y. A. Jaha, and M. A. El-Ameen. 2013. Planar momentum-dominated regime of small-scale hydrogen leakage. *International Journal of Hydrogen Energy* 38:15553–8.
283. El-Amin, M. F., and A. Subasi. 2020. Forecasting a small-scale hydrogen leakage in air using machine learning techniques. In *Proceedings of the 2nd International Conference on Computer and Information Sciences* (ICCIS), 1–6. IEEE.
284. El-Amin, M. F. 2022. Prediction of hydrogen leakage in air using random forest technique. In *Lecture Notes in Networks and Systems*, FICC 2022, vol. 439, 393–405. Springer.
285. El-Amin, M. F. 2023. Detection of hydrogen leakage using different machine learning techniques. In *Proceedings of the 20th Learning and Technology Conference*, 74–79. IEEE.
286. El-Amin, M. F. 2024. Digital twin integration for hydrogen leakage modeling and analysis. In *Proceedings of the L&T 24 Conference*. IEEE.
287. Ellabban, O., et al. 2014. Renewable energy resources: Current status, future prospects and their enabling technology. *Renewable and Sustainable Energy Reviews* 39:748–7641.
288. Fabio, D. R., D. Fabio, and P. Carlo. 2013. Profiling core-periphery network structure by random walkers. *Scientific Reports* 3:1467.

289. Fabricio, B., and Z. Liang. 2013. Fuzzy community structure detection by particle competition and cooperation. *Soft Computing* 17:659–73.

290. Feldmann, F., B. Hagemann, L. Ganzer, et al. 2016. Numerical simulation of hydrodynamic and gas mixing processes in underground hydrogen storages. *Environmental Earth Sciences* 75:1165. https://doi.org/10.1007/s12665-016-5948-z.

291. Flixborough Explosion Investigation Committee. 1974. The Flixborough disaster 1974. UK Health and Safety Executive.

292. Health and Safety Executive. 1975. *The Flixborough disaster: Report of the court of inquiry.* HMSO.

293. Fortunato, S., and M. Barthelemy. 2007. Resolution limit in community detection. *Proceedings of the National Academy of Sciences of the United States of America* 104:36–41.

294. Fortunato, S. 2010. Community detection in graphs. *Physics Reports-Review Section of Physics Letters* 486:75–174.

295. Ganzer, L., V. Reitenbach, D. Pudlo, et al. 2013. The H2STORE project - Experimental and numerical simulation approach to investigate processes in underground hydrogen reservoir storage. Paper presented at the EAGE Annual Conference & Exhibition incorporating SPE Europec. https://doi.org/10.2118/164936-MS.

296. Gbadamosi, A. O., N. S. Muhammed, S. Patil, et al. 2023. Underground hydrogen storage: A critical assessment of fluid-fluid and fluid-rock interactions. *Energy Storage* 2023:108473. https://doi.org/10.1016/j.est.2023.108473.

297. Van Genuchten, M. T., and D. R. Nielsen. 1985. On describing and predicting the hydraulic properties of unsaturated soils. *Annales Geophysicae* 3, no. 5:615–28.

298. Ghaedi, M., P. Andersen, and R. Gholami. 2024. Mixing dynamics and recovery factor during hydrogen storage in depleted gas reservoirs. *Gas Science and Engineering* 128:205382. https://doi.org/10.1016/j.jgsce.2024.205382.

299. Gregory, S. 2011. Fuzzy overlapping communities in networks. *Journal of Statistical Mechanics: Theory and Experiment* 2011:P02017.

300. Hagemann, B., M. Rasoulzadeh, M. Panfilov, et al. 2015. Mathematical modeling of unstable transport in underground hydrogen storage. *Environmental Earth Sciences* 73:6891–98. https://doi.org/10.1007/s12665-015-4414-7.

301. Hassanpouryouzband, A., K. Adie, T. Cowen, et al. 2022. Geological hydrogen storage: Geochemical reactivity of hydrogen with sandstone reservoirs. *ACS Energy Letters* 7, no. 7:2203–7.

302. Havens, T. C., J. C. Bezdek, C. Leckie, K. Ramamohanarao, and M. Palaniswami. 2013. A soft modularity function for detecting fuzzy communities in social networks. *IEEE Transactions on Fuzzy Systems* 21:1170–5.

303. Hoffmann, P. n.d. *The Hindenburg syndrome: Is hydrogen safe?* Taylor & Francis. https://www.taylorfrancis.com.
304. Hullermeier, E., and M. Rifqi. 2009. A fuzzy variant of the Rand index for comparing clustering structures. In *Proceedings IFSA/EUSFLAT Conference*, 1294–98.
305. International Hydrogen Safety Guide. 2007. *Properties and hazards of hydrogen.*
306. The European network of excellence "Safety of hydrogen as an energy carrier." 2009. Biennial report on hydrogen safety. www.hysafe.org.
307. International Energy Agency. 2023. Global hydrogen review 2023. https://www.iea.org.
308. Iordache, I., et al. 2014. Hydrogen underground storage in Romania, potential directions of development, stakeholders and general aspects. *International Journal of Hydrogen Energy* 39, no. 18:9758–96.
309. ISO/TR 15916. 2004. *Basic considerations for the safety of hydrogen systems.* Geneva: International Organization for Standardization.
310. Kanaani, M., B. Sedaee, M. Asadian-Pakfar, et al. 2023. Development of multi-objective co-optimization framework for underground hydrogen storage and carbon dioxide storage using machine learning algorithms. *Journal of Cleaner Production* 386:135785.
311. Kumar, K. R., A. Makhmutov, C. J. Spiers, et al. 2021. Geomechanical simulation of energy storage in salt formations. *Scientific Reports* 11:19640. https://doi.org/10.1038/s41598-021-99161-8.
312. Kumar, R., and H. Hajibeygi. 2022. Influence of pressure solution and evaporate heterogeneity on the geo-mechanical behavior of salt caverns. In *Mechanical behavior of salt X, 2022 proceedings.*
313. Kunz, O., and W. Wagner. 2012. The GERG-2008 wide-range equation of state for natural gases and other mixtures: An expansion of GERG-2004. *Journal of Chemical Engineering Data* 57:3032–91.
314. Lancichinetti, A., S. Fortunato, and F. Radicchi. 2008. Benchmark graphs for testing community detection algorithms. *Physical Review E* 78:046110.
315. Lancichinetti, A., and S. Fortunato. 2009. Benchmarks for testing community detection algorithms on directed and weighted graphs with overlapping communities. *Physical Review E* 80:016118.
316. Lenhard, R. J., and J. C. Parker. 1987. A model for hysteretic constitutive relations governing multiphase flow: 2. Permeability-saturation relations. *Water Resources Research* 23, no. 12:2197–206.
317. Li, J., X. Wang, and J. Eustace. 2013. Detecting overlapping communities by seed community in weighted complex networks. *Physica A* 392:6125–34.
318. Li, J., X. Shi, C. Yang, et al. 2018. Mathematical model of salt cavern leaching for gas storage in high-insoluble salt formations. *Scientific Reports* 8:372. https://doi.org/10.1038/s41598-017-18546-w.

319. Lind, C. D. 1975. What causes unconfined vapour cloud explosions? *Loss Prevention* 9:101–5.
320. Liu, J. 2010. Fuzzy modularity and fuzzy community structure in networks. *European Physical Journal B* 77:547–57.
321. Liu, W., M. Pellegrini, and X. Wang. 2014. Detecting communities based on network topology. *Scientific Reports* 4:5739.
322. Lomeland, F., E. Ebeltoft, and W. H. Thomas. 2005. A new versatile relative permeability correlation. Paper presented at the International Symposium of the Society of Core Analysts.
323. Lou, H., S. Li, and Y. Zhao. 2013. Detecting community structure using label propagation with weighted coherent neighborhood propinquity. *Physica A* 392:3095–105.
324. Lubon, K., and R. Tarkowski. 2020. Numerical simulation of hydrogen injection and withdrawal to and from a deep aquifer in NW Poland. *International Journal of Hydrogen Energy* 45:2068–83. https://doi.org/10.1016/j.ijhydene.2019.11.055.
325. McCarty, R. D., J. Hord, and H. M. Roder. 1981. *Selected properties of hydrogen*. NBS Monograph 168. Boulder, CO: National Bureau of Standards.
326. McPherson, M., et al. 2018. The role of electricity storage and hydrogen technologies in enabling global low-carbon energy transitions. *Applied Energy* 232:649–661.
327. Minougou, J. D., R. Gholami, and P. Andersen. 2023. Underground hydrogen storage in caverns: Challenges of impure salt structures. *Earth-Science Reviews*. https://doi.org/10.1016/j.earscirev.2023.104599.
328. Molkov, V. 2012. *Fundamentals of hydrogen safety engineering, part I*. www.bookboon.com.
329. NASA. 1997. *Safety standard for hydrogen and hydrogen systems. Guidelines for hydrogen system design, materials selection, operations, storage, and transportation*. Office of safety and mission assurance, Washington. http://www.hq.nasa.gov/office/codeq/doctree/canceled/871916.pdf.
330. Navaid, H. B., H. Emadi, and M. Watson. 2023. A comprehensive literature review on the challenges associated with underground hydrogen storage. *International Journal of Hydrogen Energy* 48, no. 28:10603–35. https://doi.org/10.1016/j.ijhydene.2022.11.225.
331. Nepusz, T., A. Petroczi, L. Negyssy, and F. Bazso. 2008. Fuzzy communities and the concept of bridgeness in complex networks. *Physical Review E* 77:016107.
332. Newman, M. E. J., and M. Girvan. 2004. Finding and evaluating community structure in networks. *Physical Review E* 69:026113.
333. Ozarslan, A. 2012. Large-scale hydrogen energy storage in salt caverns. *International Journal of Hydrogen Energy* 37:14265–77.
334. Pan, B., X. Yin, Y. Ju, and S. Iglauer. 2021. Underground hydrogen storage: Influencing parameters and future out-

look. *Advances in Colloid and Interface Science* 2021:102473. https://doi.org/10.1016/j.cis.2021.102473.

335. Panwar, N. L., et al. 2011. Role of renewable energy sources in environmental protection: A review. *Renewable and Sustainable Energy Reviews* 15, no. 3:1513–24.
336. Peng, D.-Y., and D. B. Robinson. 1976. A new two-constant equation of state. *Industrial & Engineering Chemistry Fundamentals* 15:59–64.
337. Perera, M. S. A. 2023. A review of underground hydrogen storage in depleted gas reservoirs: Insights into various rock-fluid interaction mechanisms and their impact on the process integrity. *Fuel* 2023:126677. https://doi.org/10.1016/j.fuel.2022.126677.
338. Psorakis, I., S. Roberts, M. Ebden, and B. Sheldon. 2011. Overlapping community detection using Bayesian non-negative matrix factorization. *Physical Review E* 83:066114.
339. Raghavan, U., R. Albert, and S. Kumara. 2007. Near linear time algorithm to detect community structures in large-scale networks. *Physical Review E* 76:036106.
340. Raza, A., M. Mahmoud, M. Arif, and S. Alafnan. 2024. Underground hydrogen storage prospects in the Kingdom of Saudi Arabia. *Fuel* 357, Part A:1296658.
341. SLR Consulting. n.d. Perception or reality: How does the safety concern of hydrogen. https://www.slrconsulting.com.
342. Soave, G. 1972. Equilibrium constants from a modified Redlich-Kwong equation of state. *Chemical Engineering Science* 27:1197–203.
343. Sobolevsky, S., and R. Campari. 2014. General optimization technique for high-quality community detection in complex networks. *Physical Review E* 90:012811.
344. Subelj, L., and M. Bajec. 2011. Robust network community detection using balanced propagation. *European Physical Journal B* 81:353–62.
345. Subelj, L., and M. Bajec. 2011. Unfolding communities in large complex networks: Combining defensive and offensive label propagation for core extraction. *Physical Review E* 83:036103.
346. Subelj, L., and M. Bajec. 2012. Ubiquitousness of link-density and link-pattern communities in real-world networks. *European Physical Journal B* 85:1–11.
347. Sun, P., L. Gao, and S. Han. 2011. Identification of overlapping and non-overlapping community structure by fuzzy clustering in complex networks. *Information Sciences* 181:1060–71.
348. Tackie-Otoo, B. N., and M. B. Haq. 2024. A comprehensive review on geo-storage of H2 in salt caverns: Prospect and research advances. *Fuel* 356:129609.
349. Tariq, Z., M. Ali, N. Yekeen, et al. 2023. Enhancing wettability prediction in the presence of organics for hydrogen geo-storage through

data-driven machine learning modeling of rock/H2/brine systems. *Fuel* 354:129354. https://doi.org/10.1016/j.fuel.2023.129354.

350. Tarkowski, R. 2019. Underground hydrogen storage: Characteristics and prospects. *Renewable and Sustainable Energy Reviews* 107:338–533.
351. US Department of Energy. 2008. Hydrogen safety training for first responders. http://hydrogen.pnl.gov/FirstResponders/.
352. Vehlow, C., T. Reinhardt, and D. Weiskopf. 2013. Visualizing fuzzy overlapping communities in networks. *IEEE Transactions on Visualization and Computer Graphics* 19:2486–95.
353. Vivian, J. E., and C. J. King. 1964. Diffusivities of slightly soluble gases in water. *AIChE Journal* 10:220–1. https://doi.org/10.1002/aic.690100217.
354. Wang, W., D. Liu, X. Liu, and L. Pan. 2013. Fuzzy overlapping community detection based on local random walk and multidimensional scaling. *Physica A* 392:6578–86.
355. Wang, X., and J. Li. 2013. Detecting communities by the core-vertex and intimate degree in complex networks. *Physica A* 392:2555–63.
356. Wang, G., G. Pickup, K. Sorbie, and E. Mackay. 2022. Numerical modelling of H2 storage with cushion gas of CO2 in subsurface porous media: Filter effects of CO2 solubility. *International Journal of Hydrogen Energy* 47, no. 67:28956–68.
357. Yekta, A. E., M. Pichavant, and P. Audigane. 2018. Evaluation of geochemical reactivity of hydrogen in sandstone: Application to geological storage. *Applied Geochemistry* 95:182–94.
358. Yekta, A. E., J.-C. Manceau, S. Gaboreau, M. Pichavant, and P. Audigane. 2018. Determination of hydrogen–water relative permeability and capillary pressure in sandstone: Application to underground hydrogen injection in sedimentary formations. *Transport in Porous Media* 122, no. 2:333–56. https://doi.org/10.1007/s11242-018-1004-7.
359. Zhang, S., R. Wang, and X. Zhang. 2007. Identification of overlapping community structure in complex networks using fuzzy c-means clustering. *Physica A* 374:483–90.
360. Zhang, Y., and D. Yeung. 2012. Overlapping community detection via bounded nonnegative matrix tri-factorization. In *Proceedings ACM SIGKDD Conference*, 606–14.
361. Zhao, X., and H. Jin. 2019. Investigation of hydrogen diffusion in supercritical water: A molecular dynamics simulation study. *International Journal of Heat and Mass Transfer* 133:718–28. https://doi.org/10.1016/j.ijheatmasstransfer.2018.12.164.
362. Zhong, H., Z. Wang, Y. Zhang, S. Suo, Y. Hong, L. Wang, and Y. Gan. 2024. Gas storage in geological formations: A comparative review on carbon dioxide and hydrogen storage. *Materials Today Sustainability* 26:100720. https://doi.org/10.1016/j.mtsust.2024.100720.

363. Zivar, D., et al. 2021. Underground hydrogen storage: A comprehensive review. *International Journal of Hydrogen Energy* 46, no. 1:798–8152.
364. Ali, A. 2021. Data-driven based machine learning models for predicting the deliverability of underground natural gas storage in salt caverns. *Energy* 229:120648. https://doi.org/10.1016/j.energy.2021.120648.
365. Altman, N. S. 1992. An introduction to kernel and nearest-neighbor nonparametric regression. *The American Statistician* 46, no. 3:175–85.
366. Alvarado, V., and E. Manrique. 2010. Enhanced oil recovery: An update review. *Energies* 3, no. 9. https://doi.org/10.3390/en3091529.
367. Alwated, B., and M. F. El-Amin. 2021. Enhanced oil recovery by nanoparticles flooding: From numerical modeling improvement to machine learning prediction. *Advances in Geo-Energy Research* 5, no. 3:297–317. https://doi.org/10.46690/ager.2021.03.06.
368. American Petroleum Institute. 1998. *Recommended practices for core analysis.* Washington, D.C.: API Publishing Services.
369. Ascher, U., S. J. Ruuth, and B. R. Wetton. 1995. Implicit-explicit methods for time-dependent partial differential equations. *SIAM Journal on Numerical Analysis* 32, no. 3:797–823.
370. Babadagli, T. 1997. Scaling of capillary imbibition during static thermal and dynamic fracture flow conditions. *Journal of Petroleum Science and Engineering* 33, no. 4:223–39.
371. Babadagli, T. 2001. Scaling of concurrent and countercurrent capillary imbibition for surfactant and polymer injection in naturally fractured reservoir. *Society of Petroleum Engineers Journal* 6, no. 4:465–78.
372. Benamar, A., N.-D. Ahfir, H. Wang, and A. Alem. 2007. Particle transport in a saturated porous medium: Pore structure effects. *Comptes Rendus Geoscience* 339, no. 10:674–81. https://doi.org/10.1016/j.crte.2007.07.012.
373. Bielinski, A. 2006. Numerical simulation of CO2 sequestration in geological formations. PhD diss., Stuttgart University.
374. Boscarino, S. 2007. Error analysis of IMEX Runge-Kutta methods derived from differential-algebraic systems. *SIAM Journal on Numerical Analysis* 45, no. 4:1600–21.
375. Bradford, S. A., S. R. Yates, M. Bettahar, and J. Simunek. 2002. Physical factors affecting the transport and fate of colloids in saturated porous media. *Water Resources Research* 38, no. 12:63-1–63-12. https://doi.org/10.1029/2002WR001340.
376. Breiman, L. 2001. Random forests. *Machine Learning* 45, no. 1:5–32.
377. Chen, Z., G. Huan, and B. Li. 2004. An improved IMPES method for two-phase flow in porous media. *Transport in Porous Media* 54, no. 3:361–76.
378. Coats, K. H. 2001. IMPES stability: Selection of stable time steps. Paper presented at the SPE Reservoir Simulation Symposium, Houston, TX, SPE-84924.

379. Cortes, C., and V. Vapnik. 1995. Support-vector networks. *Machine Learning* 20, no. 3:273–97.
380. Cullen, E., D. M. O'Carroll, E. K. Yanful, and B. Sleep. 2010. Simulation of the subsurface mobility of carbon nanoparticles at the field scale. *Advances in Water Resources* 33, no. 4:361–71. https://doi.org/10.1016/j.advwatres.2009.12.001.
381. Cushing, R. S., and D. F. Lawler. 1998. Depth filtration: Fundamental investigation through three-dimensional trajectory analysis. *Environmental Science & Technology* 32, no. 23:3793–801. https://doi.org/10.1021/es9707567.
382. El-Amin, M. F., A. Salama, and S. Sun. 2011. A generalized power-law scaling law for a two-phase imbibition in a porous medium. *Journal of Petroleum Science and Engineering* 111:159–69.
383. El-Amin, M. F., A. Salama, and S. Sun. 2015. Numerical and dimensional analysis of nanoparticles transport with two-phase flow in porous media. *Journal of Petroleum Science and Engineering* 128:53–64.
384. El-Amin, M. F., A. Salama, and S. Sun. 2017. Numerical and dimensional investigation of two-phase countercurrent imbibition in porous media. *Journal of Computational and Applied Mathematics* 242:285–96.
385. El-Amin, M. F., and A. Subasi. 2020. Predicting turbulent buoyant jet using machine learning techniques. Paper presented at the 2020 2nd International Conference on Computer and Information Sciences (ICCIS), 10. IEEE.
386. El-Amin, M. F., A. Subasi, M. M. Selim, and A. Mousa. 2021. Predicted oil recovery scaling-law using stochastic gradient boosting regression model. *Computers, Materials & Continua* 68, no. 2:2349–62.
387. El-Amin, M. F., and B. Alwated. 2023. Machine learning prediction for nanoparticles behavior in hydrocarbon reservoirs. Paper presented at the IEEE 20th Learning and Technology Conference, 80–83.
388. El-Amin, M. F., B. Alwated, and H. A. Hoteit. 2023. Machine learning prediction of nanoparticle transport with two-phase flow in porous media. *Energies* 16, no. 2:678.
389. Fan, J. 2018. Numerical study of particle transport and deposition in porous media. PhD diss., INSA de Rennes.
390. Freedman, D. A. 2009. *Statistical models: Theory and practice*. Cambridge: Cambridge University Press.
391. Friedman, J. H. 2001. Greedy function approximation: A gradient boosting machine. *Annals of Statistics* 29, no. 5:1189–232.
392. Friedman, J. H. 2002. Stochastic gradient boosting. *Computational Statistics & Data Analysis* 38, no. 4:367–78.
393. Hansen, L. K., and P. Salamon. 1990. Neural network ensembles. *IEEE Transactions on Pattern Analysis and Machine Intelligence* 12, no. 10:993–1001.

394. Hochreiter, S., and J. Schmidhuber. 1997. Long short-term memory. *Neural Computation* 9, no. 8:1735–80.
395. Jamei, M., and H. Ghafouri. 2016. A novel discontinuous Galerkin model for two-phase flow in porous media using an improved IMPES method. *International Journal of Numerical Methods for Heat & Fluid Flow* 26, no. 1:284–306.
396. Kazemi, H. 1969. Pressure transient analysis of naturally fractured reservoirs with uniform fracture distribution. *Society of Petroleum Engineers Journal* 9, no. 4:451–62. https://doi.org/10.2118/2156-A.
397. Kotsiantis, S. B. 2013. Decision trees: A recent overview. *Artificial Intelligence Review* 39, no. 4:261–83.
398. LeCun, Y., L. Bottou, Y. Bengio, and P. Haffner. 1998. Gradient-based learning applied to document recognition. *Proceedings of the IEEE* 86, no. 11:2278–324.
399. Li, K. 2007. Scaling of spontaneous imbibition data with wettability included. *Journal of Contaminant Hydrology* 89:218–30.
400. Lippmann, R. P. 1987. An introduction to computing with neural nets. *IEEE ASSP Magazine* 4, no. 2:4–22.
401. Liu, X., D. M. O'Carroll, E. J. Petersen, Q. Huang, and C. L. Anderson. 2009. Mobility of multiwalled carbon nanotubes in porous media. *Environmental Science & Technology* 43, no. 21:8153–8. https://doi.org/10.1021/es901340d.
402. Ma, S., N. R. Morrow, and X. Zhang. 1997. Generalized scaling of spontaneous imbibition data for strong water-wet systems. *Journal of Petroleum Science and Engineering* 18:165–78.
403. Mattax, C. C., and J. R. Kyte. 1962. Imbibition oil recovery from fractured water drive reservoir. *Society of Petroleum Engineers Journal* 2:177–84.
404. Murphy, M. J. 2012. Experimental analysis of electrostatic and hydrodynamic forces affecting nanoparticle retention in porous media. PhD diss., University of Texas. https://repositories.lib.utexas.edu/handle/2152/ETD-UT-2012-05-5576.
405. Nelson, P. H., and J. E. Kibler. 2003. *A catalog of porosity and permeability from core plugs in siliciclastic rocks.* Open-File Report 03-420. Denver, CO: U.S. Geological Survey. https://pubs.usgs.gov/of/2003/ofr-03-420/.
406. Park, H., N. V. Myung, H. Jung, and H. Choi. 2008. As(V) remediation using electrochemically synthesized maghemite nanoparticles. *Journal of Nanoparticle Research* 11, no. 8:1981. https://doi.org/10.1007/s11051-008-9558-x.
407. Raissi, M., P. Perdikaris, and G. E. Karniadakis. 2019. Physics-informed neural networks: A deep learning framework for solving forward and

inverse problems involving nonlinear partial differential equations. *Journal of Computational Physics* 378:686–707.

408. Rezk, M. Y., and N. K. Allam. 2019. Impact of nanotechnology on enhanced oil recovery: A mini-review. *Industrial & Engineering Chemistry Research* 58:16287–95.
409. Sheldon, J. W., and W. T. Cardwell Jr. 1959. One-dimensional, incompressible, noncapillary, two-phase fluid flow in a porous medium. *Transactions of the AIME* 216, no. 1:290–6.
410. Stone, H. L., and A. O. Gardner Jr. 1961. Analysis of gas-cap or dissolved-gas reservoirs. *TAPE AIME* 222:92–104.
411. Tavassoli, Z., R. W. Zimmerman, and M. J. Blunt. 2005. Analytic analysis of oil recovery during countercurrent imbibition in strong water-wet system. *Transport in Porous Media* 58:173–89.
412. Tong, Z., X. Xie, and N. R. Morrow. 2001. Scaling of viscosity ratio and oil recovery by imbibition from mixed-wet rocks. Paper presented at the International Symposium of the Society of Core Analysis, Edinburgh, UK, September, SCA 2001-21.
413. Van Poolen, H. 1980. *Fundamentals of enhanced oil recovery.* PennWell Books.
414. Wang, Y., Y. Li, and K. D. Pennell. 2008. Influence of electrolyte species and concentration on the aggregation and transport of fullerene nanoparticles in quartz sands. *Environmental Toxicology and Chemistry* 27, no. 9:1860–s7. https://doi.org/10.1897/08-039.1.
415. Zhang, T. 2012. Modeling of nanoparticle transport in porous media. PhD diss., University of Texas.
416. Zhou, D., L. Jia, J. Kamath, and A. R. Kovscek. 2002. Scaling of counter-current imbibition processes in low-permeability porous media. *Journal of Petroleum Science and Engineering* 33:61–74.

Index

9781032722795